Graduate Texts in Mathematics 309

Graduate Texts in Mathematics

Graduate Texts in Mathematics bridge the gap between passive study and creative understanding, offering graduate-level introductions to advanced topics in mathematics. The volumes are carefully written as teaching aids and highlight characteristic features of the theory. Although these books are frequently used as textbooks in graduate courses, they are also suitable for individual study.

Shiro Goto • Kei-ichi Watanabe

Commutative Algebra

With Appendices by Mitsuyasu Hashimoto

Springer

Shiro Goto (*deceased*)
Kawasaki-shi, Japan

Kei-ichi Watanabe
College of Humanities and Sciences
Nihon University
Setagaya-ku, Tokyo, Japan

Translated by
Kei-ichi Watanabe
College of Humanities and Sciences
Nihon University
Setagaya-ku, Tokyo
Japan

Mitsuyasu Hashimoto
Department of Mathematics
Osaka Metropolitan University
Sumiyoshi-ku, Osaka
Japan

ISSN 0072-5285 ISSN 2197-5612 (electronic)
Graduate Texts in Mathematics
ISBN 978-981-95-4226-0 ISBN 978-981-95-4227-7 (eBook)
https://doi.org/10.1007/978-981-95-4227-7

Mathematics Subject Classification: 13-01

This Springer imprint is published by the registered company Springer Nature Singapore Pte Ltd.
The registered company address is: 152 Beach Road, #21-01/04 Gateway East, Singapore 189721, Singapore

Dedicated to our wives Hiroko and Mieko,
who supported our work

Preface

This book is written to present an overview of modern commutative algebra and to provide as many techniques and ideas as possible in brief.

As an introduction to commutative algebra, M. F. Atiyah–I. G. Macdonald [AM], published in 1969, has played a great role. It has been the standard introduction to commutative algebra for a long time, even for people from other fields (number theory, algebraic geometry, singularity theory, etc.), and many people around the world have learned commutative algebra from this book. Although there is a good book by Hideyuki Matsumura [Ma2] in Japan, there are big gaps between [AM] and the state-of-the-art of modern commutative algebra—one is the world of commutative algebra based on methods of homological algebra, and the other is the background and viewpoint of algebraic geometry. It is difficult to find a book that is suitable to recommend to read next after [AM]. N. Bourbaki [Bo], D. Eisenbud [Ei2] and W. Bruns–J. Herzog [BH] are all major historical works and will surely play important roles in the development of commutative algebra, but the gap between [AM] and these works is too large to overcome without a mentor (or an appropriate book). This book is written to serve as a good guide for overcoming such a gulf. Basically, no prior knowledge is assumed, although the Appendices and Chaps. 1 to 5 are considerably compressed in their descriptions.

Around 1970, when the authors started to study commutative algebra, there was only one known example of a Gorenstein ring that was not a complete intersection, and it was the "idyllic" time. Herzog's theory of the structure of defining ideals of numerical semigroup rings generated by three elements, and the subsequent development of the theory of canonical ideals and canonical modules by E. Kunz and Herzog were very fresh, and excellent works that gave us a prediction of the subsequent development of commutative algebra.

There are various theories as to when commutative algebra was born, but there is no disputing that we can find its roots and models in the efforts on abstraction of algebra by D. Hilbert, Shozo Sono, van der Waerden, and E. Noether from the end of the nineteenth century to the beginning of the twentieth century. It is worth knowing that behind such activities, there was a fiercely concrete "age of computation," such as the research on invariant theory at the end of the nineteenth century, while very concrete geometric findings obtained by the Italian school (in algebraic geometry) and others provided a rich soil for the development of commutative algebra. After Noether, W. Krull originated the ring structure theory in the form we use today. Another work that is considered to have a great influence on the present day is that of F. S. Macaulay [Mac]. We do not know how this book was regarded at the time, but it is considered as a timeless work in the theory of homogeneous ideals in a polynomial ring or ring structure theory. Later, C. Chevalley constructed the theory of multiplicity of local rings, and I. S. Cohen [Coh] presented the structure theory of complete local rings, and the deep world encompassed by commutative algebra gradually came into view. In response to the demands of the time, O. Zariski–P. Samuel [ZS1, ZS2] was published with the purpose of providing a ring-theoretic foundation of algebraic geometry, and the framework of commutative algebra became clearer.

However, as Professor Masayoshi Nagata often mentioned before his death, there was not an independent field of "commutative algebra" in those days, but people in various fields, mainly in algebraic geometry, recorded what they thought necessary (and important) in the form of papers and books. In fact, the famous and difficult *Local Rings* [Nag] is one of the origins of commutative algebra (the others are [ZS1, ZS2]), and it is meaningful that this book and its predecessor, *Modern Algebra* [AN], were written by mathematicians whose main research field is algebraic geometry.

The basis of Nagata's *Local Rings* is obviously the structure theorem of complete local rings. In this sense, the real turning point of commutative algebra was past the middle of the twentieth century, when M. Auslander, D. A. Buchsbaum, D. Rees, D. G. Northcott, and J.-P. Serre introduced homological algebra into the field. Homological algebra is a linear algebra over rings, and it is the representation theory of rings. Thus, the significance of the development in this direction, which Serre later led from the forefront, is in an epoch-making shift in perspective, reducing ring structure theory to external representations, whereas studies on commutative rings until then had been submerged in the analysis of their internal structure. Along with its great success, Serre's work brought about a decisive change (revolution) in the subsequent development of commutative algebra. In fact, a regular local ring is a Noetherian local ring of finite global dimension, and the global dimension of a Noetherian local ring (A, m) is equal to the projective dimension of its residue field A/m. The authors do not know any proof of Serre's theorem which states that the localization of a regular local ring by a prime ideal is a regular local ring yet, except for Serre's own method.

After Serre's work, there was a rapid generational change in commutative algebra. Thereafter, giants such as Rees, Northcott, Auslander, Buchsbaum, and H. Bass shone like stars, while homological algebra methods were deepened and developed, and the first golden age of commutative algebra emerged. In retrospect, one cannot help but be amazed at how much theory can be developed purely within the framework of Noetherian local rings.

At the same time, the next decade (1960s) was the era of abstraction of algebraic geometry by A. Grothendieck. Although there is not much in this book that directly relates commutative rings to algebraic geometry, Grothendieck's work shows that it is possible to give a geometric meaning to (presumably) all arguments concerning commutative rings. Roughly speaking, a commutative ring is the set of functions (valued in fields) defined on a certain space, and in this sense commutative algebra is a local theory of "varieties." This viewpoint provides an excellent answer to the simple question of why commutative algebra can show such a rich and diverse development. Not only that, commutative algebra has benefited enormously from Grothendieck's work. For example, the invention of local cohomology is probably the greatest of all. Nevertheless, during this period, commutative algebra was tirelessly developed by the above-mentioned giants and others such as Eisenbud, Herzog, Bruns, G. Valla, R. Y. Sharp, M. Hochster, and others. The contributions of Japanese researchers during this period are also significant. The visionary work of Goto, Watanabe, and Yasuhiro Shimoda, such as studies of various graded rings, Buchsbaum rings, and ring structures of Rees algebras (Cohen–Macaulay property, Gorenstein property, etc.), started during this period and opened a new horizon in commutative algebra.

The second golden age of commutative algebra began with the appearance of Hochster. (In commutative algebra, the 1950s was the era of Nagata, Auslander, Buchsbaum, and Serre, and the 1960s was the era of Grothendieck and Bass. If we follow this along, the 1970s was the era of Hochster.) The authors encountered Hochster's work in 1970s. The Regional Conference: *Analytic Methods in Commutative Algebra* held at George Mason University (Fairfax, USA) in 1979 was the first conference on commutative algebra held in the United States. J. Sally, Eisenbud, W. V. Vasconcelos, W. Heinzer and others attended, C. Huneke made his first appearance, and the authors have been friends with them ever since. In the history of such meetings of commutative algebra, it is worth mentioning the international meeting *Commutative Algebra: Durham* (July 15–25, 1981), organized by Sharp and held for 2 weeks at Durham University (UK), and The 9th International Symposium, Division of Mathematics of the Taniguchi Foundation, *Conference on Commutative Algebra* (Katata, September 3–7, 1981), organized by Hideyuki Matsumura[1] and held in Katata in 1981. The former was the first international meeting in the field of commutative algebra, and there were many participants from all over the world, including England, Germany, Italy, France, Northern Europe, the United States, and Japan. The Taniguchi Foundation supported the meeting in Katata, and the participants were carefully selected mainly from young mathematicians. The Japanese participants were Matsumura, Masayoshi Miyanishi, Watanabe, Goto, Akira Ooishi, Tetsushi Ogoma, Shimoda, Yuji Yoshino, Jun-ichi Nishimura, and foreign participants were R. Hartshorne, Hochster, Huneke, L. Avramov, N. M. Kumar, P. Valabrega, and C. Rotthaus. In the course of discussions while staying at the same accommodation, the exchange between the authors and foreign researchers also started from this meeting. The same was true for the participants from abroad, and the Katata meeting became a common defining

[1] Hideyuki Matsumura (1930–1995).

experience for all participants, who eventually played leading roles in commutative algebra research. Japanese commutative algebra has benefited a lot from the insight and foresight of Hideyuki Matsumura [Ma2] as a pioneer.

The main issue of the second golden age, which began in the second half of the twentieth century, is the Cohen–Macaulay analysis of rings. The question of when a given ring becomes Cohen–Macaulay was studied extensively and thoroughly by Hochster, Huneke, Vasconcelos, Eisenbud, Herzog, Bruns, P. Roberts, Avramov, B. Ulrich, Goto, Watanabe, and others. The introduction of face rings by Hochster, R. P. Stanley, and G. A. Reisner led to the realization of the connection between combinatorial problems and commutative rings (in fact, it was a "representation" of combinatorics in commutative rings), which led to a rapid development. In addition, with the development of the theory of Groebner bases (which, in essence, is said to have originated at the end of the nineteenth century), the field has expanded explosively in many directions, including the development of structural analysis of monomial ideals, and continues today. The contributions of Herzog and Takayuki Hibi in this field are significant.

On the other hand, the fact that not much is known outside of the Cohen–Macaulay framework does not seem to have changed much in the past 40 years. To truly understand Cohen–Macaulay rings, it is necessary to understand the non-Cohen–Macaulay world that forms the background of Cohen–Macaulay rings, but from this viewpoint, except for Buchsbaum rings and finite local cohomology (FLC) rings introduced by W. Vogel and J. Stückrad, there is not much to see. It seems to remain as an issue for the future.

The regional conference held in 1979 had another significant impact. At *Analytic Methods in Commutative Algebra*, Hochster gave 10 intensive lectures. He proved the Briançon–Skoda theorem by making full use of Banach space theory, and at this point, the case of positive characteristic may have already been an issue in Hochster's subconscious. After a series of studies on homological conjectures by Hochster and Roberts, the theory of tight closure developed by Hochster and Huneke [HH1] in the 1990s had an explosive impact. The theory was developed in depth and seems to have had a great influence on algebraic geometry, but the essence of the theory, that is, "In rings of positive characteristics, analysis on Frobenius morphisms can be a substitute for calculus," must have been conceived a long time ago.

The authors of this book have had the good fortune of participating in and closely witnessing the development (manifestation) of commutative algebra as described above, and have had the opportunity to make a number of contributions. Although modern commutative algebra is exploding in various directions, and there are fields that do not at first glance appear to be commutative algebra, the authors are happy to find that the core structure of commutative algebra since Noether is still alive and well in such fields. In fact, the Jacobian conjecture is still unsolved, and there are still some remaining problems to be solved. Instead of hand calculations, which were done from the end of the nineteenth century to the beginning of the twentieth century, a new field of commutative algebra may be born on the basis of computer experiments. This book aims to explain almost the whole area of modern commutative algebra, leaving the realization of such expectations to the reader, and to provide a gateway to the field.

The authors wrote this book under extremely busy circumstances. There must be many omissions and misunderstandings, and we request the readers' tolerance. We hope that the readers will be able to overcome these flaws and establish their own theory of commutative rings. Needless to say, the contents of this book are only a superficial view of commutative algebra theory developed in a variety of fields. We hope that you will make efforts to find and solve your own problems in your fields of interest, while consulting the original sources.

We are greatly indebted to Naoyuki Matsuoka and Kazuma Shimomoto for their great help in writing this book. With the help of Matsuoka, we were able to describe complicated commutative diagrams, and Shimomoto helped us to prepare and select problems.

We also thank Ken-ichi Takahashi of Nippon Hyoron Sha. Although the writing of this book was interrupted many times, we would like to express our gratitude and deep respect to him for generosity and forbearance, and for not giving up until the end.

Kawasaki-shi, Japan Shiro Goto
Setagaya-ku, Tokyo, Japan Kei-ichi Watanabe
July 2011

Preface to the English Version

This book is an English translation of a book in Japanese published by Nippon Hyoron Sha in 2011.

After publishing this book, the authors, Shiro Goto and Keiichi Watanabe, wanted to translate it into English so that this book will be available to people worldwide and began preparation for translation.

But at that time both authors were quite busy and could not proceed further. Meanwhile, Shiro got ill and passed away in July 2022. After that, Mitsuyasu Hashimoto, who had long waited for the English translation of the book, proposed the second named author to translate the book. The translation was done by us, the second named author and the translator, collaboratively.

This is not a literal translation of the Japanese text and considerable updates have been done in order to improve accuracy and readability. So, this current version, we believe, has improved in several places. The Appendix on d-sequences in the Japanese version has disappeared, but its contents have been moved to the main text. Considerable supplementary material was added by the translator to Chap. 4, Sect. 4.2 and Chap. 6, Sect. 6.3. We have added Appendices D, E, F, and G, newly written by Hashimoto in order to supplement the main texts on multilinear algebra, field theory, graded rings, and purity, respectively.

Since the original version of this book was in Japanese, the introduction mentions some history of modern commutative algebra in Japan.

Special thanks are due to Anurag Singh and Futoshi Hayasaka for many valuable comments on the previous versions. The translators also thank Chen Andi and Yasutaka Nakao for pointing out errors. We, the translators, are grateful to Elizabeth Loew and Masayuki Nakamura, the editors of Springer Nature, for their help.

We dedicate this English translation to Shiro Goto, who wanted to publish the book worldwide in order to serve everyone who studies commutative algebra.

Setagaya-ku, Tokyo, Japan — Kei-ichi Watanabe
Sumiyoshi-ku, Osaka, Japan — Mitsuyasu Hashimoto
May 2025

After sending the final version of this book to the publishers in August 2025, it is with great sadness that I report that Mitsuyasu Hashimoto, who initiated this work of translation, passed away from colon cancer in December 2025.

Hashimoto and I wished to eliminate errors from this book as much as possible, and the following colleagues kindly helped check the manuscript: Naoki Endo, Kazuki Hayashi, Jun Horiuchi, Ryo Ishizuka, Kaito Kimura, Sora Miyashita, Yasutaka Nakao, Yuya Otake, and Kazuma Shimomoto; we are very grateful to them.

Above all, I remember with deepest respect and gratitude, Shiro Goto (1946–2022) and Mitsuyasu Hashimoto (1962–2025), whose vision, scholarship, and devoted efforts made this English edition possible.

Setagaya-ku, Tokyo, Japan Kei-ichi Watanabe
May 2026

Contents

Chapter 1
Rings and Modules

In this chapter, we introduce basic concepts of commutative ring theory.

In Sect. 1.1, we introduce the most fundamental objects: rings, modules, ideals, and ring homomorphisms and homomorphisms of modules.

In Sect. 1.2, we study integral domains and prime ideals, which are very important in commutative ring theory.

In Sect. 1.3, we introduce the notion of polynomial rings over a ring. The most common process of making a new ring is to make a polynomial ring and then consider a homomorphic image of it.

In Sect. 1.4, we generalize the notion of prime factorization of the natural numbers. We introduce the notion of principal ideal domain (PID) and unique factorization domain (UFD).

In Sect. 1.5, we study the fundamental concepts about modules over a ring.

In Sect. 1.6, we introduce the notion of sequences of modules and diagrams of modules. In particular, exact sequences are very important.

In Sect. 1.7, we introduce the notion of Hom and tensor product, which are the most basic tools to investigate the properties of modules.

In Sect. 1.8, we introduce localization and fields of fractions. By considering the field of fractions, we can embed an integral domain in a field.

In Sect. 1.9, we introduce the notion of $\operatorname{Spec} A$, which is a geometric object attached to the ring A. By this concept, commutative ring theory has a face as a "local theory" of algebraic geometry.

1.1 Rings, Ring Homomorphisms, and Ideals

Throughout this book, the symbols $\mathbb{Z}$, $\mathbb{Q}$, $\mathbb{R}$, $\mathbb{C}$ represent the sets of all integers, rational numbers, real numbers, and complex numbers, respectively.

S. Goto, K. Watanabe, *Commutative Algebra*, Graduate Texts in Mathematics 309,
https://doi.org/10.1007/978-981-95-4227-7_1

In the world of numbers and functions, addition and multiplication are defined. A "ring" is a set where two operations, "addition" $a + b$ and "multiplication" ab, are defined.

Definition 1.1 (Ring) A set A, equipped with two operations called the addition and the multiplication which are respectively denoted by $a + b$ and ab ($a, b \in A$), is said to be a *ring*, if the following four conditions are satisfied:

(1) A forms an abelian group under addition $+$.
(2) The associative law holds for multiplication. That is, for any elements $a, b, c \in A$, the equation $(ab)c = a(bc)$ holds.
(3) There exists some $e \in A$ such that for any $a \in A$, it holds that $ea = ae = a$.
(4) The distributive law holds. That is, for any elements $a, b, c \in A$, the equations $a(b + c) = ab + ac$ and $(a + b)c = ac + bc$ hold.

The element $e \in A$ in (3) above is unique. Indeed, if $e' \in A$ satisfies that $e'a = ae' = a$ for any $a \in A$, then $e' = e'e = e$. The element e is called the *unit element* of A and is denoted by 1 hereafter.

In general, we call the unit element of an additive group the zero element, and denote it by 0. In an additive group, the inverse of an element a is denoted by $-a$, rather than a^{-1}. These conventions are also applied to the additions of rings, and thus the words unit element and the inverse are reserved for multiplications.

A ring in which the commutative law holds for multiplication (that is, for any two elements $a, b \in A$, the equation $ab = ba$ holds) is called a *commutative ring*. $\mathbb{Z}$, $\mathbb{Q}$, $\mathbb{R}$, $\mathbb{C}$ are commutative rings under the usual addition and multiplication.

Exercise 1.1 Show that the set $A = \{(a, x) \mid a \in \mathbb{Z}, x \in \mathbb{Q}\}$ forms a commutative ring with the following addition and multiplication operations: $(a, x) + (b, y) = (a + b, x + y)$ and $(a, x)(b, y) = (ab, bx + ay)$. In this ring, $1 = (1, 0)$ and $0 = (0, 0)$.

Example 1.2 Take an integer $n \geq 1$, and let $A = \mathrm{M}_n(\mathbb{R})$ be the set of all real square matrices of order n. The set A forms a ring with matrix addition and multiplication, but if $n \geq 2$, the commutative law does not hold.

Hereafter, unless otherwise specified, "ring" refers to a commutative ring.

Let us consider the unit element 0 for the addition (it is called the *zero element*) of a ring A. If $1 = 0$ holds in A, then for any $a \in A$, $a = a1 = a0 = 0$, so $A = \{0\}$. We call such a ring a *zero ring*.

Exercise 1.2 Show that the subset $B = \{\begin{pmatrix} a & x \\ 0 & a \end{pmatrix} \mid a \in \mathbb{Z}, x \in \mathbb{Q}\}$ of the set $\mathrm{M}_2(\mathbb{Q})$ forms a commutative ring with the operations of matrix addition and multiplication. Since $1 = \begin{pmatrix} 1 & 0 \\ 0 & 1 \end{pmatrix}$ and $0 = \begin{pmatrix} 0 & 0 \\ 0 & 0 \end{pmatrix}$, we have $1 \neq 0$ in B.

Definition 1.3 (Unit, Field) Let A be a ring.

(1) When an element $a \in A$ has an inverse with respect to multiplication, that is, when there exists an element b in A such that $ab = 1$, a is said to be a *unit* of the ring A. Of course, 1 is a unit. For a unit a, its inverse is uniquely determined in the ring A, and denoted by a^{-1}.
(2) The set of all units of the ring A is denoted by $\mathrm{U}(A)$. The set $\mathrm{U}(A)$ forms a group under the multiplication of the ring A.
(3) When $A \neq 0$ and all elements other than 0 in the ring A are units, that is, when $\mathrm{U}(A) = A \setminus \{0\}$ holds, the ring A is said to be a *field*.

Note that $\mathbb{Q}$, $\mathbb{R}$, and $\mathbb{C}$ are fields, but $\mathbb{Z}$ is not a field because $\mathrm{U}(\mathbb{Z}) = \{\pm 1\}$. In the ring B of Exercise 1.2, the element $\begin{pmatrix} a & x \\ 0 & a \end{pmatrix}$ is a unit if and only if $a = \pm 1$, and we have $\begin{pmatrix} 1 & x \\ 0 & 1 \end{pmatrix}^{-1} = \begin{pmatrix} 1 & -x \\ 0 & 1 \end{pmatrix}$ and $\begin{pmatrix} -1 & x \\ 0 & -1 \end{pmatrix}^{-1} = \begin{pmatrix} -1 & -x \\ 0 & -1 \end{pmatrix}$.

Definition 1.4 (Ring Homomorphism) Let A and B be rings, not necessarily commutative. A map $\varphi : A \to B$ is said to be a *ring homomorphism* when it satisfies the following two conditions:

(1) For any elements a, b of A, $\varphi(a + b) = \varphi(a) + \varphi(b)$ and $\varphi(ab) = \varphi(a)\varphi(b)$ hold.
(2) $\varphi(1) = 1$.

A ring homomorphism that is also a bijection is said to be an *isomorphism*. When there exists at least one ring isomorphism $\varphi : A \to B$ between rings A and B, A and B are said to be *isomorphic*. When two rings A and B are isomorphic, we write $A \cong B$. The relation $\cong$ is an equivalence relation among rings. The mathematical structure of two isomorphic rings is considered to be exactly the same.

If we define the map $\varphi : A \to B$ from the ring A of Exercise 1.1 to the ring B of Exercise 1.2 by $\varphi(a, x) = \begin{pmatrix} a & x \\ 0 & a \end{pmatrix}$, then φ is a ring isomorphism. Therefore, $A \cong B$. The map $\varphi : \mathbb{C} \to \mathrm{M}_2(\mathbb{R})$, $\varphi(a + bi) = \begin{pmatrix} a & -b \\ b & a \end{pmatrix}$ $(a, b \in \mathbb{R}, i = \sqrt{-1})$ is an injective ring homomorphism, but not an isomorphism.

Definition 1.5 (Kernel of a Ring Homomorphism) For a ring homomorphism $\varphi : A \to B$, we define $\mathrm{Ker}\varphi = \{a \in A \mid \varphi(a) = 0\}$, and call this the *kernel* of φ. The subset $\mathrm{Ker}\,\varphi$ of the ring A has the following properties:

(1) $\mathrm{Ker}\,\varphi \neq \emptyset$.
(2) If $x, y \in \mathrm{Ker}\,\varphi$, $a \in A$, then $x + y, ax \in \mathrm{Ker}\,\varphi$.

Exercise 1.3 Prove that a ring homomorphism $\varphi : A \to B$ is injective if and only if $\mathrm{Ker}\,\varphi = \{0\}$.

For A as in Exercise 1.1, define $\varphi : A \to \mathbb{Z}$ by $\varphi(a, x) = a$. Then φ is a ring homomorphism and is surjective. Since $\mathrm{Ker}\,\varphi = \{(0, x) \mid x \in \mathbb{Q}\} \neq \{0\}$, φ is not injective.

Definition 1.6 A non-empty subset I of a ring A is said to be an *ideal* of A when it satisfies the following condition: *if* $x, y \in I, a \in A$, *then* $x + y, ax \in I$.

The sets A and $\{0\}$ are ideals of the ring A. The ideal $\{0\}$ is simply denoted by 0, if there is no danger of confusion. The kernel of a ring homomorphism is an ideal. If I is an ideal of the ring A, then for any $x \in I$, $-x = (-1)x \in I$, so I is a subgroup of the additive group A.

Example 1.7

(1) If we take an element a of the ring A and define $(a) = \{ax \mid x \in A\}$, the set (a) is an ideal of the ring A. Such an ideal is called a *principal ideal*. For example, the principal ideal (2) in $\mathbb{Z}$ is the set of all even numbers.
(2) More generally, for a subset S of the ring A, we define

$$(S) = \{\sum_{i=1}^{n} a_i s_i \mid n \geq 0, a_i \in A, s_i \in S\} \quad (\subseteq A).$$

Note that the sum in the right-hand side is 0 when $n = 0$. So $(S) = 0$ if $S = \emptyset$. The subset (S) of A is the smallest ideal that contains S. The ideal (S) is called the ideal of A *generated by* the set S. Given elements $a_1, a_2, \ldots, a_\ell$ of the ring A, the ideal generated by the set $S = \{a_1, a_2, \ldots, a_\ell\}$ is denoted by $(a_1, a_2, \ldots, a_\ell)$.

Definition 1.8 (Set of Generators of an Ideal) For an ideal I, a subset $S \subseteq I$ such that $I = (S)$ is called a *generating set* or a set of *generators* of the ideal I. We also say that S *generates* I. When I has a finite set as a generating set, I is said to be *finitely generated*. When I is generated by at most one element, then I is said to be a *cyclic module*.

Definition 1.9 When any ideal in A is finitely generated, A is said to be a *Noetherian* ring.

Theorem 1.10 *The following three conditions concerning a ring A are equivalent:*

(1) *A is a Noetherian ring.*
(2) *If $I_1 \subseteq I_2 \subseteq \cdots \subseteq I_i \subseteq \cdots$ is an ascending chain of ideals of A, there exists a sufficiently large number $k \geq 1$ such that $I_k = I_i$ holds for all $i \geq k$.*
(3) *Let $\mathcal{F}$ be the set of all ideals of A. Then every non-empty subset $\mathcal{S}$ of $\mathcal{F}$ contains a maximal element with respect to the inclusion relation.*

Proof (1) $\Rightarrow$(2). If we set $I = \bigcup_{i \geq 1} I_i$, then I is an ideal of the ring A. By assumption, I is finitely generated, so we can express $I = (a_1, a_2, \ldots, a_n)$ $(n \geq 1)$. If we can choose a number $k \geq 1$ such that for all $1 \leq i \leq n$, $a_i \in I_k$, then for $i \geq k$, the equation $I = I_k = I_i$ holds.

(2) $\Rightarrow$(3). If the set $\mathcal{S}$ does not contain a maximal element, then for any $I \in \mathcal{S}$, there exists at least one $J \in \mathcal{S}$ such that $I \subsetneq J$. Consequently, we can obtain a

chain of elements in the set $\mathcal{S}$ such as $I_1 \subsetneq I_2 \subsetneq \cdots \subsetneq I_i \subsetneq \cdots$. This contradicts assumption (2).

(3) $\Rightarrow$(1). Suppose there exists an ideal I in A that is not finitely generated. If we set $\mathcal{S} = \{J \mid J$ is a finitely generated ideal of A and $J \subseteq I\}$, then since $0 \in \mathcal{S}$, $\mathcal{S} \neq \emptyset$. Therefore, the set $\mathcal{S}$ contains a maximal element J. If $J \neq I$, then we can choose $a \in I$ such that $a \notin J$. If we set $K = \{j + ax \mid j \in J, x \in A\}$, then the ideal K is finitely generated and $K \subseteq I$, so $K \in \mathcal{S}$, but $J \subsetneq K$, which contradicts the maximality of the ideal J. Therefore, all ideals of A are finitely generated. □

Exercise 1.4

(1) Prove that the ring of integers $\mathbb{Z}$ is a Noetherian ring.
(2) Show that the ideal $I = \{(0, x) \mid x \in \mathbb{Q}\}$ of the ring A in Exercise 1.1 is not finitely generated (refer to Exercise 1.32).

Proposition 1.11 *A necessary and sufficient condition for a ring A to be a field is that A* $\neq 0$*, and contains no ideals other than A or* 0.

Proof Assuming that I is an ideal of the ring A, $I = A$ if and only if $I \cap \mathrm{U}(A) \neq \emptyset$. If A is a field, then $A \neq 0$. Moreover, any nonzero ideal contains a unit, and so the only ideals in the ring A are A and 0. Conversely, if $A \neq 0$ and the only ideals in A are A and 0, then for any nonzero $a \in A$, we have $(a) = A$. So we can write $1 = ax$ for some $x \in A$. So $a \in \mathrm{U}(A)$. It is assumed that A is not a zero ring, and hence A is a field. □

Corollary 1.12 *Let* $\varphi : A \to B$ *be a ring homomorphism. If A is a field and B is not a zero ring, then* φ *is injective.*

Let us describe the structure of a residue class ring.

Definition 1.13 (Residue Class Ring A/I**)** Let I be an ideal of a ring A. Since I is a subgroup of A with respect to addition, for any two elements $a, b \in A$,

$$a \equiv b \Leftrightarrow a - b \in I$$

defines an equivalence relation $\equiv$ on A. For an element $a \in A$, the equivalence class containing a is denoted by $\overline{a} = \{a + x \mid x \in I\}$. The set of residue classes $A/I = \{\overline{a} \mid a \in A\}$ forms an abelian group under addition $\overline{a} + \overline{b} = \overline{a + b}$, and by defining multiplication as $\overline{a} \cdot \overline{b} = \overline{ab}$, it forms a ring. This is called the *residue class ring* of A by the ideal I. The natural homomorphism $f : A \to A/I$, $a \mapsto \overline{a}$, is a ring homomorphism with $I = \mathrm{Ker}\, f$.

Let us state the isomorphism theorem for rings.

Theorem 1.14 *If* $\varphi : A \to B$ *is a ring homomorphism, and I is an ideal of A such that* $I \subseteq \mathrm{Ker}\,\varphi$*, then, there exists a unique ring homomorphism* $\phi : A/I \to B$ *such that* $\varphi = \phi f$*, where* $f : A \to A/I$, $f(a) = \overline{a}$ *is the natural surjection. We have* $\mathrm{Im}\,\phi = \mathrm{Im}\,\varphi$ *and* $\mathrm{Ker}\,\phi = f(\mathrm{Ker}\,\varphi)$*. In particular,* ϕ *is surjective if and only if* φ *is surjective, and* ϕ *is injective if and only if* $\mathrm{Ker}\,\varphi = I$.

Exercise 1.5 (Correspondence Theorem) Let I be an ideal of A, and let $f : A \to A/I$ be the natural surjection. Let $\mathcal{X}$ be the set of all ideals of A that contain I, and let $\mathcal{Y}$ be the set of all ideals of the residue class ring A/I. Prove the following assertions.

(1) For any $J \in \mathcal{X}$ and $K \in \mathcal{Y}$, $f(J)$ is an ideal of the ring A/I and $f^{-1}(K)$ is an ideal of the ring A that contains I. Therefore, $f(J) \in \mathcal{Y}$ and $f^{-1}(K) \in \mathcal{X}$.
(2) The maps $\Phi : \mathcal{X} \to \mathcal{Y}$, $\Phi(J) = f(J)$ and $\Psi : \mathcal{Y} \to \mathcal{X}$, $\Psi(K) = f^{-1}(K)$ are both bijective and are inverses of each other.
(3) Both maps Φ and Ψ preserve the inclusion relation of ideals.
(4) For any $K \in \mathcal{Y}$, $A/f^{-1}(K) \cong (A/I)/K$.

Definition 1.15 Let M be an ideal of a ring A. If $M \neq A$, and M is maximal among all ideals of A that are different from A with respect to the incidence relation, i.e., for any ideal I such that $M \subseteq I \subsetneq A$, it holds that $I = M$, then M is said to be a *maximal ideal* of A.

Exercise 1.6 Prove that if $p \in \mathbb{Z}$ is a prime number, then the ideal (p) of the ring $\mathbb{Z}$ is a maximal ideal.

Exercise 1.7 Let I be an ideal of a ring A. Prove the following.

(1) If A is a Noetherian ring, then A/I is also a Noetherian ring (use Exercise 1.5).
(2) A necessary and sufficient condition for I to be a maximal ideal is that the residue class ring A/I is a field (use Exercise 1.5).
(3) If A is a Noetherian ring and $A \neq I$, then there exists at least one maximal ideal M that contains I (use Theorem 1.10, compare with Theorem 1.24).

Definition 1.16 (Subring) A subset A of a ring B is said to be a subring of B when it satisfies the following two conditions:

(1) If $a, b \in A$, then $a + b, -a, ab \in A$.
(2) $1_B \in A$, where 1_B is the unit element of B.

A subset A of a ring B is a subring of B if and only if $1_B \in A$ and the set A forms a ring under the addition and multiplication of the ring B. If so, 1_B is the unit element also for the ring A. Furthermore, if $\varphi : A \to B$ is a ring homomorphism, the image of φ, $\varphi(A) = \{\varphi(a) \mid a \in A\}$, is a subring of B.

Exercise 1.8 Prove that the sets $A = \{a + b\sqrt{2} \mid a, b \in \mathbb{Q}\}$ and $B = \{a + bi \mid a, b \in \mathbb{Z}\}$ $(i = \sqrt{-1})$ are subrings of $\mathbb{R}$ and $\mathbb{C}$, respectively.

Definition 1.17 (Operations on Ideals) Let A be a ring and $\mathcal{J}$ be the set of all ideals of A.

(1) For ideals $I, J \in \mathcal{J}$, we define

$$
\begin{aligned}
&I + J = \{a + b \mid a \in I, b \in J\}, \\
&IJ = \left\{\textstyle\sum_{i=1}^{n} a_i b_i \mid n \geq 1, a_i \in I, b_i \in J\right\}, \\
&I \cap J = \{a \in A \mid a \in I \text{ and } a \in J\}, \\
&I : J = \{a \in A \mid \text{for all } x \in J, ax \in I\}
\end{aligned}
$$

and call them the sum, product, and intersection of I and J, and the quotient of I by J, respectively. They are ideals of A.

(2) The following inclusions hold:

$$I \cup J \subseteq I + J, \quad IJ \subseteq I \cap J, \quad I \subseteq I : J.$$

(3) $J \subseteq I$ is equivalent to $1 \in I : J$ and also to $I + J = I$.

(4) $(\mathcal{J}, +)$ (resp. $(\mathcal{J}, \cdot)$) is a commutative semigroup with 0 (resp. A) as the unit element, and for a finite number of ideals $I_1, I_2, \ldots, I_n \in \mathcal{J}$, the sum $\sum_{i=1}^{n} I_i$ and the product $\prod_{i=1}^{n} I_i$ are defined.

(5) For an ideal I, its powers are defined by $I^0 = A$ and $I^n = I^{n-1}I$ $(n \geq 1)$. For all $m, n \geq 0$, the equations

$$I^m I^n = I^{m+n}, \quad (I^m)^n = I^{mn}, \quad (IJ)^n = I^n J^n, \quad (I + J)^n = \sum_{i+j=n} I^i J^j$$

hold.

(6) If I, J are ideals of the ring A, the ideal IJ is generated by the set $\{ab \mid a \in I,\ b \in J\}$, and the ideal I^n is generated by the set $\{\prod_{i=1}^{n} a_i \mid a_i \in I\}$ $(n \geq 1)$.

(7) The distributive law holds between the sum and the product, that is, for all ideals I, J, and K, the equations

$$I(J + K) = IJ + IK, \quad (I + J)K = IK + JK$$

hold.

Exercise 1.9 In the ring of integers $\mathbb{Z}$, Confirm that $(4) \cap (6) = (12)$, $(4) + (6) = (2)$, $(4)(6) = (24)$, and $(6) : (4) = (3)$.

Proposition 1.18 (Modular Law) *Let I, J, and K be ideals of a ring A. If $J \subseteq I$, then*

$$(J + K) \cap I = J + (K \cap I).$$

Proof It is clear that the left side contains the right side. Take $i \in (J + K) \cap I$ and express it as $i = j + k$ $(j \in J, k \in K)$. Since $k = i - j \in I \cap K$, we have $i = j + k \in J + (I \cap K)$. Therefore, $(J + K) \cap I \subseteq J + (K \cap I)$. □

Definition 1.19 (Image and Preimage of Ideals) Let $\varphi : A \to B$ be a ring homomorphism. For an ideal I of A, the ideal of B generated by the set $\varphi(I)$ is denoted by IB. That is,

$$IB = \left\{ \sum_{i=1}^{n} \varphi(a_i)b_i \mid n \geq 1,\ a_i \in I,\ b_i \in B \right\}.$$

If I and J are ideals of A, then

$$(I + J)B = IB + JB, \quad (IJ)B = (IB)(JB).$$

If an ideal I of A is finitely generated, say $I = (a_1, a_2, \dots, a_\ell)$, then

$$IB = (\varphi(a_1), \varphi(a_2), \dots, \varphi(a_\ell))$$

holds. Therefore, IB is also finitely generated.

Let J be an ideal of B. $\varphi^{-1}(J) = \{a \in A \mid \varphi(a) \in J\}$ is called the restriction of ideal J to A, or the pullback. $\varphi^{-1}(J)$ is an ideal of A, and $\varphi^{-1}(J)B \subseteq J$ holds. If I is an ideal of A, then $I \subseteq \varphi^{-1}(IB)$ holds. Note that $\varphi^{-1}(J)$ is sometimes written as $J \cap A$, even if φ is not injective. For example, $0 \cap A$ means $\operatorname{Ker} \varphi$.

Exercise 1.10 Let $\varphi : A \to B$ be a ring homomorphism, and I, J be ideals of A. If the homomorphism φ is surjective, then $IB = \varphi(I)$. But the following statements are not true in general. Give a counterexample to each statement.

(1) $(I \cap J)B = IB \cap JB$.
(2) $\varphi(I)$ is an ideal of B.

1.20 Let $f : A \to B$ be an injective homomorphism of rings. Then we often identify A with $f(A)$ and consider A as a subring of B.

1.2 Integral Domains and Prime Ideals

Definition 1.21 Let A be a ring.

(1) Let $a \in A$. We say that a is a *zerodivisor* of A if there exists an element $x \in A$ such that $x \neq 0$ and $ax = 0$. We say that a is a *non-zerodivisor* if a is not a zerodivisor. If $A \neq 0$, then 0 is a zerodivisor, and a unit of A is a non-zerodivisor.
(2) When $A \neq 0$ and A does not contain zerodivisors other than 0, A is said to be an *integral domain* or a *domain*.
(3) Let P be an ideal of A. When $P \neq A$ and for any $a, b \in A$, if $ab \in P$ then either $a \in P$ or $b \in P$ holds, or, equivalently, if the residue class ring A/P is an integral domain, P is said to be a *prime ideal* of A.

Exercise 1.11 Let A be a ring, P a prime ideal of A, and $I_1, I_2, \dots, I_n$ ideals of A. Confirm that if $\bigcap_{i=1}^n I_i \subseteq P$, then there exists some $1 \le i \le n$ such that $I_i \subseteq P$. Therefore, if $\bigcap_{i=1}^n I_i = P$, then there exists some $1 \le i \le n$ such that $I_i = P$.

Example 1.22

(1) Let n (≥ 1) be an integer. The ideal (n) of the ring $\mathbb{Z}$ is a prime ideal if and only if n is a prime number.
(2) A field is an integral domain. Hence a maximal ideal is a prime ideal.
(3) A subring of an integral domain is an integral domain.

Exercise 1.12 Prove the following.

(1) Let $\varphi : A \to B$ be a ring homomorphism. If P is a prime ideal of B, then $\varphi^{-1}(P)$ is a prime ideal of A. (Even if P is a maximal ideal of B, $\varphi^{-1}(P)$ may not be a maximal ideal of A. Give an example.)
(2) Let I be an ideal of a ring A, and J be an ideal of A such that $I \subseteq J$. Then J is a prime (resp. maximal) ideal of A if and only if J/I is a prime (resp. maximal) ideal of the ring A/I. Therefore, there is a one-to-one correspondence between the set of all prime ideals (resp. maximal ideals) of A that contain I and the set of all prime (resp. maximal) ideals of A/I.

For an element a of a ring A, define the map $\widehat{a} : A \to A$ by $\widehat{a}(x) = ax$. Then a is a non-zerodivisor (resp. unit) of A if and only if $\widehat{a}$ is injective (resp. surjective). When the ring A is finite as a set, if $\widehat{a}$ is injective, it is also surjective.

Exercise 1.13 Show that a non-zerodivisor in a finite ring is a unit, and hence a finite integral domain is a field.

If $I \neq A$ is an ideal of a ring A, then there exists at least one maximal ideal of A that contains I. Let us prove this using *Zorn's lemma*.[1]

Let $\mathcal{X}$ be an ordered set, and C be a subset of $\mathcal{X}$. If either $x \le y$ or $y \le x$ holds for any $x, y \in C$, then C is said to form a *chain* in $\mathcal{X}$. When there exists an element $a \in \mathcal{X}$ such that $x \le a$ for all $x \in C$, the chain C is said to be *bounded above* in $\mathcal{X}$. An element $a \in \mathcal{X}$ is said to be *maximal* in $\mathcal{X}$ if $b \in \mathcal{X}$ and $a \le b$ imply $a = b$.

Using these terms, Zorn's lemma can be stated as follows:

Lemma 1.23 (Zorn's Lemma) *Let $\mathcal{X}$ be a partially ordered set. If every chain in $\mathcal{X}$ is bounded above, then the set $\mathcal{X}$ contains at least one maximal element.*

We use the lemma frequently in the following form (equivalent to the lemma). Let $\mathcal{X}$ be a non-empty partially ordered set. If every non-empty chain in $\mathcal{X}$ is bounded above, then $\mathcal{X}$ contains a maximal element.

Theorem 1.24 *If I is an ideal of a ring A with $I \neq A$, then there exists at least one maximal ideal M of A that contains I.*

---

[1] Max Zorn, 1906–1993; Zorn's lemma is a proposition equivalent to "Axiom of choice" and "Well-ordering theorem."

Proof Let $\mathcal{X} = \{J \mid J$ is an ideal of A with $I \subseteq J \subsetneq A\}$. Since $I \in \mathcal{X}$, the set $\mathcal{X}$ is not empty. Consider $\mathcal{X}$ as a partially ordered set by inclusion, and take a non-empty chain C in $\mathcal{X}$. We show that C is bounded above in $\mathcal{X}$. Let $K = \bigcup_{J \in C} J$. The set K is an ideal of A, contains the ideal I, and $1 \notin K$. Therefore, $K \in \mathcal{X}$, and since any $J \in C$ is contained in K, the chain C is bounded above in $\mathcal{X}$. Hence, by Zorn's Lemma (Lemma 1.23), the set $\mathcal{X}$ contains at least one maximal element M. This is the maximal ideal we seek. □

An element a of a ring A is a unit if and only if $(a) = A$. Therefore, the following proposition holds.

Proposition 1.25 *An element $a \in A$ is a unit if and only if $a \notin M$ for any maximal ideal M of A.*

Definition 1.26

(1) When a ring A contains exactly one maximal ideal, we say that A is a *local ring*. When A is a local ring and its only maximal ideal is $\mathfrak{m}$, we denote it as the local ring $(A, \mathfrak{m})$ or $(A, \mathfrak{m}, k)$, where $k = A/\mathfrak{m}$. The field $k = A/\mathfrak{m}$ is called the *residue field* of the local ring A.
(2) When a ring A has a finite number of maximal ideals, we say that A is a *semi-local ring*. We write "semi-local ring $(A, \mathfrak{m}_1, \mathfrak{m}_2, \ldots, \mathfrak{m}_r)$" when A is a semi-local ring and the set of maximal ideals of A is $\{\mathfrak{m}_1, \mathfrak{m}_2, \ldots, \mathfrak{m}_r\}$. Note that a zero ring cannot be local, because the definition of a local ring requires one maximal ideal. On the other hand, a zero ring is semi-local, since zero is finite.

Exercise 1.14 Let A be a ring, and $\mathfrak{m}$ be the set of all non-units of A. Show that if A is nonzero and $\mathfrak{m}$ is closed under addition, then A is a local ring with $\mathfrak{m}$ as its maximal ideal.

Definition 1.27 The intersection of all maximal ideals of a ring A is called the *Jacobson radical* or *radical* of A and is denoted by $\mathrm{J}(A)$.

Exercise 1.15 Let $a \in A$. Prove that a necessary and sufficient condition for $a \in \mathrm{J}(A)$ is that $1 - ax \in \mathrm{U}(A)$ for all $x \in A$.

Example 1.28 Let us look at some examples of Jacobson radicals.

(1) If $(A, \mathfrak{m})$ is local, then $\mathrm{J}(A) = \mathfrak{m}$.
(2) In the ring $\mathbb{Z}$, the only maximal ideals are (p) (p is a prime number). Therefore, $\mathrm{J}(\mathbb{Z}) = 0$.

Definition 1.29 Let A be a ring.

(1) For an ideal I of A, we define

$$\sqrt{I} = \{a \in A \mid \text{there exists an integer } n \geq 1 \text{ such that } a^n \in I\}$$

and call $\sqrt{I}$ the *radical* of I. Note that $\sqrt{I}$ is an ideal of A containing I, see Proposition 1.30 below.

(2) Let $a \in A$. We say that a is *nilpotent* in A if there exists an integer $n \geq 1$ such that $a^n = 0$.
(3) Note that $\sqrt{0}$ is the set of nilpotent elements of A. We call $\sqrt{0}$ the *nilradical* of A. Being a radical of an ideal, the nilradical is an ideal.
(4) If $\sqrt{I} = I$ holds, we say that the ideal I is *radical*. We say that A is *reduced* if $\sqrt{0} = 0$.

Exercise 1.16 Prove that if $a \in A$ is nilpotent, then $1 - a$ is a unit in A.

Proposition 1.30 *When A is a ring and I ($\neq A$) is an ideal of the ring A, the equation*

$$\sqrt{I} = \bigcap_{P \supseteq I} P$$

holds, where P ranges over all prime ideals of A that contain I. In particular, $\sqrt{I}$ is an ideal of A.

Proof If $P \supseteq I$ is a prime ideal and $a^n \in I$, then $a^n \in P$, and hence $a \in P$. It follows that $\sqrt{I} \subseteq \bigcap_{P \supseteq I} P$.

Suppose $f \in A$ and $f \notin \sqrt{I}$. Then we show that there exists some prime ideal P such that $I \subseteq P$ and $f \notin P$. This will show $\sqrt{I} \supseteq \bigcap_{P \supseteq I} P$. Define

$$\mathcal{X} = \{J \mid J \text{ is an ideal of } A, I \subseteq J, f \notin \sqrt{J}\}.$$

If we consider $\mathcal{X}$ as a partially ordered set with respect to inclusion, we can show that X has a maximal element by the same argument as in the proof of Theorem 1.24. Let P be a maximal element of the set X, then P is a prime ideal of the ring A. Indeed, $P \neq A$, since $f \notin \sqrt{P} \supseteq P$. Let's take $a, b \in A$ such that $a, b \notin P$. Then from the maximality of P, the ideals $(a) + P$ and $(b) + P$ are not elements of $\mathcal{X}$. Hence these ideals contain a large power of the element f. Let us express this as

$$f^n = ax + p_1, \quad f^m = by + p_2$$

($p_1, p_2 \in P, x, y \in A$). Then $f^{n+m} = (ax + p_1)(by + p_2)$. Since $p_1, p_2 \in P$, if $ab \in P$ then $f^{n+m} \in P$. But this contradicts $P \in \mathcal{X}$. Therefore, $ab \notin P$. Combining these, P is a prime ideal. By definition, $f \notin \sqrt{P} = P$ and this is what we wanted to show. □

Exercise 1.17 Let I be an ideal of a ring A, and let $J = \sqrt{I}$. Prove that if the ideal J is finitely generated, then there exists some $n \geq 1$ such that $J^n \subseteq I$.

Exercise 1.18 Prove that for ideals I, J of a ring A, we have $I + J = A$ if and only if $\sqrt{I} + \sqrt{J} = A$.

The following two theorems are famous for their simple proofs and high utility.

Theorem 1.31 (Prime Avoidance Lemma) *Let $I, P_1, P_2, \ldots, P_n$ ($n \geq 2$) be ideals of a ring A and assume that there are at most two ideals among $\{P_i\}_{1\leq i\leq n}$ that are not prime. If $I \subseteq \bigcup\limits_{i=1}^{n} P_i$, then there exists some $1 \leq i \leq n$ such that $I \subseteq P_i$.*

Proof Assume that this theorem is incorrect and consider a counterexample where the number of ideals $\{P_i\}_{1\leq i\leq n}$ is minimal. When $n = 2$, take $a_1, a_2 \in I$ such that $a_1 \notin P_2$ and $a_2 \notin P_1$. Then, since $a_i \in I \subseteq P_1 \cup P_2$, we have $a_i \in P_i$ ($i = 1, 2$). If we set $a = a_1 + a_2$, then $a = a_1 + a_2 \in I \subseteq P_1 \cup P_2$. However, if for example $a = a_1 + a_2 \in P_1$, then since $a_1 \in P_1$, it would imply that $a_2 \in P_1$, which is impossible. Hence, $n \geq 3$. Then changing the order of P_i, if necessary, we can assume that P_1 is prime. By the induction hypothesis, for any $1 \leq i \leq n$, $I \nsubseteq \cup_{j\neq i} P_j$. Hence we can take $a_i \in I$ such that $a_i \notin \cup_{j\neq i} P_j$. Of course, $a_i \in P_i$. Put $a = a_1 + \prod_{i=2}^{n} a_i$. Then, $a \in I$, but $a \notin \bigcup_{i=1}^{n} P_i$. Indeed, since $a_i \notin P_1$ for $i \geq 2$ and P_1 is prime, we have $\prod_{i=2}^{n} a_i \notin P_1$. Then since $a_1 \in P_1$, $a = a_1 + \prod_{i=2}^{n} a_i \notin P_1$. Also, for $i \geq 2$, $\prod_{i=2}^{n} a_i \in P_i$ and $a_1 \notin P_i$ imply $a \notin P_i$. This shows $I \nsubseteq \bigcup\limits_{i=1}^{n} P_i$, contradicting our hypothesis. □

Theorem 1.32 (Davis's Lemma) *Let $a \in A$, I be an ideal of a ring A, and $P_1, P_2, \ldots, P_n$ be prime ideals of A. If*

$$(a) + I \nsubseteq \bigcup_{i=1}^{n} P_i,$$

we can choose $x \in I$ such that $a + x \notin \bigcup\limits_{i=1}^{n} P_i$.

Proof Let $n = 1$. If $a \in P_1$, then our assumption $(a) + I \nsubseteq P_1$ implies $I \nsubseteq P_1$ and for $x \in I \setminus P_1$, $a + x \notin P_1$. If $a \notin P_1$ we can take $x = 0$.

Let us assume that this proposition is true up to $n - 1$ for $n \geq 2$. Without loss of generality, we can assume $P_i \nsubseteq P_n$ ($1 \leq \forall i \leq n-1$). Hence, $\prod_{i=1}^{n-1} P_i \nsubseteq P_n$. Since $(a) + I \nsubseteq \bigcup_{i=1}^{n-1} P_i$, we can choose an element $y \in I$ such that $a + y \notin \bigcup_{i\neq 1}^{n-1} P_i$. If $a + y \notin P_n$, then this $y \in I$ is the desired element. Let us assume $a + y \in P_n$. At this time, $I \nsubseteq P_n$. Indeed, if $I \subseteq P_n$, then since $a + y \in P_n$, we get $a \in P_n$, and $(a) + I \subseteq P_n$ holds. But this contradicts the assumption that $(a) + I \nsubseteq \bigcup\limits_{i=1}^{n} P_i$. Therefore, since P_n is a prime ideal, $I \cdot \prod_{i=1}^{n-1} P_i \nsubseteq P_n$. Hence, we can choose an element $z \in I \cdot \prod_{i=1}^{n-1} P_i$ such that $z \notin P_n$. As chosen, if $x = y + z$, then this $x \in I$ is the element to be found. Indeed, if $1 \leq i \leq n-1$, because $a + y \notin P_i$ and $z \in P_i$, $a + x \notin P_i$. On the other hand, because $a + y \in P_n$ and $z \notin P_n$, $a + x \notin P_n$. □

1.3 Polynomial Rings

First, let us start by organizing what the polynomial ring $A[X]$ is. Hereafter, A is a ring.

Definition 1.33 Let A and B be rings and fix a ring homomorphism $\varphi : A \to B$ (we say that B is an A-algebra and φ is its structure map in this situation, see Definition 1.116). For $a \in A$ and $b \in B$, ab means $f(a)b \in B$. An element $X \in B$ is said to be *transcendental* over A if

$$\sum_{i=0}^{n} a_i X^i = 0 \quad (a_0, \ldots, a_n \in A)$$

in B implies $a_i = 0$ $(i = 0, \ldots, n)$ for any $n \geq 0$. If X is not transcendental over A, then we say that X is *algebraic* over A.

The following theorem holds. The proof will be postponed.

Theorem 1.34 (Polynomial Ring) *Given a ring A, there exists a pair (S, X) that satisfies the following three conditions:*

(1) *S is an A-algebra, see Definition 1.116.*
(2) *$X \in S$, and X is transcendental over A.*
(3) *Every element f of the ring S can be expressed in the form $f = \sum_{i=0}^{n} a_i X^i$ by choosing n and elements $a_0, \ldots, a_n$ of the ring A.*

The ring S given in Theorem 1.34 is called a *polynomial ring* over A with *indeterminate* X, and is denoted by $S = A[X]$. As the structure map $A \to A[X]$ is injective, A is identified with a subring of $A[X]$. It is common to write $f = f(X)$ and call an element of S a *polynomial* in X over A. The representation of $f(X) \in A[X]$ in Theorem 1.34 (3) is unique. Moreover, if $f(X) \neq 0$, by choosing the representation in Theorem 1.34 (3) such that $a_n \neq 0$, the integer $n \geq 0$ is uniquely determined for $f(X)$. This n is called the *degree* of the polynomial $f(X)$, and is denoted by $\deg f(X)$ (for the polynomial 0, we define $\deg 0 = -\infty$).

Let $0 \neq f(X), g(X) \in A[X]$, and let $f(X) = \sum_{i=0}^{m} a_i X^i$, $g(X) = \sum_{i=0}^{n} b_i X^i$ $(a_m, b_n \neq 0)$. In this case, if a_m is a non-zerodivisor in the ring A, then $f(X)g(X) \neq 0$ and $\deg f(X)g(X) = \deg f(X) + \deg g(X)$. In particular, $f(X)$ is a non-zerodivisor in the ring $A[X]$. Therefore, the following holds.

Theorem 1.35 *The polynomial ring $A[X]$ over an integral domain A is an integral domain.*

Exercise 1.19 If $f(X) \in A[X]$ is a zerodivisor, show that there exists $a \in A$, $a \neq 0$ such that $a \cdot f(X) = 0$.

The essence of the polynomial ring lies in the following substitution principle.

Theorem 1.36 (Substitution Principle) *Let $A[X]$ be a polynomial ring, and $\psi : A \to T$ be a ring homomorphism. Then for each $t \in T$, there exists a unique ring homomorphism $\varphi : A[X] \to T$ that satisfies the following two conditions:*

(1) *If $a \in A$, then $\varphi(a) = \psi(a)$;*
(2) $\varphi(X) = t$.

For each $f(X) \in A[X]$, write $\varphi(f(X))$ as $f(t)$, which is called the value of $f(X)$ when X is substituted with t.

Proof Define the homomorphism $\varphi : A[X] \to T$ by $\varphi(f(X)) = \sum_{i=0}^{n} \psi(a_i)t^i$ (where $f(X) = \sum_{i=0}^{n} a_i X^i$). This is the desired homomorphism φ. The uniqueness of φ is clear. □

Remark 1.1 Let k be a *finite* ring. Let us consider the polynomial $f(X) = \prod_{a \in k}(X - a) \in k[X]$. Then for any $a \in k$, $f(a) = 0$, but as an element of $k[X]$, $f(X) \neq 0$. In other words, the polynomial $f(X)$ defines a function over the set k, and even if it is identically 0 over k, it does not necessarily mean that $f(X) = 0$ as a polynomial. In other words, $k[X] \to \mathrm{Map}(k, k)$ is not injective, if k is finite.

The uniqueness of the polynomial ring can be derived from the substitution principle.

Corollary 1.37 *If $A[X]$ and $A[Y]$ are polynomial rings, then there is a unique ring isomorphism $\varphi : A[X] \to A[Y]$ such that $\varphi(X) = Y$ and for any $a \in A$, $\varphi(a) = a$.*

Exercise 1.20 If I is an ideal of the ring A and $I \neq A$, then $A[X]/IA[X] \cong (A/I)[X]$. Hence I is a prime ideal of the ring A if and only if $IA[X]$ is a prime ideal of the polynomial ring $A[X]$. Prove this.

Let us describe the construction of the polynomial ring.

Proof (Proof of Theorem 1.34) Let $C = \{(a_0, a_1, a_2, \ldots) \mid a_i \in A\}$. Take $f, g \in C$, where $f = (a_0, a_1, a_2, \ldots)$ and $g = (b_0, b_1, b_2, \ldots)$. We define the sum and product of them as follows:

$$f + g = (a_0 + b_0, a_1 + b_1, \ldots, a_n + b_n, \ldots),$$
$$f \cdot g = (a_0 b_0, a_0 b_1 + a_1 b_0, \ldots, \sum_{i+j=n} a_i b_j, \ldots).$$

Then, C becomes a commutative ring. Put $I = \{i \in \mathbb{Z} \mid i \geq 0\}$ and define

$$T = \{f \in C \mid \text{for almost all } i \in I, a_i = 0\}.$$

Then T is a subring of the ring C. Define the map $\phi : A \to T$ by $a \mapsto (a, 0, 0, \ldots)$, and let $t = (0, 1, 0, 0, \ldots)$. Then, for any $i \in I$, $t^i = (0, \ldots, 0, 1, 0, \ldots)$ (here, 1 appears only in the ith position), and for any $f = (a_0, a_1, a_2, \ldots) \in C$, the equation $f = \sum_{i \in I} \phi(a_i)t^i$ holds. Since the map ϕ is an injective ring homomorphism, we consider A as a subring of T (1.20) and $S = T$ and $X = t$ are what we want. □

Theorem 1.38 (Hilbert's Basis Theorem) *If A is a Noetherian ring, then the polynomial ring $A[X]$ is also a Noetherian ring.*

Proof (E. Kunz) Assume that I is an ideal of $A[X]$ that is not finitely generated. Choose a sequence of elements $\{f_i\}_{i \geq 1}$ of I such that $n_i = \deg f_i$ and $f_i \notin (f_1, f_2, \ldots, f_{i-1})$, where f_i has the smallest degree among the elements of I not contained in $(f_1, f_2, \ldots, f_{i-1})$. Then, because

$$f_{i+1} \notin (f_1, f_2, \ldots, f_{i-1}),$$

we have $n_i \leq n_{i+1}$. Let a_i be the coefficient of the n_ith term of each polynomial f_i, then in the ring A, we have the chain of ideals

$$(a_1) \subseteq (a_1, a_2) \subseteq \cdots \subseteq (a_1, a_2, \ldots, a_{i-1}) \subseteq (a_1, a_2, \ldots, a_i) \subseteq \cdots.$$

Since the ring A is Noetherian, we can choose a number $i \gg 0$ such that the equation $(a_1, a_2, \ldots, a_{i-1}) = (a_1, a_2, \ldots, a_i)$ holds. We can express $a_i = \sum_{j=1}^{i-1} b_j a_j, b_j \in A$, and set

$$g = f_i - \sum_{j=1}^{i-1} b_j X^{n_i - n_j} f_j.$$

Then, since $f_i \notin (f_1, f_2, \ldots, f_{i-1})$, $g \in I \setminus (f_1, f_2, \ldots, f_{i-1})$. However, the polynomial g has degree $\leq n_i$ and its coefficient of $X_i^{n_i}$ is 0 by the choice of $b_j, 1 \leq i \leq i-1$. So $\deg g < n_i = \deg f_i$, which contradicts the choice of f_i. Therefore, if A is a Noetherian ring, then the polynomial ring $A[X]$ is also Noetherian. □

Definition 1.39 A polynomial $f \in A[X]$ is said to be *monic* if $f \neq 0$ and the coefficient of its highest-degree term is 1.

Monic polynomials are non-zerodivisors. If a monic polynomial f is given, we can perform division by f within the polynomial ring $A[X]$.

Proposition 1.40 (Division of Polynomials) *Let $f \in A[X]$ be a monic polynomial of degree n. Then for any $g \in A[X]$, there exist $q, r \in A[X]$ uniquely, satisfying the equation*

$$g = qf + r, \qquad \deg r < n.$$

Proof We prove the uniqueness. If $g = qf + r = q'f + r'$, then $(q - q')f = r' - r$. If $q \neq q'$, then $\deg[(q - q')f] = \deg(q - q') + n = \deg(r' - r) < n$, but this is impossible. We prove the existence by induction on $\deg g = s$. If $s < n$, then the conditions are satisfied when we set $q = 0$ and $r = g$. So let $s \geq n$, and let $g = aX^s + g'$ with $a \in A \setminus \{0\}$ and $\deg g' < s$. Then since $\deg(X^n - f) < n$, we have that $g = aX^{s-n}f + aX^{s-n}(X^n - f) + g'$ with $\deg(aX^{s-n}(X^n - f) + g') < s$, and the assertion follows by induction easily. □

The following theorem is known as the *factor theorem*.

Corollary 1.41 *Let $g(X) \in A[X]$ and $a \in A$. Then there exists a unique polynomial $q(X) \in A[X]$ such that $g(X) = q(X)(X - a) + g(a)$. In particular, $g(a) = 0$ if and only if $g(X)$ is divisible by $X - a$; in other words, there exists $q(X) \in A[X]$ such that $g(X) = (X - a)q(X)$. Such a $q(X)$ is unique.*

Exercise 1.21

(1) Let A be an integral domain, $a_1, a_2, \ldots, a_n$ $(n \geq 1)$ be distinct elements of A, and $f(X) \in A[X]$. Then a necessary and sufficient condition for $f(a_i) = 0$ for all $1 \leq i \leq n$ is that $f(X) = q(X) \cdot \prod_{i=1}^{n}(X - a_i)$ for some $q(X) \in A[X]$. Therefore, if $f(X) \neq 0$, then

$$\sharp\{a \in A \mid f(a) = 0\} \leq \deg f(X)$$

holds. Hence if A is infinite, then there exists $a \in A$ with $f(a) \neq 0$. Prove this.

(2) Let $A = \mathbb{Z}/(6)$. Find a distinct pair of elements $a, b \in A$ and a polynomial $f(X) \in A[X]$ such that $f(a) = f(b) = 0$, but which we cannot write as $f(X) = q(X)(X - a)(X - b)$ for any $q(X) \in A[X]$. More generally, show that if the ring A is not an integral domain, then there is always such an example.

(3) If the ring A is reduced but not an integral domain, then show that there exists a $0 \neq f(X) \in A[X]$ such that

$$\sharp\{a \in A \mid f(a) = 0\} > \deg f(X).$$

A polynomial ring with more than one indeterminates is defined inductively from the ring A, with $A_1 = A[X_1]$ and $A_n = A_{n-1}[X_n]$ $(n \geq 2)$. It can be proven by induction on n that the ring A_n has the following properties:

(1) A is a subring of A_n.

(2) Any element $f \in A_n$ can be uniquely expressed as a linear combination of the monomials $X_1^{a_1} X_2^{a_2} \cdots X_n^{a_n}$ $(a_i \geq 0)$ over A. Therefore, the elements $X_1, X_2, \ldots, X_n \in A_n$ are algebraically independent over A.

We call A_n the *polynomial ring* with indeterminates $X_1, X_2, \ldots, X_n$ over A, denoted by $A[X_1, X_2, \ldots, X_n]$.

The substitution principle of the polynomial ring $A[X_1, X_2, \ldots, X_n]$ $(n \geq 1)$ can be stated as follows in Theorem 1.42. As in the case of one variable $(n = 1)$, this theorem implies the uniqueness of ring homomorphisms in the category of polynomial rings.

Theorem 1.42 *Let $\psi : A \to T$ be a ring homomorphism and let $t_1, t_2, \ldots, t_n \in T$. Then, there exists a unique ring homomorphism $\varphi : A[X_1, X_2, \ldots, X_n] \to T$ such that $\varphi(X_i) = t_i$ and $\varphi(a) = \psi(a)$ for all $1 \le i \le n$ and for all $a \in A$.*

From Theorem 1.38, we obtain the following corollary.

Corollary 1.43 *The polynomial ring $A[X_1, X_2, \ldots, X_n]$ over a Noetherian ring A is Noetherian. In particular, the polynomial ring $k[X_1, X_2, \ldots, X_n]$ over a field k is a Noetherian ring.*

1.44 For a polynomial $f \in A[X_1, X_2, \ldots, X_n]$, write $\varphi(f)$ as $f(t_1, t_2, \ldots, t_n)$, which is the value obtained by substituting t_i for X_i. The image of the substitution map φ, $\operatorname{Im}\varphi$, is written as $A[t_1, t_2, \ldots, t_n]$ and is the smallest subring of T that contains both $\operatorname{Im}\psi$ and $\{t_i\}_{1\le i\le n}$. According to Theorem 1.43, if A is a Noetherian ring, then $A[t_1, t_2, \ldots, t_n]$ is also a Noetherian ring.

Exercise 1.22 Let k be a field.

(1) Show that for a subring R of $k[X]$ that properly contains the field k, the vector space $k[X]/R$ over k is finite-dimensional. Using this, show that R is a Noetherian ring.
(2) Is every subring of $k[X, Y]$ containing k Noetherian?

Exercise 1.23

(1) Show $\mathbb{Z}[\sqrt{-5}] = \{a + b\sqrt{-5} \mid a, b \in \mathbb{Z}\}$ and $\mathbb{Q}[\omega] = \{a + b\omega \mid a, b \in \mathbb{Q}\}$, where $\omega = \frac{-1+\sqrt{-3}}{2}$.
(2) Let $f \in A[X_1, X_2, \ldots, X_n], a_1, a_2, \ldots, a_n \in A$. In this case, $f(a_1, a_2, \ldots, a_n) = 0$ if and only if $f \in (X_i - a_i \mid 1 \le i \le n)$. Therefore, $A[X_1, X_2, \ldots, X_n]/(X_i - a_i \mid 1 \le i \le n) \cong A$.

Let us calculate an example.

Example 1.45 Let $A = k[X, Y]$ and $B = k[t]$ be polynomial rings over a field k, and consider the substitution homomorphism $\varphi : A \to B$, $\varphi(X) = t^3$, $\varphi(Y) = t^4$, $\varphi(c) = c$ $(c \in k)$. Let $\mathfrak{p} = \operatorname{Ker}\varphi$. Then, $\mathfrak{p} = (X^4 - Y^3)$, and hence

$$k[t^3, t^4] \cong k[X, Y]/(X^4 - Y^3).$$

Proof Let $f \in \mathfrak{p}$. $k[X, Y]$ is a polynomial ring over $k[X]$ with Y as an indeterminate. In this sense, $Y^3 - X^4$ is monic, so we can write

$$f = (Y^3 - X^4)g + (a + bY + cY^2)$$

for some $g \in k[X, Y]$ and $a, b, c \in k[X]$ by Proposition 1.40. Then $a(t^3)+b(t^3)t^4+c(t^3)t^8 = 0$. Since $a(t^3)$, $b(t^3)t^4$, and $c(t^3)t^8$ are linear combinations of $\{t^{3n}\}_{n\ge 0}$, $\{t^{3n+1}\}_{n\ge 0}$, and $\{t^{3n+2}\}_{n\ge 0}$ in $k[t]$, respectively, $a(t^3) = b(t^3)t^4 = c(t^3)t^8 = 0$ must hold. Therefore, $a(t^3) = b(t^3) = c(t^3) = 0$, and hence $a = b = c = 0$. It follows that $f \in (X^4 - Y^3)$. □

Exercise 1.24 Show that the ideal $I = (X^2 - Y, XY - Z)$ of the polynomial ring $A = k[X, Y, Z]$ over a field k is a prime ideal of A.

Example 1.46 Let $A = k[X, Y, Z]$ and $B = k[t]$ be polynomial rings over a field k, and consider the substitution homomorphism $\varphi : A \to B$, $\varphi(X) = t^3$, $\varphi(Y) = t^4$, $\varphi(Z) = t^5$, $\varphi(c) = c$ $(c \in k)$. Let $\mathfrak{p} = \operatorname{Ker} \varphi$. Then $\mathfrak{p}$ is a prime ideal of the ring A and $\mathfrak{p} = (X^3 - YZ,\ Y^2 - XZ,\ Z^2 - X^2Y)$, so

$$k[t^3, t^4, t^5] \cong k[X, Y, Z]/(X^3 - YZ,\ Y^2 - XZ,\ Z^2 - X^2Y).$$

1.4 Euclidean Rings, PID and UFD

In this section, we assume that all rings are integral domains. Let A be a ring and $a, b \in A$. If there exists an $x \in A$ such that $b = ax$, then we write $a \mid b$, and we say that a is a *divisor* of b, or b is a *multiple* of a. This is equivalent to saying that $b \in (a)$. It is also the same as $(b) \subseteq (a)$.

1.47 A necessary and sufficient condition for both $a \mid b$ and $b \mid a$ to hold, or equivalently, for $(a) = (b)$ to hold, is that there exists a unit u in the ring A such that $a = bu$.

Definition 1.48 An element $a \in A$ is said to be a *prime element* of the ring A when $a \neq 0$ and (a) is a prime ideal of A. Furthermore, an element $a \in A$ that is neither zero nor a unit, is said to be *irreducible* in the ring A, or that a is an *irreducible element*, if for any $b, c \in A$ such that $a = bc$, either b or c is a unit in A.

Lemma 1.49 *A prime element is an irreducible element.*

Proof If $a \in A$ is a prime element and $a = bc$, then (a) is a prime ideal of A and $bc \in (a)$, so either $b \in (a)$ or $c \in (a)$ holds. If $b \in (a)$, then expressing $b = da$ and substituting it in $a = bc$, since A is an integral domain and $a \neq 0$, from the equation $a = a(cd)$, $cd = 1$ is obtained, which implies that c is a unit in A. Similarly, b is a unit if $c \in (a)$. Therefore, all prime elements are irreducible. □

In the ring $\mathbb{Z}$, the usual definition of prime numbers is that of irreducible elements. But in general rings, irreducible elements are not necessarily prime.

Example 1.50 In the integral domain $A = \{a + b\sqrt{-5} \mid a, b \in \mathbb{Z}\}$, each of the elements $2, 3, 1 + \sqrt{-5}$ is an irreducible element. Indeed, when we define $N(\alpha) = a^2 + 5b^2$ for $\alpha = a + b\sqrt{-5} \in A$, we have that $N(\alpha\beta) = N(\alpha)N(\beta)$, $N(1) = 1$, and $N(A) \subseteq \mathbb{Z}_{\geq 0}$. As there is no $\alpha \in A$ such that $N(\alpha) = 2$ or 3, any $\alpha \in A$ such that $N(\alpha) \in \{4, 6, 9\}$ is irreducible. As we have $2 \cdot 3 = (1 + \sqrt{-5})(1 - \sqrt{-5})$, it follows that none of $2, 3, 1 + \sqrt{-5}$ is a prime element.

Exercise 1.25 Let k be a field and $k[t]$ be a polynomial ring. In the subring $k[t^2, t^3]$ of $k[t]$, confirm that the elements t^2, t^3 are irreducible but not prime.

Definition 1.51 An integral domain A is said to be a *PID* (principal ideal domain) when all of its ideals are principal ideals.

Proposition 1.52 *Let A be a PID and let a be an element of A that is neither 0 nor a unit. Then, the following conditions are equivalent:*

(1) *a is an irreducible element of A.*
(2) *(a) is a maximal ideal of A.*
(3) *a is a prime element of A.*

Proof We have only to show (1) ⇒(2). Let us take an ideal I such that $(a) \subseteq I \subsetneq A$. Since A is a PID, there exists a non-unit $b \in A$ such that $I = (b)$. Since $a \in (b)$, we can write $a = bc$. From the definition of an irreducible element, c must be a unit, so $(a) = I$. Therefore, (a) is a maximal ideal of A. □

If an element can be factored into a product of prime elements, then that factorization is essentially unique.

Lemma 1.53 *Let $p_1, p_2, \dots, p_m, q_1, q_2, \dots, q_n$ be prime elements of the ring A. If $\prod_{i=1}^m p_i = \prod_{j=1}^n q_j$, then $m = n$ and after an appropriate permutation, we obtain $(p_i) = (q_i)$ for all $1 \le i \le n$.*

Proof Induction on m. If $m = 0$, then $\prod_j q_j = 1$ is a unit, and hence $n = 0$, and the assertion holds. Assume that $m > 0$. Since $\prod_j q_j \in (p_m)$, and p_m is a prime element, there exists some j such that $q_j \in (p_m)$. After an appropriate permutation of indices, we may assume that $j = n$. Then we can write $q_n = cp_m$ for some $c \in A$. Since q is an irreducible element, we have that $c \in A^\times$, and $(q_n) = (p_m)$. So $p_1 \cdots p_{m-1} = cq_1 \cdots q_{n-1}$. If $m = 1$, then $n = 1$ and $c = 1$, and the assertion holds. If $m > 1$, then $n > 1$, and letting $q_1' = cq_1$, we have that $m - 1 = n - 1$ and $(q_i) = (p_i)$ for $2 \le i \le n-1$, and $(q_1) = (q_1') = (p_1)$, and the assertion holds. □

In an integral domain A, when any element that is neither zero nor a unit can always be expressed as a finite product of prime elements, the ring A is said to be a *UFD* (unique factorization domain). If the ring A is a UFD, then any irreducible element in A is necessarily a prime element.

Lemma 1.54 *If A is a Noetherian integral domain, then any element of the ring A that is neither zero nor a unit can always be expressed as a finite product of irreducible elements.*

Proof Suppose there is an element $a \in A \setminus (\{0\} \cup \mathrm{U}(A))$ that does not have such a representation. Let $\mathcal{S}$ be the set of principal ideals (a) such that $(a) \neq (0)$, $(a) \neq (1)$, and a cannot be expressed as a finite product of irreducible elements. Take $a \in A$ so that the ideal (a) is maximal with respect to inclusion within the set $\mathcal{S}$ (Theorem 1.10). Then, since a is reducible, we can choose $b, c \in A$ such that $a = bc$ and neither b nor c are units. Then $(a) \subsetneq (b)$ and $(a) \subsetneq (c)$. Therefore, by the maximality of the ideal (a) within $\mathcal{S}$, both b and c are products of a finite number of irreducible elements. Hence $a = bc$ is also such a product, which contradicts the choice of the element a. □

Since a PID is a Noetherian ring, the following is obtained from Proposition 1.52 and Lemma 1.54.

Proposition 1.55 *A PID is a UFD.*

The Euclidean algorithm is a general method for finding the greatest common divisor of two integers. The concept of a Euclidean ring, which we will describe below, is a generalization of this.

Definition 1.56 Let $A^* = A \setminus \{0\}$. We say that A is a *Euclidean ring* if there exists a function $\varphi : A^* \to \mathbb{Z}$ satisfying the following conditions (1), (2):

(1) For all $a \in A^*$, $\varphi(a) \geq 0$.
(2) Given two elements $a, b \in A$ $(a \neq 0)$, there exist elements $q, r \in A$ that satisfy the equation $b = aq + r$ where $r = 0$ or, if $r \neq 0$, $\varphi(r) < \varphi(a)$ can be chosen.

The ring of integers $\mathbb{Z}$ becomes a Euclidean ring with the map $\varphi : \mathbb{Z}^* \to \mathbb{Z}$, $\varphi(a) = |a|$. The ring of polynomials $A = k[X]$ over a field k becomes a Euclidean ring with the map $\varphi : A \setminus \{0\} \to \mathbb{Z}$, $\varphi(f) = \deg f$. Also, the ring $A = \mathbb{Z}[\sqrt{-1}] = \{a+bi \mid a, b \in \mathbb{Z}\}$ becomes a Euclidean ring with the map $\varphi : A^* \to \mathbb{Z}$, $\varphi(a+bi) = a^2 + b^2$.

Exercise 1.26 Consider the ring $A = \mathbb{Z}[\sqrt{-1}]$ of Gaussian integers.

(1) Show that 7 is a prime element in the ring A.
(2) For 5 and 13, find their prime factorizations in the ring A.
(3) Show that a prime number $p \in \mathbb{Z}$ is also a prime element in $\mathbb{Z}[\sqrt{-1}]$ if and only if $p \equiv 3 \pmod 4$.

Proposition 1.57 *A Euclidean ring is a PID and hence a UFD.*

Proof Let $I \neq 0$ be an ideal of the Euclidean ring A. We seek an element $a \in A$ such that $I = (a)$. Let $n = \min\{\varphi(a) \mid 0 \neq a \in I\}$, and take an element $a \neq 0$ of I such that $n = \varphi(a)$. We will show that $I = (a)$. For any element $b \in I$, since A is a Euclidean ring, we can take elements $q, r \in A$ such that the equation $b = aq + r$ holds, and $r = 0$ or $\varphi(r) < \varphi(a)$. Since $r = b - aq \in I$, we must have $r = 0$ from the minimality of $\varphi(a)$ and we get $b = aq \in (a)$, so $I = (a)$. Therefore, a Euclidean ring is a PID. □

We define the concept of *field of fractions* of an integral domain (see Definition 1.128 for a generalization of this concept).

Definition 1.58 If A is an integral domain, we can construct a field F that contains A as a subring. If we set $K = \{\frac{b}{a} \mid a, b \in A, a \neq 0\}$ within the field F, K is a subring of F and forms a field (note that $\frac{b}{a} = ba^{-1}$, where a^{-1} represents the inverse of a considered in the field F). Therefore, for an element b of A, $\frac{b}{1} = b$, hence $A \subseteq K \subseteq F$. This K is called the *field of fractions* of the integral domain A.

Let us prove that if A is a UFD, then the polynomial ring $A[X]$ is also a UFD.

Hereafter, let A be a UFD and K its field of fractions. Given elements $a_1, a_2, \ldots, a_n \in A$, there exists a $d \in A$ that satisfies the following two conditions, disregarding unit differences in A. This d is called the *GCD (greatest common divisor)* of $a_1, a_2, \ldots, a_n$.

(1) For all $1 \leq i \leq n$, $d \mid a_i$.
(2) If $e \in A$ and $e \mid a_i$ for all $1 \leq i \leq n$, then $e \mid d$.

For $f = a_0 + a_1X + \cdots + a_nX^n \in A[X]$, $f \neq 0$, the GCD of $a_0, a_1, \ldots, a_n$ is denoted by $c(f)$. When $c(f) = 1$, the polynomial $f \in A[X]$ is said to be *primitive*. The primitiveness of a polynomial f is equivalent to the fact that no prime element p of the ring A divides f in the ring $A[X]$.

Since $A[X]/pA[X] \cong (A/(p))\,[X]$ (Exercise 1.20), if $p \in A$ is a prime element of the ring A, then p is also a prime element in the polynomial ring $A[X]$. Therefore, the following can be obtained.

Lemma 1.59 *Every prime element of a ring A is also a prime element in the polynomial ring $A[X]$.*

Proposition 1.60 (Gauss's Lemma) *Let A be a UFD, K its field of fractions, and $f, g \in A[X]$. The following hold true:*

(1) *If f, g are primitive, then their product fg is also primitive.*
(2) *Suppose f is primitive. If f divides g in the polynomial ring $K[X]$, then f also divides g in $A[X]$.*

Proof

(1) If the product fg is not primitive, we can choose a prime element p of the ring A such that it divides all coefficients of the polynomial fg. The ideal $pA[X]$ is a prime ideal in the polynomial ring $A[X]$ and since $fg \in pA[X]$, either $f \in pA[X]$ or $g \in pA[X]$. However, since both f and g are primitive, this is impossible.
(2) It is sufficient to assume $g \neq 0$. In the ring $K[X]$, we have the equation $g = f\varphi$ for some $\varphi \in K[X]$. Take the common denominator $c \in A$ of the coefficients of φ, and let d be the greatest common divisor of the coefficients of the polynomial $c\varphi \in A[X]$. If we take $h \in A[X]$ such that the equation $c\varphi = dh$ holds, then the polynomial h is primitive. Since fh is primitive by (1), from the equation $cg = dfh$, we obtain that d is the greatest common divisor of the coefficients of dfh, and c is a common divisor of the coefficients of dfh. Therefore, $c \mid d$ and we have $\varphi = \frac{d}{c}h \in A[X]$.

□

Theorem 1.61 *If A is a UFD, then the polynomial ring $A[X]$ is also a UFD.*

Proof $K[X]$ is a Euclidean ring, and hence a UFD (Theorem 1.57). Let $f \in A[X]$, f is primitive and $f = \varphi_1\varphi_2\cdots\varphi_n$ be the factorization of f into irreducible elements in $K[X]$. For each φ_i, take the common denominator $c_i \in A$ of the coefficients of φ_i, and let d_i be the greatest common divisor of the coefficients of the polynomial $c_i\varphi_i \in A[X]$. Then, we can choose a primitive polynomial $f_i \in A[X]$ such that the equation $c_i\varphi_i = d_i f_i$ holds.

Hence, $f \cdot \prod_{i=1}^n c_i = \prod_{i=1}^n d_i \cdot \prod_{i=1}^n f_i$. Here, since all f_i are primitive, by Gauss's lemma (1), the polynomial $\prod_{i=1}^n f_i$ is also primitive, and it is understood that the original $\prod_{i=1}^n d_i$ is the greatest common divisor of the coefficients of the polynomial $\prod_{i=1}^n d_i \cdot \prod_{i=1}^n f_i$. On the other hand, since the polynomial f is primitive, $\prod_{i=1}^n c_i$ is also the greatest common divisor of the coefficients of the polynomial $f \cdot \prod_{i=1}^n c_i$. Hence, in the ring A, the equality of ideals $(\prod_{i=1}^n c_i) = (\prod_{i=1}^n d_i)$ is obtained, and $f = u \cdot \prod_{i=1}^n f_i$ by the unit u of A. By Gauss's lemma (Proposition 1.60), (2), we have

$$f_i A[X] = f_i K[X] \cap A[X] = \varphi_i K[X] \cap A[X].$$

Since φ_i is a prime element in $K[X]$, it follows that f_i is a prime element in the ring $A[X]$. Hence, if A is a UFD, then the polynomial ring $A[X]$ is also a UFD. □

Corollary 1.62 *If A is a UFD, then the polynomial ring $A[X_1, X_2, \ldots, X_n]$ is also a UFD. Hence, the polynomial ring $k[X_1, X_2, \ldots, X_n]$ over a field k is a UFD.*

Lemma 1.63 *Let A be a UFD and S be a multiplicatively closed subset that does not contain 0. Then the localization $S^{-1}A$ (see Definition 1.124) is a UFD.*

Proof Left to the reader. □

1.5 Modules

Module theory is the representation theory of rings. One way to understand the structure of a given ring is to elucidate how various ideals are contained in the ring, but to do so, only the internal structure analysis of the ring is insufficient, and module theory, as a representation theory of rings, is necessary.

Let us first summarize the basics of module theory. Hereafter, let A be a ring.

Definition 1.64 (A-Module) We say that M is an A-module if M is an additive group (i.e., an Abelian group whose operation is denoted by $+$) and there is an action of the ring A, that is, a map $\mu : A \times M \to M$, $\mu(a, x) = ax$ is given, and for any $a, b \in A$, $x, y \in M$, the following equations hold:

(1) $(a + b)x = ax + bx$,
(2) $a(x + y) = ax + ay$,
(3) $a(bx) = (ab)x$,
(4) $1x = x$.

Note that, if $a \in A$ and $x, y \in M$, then $a0 = 0x = 0$, $a(-x) = (-a)x = -ax$, $a(x - y) = ax - ay$, and $(a - b)x = ax - bx$ hold. We only show $a0 = 0$. We have: $a0 = a0 + 0 = a0 + (a0 - a0) = (a0 + a0) - a0 = a(0 + 0) - a0 = a0 - a0 = 0$.

Example 1.65

(1) A $\mathbb{Z}$-module is nothing but an additive group.
(2) Modules over a field are called vector spaces.
(3) The ring A is an A-module under its multiplication.
(4) If $\{M_i\}_{i\in I}$ is a family of A-modules, then the direct product $\prod_{i\in I} M_i$ and the direct sum $\bigoplus_{i\in I} M_i$ as Abelian groups become A-modules by the operation $a\cdot\{m_i\}_{i\in I} = \{am_i\}_{i\in I}$. If I is a finite set, then $\prod_{i\in I} M_i = \bigoplus_{i\in I} M_i$.
(5) If M is an A-module, we denote the n-fold direct sum of M as M^n. Elements of M^n are often represented as column vectors: $M^n = \left\{ \begin{pmatrix} x_1 \\ x_2 \\ \vdots \\ x_n \end{pmatrix} \middle|\ x_i \in M \right\}$.
(6) If $\varphi : A \to B$ is a ring homomorphism and L is a B-module, then L becomes an A-module by defining the action of the ring A as $a\cdot x = \varphi(a)x$ $(a \in A, x \in L)$. Consequently, B is an A-module.

Definition 1.66 (A-Module Homomorphism, Linear Map) Let M, M' be A-modules, and $\varphi : M \to M'$ be a map. The map φ is called an A-homomorphism (or an A-linear map) if for any $x, y \in M$ and $a \in A$, the following equations hold:

(1) $\varphi(x + y) = \varphi(x) + \varphi(y)$;
(2) $\varphi(ax) = a\varphi(x)$.

Exercise 1.27 If M is an A-module, then for any $a \in A$, the map $\widehat{a} : M \to M$, $\widehat{a}(x) = ax$ is A-linear.[2] Such a linear map is sometimes called a *homothety*. If a is a unit of the ring A, then $\widehat{a} : M \to M$ is a bijection. Verify this.

1.67 Let $m, n \geq 1$ be natural numbers. Let $\varphi : A^n \to A^m$ be an A-linear map. If $\boldsymbol{e}_1, \ldots, \boldsymbol{e}_n$ is the standard basis of A^n and P is the (m, n)-matrix whose jth column is $\varphi(\boldsymbol{e}_j)$, then we have $\varphi(\boldsymbol{x}) = P\boldsymbol{x}$ for any $\boldsymbol{x} \in A^n$. Just like in the case of vector spaces, A-linear maps are diverse. In addition to this, not all A-modules take the form A^n in a general ring, so a more complex and rich world unfolds.

1.68 The composition of linear maps is linear. For an A-module M, the identity map 1_M is linear.

A linear map $\varphi : M \to M'$ between A-modules is said to be an *isomorphism* when it is bijective. When there exists at least one isomorphism $\varphi : M \to M'$ between A-modules M, M', we say that M and M' are *isomorphic*, and we write $M \cong M'$. The inverse of an isomorphism $\varphi : M \to M'$ of A-modules, $\varphi^{-1} : M' \to M$, is linear, hence $\cong$ is an equivalence relation among A-modules.

Definition 1.69 (A-Submodule) Let M be an A-module. A non-empty subset N of M is said to be an A-*submodule* of M if for any $x, y \in N$ and $a \in A$, we have

[2] The commutativity of the ring is necessary.

$x + y, ax \in N$. A submodule of A is a subgroup of M with respect to addition, and becomes an A-module under the action of the ring A. If N is a submodule of M, the inclusion $\iota : N \to M$ is A-linear.

Note that M and $\{0\}$ are A-submodules of M. The A-submodule $\{0\}$ is often simply denoted by 0, if there is no danger of confusion. An ideal of A is an A-submodule of the ring A. In this sense, module theory is a natural extension of ideal theory. Let $\varphi : M \to M'$ be an A-linear map between A-modules. We define $\operatorname{Ker}\varphi = \{x \in M \mid \varphi(x) = 0\}$, and call it the *kernel* of φ. $\operatorname{Ker}\varphi$ is a submodule of M, and $\operatorname{Im}\varphi$ is a submodule of M'. φ is injective if and only if $\operatorname{Ker}\varphi = 0$.

Exercise 1.28 Let $\varphi : M \to M'$ be a linear map, X be a submodule of M, and X' be a submodule of M'. Verify that the image of X under φ, denoted by $\varphi(X)$, is a submodule of M', and the preimage of X' under φ, denoted by $\varphi^{-1}(X')$, is a submodule of M.

Definition 1.70 (Quotient Module) Let M be an A-module and N be a submodule of M. The quotient group M/N becomes an A-module under the operation $a \cdot \overline{x} = \overline{ax}$ $(a \in A, x \in M)$. Here, $\overline{x}$ denotes the residue class of x modulo N. We call this A-module M/N the *quotient module* of M by N. The natural surjection $f : M \to M/N$, $f(x) = \overline{x}$ is A-linear and satisfies $\operatorname{Ker} f = N$.

Let us state the fundamental theorem on homomorphisms, the isomorphism theorem, and the correspondence theorem for modules.

Theorem 1.71 (Fundamental Theorem on Homomorphisms) *If $\varphi : M \to M'$ is an A-linear map between A-modules and if N is an A-submodule of M such that $N \subseteq \operatorname{Ker}\varphi$, then there exists a unique A-linear map $\phi : M/N \to M'$ such that $\varphi = \phi f$. Here, $f : M \to M/N$, $f(x) = \overline{x}$ is the natural surjection. Moreover, we have $\operatorname{Im}\phi = \operatorname{Im}\varphi$, and $\operatorname{Ker}\phi = f(\operatorname{Ker}\varphi)$. In particular, ϕ is an injection if and only if $N = \operatorname{Ker}\varphi$.*

Corollary 1.72 (Isomorphism Theorem) *If $\varphi : M \to M'$ is A-linear, then $M/\operatorname{Ker}\varphi \cong \operatorname{Im}\varphi$ as A-modules.*

Exercise 1.29 Let M be an A-module, N be an A-submodule of M and $f : M \to M/N$ be the natural surjection. Let $\mathcal{X}$ be the set of all A-submodules of M that contain N, and $\mathcal{Y}$ be the set of all A-submodules of the quotient module M/N. Prove the following:

(1) For any $J \in \mathcal{X}$ and $K \in \mathcal{Y}$, the image of J under f, denoted by $f(J)$, is an A-submodule of M/N, and the preimage of K under f, denoted by $f^{-1}(K)$, is an A-submodule of M that contains N. Therefore, $f(J) \in \mathcal{Y}$ and $f^{-1}(K) \in \mathcal{X}$.
(2) The map $\Phi : \mathcal{X} \to \mathcal{Y}$, $\Phi(J) = f(J)$ and the map $\Psi : \mathcal{Y} \to \mathcal{X}$, $\Psi(K) = f^{-1}(K)$ are both bijective, and they are inverses of each other.
(3) Both maps Φ, Ψ preserve the inclusion relation.
(4) For any $K \in \mathcal{Y}$, $M/f^{-1}(K) \cong (M/N)/K$.

Let M be an A-module, L and N be submodules of M, and $\mathfrak{a}$ be an ideal of the ring A. Then $L + N = \{x + y \mid x \in L, y \in N\}$ and $L \cap N$ are also submodules of M, and

$$\mathfrak{a}N = \{a_1x_1 + a_2x_2 + \ldots + a_nx_n \mid n \geq 1, a_i \in \mathfrak{a}, x_i \in N\}$$

is a submodule of M. In general, when a family of submodules $\{N_i\}_{i\in I}$ of M is given,

$$\sum_{i\in I} N_i = \{\sum_{i\in I} x_i \mid x_i \in N_i \text{ and for all but finitely many } i, x_i = 0\}$$

is called the *sum* of $\{N_i\}_{i\in I}$ and is a submodule of M. The intersection $\bigcap_{i\in I} N_i$ is a submodule of M, where we understand that $\bigcap_{i\in I} N_i = M$ if I is the empty set.

Exercise 1.30 Let M be an A-module and L, N be submodules of M. Prove that $L/(L \cap N) \cong (L + N)/N$.

Exercise 1.31 Let M be an A-module and $\mathfrak{a}$ be an ideal of A. Verify the following assertions:

(1) If $\mathfrak{a}M = 0$, then M becomes an $A/\mathfrak{a}$-module by the following operation: $\overline{a}x = ax$, $(a \in A, x \in M)$. Here, $\overline{a}$ represents the image of the element $a \in A$ in the ring $A/\mathfrak{a}$.
(2) Since $\mathfrak{a} \cdot (M/\mathfrak{a}M) = 0$, $M/\mathfrak{a}M$ is also an $A/\mathfrak{a}$-module.
(3) If $\varphi : M \to M'$ is an A-linear map, then φ induces an $A/\mathfrak{a}$-linear map

$$\overline{\varphi} : M/\mathfrak{a}M \to M'/\mathfrak{a}M', \quad \overline{\varphi}(\overline{x}) = \overline{\varphi(x)}$$

(here, $\overline{x}$ and $\overline{\varphi(x)}$ represent the images of $x \in M$ and $\varphi(x) \in M'$ in $M/\mathfrak{a}M$ and $M'/\mathfrak{a}M'$, respectively).

Definition 1.73 (Generator, Finitely Generated Modules, Finitely Represented Modules) Let M be an A-module.

(1) For an element x of M, define $Ax = \{ax \mid a \in A\}$. Ax is an A-submodule of M. For a subset S of M, define

$$AS = \sum_{s\in S} As = \{m \in M \mid \exists r \geq 0\ \exists s_1, \ldots, s_r \in S$$
$$\exists a_1, \ldots, a_r \in A\ m = a_1s_1 + \cdots + a_rs_r\},$$

which is called the A-submodule of M generated by S. In fact, AS is the smallest A-submodule of M that contains the subset S. If $AS = M$, we say that S is a *generator* of M. We also say that S generates M.

(2) When M can be generated by a finite number of elements, that is, when a finite subset S can be chosen such that $M = AS$, M is said to be a *finitely generated* A-module.

Exercise 1.32 (cf. Exercise 1.4) Verify that $\mathbb{Q}$ is not finitely generated as a $\mathbb{Z}$-module.

Let M be an A-module. Consider a family of A-modules $\{M_i\}_{i \in I}$ indexed by a set I, where each $M_i = M$ for all $i \in I$. The direct sum $\bigoplus_{i \in I} M_i$ is denoted by $M^{(I)}$ (when $I = \emptyset$, we define $M^{(I)} = 0$).

Definition 1.74 An A-module M is said to be a *free A-module* when there exists a set I such that $M \cong A^{(I)}$.

Definition 1.75 Let M be an A-module and $x_1, \ldots, x_n$ be elements of M. If $a_1x_1 + \cdots + a_nx_n \neq 0$ for $(a_1, \ldots, a_n) \in A^n \setminus \{(0, \ldots, 0\}$, we say that $x_1, \ldots, x_n$ is linearly independent. A family of elements $\{x_i\}_{i \in I}$ of M is said to be linearly independent if every finitely many distinct elements $x_{i_1}, \ldots, x_{i_n}$ of this family is linearly independent. We say that $\{x_i\}_{i \in I}$ is a *free basis* of M if it is linearly independent and generates M.

In the case $M = A^{(I)}$, we denote by $\mathbf{e}_i$ the element of $A^{(I)}$ whose jth component is $\delta_{i,j}$. The set $\{\mathbf{e}_i\}_{i \in I}$ is called the *standard free basis* of $A^{(I)}$.

Proposition 1.76 *Let M be an A-module. Given a family of elements $\{x_i\}_{i \in I}$ in M, there exists a unique A-linear map $\varphi : A^{(I)} \to M$ such that $\varphi(\mathbf{e}_i) = x_i$ for any $i \in I$. The map φ is injective (resp. surjective, isomorphic) if and only if $\{x_i\}_{i \in I}$ is linearly independent (resp. generates M, is a free basis of M). Hence, the following two conditions regarding M are equivalent:*

(1) *M is a free A-module.*
(2) *M has a free basis.*

A general A-module can be obtained as a quotient of a free A-module.

Corollary 1.77 *Any A-module M is a homomorphic image of some free A-module F. In other words, for given M, we can choose a free A-module F and its A-submodule K such that $M \cong F/K$.*

Proof For M, if we take $F = A^{(M)}$ and the A-linear map $\varphi : F \to M$ such that $\varphi(\mathbf{e_m}) = m$ holds for each $m \in M$, then $M \cong F/\mathrm{Ker}\,\varphi$ because φ is surjective. □

Exercise 1.33 Given a set S, confirm that we can construct a free A-module F such that $S \subseteq F$ and S is its free basis.

Every vector space over a field has a free basis (fundamental theorem of abstract vector spaces). The converse is also true.

Proposition 1.78 *If every A-module is a free A-module, then A is a field.*

Proof Let $\mathfrak{a}$ be an ideal of A with $\mathfrak{a} \neq A$. Since $A/\mathfrak{a}$ is a free A-module, it has a free basis $\{\overline{x_i}\}_{i \in I}$ ($x_i \in A$). Here, since $A/\mathfrak{a} \neq 0$, $I \neq \emptyset$. For any $i \in I$ and $a \in \mathfrak{a}$, we have $a \cdot \overline{x_i} = \overline{ax_i} = 0$, implying $a = 0$. Therefore, the ring A contains only the ideals A and 0. □

Exercise 1.34 Let $\mathfrak{a}$ be an ideal of a ring A. Show that unless $\mathfrak{a}$ is principal, $\mathfrak{a}$ is not a free A-module.

Let M be an A-module and given elements $x_1, x_2, \ldots, x_n \in M$, we can define $\varphi : A^n \to M$ such that $\varphi(\mathbf{e}_i) = x_i$ $(1 \leq i \leq n)$ (Proposition 1.76). Then φ is surjective if and only if M is generated by $x_1, x_2, \ldots, x_n$. Hence the condition that M is finitely generated is equivalent to saying that M is a homomorphic image of a finitely generated free A-module.

Definition 1.79 (Noetherian Module) Let M be an A-module. If every submodule of M is a finitely generated A-module, M is said to be a *Noetherian* A-module.

Of course, a Noetherian A-module is finitely generated. The proof of the following theorem is exactly the same as that of Theorem 1.10.

Proposition 1.80 *The following three conditions on an A-module M are equivalent:*

(1) *M is a Noetherian A-module.*
(2) *If $N_1 \subseteq N_2 \subseteq \cdots \subseteq N_i \subseteq \cdots$ is a sequence of A-submodules of M, then there exists a number $k \geq 1$ such that $N_k = N_i$ holds for all $i \geq k$.*
(3) *Let $\mathcal{F}$ be the set of all A-submodules of M. Then any non-empty subset $\mathcal{S}$ of $\mathcal{F}$ contains a maximal element with respect to the inclusion relation.*

Theorem 1.81 *The following two conditions for a ring A are equivalent:*

(1) *A is a Noetherian ring.*
(2) *All finitely generated A-modules are Noetherian A-modules.*

Proof (2) $\Rightarrow$(1) Since A is a finitely generated A-module, all ideals of the ring A are finitely generated.

(1) $\Rightarrow$(2) If M is a finitely generated A-module, then we can take a surjective A-linear map $\varphi : A^n \to M$. Then any A-submodule N of M is the homomorphic image of the submodule $\varphi^{-1}(N)$ of A^n. Therefore, it is sufficient to show that all submodules of A^n are finitely generated. This is the definition of a Noetherian ring when $n = 1$. Assume it is true up to $n - 1$ for $n > 1$. Consider the linear map $p : A^n \to A^{n-1}$, where $p(a_1, a_2, \ldots, a_n) = (a_2, a_3, \ldots, a_n)$. Then p is surjective, and $\operatorname{Ker} p \cong A$. Let N be an A-submodule of A^n. The submodule $p(N)$ of A^{n-1} is finitely generated by the induction hypothesis. On the other hand, $N \cap \operatorname{Ker} p$ is isomorphic to an ideal of A and is finitely generated. If we take elements $x_1, x_2, \ldots, x_k$ and $y_1, y_2, \ldots, y_\ell$ such that $p(N) = \sum_{i=1}^{k} Ap(x_i)$ and $N \cap \operatorname{Ker} p = \sum_{j=1}^{\ell} Ay_j$, then $N = \sum_{i=1}^{k} Ax_i + \sum_{j=1}^{\ell} Ay_j$.

In fact, if we take $x \in N$, since $p(x) \in p(N)$, we can express $p(x) = \sum_{i=1}^{k} a_i p(x_i)$ where $a_i \in A$. Since $x - \sum_{i=1}^{k} a_i x_i \in N \cap \operatorname{Ker} p$, we have $x - \sum_{i=1}^{k} a_i x_i = \sum_{j=1}^{\ell} b_j y_j$ where $b_j \in A$, and thus $x \in \sum_{i=1}^{k} Ax_i + \sum_{j=1}^{\ell} Ay_j$. Therefore, N is finitely generated. □

Corollary 1.82 (Corollary of the Proof of Theorem 1.81) *Let A be a PID and N be a submodule of the free A-module A^n. Then N is a free A-module and has a free basis consisting of at most n elements.*

Proof The proof is similar to Theorem 1.81 using induction on n. Since A is a PID, it is trivially true for $n = 1$. Assume that the statement is true for $n - 1$ where $n > 1$. Let N be a submodule of A^n. Since $p(N)$ is a free A-module by our induction hypothesis, in the notation of the proof of Theorem 1.81, we can take $x_1, x_2, \ldots, x_k \in N$ $(k \leq n-1)$ so that $p(x_1), p(x_2), \ldots, p(x_k)$ form a free basis of $p(N)$. Since $N \cap \operatorname{Ker} p$ is isomorphic to an ideal of the ring A, either $N \cap \operatorname{Ker} p = 0$ or $N \cap \operatorname{Ker} p = (y)$. Then it is easy to see that if $N \cap \operatorname{Ker} p = 0$,$\{x_1, x_2, \ldots, x_k\}$ is a free basis of N and if $N \cap \operatorname{Ker} p = (y)$, then $\{y, x_1, x_2, \ldots, x_k\}$ is a free basis of N. Therefore, N is a free A-module with a basis of at most n elements. □

Theorem 1.83 *If A is a PID, then any A-submodule of a free A-module is a free module.*

Proof (S. Lang) Let F be a free A-module with a free basis B. Let M be a submodule of F, and we want to show that M is free. For $C \subseteq B$, we set $F_C = \sum_{c \in C} Ac$, and $M_C = F_C \cap M$. We define

$$\mathcal{S} := \{(C, X) \mid C \subseteq B,\ X \subseteq M_C,\ X \text{ is a free basis of } M_C\}.$$

For $(C, X), (D, Y) \in \mathcal{S}$, we say that $(C, X) \leq (D, Y)$ if $C \subseteq D$ and $X \subseteq Y$. It is easy to see that $\mathcal{S}$ satisfies the assumption of Zorn's lemma, and hence $\mathcal{S}$ has a maximal element, say, (C, X). It suffices to show that $C = B$ to prove the theorem. Assume the contrary. Then we can take $b \in B \setminus C$. We set $D = C \cup \{b\}$. If $M_D = M_C$, then $(C, X) < (D, X) \in \mathcal{S}$, and this contradicts the maximality of (C, X). As the kernel of the composite map

$$M_D \hookrightarrow F_D \to F_D/F_C \cong Ab \cong A$$

is M_C, we have that M_D/M_C is isomorphic to a nonzero ideal of A. Since A is a PID, $M_D/M_C \cong A$. If we take $y \in M_D$ such that the image of y in M_D/M_C is a free generator, then $Y = X \cup \{y\}$ is a free basis of M_D, and $(C, X) < (D, Y) \in \mathcal{S}$. This is also a contradiction, and we are done. □

Exercise 1.35 Let $\varphi : A \to B$ be a ring homomorphism. Prove the following.

(1) Let L be a B-module. If L is a finitely generated A-module, then L is also a finite generated B-module.
(2) If A is a Noetherian ring and B is a finitely generated A-module, then B is a Noetherian ring.

Definition 1.84 A finitely generated A-module M is said to have a *finite presentation* when it can be expressed as $M \cong F/K$, where F is a finitely generated free A-module and K is a finitely generated A-submodule of F.

Corollary 1.85 *If the ring A is Noetherian, then all finitely generated A-modules have a finite presentation.*

The following proposition is often referred to as *Nakayama's*[3] *lemma*, but in Masayoshi Nagata's[4] "Local Rings [Nag]," it is referred to as the *lemma of Krull*[5]*–Azumaya.*[6] For simplicity, we will abbreviate it as *NAK* in this book.

Proposition 1.86 (NAK) *Let M be a finitely generated A-module and let N be an A-submodule of M. Let $J = \mathrm{J}(A)$. If $M = N + JM$, then $M = N$. Furthermore, if $M = JM$, then $M = 0$.*

Proof If we set $\overline{M} = M/N$ and $M = N + JM$, then $\overline{M} = J\overline{M}$. Since $\overline{M}$ is finitely generated, it is sufficient to show the latter statement. Since M is finitely generated, we can select a finite set of elements $x_1, x_2, \ldots, x_n$ in M and express $M = \sum_{i=1}^{n} Ax_i$. Then, $x_n \in JM = \{a_1x_1 + a_2x_2 + \ldots + a_nx_n \mid a_i \in J\}$, so we can write $x_n = \sum_{i=1}^{n} a_ix_i$ $(a_i \in J)$. Therefore, $(1 - a_n)x_n = \sum_{i=1}^{n-1} a_ix_i$ and $1-a_n$ is a unit of the ring A (from Exercise 1.15), so $x_n \in \sum_{i=1}^{n-1} Ax_i$. Consequently, $M = \sum_{i=1}^{n-1} Ax_i$. This argument can be repeated any number of times, so we obtain $M = 0$. □

Let $(A, \mathfrak{m})$ be a local ring and M be an A-module. We say that elements $x_1, \ldots, x_n$ of M form a *minimal set of generators* of M if $x_1, \ldots, x_n$ generates M, but any proper subset of $\{x_1, \ldots, x_n\}$ does not. Then $M/\mathfrak{m}M$ is a vector space over the residue field $A/\mathfrak{m}$. We denote the dimension of $M/\mathfrak{m}M$ as a vector space by $\dim_{A/\mathfrak{m}} M/\mathfrak{m}M$. Note that a minimal set of generators of a vector space over a field is nothing but a basis (that is, a linearly independent set of generators).

Corollary 1.87 *Let $(A, \mathfrak{m})$ be a local ring, M a finitely generated A-module, and $x_1, \ldots, x_n \in M$.*

(1) *The elements $x_1, \ldots, x_n$ generate M if and only if their images generate the $A/\mathfrak{m}$-vector space $M/\mathfrak{m}M$.*
(2) *$x_1, \ldots, x_n$ form a minimal set of generators of M if and only if their images form an $A/\mathfrak{m}$-basis of the vector space $M/\mathfrak{m}M$.*
(3) *If $x_1, \ldots, x_n$ and $y_1, \ldots, y_m$ form a minimal set of generators of M, then $n = m = \dim_{A/\mathfrak{m}} M/\mathfrak{m}M$.*

---

[3] Tadashi Nakayama, 1912–1964.

[4] Masayoshi Nagata, 1927–2008.

[5] Wolfgang Krull, 1899–1971.

[6] Goro Azumaya, 1920–2010.

Proof

(1) The 'only if' part is obvious. Conversely, if the images of $x_1, \ldots, x_n$ generate $M/\mathfrak{m}M$, then we have $M = \sum_{i=1}^n Ax_i + \mathfrak{m}M$, and it follows that $M = \sum_{i=1}^n Ax_i$ by Theorem 1.86.
(2) follows immediately from (1), and (3) follows immediately from (2).

□

Exercise 1.36 Let $(A, \mathfrak{m})$ be a local ring. Consider an A-linear map $\varphi : M \to M'$ between A-modules. If $\mathfrak{a}$ is an ideal of the ring A with $\mathfrak{a} \neq A$, let $\overline{\varphi} : M/\mathfrak{a}M \to M'/\mathfrak{a}M'$ be the homomorphism of $A/\mathfrak{a}$-modules defined by $\overline{\varphi}(\overline{x}) = \overline{\varphi(x)}$. If M' is finitely generated, verify that φ is surjective if and only if $\overline{\varphi}$ is surjective.

Definition 1.88 (Annihilator) Let M be an A-module.

(1) Let N, L be submodules of M. We define

$$N :_A L = \{a \in A \mid \text{for all } x \in L, ax \in N\}.$$

$N :_A L$ is an ideal of A. When L is generated by a single element x, it is common to write $N :_A L$ as $N :_A x$.
(2) We write $\mathrm{Ann}_A(M) = 0 :_A M$, and call it the *annihilator* of the module M. An A-module M such that $\mathrm{Ann}_A M = 0$ is said to be *faithful*.

Exercise 1.37 $M = \bigoplus_{p:\text{prime}} \mathbb{Z}/(p)$ and $\mathbb{Q}/\mathbb{Z}$ are faithful $\mathbb{Z}$-modules. Verify this.

Exercise 1.38 Let $\varphi : A \to B$ be a ring homomorphism. Show that B is a faithful A-module if and only if φ is injective.

Exercise 1.39 Let M be an A-module and L, N be submodules of M. Verify the following assertions:

(1) $N :_A L = \mathrm{Ann}_A((L + N)/N)$.
(2) $L \subseteq N$ and $N :_A L = A$ are equivalent.
(3) If we set $\mathfrak{a} = \mathrm{Ann}_A M$, then M is a faithful $A/\mathfrak{a}$-module.
(4) If $M = \sum_{i=1}^n Ax_i$ is finitely generated, and we consider the homomorphism $\varphi : A \to M^n$ defined by $\varphi(a) = (ax_1, ax_2, \ldots, ax_n)$, then φ is A-linear and $\mathrm{Ann}_A M = \bigcap_{i=1}^n [0 :_A x_i] = \mathrm{Ker}\,\varphi$.

1.89(Idealization) Let us explain the principle of idealization. Let A be a ring and M be an A-module. Then we can consider $B = A \oplus M$ as a ring by the following product:

$$(a, x)(b, y) = (ab, ay + bx).$$

This ring B is called the *idealization* of M over A, and we write $B = A \ltimes M$. The homomorphisms $f : A \to B,\ f(a) = (a, 0)$ and $p : B \to A, p(a, x) = a$ are ring homomorphisms and $p \circ f = 1_A$. Let $\mathfrak{a} = 0 \times M$. Since $\mathfrak{a} = \mathrm{Ker}\, p$, $\mathfrak{a}$ is an ideal of the ring B. When we view the A-module M as a B-module through the morphism $p : B \to A$, we write ${}_pM$ for distinction, and the homomorphism $\alpha : {}_pM \to \mathfrak{a}$, $\alpha(x) = (0, x)$ is an isomorphism of B-modules, so through the isomorphism α, the A-module M is identified with the ideal $\mathfrak{a}$ of the ring B. This is why B is called the idealization of M.

Through the idealization B, the structural analysis of the module M is essentially reduced to that of the ring B.

Theorem 1.90 (Principle of Idealization and Noetherian Property) *The following properties hold for the idealization $B = A \ltimes M$:*

(1) $\mathfrak{a}^2 = 0$.
(2) *The ring B is Noetherian if and only if A is Noetherian and the A-module M is finitely generated.*
(3) *If we denote the set of all prime ideals of the ring B as* Spec B *and the set of all prime ideals of the ring A as* Spec A, *then* $\operatorname{Spec} B = \{\mathfrak{p} \times M \mid \mathfrak{p} \in \operatorname{Spec} A\}$.

Proof

(1) For any $x, y \in M$, $(0, x)(0, y) = (0, 0)$.
(2) If B is Noetherian, then A, which is a homomorphic image of B, is also Noetherian. Since ${}_pM \cong \mathfrak{a}$, ${}_pM$ is a finitely generated B-module. Therefore, M is also finitely generated as an A-module. Conversely, if A is a Noetherian ring and the A-module M is finitely generated, then through the ring homomorphism $f : A \to B$, if we consider B as an A-module, B is simply the direct sum of A-modules A and M. Therefore, the A-module B is finitely generated. Because ideals of B are also A-modules, they are necessarily finitely generated as A-modules, and also finitely generated as B-modules. Hence the ring B is Noetherian.
(3) For each $\mathfrak{p} \in \operatorname{Spec} A$, because $\mathfrak{p} \times M = p^{-1}(\mathfrak{p})$, $\mathfrak{p} \times M \in \operatorname{Spec} B$. Because $\mathfrak{a}^2 = 0$, any $P \in \operatorname{Spec} B$ also contains $\mathfrak{a} = \operatorname{Ker} p$. Since the ring homomorphism $p : B \to A$ is surjective, the equation $\operatorname{Spec} B = \{\mathfrak{p} \times M \mid \mathfrak{p} \in \operatorname{Spec} A\}$ holds.

□

Corollary 1.91 *Let M be a finitely generated A-module. If $(A, \mathfrak{m})$ is a Noetherian local ring, then the idealization $B = A \ltimes M$ is also a Noetherian local ring.*

1.6 Sequences of Modules and Exact Sequences

Let A be a ring.

Definition 1.92 (Exactness and Short Exact Sequences) We say that $X \xrightarrow{\alpha} Y \xrightarrow{\beta} Z$ is a sequence of A-modules if X, Y, Z are A-modules and α, β are A-linear maps. When a sequence of A-modules $X \xrightarrow{\alpha} Y \xrightarrow{\beta} Z$ is given, we say that this sequence is *exact* at Y if $\operatorname{Im} \alpha = \operatorname{Ker} \beta$. An A-linear map $\alpha : X \to Y$ is said to be a *zero* *map* when $\alpha(x) = 0$ for all $x \in X$, and we write $\alpha = 0$. A sequence of the form $0 \to X \xrightarrow{\alpha} Y \xrightarrow{\beta} Z \to 0$ is called a *short exact sequence* when it is exact at X, Y, Z. This condition means that α is injective, β is surjective, and it is exact at Y.

We also define the exactness of longer sequences.

Definition 1.93 (Exact Sequence) Consider a sequence

$$\cdots \to M_{i+1} \xrightarrow{f_{i+1}} M_i \xrightarrow{f_i} M_{i-1} \to \cdots \qquad (\natural)$$

of A-modules and A-linear maps. When $f_i \circ f_{i+1} = 0$ (or equivalently, $\operatorname{Im} f_{i+1} \subseteq \operatorname{Ker} f_i$) holds for all i, then this sequence is said to form a *chain complex*, and when $\operatorname{Im} f_{i+1} = \operatorname{Ker} f_i$ holds for all i, this sequence is said to be an *exact sequence*. The chain complex $(\natural)$ can be decomposed into a bunch of complexes

$$0 \to \operatorname{Im} f_{i+1} \xrightarrow{\iota_i} M_i \xrightarrow{f_i'} \operatorname{Im} f_i \to 0, \qquad (\flat_i)$$

where $\iota_i : \operatorname{Im} f_{i+1} \to M_i$ is the inclusion, and $f_i' : M \to \operatorname{Im} f_i$ is the map induced by f_i. Defining $f_i = \iota_{i-1} f_i'$, we can recover the complex $(\natural)$ from the collection of complexes $\{(\flat_i)\}_{i \in \mathbb{Z}}$

The chain complex $(\natural)$ is exact if and only if $(\flat_i)$ is a short exact sequence for every i.

For an A-linear map $\varphi : M \to M'$, we define $\operatorname{Coker}\varphi = M'/\operatorname{Im}\varphi$, and call it the *cokernel* of the linear map φ. Given an A-linear map $\varphi : M \to M'$ of A-modules, there is a natural exact sequence

$$0 \to \operatorname{Ker}\varphi \xrightarrow{\iota} M \xrightarrow{\varphi} M' \xrightarrow{\varepsilon} \operatorname{Coker}\varphi \to 0,$$

where $\iota : \operatorname{Ker}\varphi \to M$ is the inclusion map and $\varepsilon : M' \to \operatorname{Coker}\varphi$ is the natural surjection. Then, the pair of A-modules and A-module homomorphisms $(\operatorname{Ker}\varphi, \iota)$ and $(\operatorname{Coker}\varphi, \varepsilon)$ are characterized as *universal* under these conditions. In other words, we have the following proposition.

Proposition 1.94 *Let* $\varphi : M \to M'$ *be as above.*

(1) *If* $\psi : M' \to L$ *is an A-linear map satisfying* $\psi \circ \varphi = 0$, *then* ψ *uniquely factors through* $\operatorname{Coker}\varphi$. *That is, an A-linear map* $\xi : \operatorname{Coker}\varphi \to L$ *is uniquely determined, so that the following diagram commutes:*

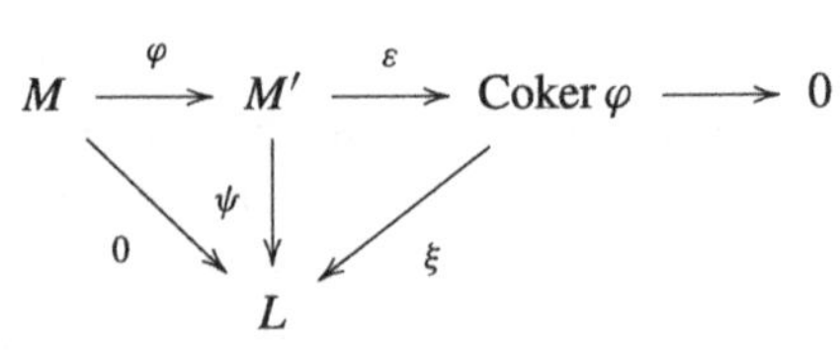

(2) *If an* A*-linear map* $\psi : N \to M$ *satisfies* $\varphi \circ \psi = 0$*, then* ψ *uniquely factors through* $\operatorname{Ker}\varphi$*. That is, an* A*-linear map* $\xi : N \to \operatorname{Ker}\varphi$ *is uniquely determined, and the following diagram commutes:*

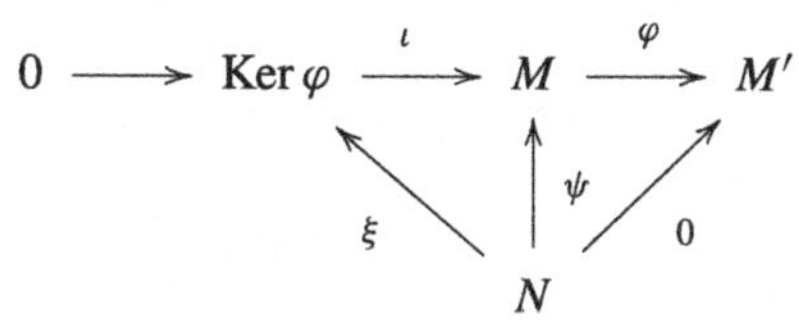

Note that Proposition 1.94, (1) is a restatement of Theorem 1.71.

Let us discuss the splitting of short exact sequences.

1.95 Let X, Y, and Z be A-modules, and

$$X \underset{f}{\overset{\alpha}{\rightleftarrows}} Y \underset{g}{\overset{\beta}{\rightleftarrows}} Z$$

be a diagram of A-modules. We say that the diagram (#) forms a *biproduct* if

$$f\alpha = 1_X, \quad \beta g = 1_Z, \quad \text{and} \quad g\beta + \alpha f = 1_Y.$$

If a biproduct (#) is given, then α and g are split injective, and β and f are split surjective (see Appendices, A.2). Moreover, $\beta\alpha f = \beta(1_Y - g\beta) = \beta - \beta g\beta = \beta - \beta = 0$. So $\beta\alpha = \beta\alpha f\alpha = 0\alpha = 0$, and hence $\operatorname{Im}\alpha \subseteq \operatorname{Ker}\beta$. On the other hand, if $y \in Y$ and $\beta(y) = 0$, then $y = (g\beta + \alpha f)(y) = \alpha(fy) \in \operatorname{Im}\alpha$, and hence we have $\operatorname{Im}\alpha = \operatorname{Ker}\beta$. It follows that the sequence

$$0 \to X \overset{\alpha}{\to} Y \overset{\beta}{\to} Z \to 0$$

is exact. Similarly, the sequence

$$0 \to Z \overset{g}{\to} Y \overset{f}{\to} X \to 0$$

is also exact. Moreover, we have $Y = \operatorname{Im}\alpha \oplus \operatorname{Im} g \cong X \oplus Z$, see Lemma C.1.

If Y is an A-module and X and Z are A-submodules of Y such that $Y = X \oplus Z$, then we define $\alpha : X \to Y$ and $g : Z \to Y$ to be the inclusions, $f : Y \to X$ be the map $Y \to Y/Z = (X + Z)/Z \cong X/(X \cap Z) \cong X$, and $\beta : Y \to Z$ be the map $Y \to Y/X = (X + Z)/X \cong Z/(X \cap Z) \cong Z$. Namely, $\alpha(x) = x$, $g(z) = z$, $f(x + z) = x$, and $\beta(x + z) = z$. Then it is easy to see that the diagram (#) forms a biproduct. In conclusion, making a biproduct (#) is the same as giving an isomorphism $Y \cong X \oplus Z$.

Proposition 1.96 *Let* $0 \to X \xrightarrow{\alpha} Y \xrightarrow{\beta} Z \to 0$ *be a short exact sequence. The following conditions are equivalent.*

(1) α *is a split injection. Namely, there exists an* A*-linear map* $f : Y \to X$ *satisfying* $f \circ \alpha = 1_X$.
(2) β *is a split surjection. Namely, there exists an* A*-linear map* $g : Z \to Y$ *satisfying the equation* $\beta \circ g = 1_Z$.
(3) *There exist* A*-linear maps* $f : Y \to X$ *and* $g : Z \to Y$ *such that* (#) *forms a biproduct.*

If condition (3) is satisfied, then we have $Y = \operatorname{Im}\alpha \oplus \operatorname{Im} g \cong X \oplus Z$ *as* A*-modules.*

If the equivalent conditions in the proposition are satisfied, we say that the short exact sequence $0 \to X \xrightarrow{\alpha} Y \xrightarrow{\beta} Z \to 0$ *splits*.

Proof (1) ⇒(3). Since $f\alpha = 1_X$, we have $(1_Y - \alpha f)\alpha = \alpha - \alpha f \alpha = \alpha - \alpha = 0$. By Proposition 1.94, (1), there exists a unique map $g : Z \to Y$ such that $g\beta = 1_Y - \alpha f$. So $g\beta + \alpha f = 1_Y$. Since $\beta g \beta = \beta(1_Y - \alpha f) = \beta = 1_Z \beta$ and β is surjective, we have that $\beta g = 1_Z$.

(2) ⇒(3). Since $\beta(1_Y - g\beta) = \beta - \beta = 0$, there exists a unique map $f : Y \to X$ such that $\alpha f = 1_Y - g\beta$ by Proposition 1.94, (2). Hence $1_Y = g\beta + \alpha f$. Since $\alpha(1_X - f\alpha) = \alpha - (1_Y - g\beta)\alpha = \alpha - \alpha = 0$ and α is injective, $1_X - f\alpha = 0$.

(3) ⇒(1) and (3) ⇒(2) are obvious. □

The following lemma, known as the *snake lemma*, has many applications.

Theorem 1.97 (Snake Lemma) *Consider the following diagram with two exact rows:*

$$\begin{array}{ccccccc} & & L_1 & \xrightarrow{\alpha} & M_1 & \xrightarrow{\beta} & N_1 & \longrightarrow & 0 \\ & & \downarrow{\scriptstyle f} & & \downarrow{\scriptstyle g} & & \downarrow{\scriptstyle h} & & \\ 0 & \longrightarrow & L_2 & \xrightarrow{\varphi} & M_2 & \xrightarrow{\psi} & N_2 & & \end{array}$$

Assume that this diagram is commutative, that is, $\varphi \circ f = g \circ \alpha$ *and* $\psi \circ g = h \circ \beta$. *In this case, there exists a long exact sequence of the following form:*

$$\operatorname{Ker} f \xrightarrow{\alpha'} \operatorname{Ker} g \xrightarrow{\beta'} \operatorname{Ker} h \xrightarrow{\Delta} \operatorname{Coker} f \xrightarrow{\overline{\varphi}} \operatorname{Coker} g \xrightarrow{\overline{\beta}} \operatorname{Coker} h.$$

Here, the morphisms $\alpha', \beta', \overline{\varphi}, \overline{\psi}$ *are naturally induced linear maps from* $\alpha, \beta, \varphi, \psi$, *respectively.*

Proof The *connecting homomorphism*, denoted by $\Delta : \operatorname{Ker} h \to \operatorname{Coker} f$, is defined as follows. For any $x \in \operatorname{Ker} h$, since the map β is surjective, we can take $y \in M_1$ so that $x = \beta(y)$. If we set $z = g(y) \in M_2$, then from the commutativity of the diagram, we have $\psi(z) = h(\beta(y)) = h(x) = 0$. Then, the exactness of the row guarantees that we can take $u \in L_2$, such that $z = \phi(u)$. We define $\Delta(x) = \overline{u}$ using this u. Here, $\overline{u}$ represents the image of u in $\operatorname{Coker} f$. While there is freedom in choosing the element $y \in M_1$ for a given $x \in \operatorname{Ker} h$, once y is specified, $u \in L_2$ is uniquely determined by y. The fact that this definition of $\overline{u}$ does not depend on the choice of $y \in M_1$ can be shown as follows. If $y' \in M_1$ satisfies $\beta(y') = x$, and we take $u' \in L_2$ such that $g(y') = \varphi(u')$, then because $\beta(y - y') = 0$, there exists $w \in L_1$ such that $\alpha(w) = y - y'$. Then we have $\varphi(u - u') = g(y) - g(y') = g(\alpha(w)) = \varphi(f(w))$. Since the map φ is injective, we have $u - u' = f(w)$, and thus $\overline{u} = \overline{u'}$ in $\operatorname{Coker} f$. The proof of exactness of the sequence

$$\operatorname{Ker} f \xrightarrow{\alpha'} \operatorname{Ker} g \xrightarrow{\beta'} \operatorname{Ker} h \xrightarrow{\Delta} \operatorname{Coker} f \xrightarrow{\overline{\varphi}} \operatorname{Coker} g \xrightarrow{\overline{\beta}} \operatorname{Coker} h$$

is left to the reader. Please enjoy it. □

Exercise 1.40 Consider an exact sequence of A-modules

$$0 \to L \to M \to N \to 0.$$

For any element $x \in A$, show that the exact sequence of A-modules

$$0 \to 0 :_L x \to 0 :_M x \to 0 :_N x \to L/xL \to M/xM \to N/xN \to 0$$

can be obtained. Here, we write $0 :_M x = \{m \in M \mid xm = 0\}$.

1.98 For a chain complex

$$C : \cdots \to X_{i+1} \xrightarrow{\partial^C_{i+1}} X_i \xrightarrow{\partial^C_i} X_{i-1} \to \cdots$$

we define $\mathrm{H}_i(C) = \operatorname{Ker} \partial^C_i / \operatorname{Im} \partial^C_{i+1}$, and call it the ith *homology* of C. When $\mathcal{D} : \cdots \to Y_{i+1} \xrightarrow{\partial^{\mathcal{D}}_{i+1}} Y_i \xrightarrow{\partial^{\mathcal{D}}_i} \cdots$ is a chain complex, a family of linear maps $f = \{f_i : X_i \to Y_i\}_{i \in \mathbb{Z}}$ is said to be a *morphism of chain complexes* if the following diagram commutes for all i:

$$\begin{array}{ccc} X_{i+1} & \xrightarrow{\partial^C_{i+1}} & X_i \\ {\scriptstyle f_{i+1}}\downarrow & & \downarrow{\scriptstyle f_i} \\ Y_{i+1} & \xrightarrow{\partial^{\mathcal{D}}_{i+1}} & Y_i \end{array}$$

and we write

$$f : C \to \mathcal{D}.$$

If $f : C \to \mathcal{D}$ is a morphism of chain complexes, then for each i, we have

$$f_i(\mathrm{Ker}\,\partial_i^C) \subseteq \mathrm{Ker}\,\partial_i^{\mathcal{D}}, \quad f_i(\mathrm{Im}\,\partial_{i+1}^C) \subseteq \mathrm{Im}\,\partial_{i+1}^{\mathcal{D}}.$$

Therefore, we obtain an A-linear map between homologies

$$f : \mathrm{H}_i(C) \to \mathrm{H}_i(\mathcal{D}), \quad \overline{z} \mapsto \overline{f_i(z)}$$

where $z \in \mathrm{Ker}\,\partial_i^C$ and $\overline{z}$, $\overline{f_i(z)}$ represent the images of z and $f_i(z)$ in $\mathrm{H}_i(C)$, $\mathrm{H}_i(\mathcal{D})$ respectively (see Proposition 1.94 (1)).

Definition 1.99 Let $C, \mathcal{D}$ be as above and let $\mathcal{E}$ be $\mathcal{E} : \cdots \to Z_{i+1} \xrightarrow{\partial_{i+1}^{\mathcal{E}}} Z_i \xrightarrow{\partial_i^{\mathcal{E}}} Z_{i-1} \to \cdots$.

A sequence of chain complexes and their morphisms $0 \to C \xrightarrow{f} \mathcal{D} \xrightarrow{g} \mathcal{E} \to 0$ is said to be *exact* if for all $i \in \mathbb{Z}$, the sequence

$$0 \to X_i \xrightarrow{f_i} Y_i \xrightarrow{g_i} Z_i \to 0$$

is exact.

The following theorem is fundamental in homological algebra.

Theorem 1.100 (Long Exact Sequence of Homology) *Given a short exact sequence of chain complexes* $0 \to C \xrightarrow{f} \mathcal{D} \xrightarrow{g} \mathcal{E} \to 0$, *we obtain a long exact sequence of homologies*

$$\cdots \to \mathrm{H}_i(C) \xrightarrow{f} \mathrm{H}_i(\mathcal{D}) \xrightarrow{g} \mathrm{H}_i(\mathcal{E}) \xrightarrow{\Delta} \mathrm{H}_{i-1}(C) \xrightarrow{f} \mathrm{H}_{i-1}(\mathcal{D}) \xrightarrow{g} \mathrm{H}_{i-1}(\mathcal{E}) \to \cdots$$

Proof By assumption, the diagram

$$\begin{array}{ccccccccc} 0 & \longrightarrow & X_i & \xrightarrow{f_i} & Y_i & \xrightarrow{g_i} & Z_i & \longrightarrow & 0 \\ & & \downarrow{\scriptstyle \partial_i^C} & & \downarrow{\scriptstyle \partial_i^{\mathcal{D}}} & & \downarrow{\scriptstyle \partial_i^{\mathcal{E}}} & & \\ 0 & \longrightarrow & X_{i-1} & \xrightarrow{f_{i-1}} & Y_{i-1} & \xrightarrow{g_{i-1}} & Z_{i-1} & \longrightarrow & 0 \end{array}$$

is commutative with exact rows. By applying the snake lemma, we obtain the exact sequence

$$\begin{aligned} 0 \to \mathrm{Ker}\,\partial_i^C \to \mathrm{Ker}\,\partial_i^{\mathcal{D}} \to \mathrm{Ker}\,\partial_i^{\mathcal{E}} & \\ \to X_{i-1}/\mathrm{Im}\,\partial_i^C \to Y_{i-1}/\mathrm{Im}\,\partial_i^{\mathcal{D}} \to Z_{i-1}/\mathrm{Im}\,\partial_i^{\mathcal{E}} \to 0. \end{aligned}$$

As a result, we obtain the commutative diagram with two exact rows

$$
\begin{array}{ccccccc}
 & X_i/\operatorname{Im}\partial_i^{\mathcal{C}} & \longrightarrow & Y_i/\operatorname{Im}\partial_i^{\mathcal{D}} & \longrightarrow & Z_i/\operatorname{Im}\partial_i^{\mathcal{E}} & \longrightarrow 0 \\
 & \downarrow{\scriptstyle \partial_i^{\mathcal{C}}} & & \downarrow{\scriptstyle \partial_i^{\mathcal{D}}} & & \downarrow{\scriptstyle \partial_i^{\mathcal{E}}} & \\
0 \longrightarrow & \operatorname{Ker}\partial_{i-1}^{\mathcal{C}} & \longrightarrow & \operatorname{Ker}\partial_{i-1}^{\mathcal{D}} & \longrightarrow & \operatorname{Ker}\partial_{i-1}^{\mathcal{E}} &
\end{array}
$$

From Proposition 1.94 (1), we obtain the long exact sequence we seek by applying the snake lemma once more:

$$\cdots \to \mathrm{H}_i(\mathcal{C}) \xrightarrow{f} \mathrm{H}_i(\mathcal{D}) \xrightarrow{g} \mathrm{H}_i(\mathcal{E}) \xrightarrow{\Delta} \mathrm{H}_{i-1}(\mathcal{C}) \xrightarrow{f} \mathrm{H}_{i-1}(\mathcal{D}) \xrightarrow{g} \mathrm{H}_{i-1}(\mathcal{E}) \to \cdots$$

□

It is easy to see that for a commutative diagram of complexes of A-modules

$$
\begin{array}{ccccccccc}
0 & \longrightarrow & \mathcal{C} & \longrightarrow & \mathcal{D} & \longrightarrow & \mathcal{E} & \longrightarrow & 0 \\
 & & \downarrow{\scriptstyle f} & & \downarrow{\scriptstyle g} & & \downarrow{\scriptstyle h} & & \\
0 & \longrightarrow & \mathcal{C}' & \longrightarrow & \mathcal{D}' & \longrightarrow & \mathcal{E}' & \longrightarrow & 0
\end{array}
$$

with exact rows, the diagram

$$
\begin{array}{ccc}
\mathrm{H}_i(\mathcal{E}) & \xrightarrow{\Delta} & \mathrm{H}_{i-1}(\mathcal{C}) \\
\downarrow{\scriptstyle \mathrm{H}_i(h)} & & \downarrow{\scriptstyle \mathrm{H}_{i-1}(f)} \\
\mathrm{H}_i(\mathcal{E}') & \xrightarrow{\Delta'} & \mathrm{H}_{i-1}(\mathcal{C}')
\end{array}
$$

is commutative, where Δ and Δ' are the connecting homomorphisms.

Lemma 1.101 (Five Lemma) *Let*

$$
\begin{array}{ccccccccc}
M_1 & \xrightarrow{f_1} & M_2 & \xrightarrow{f_2} & M_3 & \xrightarrow{f_3} & M_4 & \xrightarrow{f_4} & M_5 \\
\downarrow{\scriptstyle \varphi_1} & & \downarrow{\scriptstyle \varphi_2} & & \downarrow{\scriptstyle \varphi_3} & & \downarrow{\scriptstyle \varphi_4} & & \downarrow{\scriptstyle \varphi_5} \\
N_1 & \xrightarrow{g_1} & N_2 & \xrightarrow{g_2} & N_3 & \xrightarrow{g_3} & N_4 & \xrightarrow{g_4} & N_5
\end{array}
\tag{1.1}
$$

be a commutative diagram of A-modules whose rows are exact. Then the following hold:

(1) *If φ_1 is surjective and φ_2 and φ_4 are injective, then φ_3 is injective.*
(2) *If φ_5 is injective and φ_2 and φ_4 are surjective, then φ_3 is surjective.*

Proof (1) Let $x \in M_3$ and assume $\varphi_3(x) = 0$. Since $\varphi_4(f_3(x)) = 0$ and φ_4 is injective, $f_3(x) = 0$ and hence $x = f_2(y)$ for some $y \in M_2$. Then since $g_2(\varphi_2(y)) = \varphi_3(x) = 0$, we must have $\varphi_2(y) = g_1(z)$ for some $z \in N_1$ and since φ_1 is surjective, $z = \varphi_1(u)$ for some $u \in M_1$. Then since φ_2 is injective, we have $y = f_1(u)$ and hence $x = f_2(y) = f_2(f_1(u)) = 0$.

The proof of (2) is similar, and is left to the reader. □

1.7 Hom and ⊗

Let A be a ring.

1.102 Let M, M', N, N' be A-modules. We denote the set of all A-linear maps $M \to N$ by $\mathrm{Hom}_A(M, N)$. $\mathrm{Hom}_A(M, N)$ becomes an A-module with the following sum and action of A: for $f, g \in \mathrm{Hom}_A(M, N)$ and $a \in A$, we define

$$(f + g)(x) = f(x) + g(x), \qquad (af)(x) = af(x).$$

If A-linear maps $u : M \to M'$, $v : N \to N'$ are given, A-linear maps

$$u^* : \mathrm{Hom}_A(M', N) \to \mathrm{Hom}_A(M, N), \qquad u^*(f) = f \circ u,$$

$$v_* : \mathrm{Hom}_A(M, N) \to \mathrm{Hom}_A(M, N'), \qquad v_*(f) = v \circ f$$

are defined. $\varphi : A \to B$ is a ring homomorphism and if L is a B-module and M is an A-module, then the A-module $\mathrm{Hom}_A(L, M)$ becomes a B-module with the following action: for $f \in \mathrm{Hom}_A(L, M)$, $b \in B$, and $x \in L$, $(bf)(x) = f(bx)$. And if $u : M \to M'$ is an A-linear map and $v : L \to L'$ is a B-linear map, then

$$u^* : \mathrm{Hom}_A(L', N) \to \mathrm{Hom}_A(L, N), \qquad u^*(f) = f \circ u$$

and

$$v_* : \mathrm{Hom}_A(L, N) \to \mathrm{Hom}_A(L, N'), \qquad v_*(f) = v \circ f$$

are B-linear maps.

The A-module $\mathrm{Hom}_A(M, M)$ is often denoted by $\mathrm{End}_A M$, and is called the *endomorphism ring* of M. In fact, the composition gives its A-algebra structure, which is not necessarily commutative.

Exercise 1.41 Let A and B be commutative rings. Prove the following.

(1) Let M be an additive group. Then M has a unique $\mathbb{Z}$-module structure. So a $\mathbb{Z}$-module and an additive group are one and the same thing.

(2) Let $A \to B$ be a ring homomorphism, and M be an A-module. To give M a B-module structure is the same as to give an A-algebra map $\varphi : B \to \operatorname{End}_A M$. The correspondence is given by $bm = \varphi(b)(m)$ for $b \in B$ and $m \in M$. The map φ is injective if and only if M is a faithful B-module.
(3) The A-module A yields an isomorphism $A \to \operatorname{End}_A A$ $(a \mapsto (x \mapsto ax))$.
(4) There is a unique ring homomorphism $u_A : \mathbb{Z} \to A$.

Proposition 1.103 *Let M be an A-module. The following assertions are true.*

(1) $\operatorname{Hom}_A(A, M) \cong M$.
(2) *If $\mathfrak{a}$ is an ideal of the ring A, then as an $A/\mathfrak{a}$-module,* $\operatorname{Hom}_A(A/\mathfrak{a}, M) \cong 0 :_M \mathfrak{a}$.
(3) *If $\{M_i\}_{i\in I}$ is a family of A-modules, then for any A-module X,*

$$\operatorname{Hom}_A(\bigoplus_{i\in I} M_i, X) \cong \prod_{i\in I} \operatorname{Hom}_A(M_i, X),$$

$$\operatorname{Hom}_A(X, \prod_{i\in I} M_i) \cong \prod_{i\in I} \operatorname{Hom}_A(X, M_i).$$

Hence for any set I, we have $\operatorname{Hom}_A(A^{(I)}, M) \cong M^I$. *Here, for an A-module N, N^I denotes the direct product of the family of A-modules $\{X_i = N\}_{i\in I}$, i.e., $\prod_{i\in I} X_i$.*
(4) *If the A-module X is finitely generated, then we have*

$$\operatorname{Hom}_A(X, \bigoplus_{i\in I} M_i) \cong \bigoplus_{i\in I} \operatorname{Hom}_A(X, M_i).$$

Proof

(1) The homomorphism $\operatorname{Hom}_A(A, M) \to M$ defined by $f \mapsto f(1)$ gives the isomorphism.
(2) The homomorphism $\operatorname{Hom}_A(A/\mathfrak{a}, M) \to M$ defined by $f \mapsto f(\overline{1})$ gives the isomorphism.
(3) See Proposition A.1 in Appendix A.
(4) Follows from the fact that there exists a finite subset $J \subseteq I$ such that for each $f \in \operatorname{Hom}_A(X, \bigoplus_{i\in I} M_i)$, we have $f(X) \subseteq \bigoplus_{j\in J} M_j$.

□

Proposition 1.104

(1) *If $0 \to X \xrightarrow{f} Y \xrightarrow{g} Z$ is an exact sequence of A-modules and M is an A-module, then the sequence of A-modules*

$$0 \to \operatorname{Hom}_A(M, X) \xrightarrow{f_*} \operatorname{Hom}_A(M, Y) \xrightarrow{g_*} \operatorname{Hom}_A(M, Z)$$

is exact.

(2) *If* $X \xrightarrow{f} Y \xrightarrow{g} Z \to 0$ *is an exact sequence of A-modules and N is an A-module, then the sequence of A-modules*

$$0 \to \mathrm{Hom}_A(Z, N) \xrightarrow{g^*} \mathrm{Hom}_A(Y, N) \xrightarrow{f^*} \mathrm{Hom}_A(X, N)$$

is exact.

Proof (1) Since f is injective, for $u \in \mathrm{Hom}_A(M, X)$, if $fu = f_*(u) = 0$, then $u = 0$. Since $g_* f_* = (gf)_* = 0$, $\mathrm{Im}\, f_* \subseteq \mathrm{Ker}\, g_*$. If $v \in \mathrm{Ker}\, g_*$, then $gv = 0$, so from Proposition 1.94 (2), there exists $u \in \mathrm{Hom}_A(M, X)$ such that $v = fu = f_*(u)$. Therefore, the sequence

$$0 \to \mathrm{Hom}_A(M, X) \xrightarrow{f_*} \mathrm{Hom}_A(M, Y) \xrightarrow{g_*} \mathrm{Hom}_A(M, Z)$$

is exact. (2) is similar. □

Exercise 1.42 Prove that the converse of Proposition 1.104 is also true. That is:

(1) Let $0 \to X \xrightarrow{f} Y \xrightarrow{g} Z$ be a sequence of A-modules. If the sequence of A-modules

$$0 \to \mathrm{Hom}_A(M, X) \xrightarrow{f_*} \mathrm{Hom}_A(M, Y) \xrightarrow{g_*} \mathrm{Hom}_A(M, Z)$$

is exact for any A-module M, then the sequence $0 \to X \xrightarrow{f} Y \xrightarrow{g} Z$ is exact.

(2) Let $X \xrightarrow{f} Y \xrightarrow{g} Z \to 0$ be a sequence of A-modules. If the sequence of A-modules

$$0 \to \mathrm{Hom}_A(Z, N) \xrightarrow{g^*} \mathrm{Hom}_A(Y, N) \xrightarrow{f^*} \mathrm{Hom}_A(X, N)$$

is exact for any A-module N, then the sequence $X \xrightarrow{f} Y \xrightarrow{g} Z \to 0$ is exact.

Definition 1.105 Let P be an A-module. We say that P is *projective*, or is a *projective A-module*, if there exist a free A-module F and two linear maps $P \xrightarrow{\alpha} F \xrightarrow{\beta} P$ such that $\beta \circ \alpha = 1_P$ holds.

Proposition 1.106 *Let P be an A-module. Then the following conditions are equivalent:*

(1) *P is projective.*

(2) *For any exact sequence* $0 \to X \xrightarrow{f} Y \xrightarrow{g} Z \to 0$ *of A-modules, the sequence*

$$0 \to \mathrm{Hom}_A(P, X) \xrightarrow{f_*} \mathrm{Hom}_A(P, Y) \xrightarrow{g_*} \mathrm{Hom}_A(P, Z) \to 0$$

is also exact.

(3) *If the sequence* $Y \xrightarrow{g} Z \to 0$ *of A-modules is exact, then*

$$\mathrm{Hom}_A(P, Y) \xrightarrow{g_*} \mathrm{Hom}_A(P, Z) \to 0$$

is also exact.

Proof From Proposition 1.104 (1), it is sufficient to show (1) ⇔(3).

(1) ⇒(3). We take a free A-module F and linear maps $P \xrightarrow{\alpha} F \xrightarrow{\beta} P$ such that $\beta\alpha = 1_P$. For any $u \in \mathrm{Hom}_A(P, Z)$, we want to show that there exists $w \in \mathrm{Hom}_A(P, Y)$ such that $u = gw$ holds. As we have $\beta\alpha = 1_P$, it is sufficient if there exists $v \in \mathrm{Hom}_A(F, Y)$ such that $u\beta = gv$. In other words, we may assume that $P = F$. Take a free basis $\{\mathbf{e}_i\}_{i\in I}$ of the free module F. Then, because g is surjective, for each $i \in I$, there exists $y_i \in Y$ such that $u(\mathbf{e}_i) = g(y_i)$. For the family of elements $\{y_i\}_{i\in I}$ of Y, if we define the A-linear map $v : F \to Y$ by $v(\mathbf{e}_i) = y_i$ for each $i \in I$, then this v is the desired map.

(3) ⇒(1). If we take a free A-module F and a surjective A-linear map $\beta : F \to P$, then for $1_P : P \to P$, we can find an A-linear map $\alpha : P \to F$ such that $\beta\alpha = 1_P$. □

Let M, N, and P be A-modules, and $f : M \times N \to P$ be a map.

Definition 1.107 We say that $f : M \times N \to P$ is *A-bilinear* if for any $x, x' \in M$, $y, y' \in N$ and $a \in A$, the equations

(1) $f(x + x', y) = f(x, y) + f(x', y),\ f(x, y + y') = f(x, y) + f(x, y')$,
(2) $f(ax, y) = f(x, ay) = af(x, y)$

hold. This condition is equivalent to saying that if the element $y \in N$ is fixed, the map $f_y : M \to P$, $f_y(x) = f(x, y)$ is A-linear, and if the element $x \in M$ is fixed, the map $f_x : N \to P$, $f_x(y) = f(x, y)$ is A-linear.

We define $\mathrm{Bil}_A(M, N; P) = \{f : M \times N \to P \mid f \text{ is bilinear}\}$. $\mathrm{Bil}_A(M, N; P)$ becomes an A-module by the following sum and action: for $f, g \in \mathrm{Bil}_A(M, N; P)$ and $a \in A$,

$$(f + g)(x, y) = f(x, y) + g(x, y), \qquad (af)(x, y) = af(x, y).$$

Lemma 1.108 *The map* $\Phi : \mathrm{Bil}_A(M, N; P) \to \mathrm{Hom}_A(M, \mathrm{Hom}_A(N, P))$ *defined by* $\Phi(f)(m) = f(m, -)$ *is an isomorphism of A-modules.*

Proof By the definition of bilinear maps, $\Phi(f)(m) = f(m, -)$ is in $\mathrm{Hom}_A(N, P)$. Since

$$\begin{aligned}\Phi(f)(am + a'm')(n) &= f(am + a'm', n) = af(m, n) + a'f(m', n)\\ &= (a\Phi(f)(m) + a'\Phi(f)(m'))(n)\end{aligned}$$

for $m, m' \in M$, $a, a' \in A$ and $n \in N$, we have that $\Phi(f) \in \mathrm{Hom}_A(M, \mathrm{Hom}_A(N, P))$. It is easy to see that Φ is A-linear. Now we define $\Psi : \mathrm{Hom}_A(M, \mathrm{Hom}_A(N, P)) \to \mathrm{Bil}_A(M, N; P)$ by $\Psi(h)(m, n) = h(m)(n)$ for $h \in \mathrm{Hom}_A(M, \mathrm{Hom}_A(N, P))$, $m \in M$ and $n \in N$. As $h(m)$ is linear for $m \in M$ and h is linear, it is easy to see that $\Psi(h)$ certainly lies in $\mathrm{Bil}_A(M, N; P)$. Since $\Phi(\Psi(h))(m)(n) = \Psi(h)(m, n) = h(m)(n)$, we have $\Phi \circ \Psi = \mathrm{id}$. As $\Psi(\Phi(f))(m, n) = \Phi(f)(m)(n) = f(m, n)$, we have $\Psi \circ \Phi = \mathrm{id}$, and Φ is bijective. Hence Φ is an A-linear isomorphism. □

Let us describe the definition and construction of a tensor product.

Theorem–Definition 1.109 Let M, N be A-modules. Then there exists a pair (T, f) of an A-module T and an A-bilinear map f such that for any given pair (P, g) of an A-module P and an A-bilinear map $(g : M \times N \to P)$, there exists a unique A-linear map $\varphi : T \to P$ such that the diagram

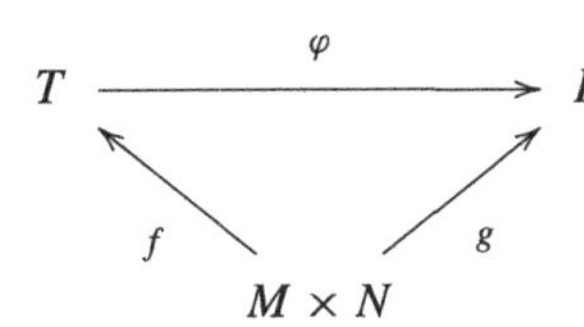

is commutative, that is, $g = \varphi f$. As the pair (T, f) is uniquely determined up to isomorphisms, we write $T = M \otimes_A N$ (often simply written as $M \otimes N$), and call it the *tensor product* of M and N.

Proof First, we show the uniqueness of the pair (T, f). If the pairs (T, f) and (T', f') satisfying the conditions of the theorem are given, then the maps $\varphi : T \to T'$ and $\varphi' : T' \to T$ are uniquely determined, and the equations $f' = \varphi f$ and $f = \varphi' f'$ hold.

When we consider the diagram

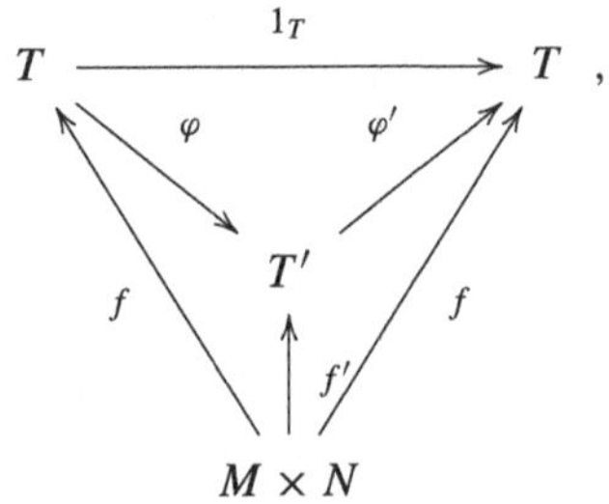

we have that $(\varphi'\varphi)f = f = 1_T f$. From the uniqueness of the A-linear map $\alpha : T \to T$ satisfying the equation $\alpha f = f$, we get the equation $\varphi'\varphi = 1_T$. Similarly, we have $\varphi\varphi' = 1_{T'}$. Therefore, the map φ is an isomorphism.

Now we demonstrate the existence of the pair (T, φ). Let F be a free A-module with the set $M \times N$ as its basis (refer to Exercise 1.33), and let C be an A-submodule of F generated by the subset

$$Z = \left\{ \begin{array}{c} (x+x', y) - (x, y) - (x', y) \\ (x, y+y') - (x, y) - (x, y') \\ (ax, y) - a(x, y) \\ (x, ay) - a(x, y) \end{array} \middle| \; x, x' \in M, \quad y, y' \in N, \quad a \in A \right\}$$

of F. We set $T = F/C$. For each $(x, y) \in M \times N$, we denote the image of (x, y) in T by $x \otimes y$. Then, by the definition of the A-module T, for any $x, x' \in M$, $y, y' \in N$, $a \in A$, the equations

$$\begin{aligned} (x+x') \otimes y &= x \otimes y + x' \otimes y, \\ x \otimes (y+y') &= x \otimes y + x \otimes y', \\ (ax) \otimes y &= x \otimes (ay) = a(x \otimes y) \end{aligned}$$

hold. Therefore, the map $f : M \times N \to T$, $(x, y) \mapsto x \otimes y$ is A-bilinear. Given a pair (P, g), where P is an A-module and $g : M \times N \to P$ is an A-bilinear map, since $M \times N$ is a basis of the free module F, there exists an A-linear map $\alpha : F \to P$ such that $\alpha(x, y) = g(x, y)$ for any $(x, y) \in M \times N$ (refer to Proposition 1.76). Since g is A-bilinear, $C \subseteq \operatorname{Ker}\alpha$, and there exists an A-linear map $\varphi : T \to P$ satisfying $g = \varphi f$, which is uniquely determined. □

From the definition of the tensor product $M \otimes_A N$, we have

$$\operatorname{Bil}_A(M, N; P) = \operatorname{Hom}_A(M \otimes_A N, P)$$

as A-modules. Therefore, from Proposition 1.108, we immediately get the following corollary. In fact, the tensor product $M \otimes_A N$ is constructed as the A-module such that this isomorphism Corollary 1.110 holds.

Corollary 1.110 *For any A-module P, we have*

$$\operatorname{Hom}_A(M \otimes_A N, P) = \operatorname{Hom}_A(M, \operatorname{Hom}_A(N, P)).$$

1.111 Just as with polynomial rings, what is important for the tensor product is its universality Theorem–Definition 1.109. There is no problem or difficulty even if we forget about the construction. However, we would like to record the following two facts that directly follow from the construction.

(1) Every element of $M \otimes_A N$ is a finite sum of elements of the form $x \otimes y$ $(x \in M,\ y \in N)$.
(2) Let $x_i \in M$, $y_i \in N$ $(1 \le i \le n)$. If $\sum_{i=1}^n x_i \otimes y_i = 0$ in $M \otimes_A N$, then we can choose a finitely generated A-submodule M_0 of M and a finitely generated A-submodule N_0 of N such that $x_i \in M_0$, $y_i \in N_0$ $(1 \le i \le n)$, and $\sum_{i=1}^n x_i \otimes y_i = 0$ also in $M_0 \otimes_A N_0$.

Proof (2) Retain the notation of the proof of Theorem–Definition 1.109. If $\sum_{i=1}^n x_i \otimes y_i = 0$ in $M \otimes_A N$, then $\sum_{i=1}^n (x_i, y_i) \in C$ holds in the free A-module F, so the element $z = \sum_{i=1}^{n}(x_i, y_i) \in F$ can be expressed as an A-linear combination of a finite number of elements $\{z_j\}_{1\leq j\leq m}$ that generate the A-submodule C. The finitely generated A-submodule M_0 of M and A-submodule N_0 of N can be chosen to contain all of the $\{z_j\}_{1\leq j\leq m}$ necessary to express the element z in the construction of $M_0 \otimes_A N_0$ (where $M_0 \times N_0$ is taken as the free basis) in the free A-module F_0. In addition, M_0 should contain all elements $\{x_i\}_{1\leq i\leq n}$ and N_0 should contain all elements $\{y_i\}_{1\leq i\leq n}$. □

Let M, M', N, and N' be A-modules, and $f : M \to M'$ and $g : N \to N'$ be linear maps. As the map

$$M \times N \to M' \otimes_A N', \qquad (x, y) \mapsto f(x) \otimes g(y)$$

is bilinear, it induces an A-linear map $f \otimes g : M \otimes_A N \to M' \otimes_A N'$ such that the equation

$$(f \otimes g)(x \otimes y) = f(x) \otimes g(y)$$

holds for all $x \in M$ and $y \in N$. When $N = N'$, $g = 1_N$, we simply write $f \otimes 1_N : M \otimes_A N \to M' \otimes_A N, x \otimes y \mapsto f(x) \otimes y$ as $f \otimes N$. Similarly, $M \otimes g$ denotes $1_M \otimes g : M \otimes_A N \to M \otimes_A N', x \otimes y \mapsto x \otimes g(y)$.

Proposition 1.112 *The following are true. The maps from the A-modules on the left-hand sides to the A-modules on the right-hand sides are isomorphisms, and what the corresponding elements are is as described.*

(1) $A \otimes_A M \overset{\cong}{\to} M,\ 1 \otimes m \mapsto m.$
(2) $M \otimes_A N \overset{\cong}{\to} N \otimes_A M,\ m \otimes n \mapsto n \otimes m.$
(3) $(L \otimes_A M) \otimes_A N \overset{\cong}{\to} L \otimes_A (M \otimes_A N),\ (\ell \otimes m) \otimes n \mapsto \ell \otimes (m \otimes n).$
(4) $L \otimes_A (M \oplus N) \overset{\cong}{\to} (L \otimes_A M) \oplus (L \otimes_A N),\ \ell \otimes (m, n) \mapsto (\ell \otimes m, \ell \otimes n).$
(5) *In general, for an A-module L and a family of A-modules* $\{M_i\}_{i\in I}$*, we have*

$$L \otimes_A (\bigoplus_{i\in I} M_i) \overset{\cong}{\to} \bigoplus_{i\in I}(L \otimes_A M_i),\ \ \ell \otimes (x_i)_{i\in I} \mapsto (\ell \otimes x_i)_{i\in I}.$$

Proof To confirm this kind of fact, there is a common pattern. We would like you to learn this basic pattern through the following proof.

(1) Since $A \times M \to M$, $(a, x) \mapsto ax$ is bilinear, there is a unique linear map $\alpha : A \otimes_A M \to M$ such that for all $a \in A$ and $m \in M$, $\alpha(a \otimes m) = am$. The linear map $\beta : M \to A \otimes_A M$ given by $m \mapsto 1 \otimes m$ is the inverse of α.
(2) $M \times N \to N \otimes_A M$, $(m, n) \mapsto n \otimes m$ is bilinear, so it induces the unique A-linear map $\alpha : M \otimes_A N \to N \otimes_A M$ such that $\alpha(m \otimes n) = n \otimes m$. Similarly, there is a unique A-linear map $\beta : N \otimes_A M \to M \otimes_A N$ such that

$$\beta(n \otimes m) = m \otimes n$$

for $m \in M$ and $n \in N$. We need to confirm that α and β are inverses of each other. Namely, we should show that $\alpha\beta = 1_{N\otimes_A M}$ and $\beta\alpha = 1_{M\otimes_A N}$. As generators for $M \otimes_A N$ and $N \otimes_A M$ as A-modules, we can take $m \otimes n$ and $n \otimes m$ $(m \in M, n \in N)$, respectively. Therefore, it is sufficient to verify that the images under $\alpha\beta$ and $1_{M\otimes_A N}$ agree for these special elements, and that the images under $\beta\alpha$ and $1_{N\otimes_A M}$ coincide for these special elements, but it is clearly so.

(3) Fix $n \in N$. The bilinear map $\widehat{n} : L \times M \to L \otimes_A (M \otimes_A N)$, $(\ell, m) \mapsto \ell \otimes (m \otimes n)$ can be extended to a linear map $\tilde{n} : L \otimes_A M \to L \otimes_A (M \otimes_A N)$. The map

$$(L \otimes_A M) \times N \to L \otimes_A (M \otimes_A N), \qquad (x, n) \mapsto \tilde{n}(x)$$

is bilinear, so it can be extended to a linear map $\alpha : (L \otimes_A M) \otimes_A N \to L \otimes_A (M \otimes_A N)$. For any $\ell \in L,\ m \in M,\ n \in N$, we have

$$\alpha((\ell \otimes m) \otimes n) = \ell \otimes (m \otimes n).$$

By the same argument, there is a linear map $\beta : L \otimes_A (M \otimes_A N) \to (L \otimes_A M) \otimes_A N$ such that for any $\ell \in L,\ m \in M,\ n \in N$, we have

$$\beta(\ell \otimes (m \otimes n)) = (\ell \otimes m) \otimes n.$$

Furthermore, α and β are inverses of each other.

(4) The map $L \times (M \oplus N) \to (L \otimes_A M) \oplus (L \otimes_A N)$, $(\ell, (m, n)) \mapsto (\ell \otimes m, \ell \otimes n)$ is bilinear, so it can be extended to a linear map $\alpha : L \otimes_A (M \oplus N) \to (L \otimes_A M) \oplus (L \otimes_A N)$. From the natural maps

$$i : M \to M \oplus N,\ m \mapsto (m, 0), \qquad j : N \to M \oplus N,\ n \mapsto (0, n),$$

we get the linear maps $L \otimes i : L \otimes_A M \to L \otimes_A (M \oplus N)$ and $L \otimes j : L \otimes_A N \to L \otimes_A (M \oplus N)$. The linear map

$$\beta : (L \otimes_A M) \oplus (L \otimes_A N) \to L \otimes_A (M \oplus N),\ (u, v) \mapsto (L \otimes i)(u) + (L \otimes j)(v)$$

induced by $L \otimes i$ and $L \otimes j$ is the inverse of α.

(5) Even if the index set I is infinite, exactly the same argument as in (4) works (see Appendix A).

□

Exercise 1.43 Confirm that the following assertions hold:

(1) $\mathbb{Z}/(m) \otimes_{\mathbb{Z}} \mathbb{Z}/(n) \cong \mathbb{Z}/(d)$. Here, d is the greatest common divisor of the natural numbers m and n.

(2) $\mathbb{Q} \otimes_{\mathbb{Z}} \mathbb{Z}/(m) = 0$. Here, m is a positive integer.
(3) As an $\mathbb{R}$-module, $\mathbb{C} \otimes_{\mathbb{R}} \mathbb{C} \cong \mathbb{C} \oplus \mathbb{C}$.

Proposition 1.113 *If $X \xrightarrow{f} Y \xrightarrow{g} Z \to 0$ is an exact sequence of A-modules, then for any A-module M, we get an exact sequence of A-modules $M \otimes_A X \xrightarrow{M\otimes f} M \otimes_A Y \xrightarrow{M\otimes g} M \otimes_A Z \to 0$.*

Proof It is clear that $(M \otimes g)(M \otimes f) = M \otimes (gf) = 0$ and that $M \otimes g$ is surjective. We show that $\mathrm{Ker}\,(M \otimes g) \subseteq \mathrm{Im}\,(M \otimes f)$. Because $(M \otimes g)(M \otimes f) = 0$, by Proposition 1.94, there is a linear map $\alpha : \mathrm{Coker}\,(M \otimes f) \to M \otimes_A Z$ such that $\alpha\varepsilon = M \otimes g$, where $\varepsilon : M \otimes_A Y \to \mathrm{Coker}\,(M \otimes f)$ is a natural surjection. On the other hand, if we take $(m, z) \in M \times Z$ and write $z = g(y)$ $(y \in Y)$, then $\overline{m \otimes y} \in \mathrm{Coker}\,(M \otimes f)$ is defined (where $\overline{m \otimes y}$ represents the image of $m \otimes y$ in $\mathrm{Coker}\,(M \otimes f)$). In fact, if $z = g(y) = g(y')$ for some $y' \in Y$, then expressing $y - y' = f(x)$ $(x \in X)$, we have $m \otimes y - m \otimes y' = m \otimes f(x) = (M \otimes f)(m \otimes x) \in \mathrm{Im}\,(M \otimes f)$, hence $\overline{m \otimes y} = \overline{m \otimes y'}$. The map $M \times Z \to \mathrm{Coker}\,(M \otimes_A f)$, $(m, z) \mapsto \overline{m \otimes y}$ $(y \in Y, z = g(y))$ is bilinear, so it induces a linear map $\beta : M \otimes_A Z \to \mathrm{Coker}\,(M \otimes f)$, and it is easy to verify that it is the inverse of α. Therefore, $\alpha : \mathrm{Coker}\,(M \otimes_A f) \to M \otimes_A Z$ is an isomorphism of A-modules, and hence $\mathrm{Ker}\,(M \otimes g) = \mathrm{Im}\,(M \otimes f)$ follows. □

Exercise 1.44 Prove Proposition 1.113 using Exercise 1.42 (2) and Corollary 1.110.

Example 1.114 Let M be an A-module.

(1) If $\mathfrak{a}$ is an ideal of the ring A, then $A/\mathfrak{a} \otimes_A M \cong M/\mathfrak{a}M$. In fact, from the short exact sequence

$$0 \to \mathfrak{a} \xrightarrow{\iota} A \to A/\mathfrak{a} \to 0$$

(where $\iota : \mathfrak{a} \to A$ is the inclusion), we get the exact sequence of A-modules

$$\mathfrak{a} \otimes_A M \xrightarrow{\iota\otimes M} A \otimes_A M \to A/\mathfrak{a} \otimes_A M \to 0$$

(Proposition 1.113). If we identify $A \otimes_A M$ with M (see Proposition 1.112 (1)), then we have $\mathrm{Im}(\iota \otimes M) = \mathfrak{a}M$, and hence $A/\mathfrak{a} \otimes_A M \xrightarrow{\sim} M/\mathfrak{a}M$, $1 \otimes m \mapsto \overline{m}$ is obtained.

(2) Let M be an A-module with a finite presentation, and take an exact sequence $A^n \xrightarrow{u} A^m \to M \to 0$ of A-modules. Assume that the map u is given by the (m, n)-matrix $T = (t_{ij})$. Then for an A-module N, when we identify $A^n \otimes N = N^n$, $A^m \otimes_A N = N^m$ (Proposition 1.112 (1), (2), (5)), and set $\phi = u \otimes N$, then the exact sequence

$$N^n \xrightarrow{\phi} N^m \to M \otimes_A N \to 0$$

is obtained. Therefore, $M \otimes_A N \cong \operatorname{Coker}\phi$, and if the element $\mathbf{x}$ of N^n is represented by a vector of length n with elements of N, $\phi(\mathbf{x}) = T\mathbf{x}$ (product of a row matrix and a column vector) is given.

Exercise 1.45 Even if $f : X \to Y$ is an injective A-linear map between A-modules, the morphism

$$M \otimes_A X \xrightarrow{M\otimes f} M \otimes_A Y$$

is not necessarily injective. Give an example.

Theorem 1.115 *For an A-module N, the following conditions are equivalent:*

(1) *For any short exact sequence of A-modules $0 \to X \xrightarrow{f} Y \xrightarrow{g} Z \to 0$, the sequence of A-modules*

$$0 \to X \otimes_A N \xrightarrow{f\otimes N} Y \otimes_A N \xrightarrow{g\otimes N} Z \otimes_A N \to 0$$

is exact.

(2) *If $0 \to X \xrightarrow{f} Y$ is an exact sequence of A-modules, then the sequence*

$$0 \to X \otimes_A N \xrightarrow{f\otimes N} Y \otimes_A N$$

is also exact.

(3) *For any exact sequence $0 \to X \xrightarrow{f} Y$ of finitely generated A-modules, the sequence*

$$0 \to X \otimes_A N \xrightarrow{f\otimes N} Y \otimes_A N$$

is exact.

A module N that satisfies the conditions of this theorem is said to be *flat*.

Proof (3) ⇒ (2) Take an element $\varphi \in \operatorname{Ker}(f \otimes N)$, and express it as $\varphi = \sum_{i=1}^{n} x_i \otimes n_i$ $(x_i \in X, n_i \in N)$. Then $\sum_{i=1}^{n} f(x_i) \otimes n_i = 0$. Let Y' be a submodule of Y such that $f(x_i) \in Y'$ $(1 \le i \le n)$ and the equation $\sum_{i=1}^{n} f(x_i) \otimes n_i = 0$ holds in $Y' \otimes_A N$ (1.111 (2)). If we set $X' = \sum_{i=1}^{n} Ax_i$, the map f induces an injection $f' : X' \to Y'$, $f'(x) = f(x)$, and $\sum_{i=1}^{n} x_i \otimes n_i \in \operatorname{Ker}(f' \otimes N')$ holds. Since X' and Y' are finitely generated, $\sum_{i=1}^{n} x_i \otimes n_i = 0$ is obtained in $X' \otimes_A N$, and hence, the equation $\varphi = \sum_{i=1}^{n} x_i \otimes n_i = 0$ also holds in $X \otimes_A N$. Therefore, $f \otimes N$ is injective.

The implication (2) ⇒ (1) is a consequence of Proposition 1.113, and (1) ⇒ (3) is easy. □

Let us discuss the extension of the scalar multiplication.

Definition 1.116 (**A-Algebra**) Let A be a commutative ring, and B be a ring which is not necessarily commutative. Given a ring homomorphism $\varphi : A \to B$ such that $b\varphi(a) = \varphi(a)b$ for any $a \in A$ and $b \in B$ (note that this condition is automatic if B is commutative), we say that B is an *A-algebra* and call φ the *structure map* of the A-algebra B. When we say that a ring B is an A-algebra, in the context of a certain argument, the structure map $\varphi : A \to B$ is fixed. When B and C are A-algebras with structure maps $\varphi_B : A \to B$ and $\varphi_C : A \to C$ respectively, then a ring homomorphism $f : B \to C$ satisfying $f \circ \varphi_B = \varphi_C$ is called a *homomorphism of A-algebras*. If an A-algebra B is commutative as a ring, then we say that B is a commutative A-algebra.

1.117 Note that if B is an A-algebra, then B is an A-module in a natural way, and we have

$$a(b_1b_2) = (ab_1)b_2 = b_1(ab_2) \qquad (a \in A,\ b_1, b_2 \in B). \tag{1.2}$$

Conversely, if B is a ring, not necessarily commutative, B is also an A-module (the addition of the ring B and that of the A-module B are the same), and (1.2) is satisfied, then defining $\varphi : A \to B$ by $\varphi(a) = a \cdot 1_B$, we get a ring homomorphism $A \to B$, and hence B is an A-algebra. This correspondence is one-to-one, and so given a ring B, not necessarily commutative, to give an A-algebra structure to B is the same as to give an A-module structure to B which satisfies (1.2). From this point of view, a homomorphism of A-algebras is nothing but a homomorphism of rings which is also A-linear.

In what follows, until the end of this book, an algebra means a commutative algebra, unless otherwise specified.

Let $\varphi : A \to B$ be a ring homomorphism. A subset C of B is said to be an *A-subalgebra* of B if $\varphi(A) \subseteq C$ and C is a subring of B. This is equivalent to saying that C is both a subring of B and A-submodule of B. Let S be a subset of B. The smallest A-subalgebra of B containing S is called the A-subalgebra *generated* by S, and is denoted by $A[S]$. It is nothing but the smallest subring of B containing $\varphi(A) \cup \{S\}$. If $S = \{b_1, \dots, b_n\}$, then $A[S]$ agrees with $A[b_1, \dots, b_n]$ defined in 1.44. If there exists some finite subset S of B such that $B = A[S]$, then we say that B is a *finitely generated A-algebra* or an A-algebra of *finite type*. This is equivalent to saying that there is a surjective A-algebra map $A[X_1, X_2, \dots, X_n] \to B$.

1.118 Let B be an A-algebra. Let L be a B-module, and let M be an A-module. In this case, the A-module $L \otimes_A M$ becomes a B-module by the following action:

$$b(\ell \otimes m) = (b\ell) \otimes m \qquad (b \in B,\ \ell \in L,\ m \in M).$$

In fact, when $b \in B$ is fixed, the map $L \times M \to L \otimes_A M,\ (\ell, m) \mapsto (b\ell) \otimes m$ is a bilinear map of A-modules, so it induces a linear map $\widehat{b} : L \otimes_A M \to L \otimes_A M$ of A-modules. The action of the ring B is defined by

$$B \times L \otimes_A M \to L \otimes_A M, \qquad (b, x) \mapsto \widehat{b}(x).$$

We call $B \otimes_A M$ the *scalar extension* of M to B.

If $\alpha : M \to N$ is an A-linear map, then the A-linear map induced by α,

$$L \otimes \alpha : L \otimes_A M \to L \otimes_A N$$

is B-linear. When $L = B$, the map $f : M \to B \otimes_A M, m \mapsto 1 \otimes m$ is A-linear.

Proposition 1.119 *Let L be a B-module and M an A-module. Let $g : M \to L$ be an A-linear map. Then, there exists a unique B-linear map $\alpha : B \otimes_A M \to L$ such that $g = \alpha f$. In other words, the map*

$$\mathrm{Hom}_B(B \otimes_A M, L) \to \mathrm{Hom}_A(M, L), \quad \alpha \mapsto \alpha f$$

is a bijection. Here, $f : M \to B \otimes_A M,\ m \mapsto 1 \otimes m$ is the canonical map.

Proof The map $B \times M \to L, (b, m) \mapsto bg(m)$ is a bilinear map as A-modules. Therefore, it uniquely induces an A-linear map $\alpha : B \otimes_A M \to L$ such that $g = \alpha f$. It is easy to verify that this α is also B-linear. □

Exercise 1.46 Let B be an A-algebra, C a B-algebra, and M, N be A-modules. Prove the following statement:

(1) As a B-module,

$$B \otimes_A (M \otimes_A N) \to (B \otimes_A M) \otimes_B (B \otimes_A N), \qquad 1 \otimes (m \otimes n) \mapsto (1 \otimes m) \otimes (1 \otimes n)$$

is an isomorphism.

(2) As a C-module,

$$C \otimes_B (B \otimes_A M) \to C \otimes_A M, \qquad 1 \otimes (1 \otimes m) \mapsto 1 \otimes m$$

is an isomorphism.

Let B be an A-algebra, and L, M, and N be A-modules. The map

$$\mathrm{Hom}_A(M, N) \to \mathrm{Hom}_A(B \otimes_A M, B \otimes_A N), \qquad f \mapsto B \otimes f$$

is A-linear, so it induces a B-linear map

$$\theta_{M,N} : B \otimes_A \mathrm{Hom}_A(M, N) \to \mathrm{Hom}_B(B \otimes_A M, B \otimes_A N).$$

Proposition 1.120 *If B is flat as an A-module and M is a finitely presented A-module, then for any A-module N, $\theta_{M,N} : B \otimes_A \mathrm{Hom}_A(M, N) \to \mathrm{Hom}_B(B \otimes_A M, B \otimes_A N)$ is an isomorphism.*

Proof Take a finite presentation of M as $A^n \to A^m \to M \to 0$. Applying $\mathrm{Hom}_A(*, N)$ gives an exact sequence of A-modules

$$0 \to \mathrm{Hom}_A(M, N) \to \mathrm{Hom}_A(A^m, N) \to \mathrm{Hom}_A(A^n, N).$$

Applying $B \otimes_A *$ gives an exact sequence of B-modules

$$0 \to B \otimes_A \mathrm{Hom}_A(M, N) \to B \otimes_A \mathrm{Hom}_A(A^m, N) \to B \otimes_A \mathrm{Hom}_A(A^n, N)$$

since B is flat as an A-module. On the other hand, from the exact sequence of B-modules

$$B \otimes_A A^n \to B \otimes_A A^m \to B \otimes_A M \to 0,$$

applying $\mathrm{Hom}_B(*, B \otimes_A N)$ gives the exact sequence

$$0 \to \mathrm{Hom}_B(B \otimes_A M, B \otimes_A N) \to \mathrm{Hom}_B(B \otimes_A A^m, B \otimes_A N) \to \mathrm{Hom}_B(B \otimes_A A^n, B \otimes_A N).$$

Therefore, a commutative diagram with exact rows of B-modules

$$\begin{array}{ccccc}
0 \longrightarrow B \otimes_A \mathrm{Hom}_A(M, N) & \longrightarrow & B \otimes_A \mathrm{Hom}_A(A^m, N) & \longrightarrow & B \otimes_A \mathrm{Hom}_A(A^n, N) \\
\downarrow \theta_{M,N} & & \downarrow \theta_{A^m,N} & & \downarrow \theta_{A^n,N} \\
0 \to \mathrm{Hom}_B(B \otimes_A M, B \otimes_A N) & \longrightarrow & \mathrm{Hom}_B(B \otimes_A A^m, B \otimes_A N) & \longrightarrow & \mathrm{Hom}_B(B \otimes_A A^n, B \otimes_A N)
\end{array}$$

is obtained. Furthermore, to know that the desired map $\theta_{M,N}$ is an isomorphism, it is sufficient to show that both $\theta_{A^m,N}$ and $\theta_{A^n,N}$ are isomorphisms. In other words, we can assume $M = A^n$, and by induction on n, we can reduce to the case $n = 1$. If $n = 1$, then $B \otimes_A \mathrm{Hom}_A(A, N) = B \otimes_A N = \mathrm{Hom}_B(B, B \otimes_A N)$, so indeed $\theta_{A,N}$ is an isomorphism. □

Exercise 1.47 Let B be an A-algebra, L, M be B-modules, and N be an A-module. Verify that as B-modules,

$$\mathrm{Hom}_B(L, \mathrm{Hom}_A(M, N)) \cong \mathrm{Hom}_A(L \otimes_B M, N).$$

Exercise 1.48 Let M and N be finitely generated modules over a local ring A. Show that if $M \otimes_A N = 0$, then $M = 0$ or $N = 0$.

Definition 1.121 (Tensor Product of A-Algebras) If B and C are A-algebras, the A-module $B \otimes_A C$ is equipped with the multiplication given by

$$(b \otimes c)(b' \otimes c') = bb' \otimes cc', \quad (b, b' \in B, c, c' \in C)$$

and becomes an A-algebra by the map $\varphi_{B \otimes_A C} : A \to B \otimes_A C, a \mapsto \varphi_B(a) \otimes 1 = 1 \otimes \varphi_C(a)$. Here, φ_B and φ_C denote the structure maps of B and C as A-algebras, respectively. We call $B \otimes_A C$ the *tensor product* of the A-algebras B and C.

Exercise 1.49 Let B, C, and D be A-algebras. Given morphisms of A-algebras $f : B \to D$ and $g : C \to D$, prove that there exists a unique morphism of A-algebras $h : B \otimes_A C \to D$ such that for any $b \in B$ and $c \in C$, $h(b \otimes 1) = f(b)$ and $h(1 \otimes c) = g(c)$.

Exercise 1.50 Let B be an A-algebra, and $A[X_1, X_2, \ldots, X_n]$ and $B[X_1, X_2, \ldots, X_n]$ be polynomial rings. Prove the following:

(1) $B \otimes_A A[X_1, \ldots, X_n] \cong B[X_1, \ldots, X_n]$.
(2) If I is an ideal of the ring $A[X_1, \ldots, X_n]$, then

$$B \otimes_A (A[X_1, \ldots, X_n]/I) \cong B[X_1, \ldots, X_n]/I \cdot B[X_1, \ldots, X_n].$$

Exercise 1.51 Let k be a field, and let A and B be finitely generated k-algebras.

(1) Prove that $A \otimes_k B$ is a finitely generated k-algebra.
(2) Let k be an algebraically closed field. If A and B are integral domains, prove that $A \otimes_k B$ is also an integral domain.

Let A be a ring. If there is a positive integer n such that $n \cdot 1_A = 1_A + 1_A + \cdots + 1_A = 0_A$ (1_A appears n times), then we define the *characteristic* of A, denoted by $\mathrm{char}(A)$, to be the smallest positive integer among such. If there is no such positive integer, then we define $\mathrm{char}(A)$ to be 0. If $u_A : \mathbb{Z} \to A$ is the unique ring homomorphism, then we can write $\mathrm{Ker}\, u_A = (n)$ for a unique non-negative integer n, since $\mathbb{Z}$ is a Euclidean ring. Note that this n agrees with $\mathrm{char}(A)$. So A contains a subring isomorphic to $\mathbb{Z}/(n)$ with $n = \mathrm{char}(A)$. It follows that if A is a domain, then $n = \mathrm{char}(A)$ is a prime number or 0.

Exercise 1.52 Answer the following questions:

(1) Let k and k' be subfields of the complex number field $\mathbb{C}$. If k' is an extension field of k and the extension degree $n = [k' : k]$ is finite, show that $\mathbb{C} \otimes_k k' \cong \overbrace{\mathbb{C} \times \mathbb{C} \times \ldots \times \mathbb{C}}^{n \text{ times}}$.
(2) Let k be a field of characteristic $p > 0$, and let $k' = k[X]/(X^p - a)$. Here, $a \in k$ and we assume that there does not exist $b \in k$ such that $b^p = a$. In this case, confirm that the ring k' is an extension field of k and that $k' \otimes_k k' \cong k'[Y]/(Y^p)$.

1.8 Localizations and Fields of Fractions

Commutative algebra is deeply connected with algebraic geometry, and elements of a ring A are often thought of as functions defined on a space X. In such cases, if $f \in A$ is such that $f(x) \neq 0$ for all the points x in X, then $1/f$ can also be considered as a function on X. However, even if there are points x where $f(x) = 0$, $1/f$ can still be considered as a function on the subset $U = \{x \in X \mid f(x) \neq 0\}$ of X. The term "localization" is a term that means "to limit the space" or "to consider in some neighborhood of a certain point." In ring theory, "localization" becomes "adding denominators."

Definition 1.122 A subset S of a ring A is said to be a *multiplicatively closed subset* of A when it satisfies the following conditions:[7]

(1) $1 \in S$;
(2) If $s, t \in S$, then $st \in S$.

We will give some examples of multiplicatively closed subsets.

Example 1.123

(1) $S = \{a \in A \mid a$ is a non-zerodivisor of ring $A\}$.
(2) $S = \{f^n \mid n \geq 0\}$.
(3) Take a family of prime ideals $\{P_i\}_{i \in I}$ of A, and set $S = A \setminus \bigcup_{i \in I} P_i$. In particular, take a prime ideal P, and set $S = A \setminus P$.

Definition 1.124 (Localization $S^{-1}A$) Let S be a multiplicatively closed subset of A.

(1) Let $X = S \times A$. For any two elements $(s, a), (t, b) \in X$, we say that $(s, a) \sim (t, b)$ if there exists an element $u \in S$ such that $u(sb - ta) = 0$. The relation $\sim$ is an equivalence relation on the set $X = S \times A$. We write the quotient set $X/\!\sim$ as $S^{-1}A$, and for each element $(s, a) \in X$, we denote the equivalence class containing (s, a) as a/s. In other words, $S^{-1}A = \{a/s \mid a \in A, s \in S\}$ where $a/s = b/t \Leftrightarrow \exists u \in S, u(sb - ta) = 0$.
    We call $S^{-1}A$ (including the ring structure described below) the *localization of A at S*.
(2) In $S^{-1}A$, we define the addition $+$ by $a/s + b/t = (ta + sb)/st$ and the multiplication $\cdot$ by $(a/s)\cdot(b/t) = ab/st$. With these definitions, $S^{-1}A$ forms a ring. In the ring $S^{-1}A$, we have $1 = 1/1 = s/s$, $0 = 0/1 = 0/s$, and $-a/s = (-a)/s$.
(3) If we define the map $i_S : A \to S^{-1}A$ as $i_S(a) = a/1$, then i_S becomes a ring homomorphism. We call i_S the *natural map to the localization*.

Exercise 1.53 When S is a multiplicatively closed subset of a ring A, verify the following properties:

(1) For $s, t \in S, a \in A$,

$$a/s = ta/ts, \quad 0/s = 0/1, \quad sa/s = a/1.$$

(2) The addition and the multiplication of $S^{-1}A$ defined above are certainly well-defined. That is to say, if $a/s = a'/s'$ and $b/t = b'/t'$, then $(at + bs)/st = (a't' + b's')/s't'$ and $ab/st = a'b'/s't'$.

---

[7] In the Japanese original of this book, the condition $0 \notin S$ was required for S to be a multiplicatively closed subset. When we would require the condition, then there would be an example of a ring homomorphism $f : A \to B$ and a "multiplicatively closed subset" S of A such that $f(S)$ were not "multiplicatively closed."

From the definition of $S^{-1}A$, the following assertions hold true.

Lemma 1.125

(1) *An element of* $i_S(S)$ *is a unit in* $S^{-1}A$.
(2) *Any element* α *of* $S^{-1}A$ *can always be expressed as* $\alpha = i_S(a) \cdot i_S(s)^{-1}$ *for some* $a \in A$ *and* $s \in S$.
(3) $\operatorname{Ker} i_S = \{a \in A \mid$ *there exists* $s \in S$ *such that* $sa = 0\}$. *In other words,* $S^{-1}A$ *is obtained by adding the inverses of the elements of* S *to* A*, and ignoring those that become* 0 *when multiplied by some element of* S.
(4) *The map* i_S *is injective if and only if* S *does not have a zerodivisor.*

$S^{-1}A$ has the following "universal property":

Proposition 1.126 *Let* $g : A \to B$ *be a ring homomorphism such that* $g(S) \subseteq \mathrm{U}(B)$. *Then there exists a unique ring homomorphism* $h : S^{-1}A \to B$ *that satisfies* $g = hi_S$.

Proof If $h : S^{-1}A \to B$ satisfies the equality $g = hi_S$, then for any $a/s \in S^{-1}A$, we have $h(a/s)g(s) = h(a/s)h(s/1) = h(a/1) = g(a)$, and hence $h(a/s) = g(a) \cdot g(s)^{-1}$. Therefore, it is uniquely determined by g. Taking an element $\alpha \in S^{-1}A$, and expressing it as $\alpha = a/s = a'/s'$, there exists a $u \in S$ such that the equation $u(sa' - s'a) = 0$ holds. Since $g(u)\big(g(s)g(a') - g(s')g(a)\big) = 0$ and $g(u) \in \mathrm{U}(B)$, the equation $g(s)g(a') = g(s')g(a)$ holds in the ring B, and we obtain $g(a)g(s)^{-1} = g(a')g(s')^{-1}$. Therefore, the map $h : S^{-1}A \to B$, defined by $h(a/s) = g(a)g(s)^{-1}$, is well-defined. It is straightforward to check that this h is a homomorphism of rings. If $a \in A$, then $(hi_S)(a) = h(i_S(a)) = h(a/1) = g(a)g(1)^{-1} = g(a)$, so we obtain $g = hi_S$. □

The following holds for ideals and prime ideals of $S^{-1}A$.

Proposition 1.127 *Let* S *be a multiplicatively closed subset of* A*, and* $i_S : A \to S^{-1}A$ *be the natural homomorphism. Then the following hold:*

(1) *An ideal* J *of* $S^{-1}A$ *satisfies* $J = i_S^{-1}(J)(S^{-1}A)$. *If* I *is an ideal of* A*, then we denote by* $S^{-1}I$ *the ideal generated by* $i_s(I)$.
(2) *There is a one-to-one correspondence between the set of prime ideals of* $S^{-1}A$ *and the set of prime ideals of* A *that do not intersect* S. *That is, if* P *is a prime ideal of* A *such that* $P \cap S = \emptyset$*, then* $PS^{-1}A$ *is a prime ideal of* $S^{-1}A$*, and* $P = i_S^{-1}(P(S^{-1}A))$. *Conversely, if* Q *is a prime ideal of* $S^{-1}A$*, then* $i_S^{-1}(Q)$ *is a prime ideal of* A *that does not intersect* S*, and* $Q = i_S^{-1}(Q)(S^{-1}A)$.

Proof

(1) Let J be any ideal of $S^{-1}A$, and let a/s be any element of J. Then $i_S(a) = (s/1) \cdot (a/s) \in J$, and hence $a \in i_S^{-1}(J)$. So $a/s = (1/s) \cdot i_S(a) \in i_S^{-1}(J)(S^{-1}A)$.
(2) For an ideal I of A, it is easy to see that

$$I(S^{-1}A) = \{a/s \mid a \in I\}.$$

Let P be a prime ideal of A such that $P \cap S = \emptyset$. Then $S^{-1}P$ is a prime ideal of $S^{-1}A$. In fact, if $(x/s) \cdot (y/s') = a/s'' \in S^{-1}P$ with $a \in P$, then for some $u \in S$, we have $u(xys'' - ass') = 0$. Then $uxys'' \in P$, but since $u, s'' \notin P$, either x or y must be an element of P. As $a/1 \in P(S^{-1}A)$ if and only if $sa \in P$ for some $s \in S$ and $P \cap S = \emptyset$, we have that $i_S^{-1}(S^{-1}P) = P$. For a prime ideal Q of $S^{-1}A$, set $P = i_S^{-1}(Q)$. Then P is a prime ideal. By (1), we have $Q = P(S^{-1}A)$. In particular, $i_S(P)$ cannot contain a unit of $S^{-1}A$, and hence $P \cap S = \emptyset$.

□

In the case where every element of the set S is a non-zerodivisor of the ring A, the natural homomorphism $i_S : A \to B = S^{-1}A$ is injective, so A can be considered as a subring of B.

Definition 1.128 (Total Quotient Ring and Field of Fractions) When S is the set of all non-zerodivisors of the ring A, we write $S^{-1}A = \mathrm{Q}(A)$, and call $\mathrm{Q}(A)$ the *total quotient ring* of A. Since $i_S : A \to S^{-1}A$ is injective, we consider A as a subring of $\mathrm{Q}(A)$ through this map.

If A is an integral domain, then $\mathrm{Q}(A)$ forms a field. We call $\mathrm{Q}(A)$ the *field of fractions* of the integral domain A.

Let A be a ring, and T be a subset of A. The set of prime ideals of A which contain T is denoted by $\mathrm{V}(T)$. For an ideal I of A, a minimal element of $\mathrm{V}(I)$ is called a *minimal prime* (ideal) of I. A minimal prime of (0) is nothing but a minimal element of the set of prime ideals of A. It is called a minimal prime of the ring A.

Exercise 1.54 Confirm that the following assertions are true:

(1) A non-zerodivisor of $\mathrm{Q}(A)$ is a unit of $\mathrm{Q}(A)$.
(2) If A is reduced and the minimal prime ideals of A are $P_1, \ldots, P_s$, then

$$\mathrm{Q}(A) \cong \mathrm{Q}(A/P_1) \times \mathrm{Q}(A/P_2) \times \cdots \times \mathrm{Q}(A/P_s).$$

Localization can also be defined for A-modules.

Definition 1.129 Let S be a multiplicative closed subset in the ring A, and M be an A-module.

(1) $S^{-1}M = \{x/s \mid x \in M, s \in S\}$ where $x/s = y/t \Leftrightarrow \exists u \in S, u(sy - tx) = 0$. $S^{-1}M$ is called the *localization* of M by S.
(2) In $S^{-1}M$, we define the addition $+$ as $x/s + y/t = (tx + sy)/st$ $(x, y \in M,\ s, t \in S)$, and the action of $S^{-1}A$ on $S^{-1}M$ as $(a/s)\cdot(x/t) = ax/st$ $(a \in A)$. By doing so, $S^{-1}M$ becomes an $S^{-1}A$-module.

Theorem 1.130 *If M is a Noetherian A-module, then the localization $S^{-1}M$ is a Noetherian $S^{-1}A$-module.*

Proof Let L be an $S^{-1}A$-submodule of $S^{-1}M$, and we set $N = L \cap M$. That is, $N = \{x \in M \mid \frac{x}{1} \in L\}$. We show that $L = S^{-1}N$. If $x \in N$, then for any $s \in S$, $x/s = (1/s)\cdot(x/1)$, so $S^{-1}N \subseteq L$. If we take $y \in L$ and write $y = x/s$ $(x \in M,\ s \in S)$, then $x/1 = (s/1)\cdot(x/s) \in L$, so $x \in N$. Therefore, $L = S^{-1}N$ holds. Now, since N is an A-submodule of M and M is a Noetherian A-module, N is finitely generated. Therefore, if we write $N = \sum_{i=1}^{n} Ax_i$ $(x_i \in M)$, then $L = \sum_{i=1}^{n} S^{-1}A\cdot\frac{x_i}{1}$, and it follows that L is a finitely generated $S^{-1}A$-module. Therefore, $S^{-1}M$ is a Noetherian $S^{-1}A$-module. □

Corollary 1.131 *If A is a Noetherian ring, then the localization $S^{-1}A$ is also a Noetherian ring.*

$S^{-1}M$ can be described as a tensor product.

Theorem 1.132 *Let S be a multiplicatively closed subset of A. For any A-module M,*

(1) *Define $\phi : S^{-1}A \otimes_A M \to S^{-1}M$ by $\phi((a/s) \otimes x) = ax/s$. Then ϕ is an isomorphism of $S^{-1}A$-modules.*
(2) *For an exact sequence $0 \to M' \to M \to M'' \to 0$ of A-modules, the sequence $0 \to S^{-1}M' \to S^{-1}M \to S^{-1}M'' \to 0$ is still an exact sequence. Therefore, $S^{-1}A$ is a flat A-algebra.*

Proof

(1) It is clear from the definition that the map ϕ is a surjection. To show the injectivity, let $y = \sum_{i=1}^{n}(a_i/s_i) \otimes x_i \in \mathrm{Ker}\,(\phi)$. First, if we set $s = s_1 \cdots s_n$, then $a_i/s_i = a_i s_1 \cdots s_{i-1}s_{i+1} \cdots s_n/s$, so we may assume that the denominator is a common element s. So let $s_1 = \ldots = s_n = s$. Since $\phi(y) = 0$, there exists $u \in S$ such that $u\sum a_i x_i = 0$. Therefore, $y = (1/us) \otimes (u\sum a_i x_i) = 0$ and hence ϕ is injective.
(2) From the right exactness of the tensor product in Proposition 1.113, it is sufficient to show that $S^{-1}M' \to S^{-1}M$ is injective, which is clear from the definition.

□

Definition 1.133 Let M be an A-module, and A be an integral domain.

(1) We say that $x \in M$ is a *torsion element* if there exists some $a \in A \setminus \{0\}$ such that $ax = 0$ ($0 \in M$ is always a torsion element).
(2) We denote the set of torsion elements of M by $\mathrm{t}(M)$, and call it the *torsion part* of M. Note that $\mathrm{t}(M)$ is a submodule of M.
(3) When $\mathrm{t}(M) = 0$, we say that M is *torsion-free*. When $M = \mathrm{t}(M)$, we say that M is a *torsion module*. Note that $\mathrm{t}(M)$ is the unique torsion submodule of M such that $M/\mathrm{t}(M)$ is torsion-free. Note also that $\mathrm{t}(M)$ is the kernel of the canonical map $M \to M \otimes_A \mathrm{Q}(A)$. In particular, M is torsion-free if and only if the map is injective.

1.134(Rank of a Module) Let A be a ring, M an A-module, and r a non-negative integer. Set $Q = \mathrm{Q}(A)$. We say that M is of *rank* r, and denote by $\operatorname{rank} M = r$, if $M \otimes_A Q \cong Q^r$. If A is an integral domain, then every finitely generated A-module M has a rank. Assume that $A \neq 0$. If M has a rank, then $\operatorname{rank} M$ of M is unique. Every finitely generated free module F has a rank, and $F \cong A^{\operatorname{rank} F}$. It is easy to see that M has rank r if and only if there is a free submodule F of M of rank r such that M/F is a torsion module.

1.9 The Spectrum Spec A and the Zariski Topology

When discussing commutative rings, "geometric" insights are often useful. For a ring A, we consider a "space" called Spec A, and introduce a topology called the "Zariski topology" on it.

Definition 1.135 For a ring A:

(1) We denote the set of prime ideals of A by $\operatorname{Spec}(A)$.
(2) For a subset $T \subseteq A$, we define

$$\mathrm{V}(T) = \{\mathfrak{p} \in \operatorname{Spec} A \mid \mathfrak{p} \supseteq T\}, \quad D(T) = \operatorname{Spec} A \setminus \mathrm{V}(T).$$

(3) We denote the set of maximal ideals of A as $\operatorname{Max}(A)$.

Remark 1.2 Let I be the ideal generated by the subset $T \subseteq A$. When $T \subseteq \mathfrak{p}$, $\mathfrak{p}$ contains I, so $\mathrm{V}(T) = \mathrm{V}(I)$.

Lemma 1.136 $\mathrm{V}(T)$ *has the following properties:*

(1) $\mathrm{V}((0)) = \operatorname{Spec} A$, $\mathrm{V}((1)) = \emptyset$.
(2) $\mathrm{V}(\bigcup_{\lambda \in \Lambda} T_\lambda) = \bigcap_{\lambda \in \Lambda} \mathrm{V}(T_\lambda)$.
(3) *For ideals* I, J, $\mathrm{V}(IJ) = \mathrm{V}(I \cap J) = \mathrm{V}(I) \cup \mathrm{V}(J)$.
(4) $\sqrt{I} \subseteq \sqrt{J}$ *if and only if* $\mathrm{V}(I) \supseteq \mathrm{V}(J)$. *In particular,* $\mathrm{V}(I) = \mathrm{V}(J) \Leftrightarrow \sqrt{I} = \sqrt{J}$. *So* $\mathrm{V}(I) = \mathrm{V}(\sqrt{I})$. *Also, if* $I \subseteq J$, *then* $\mathrm{V}(I) \supseteq \mathrm{V}(J)$.

Proof (1), (2) are clear from the definition.

We prove (4). Since $\sqrt{I} = \bigcap_{\mathfrak{p} \in \mathrm{V}(I)} \mathfrak{p}$, we have that $\mathrm{V}(I) \supseteq \mathrm{V}(J)$ implies $\sqrt{I} \subseteq \sqrt{J}$. Conversely, if $\sqrt{I} \subseteq \sqrt{J}$ and $\mathfrak{p} \in \mathrm{V}(J)$, then

$\mathfrak{p} \supseteq \bigcap_{\mathfrak{q} \in \mathrm{V}(J)} \mathfrak{q} = \sqrt{J} \supseteq \sqrt{I} \supseteq I$, and $\mathfrak{p} \in \mathrm{V}(I)$. We have proved the first assertion. The rest of the assertions are easy.

Finally, we prove (3). In view of (4), $\mathrm{V}(IJ) \supseteq \mathrm{V}(I \cap J) \supseteq \mathrm{V}(I) \cup \mathrm{V}(J)$ is trivial. If $\mathfrak{p} \in \mathrm{V}(IJ) \setminus (\mathrm{V}(I) \cup \mathrm{V}(J))$, then we can take $x \in I \setminus \mathfrak{p}$ and $y \in J \setminus \mathfrak{p}$. This shows $xy \in IJ \setminus \mathfrak{p}$, and this contradicts $\mathfrak{p} \in \mathrm{V}(IJ)$. Hence $\mathrm{V}(IJ) \subseteq \mathrm{V}(I) \cup \mathrm{V}(J)$, as desired. □

By this lemma, we can introduce a topology called the *Zariski topology* on Spec A.

Definition 1.137 The topology on Spec A, where the subsets of the form $\mathrm{V}(T)$ are closed, is called the *Zariski topology*[8].

Example 1.138 Let us see what Spec $(\mathbb{Z})$ looks like. As a set, Spec $(\mathbb{Z})$ is

$$\{(0)\} \cup \{(p) \mid p \text{ is a prime number}\}.$$

Next, we look at the closed subsets of Spec $(\mathbb{Z})$. The ideals of $\mathbb{Z}$ are principal, and of the form (n). When $n \neq 0$, $\mathrm{V}((n)) = \{p \mid p \text{ is a prime number and divides } n\}$ is a finite set. Since $\mathrm{V}((0)) = \mathrm{Spec}\,(\mathbb{Z})$, the non-empty closed subsets of Spec $(\mathbb{Z})$ are only finite subsets of $\mathrm{Spec}\,(\mathbb{Z}) \setminus \{(0)\}$, and the whole space Spec $(\mathbb{Z})$.

Remark 1.3 The open subsets of Spec A are of the form $D(I) = \mathrm{Spec}\,(A) \setminus \mathrm{V}(I)$, and since $\mathrm{V}(I) = \bigcap_{f \in I} \mathrm{V}(f)$,

$$D(I) = \mathrm{Spec}\, A \setminus \mathrm{V}(I) = \bigcup_{f \in I} D(f).$$

So $\{D(f) \mid f \in A\}$ is an open basis of Spec A.

A ring homomorphism $\varphi : A \to B$ has a geometric meaning when viewed in Spec.

Proposition 1.139 *When $\varphi : A \to B$ is a ring homomorphism, if we define ${}^a\varphi$: Spec $B \to$ Spec A by ${}^a\varphi(\mathfrak{q}) = \varphi^{-1}(\mathfrak{q})$, then $({}^a\varphi)^{-1}(D(a)) = D(\varphi(a))$ $(a \in A)$. In particular, ${}^a\varphi$ is a continuous map.*

Proof When $\mathfrak{q}$ is a prime ideal of B, $\varphi^{-1}(\mathfrak{q})$ is a prime ideal of A. Also, for $\mathfrak{q} \in$ Spec B, $\mathfrak{q} \in D(\varphi(a)) \Leftrightarrow \varphi(a) \notin \mathfrak{q} \Leftrightarrow a \notin \varphi^{-1}(\mathfrak{q}) \Leftrightarrow {}^a\varphi(\mathfrak{q}) \in D(a)$. □

Let A be a ring, and $f \in A$. Then $S^{-1}A$ with $S = \{1, f, f^2, \ldots\}$ is denoted by A_f. The residue ring A/I and the localization A_f respectively correspond to closed and open subvarieties.

Proposition 1.140 *Let I be an ideal of A, and S be a multiplicatively closed subset of A. Let $\pi : A \to A/I$ and $i_S : A \to S^{-1}A$ be the canonical surjection and the canonical map, respectively. Then ${}^a\pi$: Spec $(A/I) \to$ Spec (A) and ${}^a i_S$: Spec $(S^{-1}A) \to$ Spec A give homeomorphisms onto the subspaces (with the induced topology) $\mathrm{V}(I)$ and $\{\mathfrak{p} \in \mathrm{Spec}\, A \mid \mathfrak{p} \cap S = \emptyset\}$ of Spec A, respectively. In particular, for $f \in A$, we have that Spec $A_f \to$ Spec A induced by the canonical map $A \to A_f$ induces a homeomorphism from Spec A_f to $D(f)$.*

Proof By Exercise 1.12 and Proposition 1.127, ${}^a\pi$: Spec $A/I \to \mathrm{V}(I)$ and ${}^a i$: Spec $(S^{-1}A) \to \{\mathfrak{p} \in \mathrm{Spec}\, A \mid \mathfrak{p} \cap S = \emptyset\}$ are bijective. As a closed subset of Spec A/I is of the form $\mathrm{V}(J)$ with J an ideal of A/I, $\mathrm{V}(J) = \mathrm{V}(\pi^{-1}(J)(A/I)) = ({}^a\pi)^{-1}(\mathrm{V}(\pi^{-1}(J)))$, and ${}^a\pi$ is continuous, we have that ${}^a\pi$: Spec $A/I \to \mathrm{V}(I)$

[8] Oscar Zariski, 1899–1986.

is a homeomorphism. ${}^a i_S$ is a homeomorphism by a similar reasoning. When $S = \{f^n \mid n \geq 0\}$, the conditions $\mathfrak{p} \cap S = \emptyset$ and $f \notin \mathfrak{p}$ are equivalent for a prime ideal $\mathfrak{p}$. So the last assertion holds. □

Exercise 1.55 Show that the topological space Spec A is quasi-compact, that is, for any open cover Spec $A = \bigcup_{\lambda \in \Lambda} U_\lambda$, there exists a finite subcover of Spec A.

Exercise 1.56 Show that if A is an integral domain that is not a field, then Spec A is not a Hausdorff space.

As we have seen, a prime ideal $\mathfrak{p}$ is a point of the space Spec A. We define the local ring of the point.

Definition 1.141 For a prime ideal $\mathfrak{p}$ of a ring A, let $S = A \setminus \mathfrak{p}$,

(1) Define $A_\mathfrak{p} = S^{-1}A$, which is called the localization of A at $\mathfrak{p}$, also known as the local ring at the point $\mathfrak{p}$. According to Proposition 1.127, the prime ideals of $A_\mathfrak{p}$ are in one-to-one correspondence with the prime ideals contained in $\mathfrak{p}$, and hence $A_\mathfrak{p}$ is a local ring with the maximal ideal $\mathfrak{p}A_\mathfrak{p}$. We denote the natural map to the localization by $i_\mathfrak{p} : A \to A_\mathfrak{p}$.
(2) For an A-module M, we write $M_\mathfrak{p} = S^{-1}M = M \otimes_A A_\mathfrak{p}$.

Proposition 1.142 *The following conditions for an A-module M are equivalent:*

(1) $M = 0$.
(2) $\forall \mathfrak{p} \in \operatorname{Spec} A,\ M_\mathfrak{p} = 0$.
(3) $\forall \mathfrak{m} \in \operatorname{Max} A,\ M_\mathfrak{m} = 0$.

From this property, a morphism $u : M \to N$ of A-modules is injective (resp. surjective) if and only if for any maximal ideal $\mathfrak{m}$, $u_\mathfrak{m} : M_\mathfrak{m} \to N_\mathfrak{m}$ is injective (resp. surjective).

Proof It suffices to show the implication (3) ⇒(1). When $M \neq 0$, take $x \in M, x \neq 0$. Since $\operatorname{Ann}_A(x) \neq A$, we can take a maximal ideal $\mathfrak{m}$ such that $\operatorname{Ann}_A(x) \subseteq \mathfrak{m}$. Then $x/1 \in M_\mathfrak{m}$ is not 0. The latter half is obtained from the fact that u is injective ⇔ Ker $u = 0$ (resp. surjective ⇔ Coker $u = 0$) and the localization is exact. □

Proposition 1.142 is an example of properties that allow the principle that "the whole space can be understood through the localizations at all its points." We call such a property a *local property*. More precisely, when $\mathbb{P}(A, M)$ is a property, where A is a ring and M is an A-module, we say that $\mathbb{P}$ is *local* if $\mathbb{P}(A, M)$ holds if and only if $\mathbb{P}(A_\mathfrak{p}, M_\mathfrak{p})$ holds for any $\mathfrak{p} \in \operatorname{Spec} A$. The proposition says that the property $\mathbb{P}(A, M)$: "the A-module M is zero" is local.

Exercise 1.57 Let M be a finitely generated A-module, and let $x_1, x_2, \ldots, x_n \in M$.

(1) Let $\mathfrak{p} \in \operatorname{Spec} A$, and assume that $\frac{x_1}{1}, \frac{x_2}{1}, \ldots, \frac{x_n}{1} \in M_\mathfrak{p}$ generate $M_\mathfrak{p}$. Show that there exists an open neighborhood U of $\mathfrak{p}$ such that for each $\mathfrak{q} \in U$, $\frac{x_1}{1}, \frac{x_2}{1}, \ldots, \frac{x_n}{1} \in M_\mathfrak{q}$ generate $M_\mathfrak{q}$.

(2) Using this, show that the function $\dim_{\kappa(\mathfrak{p})} \kappa(\mathfrak{p}) \otimes_A M$ is an upper semi-continuous function on $\mathrm{Spec}(A)$, where $\kappa(\mathfrak{p}) = A_\mathfrak{p}/\mathfrak{p}A_\mathfrak{p}$ is the residue field of the local ring $A_\mathfrak{p}$, and a function $f : X \to \mathbb{R}$ on a topological space X is said to be *upper semi-continuous* if $\{x \in X \mid f(x) < c\}$ is open for each $c \in \mathbb{R}$.

Exercise 1.58 Confirm that the following three conditions are equivalent for the ring A.

(1) Spec A is not connected.
(2) A has idempotent elements (elements $e \in A$ such that $e^2 = e$) other than 0 and 1.
(3) A has a non-trivial direct product structure ($A \cong A_1 \times A_2$, where A_i is not a zero ring).

1.143 In algebraic geometry, Spec A becomes an "affine scheme" as we consider the "structure sheaf" on the topological space that we have defined in this section, and commutative algebra becomes the local theory of algebraic geometry. The "structure sheaf" makes a ring correspond to each open subset of Spec A, and for each point $\mathfrak{p}$ of Spec A, $A_\mathfrak{p}$ is the "ring of germs of functions" around the point $\mathfrak{p} \in \mathrm{Spec}\, A$. A sheaf on Spec A is also associated with an A-module M. For example, the concept of "vector bundles" dealt with in geometry can also be described by locally free sheaves corresponding to finitely generated projective modules. In this sense, the following concept of the "support" of a module is defined as follows.

Definition 1.144 For the A-module M, we define

$$\mathrm{Supp}_A M = \{\mathfrak{p} \in \mathrm{Spec}\, A \mid M_\mathfrak{p} \neq 0\},$$

and call it the *support* of M.

When $\mathfrak{p} \subseteq \mathfrak{q}$ are elements of Spec A, since $M_\mathfrak{p}$ is a localization of $M_\mathfrak{q}$, if $\mathfrak{p} \in \mathrm{Supp} M$ then $\mathfrak{q} \in \mathrm{Supp} M$ holds. In other words, Supp$M$ is closed under specialization.

The set of minimal elements of $\mathrm{Supp}_A M$ is denoted by $\mathrm{Min}_A\, M$. Its elements are called *minimal primes* of M (this is a natural generalization of the notion of a minimal prime of A).

When M is a finitely generated module, Supp$M$ becomes a closed set.

Proposition 1.145 *If M is a finitely generated A-module, then we have*

$$\mathrm{Supp} M = \mathrm{V}(\mathrm{Ann}_A\, M).$$

Proof Let $\mathfrak{p} \in \mathrm{Spec}\, A$, and suppose $\mathfrak{p}$ does not contain $\mathrm{Ann}_A\, M$. Then there exists $s \in \mathrm{Ann}_A\, M$, $s \notin \mathfrak{p}$, so $M_\mathfrak{p} = 0$, and $\mathfrak{p} \notin \mathrm{Supp} M$. Conversely, suppose $M_\mathfrak{p} = 0$, and let the generators of M be $x_1, x_2, \ldots, x_n$. Then for each x_i, there exists an element $s_i \in A \setminus \mathfrak{p}$ such that $s_i x_i = 0$. If we set $s = s_1 s_2 \cdots s_n$, then $sM = 0$ and $s \notin \mathfrak{p}$. Therefore, $\mathrm{Ann}_A\, M \not\subseteq \mathfrak{p}$ has been shown. □

Note that an element of $\mathrm{Min}_A(A/I)$ is nothing but a minimal prime of the ideal I, that is, a minimal element of $\mathrm{V}(I)$, by the proposition. In particular, the set of the minimal prime ideals of the ring A is nothing but $\mathrm{Min}_A A$.

We define the following concepts of topological space. These concepts are also important in algebraic context.

Definition 1.146 (Irreducible Space and Noetherian Space) Let X be a topological space.

(1) We say that X is *irreducible* if X is non-empty, and X cannot be expressed as $X = Y_1 \cup Y_2$, a union of two closed proper subsets. Note that X is irreducible if and only if X is non-empty and any non-empty open subset of X is dense. This is also equivalent to saying that any two non-empty open subsets of X intersect.
(2) When the collection of open subsets of X satisfies the maximal condition (the collection of closed subsets satisfies the minimal condition), we say that X is a *Noetherian space*. This is equivalent to saying that every open subset of X is quasi-compact, and is also equivalent to saying that every subset of X is quasi-compact.
(3) A point $x \in X$ is said to be a *generic point* of X if the singleton $\{x\}$ is dense in X. Note that $x \in X$ is a generic point of X if and only if every non-empty open subset of X contains x. In particular, if X has a generic point, then X is irreducible.

Example 1.147 For a ring A and a prime ideal $\mathfrak{p}$ of A, $\mathfrak{p} \in \mathrm{V}(\mathfrak{p})$ is the unique generic point of $\mathrm{V}(\mathfrak{p}) \subseteq \operatorname{Spec} A$. In particular, $\mathrm{V}(\mathfrak{p})$ is irreducible. Indeed, if $\mathrm{V}(I) \subsetneq \mathrm{V}(\mathfrak{p})$, then since $\mathrm{V}(\mathfrak{p}) \not\subseteq \mathrm{V}(I)$, we have $\mathfrak{p} \notin \mathrm{V}(I)$. Thus $\mathfrak{p}$ is a generic point of $\mathrm{V}(\mathfrak{p})$. If $\mathfrak{q}$ is another generic point of $\mathrm{V}(\mathfrak{p})$, then since it is also a generic point of $\mathrm{V}(\mathfrak{q})$, $\mathrm{V}(\mathfrak{p}) = \mathrm{V}(\mathfrak{q})$, and hence $\mathfrak{q} = \mathfrak{p}$. When A is a Noetherian ring, a descending chain $(*)$: $\mathrm{V}(I_1) \supseteq \mathrm{V}(I_2) \supseteq \cdots$ of closed subsets of $\operatorname{Spec} A$ yields an ascending chain of radical ideals $\sqrt{I_1} \subseteq \sqrt{I_2} \subseteq \cdots$ of A, which must be eventually constant. By Lemma 1.136, $(*)$ is eventually constant. So $\operatorname{Spec} A$ becomes a Noetherian space. Furthermore, by definition, a subspace of a Noetherian space is a Noetherian space.

Exercise 1.59 Construct a ring A that is not a Noetherian ring, but $\operatorname{Spec} A$ is a Noetherian space.

From the above considerations, the following proposition is obtained.

Proposition–Definition 1.148 Let A be a ring and let $X = \operatorname{Spec} A$.

(1) An irreducible closed subset Y of X can be written as $Y = \mathrm{V}(\mathfrak{p})$ for some prime ideal $\mathfrak{p}$.
(2) A maximal irreducible closed subset of X is called an *irreducible component* of X.

The irreducible components of X and the minimal prime ideals of A are in one-to-one correspondence.

Proposition 1.149 (Cf. Theorem 2.26) *A Noetherian space has only finitely many irreducible components, and it is the union of them. In particular, a Noetherian ring possesses only finitely many minimal prime ideals. If A is a Noetherian ring and M is a finitely generated A-module, then* $\operatorname{Min}_A M$ *is a finite set.*

Proof We prove the first assertion. Let X be a Noetherian space. First, we prove that X can be written as the union of finitely many irreducible closed subsets of it. Assume the contrary. Then by the Noetherian property, there exists a minimal element Y of the set of closed subsets of X that cannot be written as a union of finitely many irreducible closed subsets. The empty set is the union of zero many irreducible closed subsets, and hence $Y \neq \emptyset$. Since an irreducible topological space Z is the union of one irreducible closed subset (that is, Z itself), Y is not irreducible. So $Y = Y_1 \cup Y_2$ can be written as the union of smaller closed subsets Y_1, Y_2. Then, by the minimality of Y, Y_1, Y_2 can be written as the union of finitely many irreducible closed subsets, and hence Y itself can also be written as the union of finitely many irreducible closed subsets, and this contradicts the choice of Y. So we can write $X = Y_1 \cup \cdots \cup Y_r$ with Y_i irreducible. We may assume that the expression does not have a redundancy, that is, $\bigcup_{i \neq j} Y_i \neq X$ for any $1 \leq j \leq r$. For any irreducible closed subset Y of X, we have $Y = \bigcup_i (Y \cap Y_i)$, and hence $Y \subseteq Y_i$ for some i. In particular, if Y is an irreducible component, then $Y = Y_i$ for some i. If Y_i is not an irreducible component, then $Y_i \subsetneq Y$ for some irreducible closed subset Y, and $Y \subseteq Y_j$ for some j. But this contradicts the fact that the expression is irredundant. So each Y_i is an irreducible component. So X is the union of its irreducible components $Y_1, \ldots, Y_r$, and there is no other irreducible component. This proves the first assertion. The last assertion follows from Proposition–Definition 1.148. □

Exercise 1.60 (1) Show that an irreducible Hausdorff space is one point. (2) If a Noetherian topological space X is Hausdorff, then X is a finite set.

Proposition 1.150 *Let A be a ring, and $P \in$* Spec A. *Then there exists a minimal prime ideal of A contained in P.*

Proof Set $\Omega = \{Q \in \operatorname{Spec} A \mid Q \subseteq P\}$. Then $P \in \Omega$, and hence Ω is non-empty. Let $(Q_\lambda)_{\lambda \in \Lambda}$ be a non-empty chain of Ω. Then $Q = \bigcap Q_\lambda$ is a prime ideal. Since $\Lambda \neq \emptyset$, we have $Q \neq A$. Let $a \in A \setminus Q$, $b \in A$, and $ab \in Q$. Then there exists some $\lambda \in \Lambda$ such that $a \notin Q_\mu$ for any $Q_\mu \subseteq Q_\lambda$. Since $ab \in Q_\mu$, we have $b \in \bigcap Q_\mu = Q$. Hence Q is a prime ideal. By Zorn's lemma, the assertion follows. □

Corollary 1.151 *Let A be a ring, and I be an ideal of A. Then $\sqrt{I} = \bigcap_{P \in \operatorname{Min}_A(A/I)} P$, where the intersection runs through the set of minimal prime ideals of I.*

Chapter 2
Noetherian Rings and Dimension

In this chapter, we develop the theory of Noetherian and Artinian rings. We introduce the notion of associated prime ideals for modules over Noetherian rings and then primary decomposition of ideals. We also discuss dimension theory of Noetherian rings, which shows that commutative ring theory is deeply connected to algebraic geometry. Krull's altitude theorem is essential.

2.1 Noetherian Rings

First, let us recall important results we have discussed on Noetherian rings.

2.1 We say that a ring A is a Noetherian[1] ring if every ideal of A is finitely generated (Definition 1.9, Theorem 1.10). Let A be a Noetherian ring. Then the following statements are true:

(1) (Exercise 1.7) If I is an ideal of A, then the residue class ring A/I is also Noetherian.
(2) (Corollary 1.131) A localization $S^{-1}A$ of A is also a Noetherian ring.
(3) (Corollary 1.43) A polynomial ring $A[X_1, X_2, \ldots, X_n]$ over A is also Noetherian.

From (3), if B is a finitely generated A-algebra, then B is also Noetherian. If M is a Noetherian A-module, then $S^{-1}M$ is a Noetherian $S^{-1}A$-module (Theorem 1.130).

We show that if every prime ideal of A is finitely generated, then A is Noetherian (Cohen's theorem).

[1] Emmy Noether, 1882–1935.

S. Goto, K. Watanabe, *Commutative Algebra*, Graduate Texts in Mathematics 309,
https://doi.org/10.1007/978-981-95-4227-7_2

Theorem 2.2 (Cohen's Theorem) *Consider the set $\mathcal{S}$ of ideals of A which are not finitely generated, and let I be a maximal element of $\mathcal{S}$ with respect to inclusion relation. Then I is a prime ideal.*

Proof Assume the contrary. Since $I \neq A$, we can take $x, y \in A$, $x, y \notin I$ and $xy \in I$. Then since $I \subsetneq I + (x)$ and I is maximal in $\mathcal{S}$, the ideal $I + (x)$ is finitely generated. Also, if we put $J = I : x$, then since $I \subsetneq I + (y) \subseteq J$, J is also finitely generated. Since we have $(x) \cap I = xJ$, we have an exact sequence of A-modules

$$0 \to xJ \to I \to [I + (x)]/(x) \to 0.$$

Since J and $I + (x)$ are finitely generated, we have that I is also finitely generated, and this contradicts $I \in \mathcal{S}$. □

Corollary 2.3 (Cohen) *If every prime ideal of a ring A is finitely generated, then A is a Noetherian ring.*

Proof Let $\mathcal{S}$ be the set of non-finitely generated ideals, as in Theorem 2.2. Assume the corollary is not true, and $\mathcal{S}$ is non-empty. Then $\mathcal{S}$ has a maximal element. Indeed, if we take a non-empty chain C in $\mathcal{S}$, then $I = \bigcup_{J \in \mathcal{S}} J$ is *not* finitely generated. Hence we can take a maximal element I in $\mathcal{S}$ by Zorn's lemma. By Theorem 2.2, I is a prime ideal. By assumption, I is finitely generated, and this is a contradiction. □

2.2 Artinian Rings and Composition Series

We defined Noetherian rings by the Ascending Chain Condition (ACC). We define Artinian rings by the Descending Chain Condition (DCC).

Proposition 2.4 *The following conditions for a ring A are equivalent to each other:*

(1) **(DCC)** *If $I_1 \supseteq I_2 \supseteq \cdots \supseteq I_i \supseteq \cdots$ is a descending chain of ideals of A, then for sufficiently big N, we have $I_N = I_i$ for every $i \geq N$.*
(2) **(Minimality Condition)** *If $\mathcal{S}$ is a non-empty set of ideals of A, then there is a minimal element in $\mathcal{S}$ with respect to inclusion.*

The proof is a formality and hence is the same as Theorem 1.10. We say that A is *Artinian*[2] if the equivalent conditions in Proposition 2.4 are satisfied. The following is also proved.

Proposition 2.5 *Let A be an Artinian ring. If $\varphi : A \to B$ is a ring homomorphism such that $(J \cap A)B = J$ for every ideal J of B, then B is also Artinian. In particular, we have:*

(1) *For every ideal I of A, A/I is also an Artinian ring.*

[2] Emil Artin, 1898–1962.

(2) *A localization $S^{-1}A$ is an Artinian ring for every multiplicatively closed subset S of A.*

However, Artinian rings have properties very different from Noetherian rings.

Lemma 2.6 *If A is an Artinian ring that is an integral domain, then A is a field.*

Proof If A is an Artinian integral domain, then for every nonzero element $a \in A$, the chain $\{(a^n)\}_{n\geq 1}$ is a descending chain. Then by (DCC) for $n \gg 1$, we have $(a^n) = (a^{n+1})$. Hence we have $a^n = xa^{n+1}$ for some $x \in A$, which induces $a^n(1 - ax) = 0$, namely $1 = ax$, which shows that A is a field. □

Proposition 2.7 *If A is an Artinian ring, then every prime ideal of A is a maximal ideal and* Spec A *is a finite set.*

Proof If P is a prime ideal of A, then A/P is an Artinian domain. Hence A/P is a field by Lemma 2.6, which means P is a maximal ideal.

If $\operatorname{Spec} A = \operatorname{Max} A$ is an infinite set, then taking an infinite set $\{\mathfrak{m}_i\}_{i\geq 1}$ of different maximal ideals, we can make an infinite descending chain of ideals

$$\mathfrak{m}_1 \supsetneq \mathfrak{m}_1 \cap \mathfrak{m}_2 \supsetneq \mathfrak{m}_1 \cap \mathfrak{m}_2 \cap \mathfrak{m}_3 \supsetneq \cdots \supsetneq \bigcap_{i=1}^{n} \mathfrak{m}_i \supsetneq \cdots .$$

Indeed, if for some n, we have $\bigcap_{i=1}^{n} \mathfrak{m}_i = \bigcap_{i=1}^{n+1} \mathfrak{m}_i$, then we have $\bigcap_{i=1}^{n} \mathfrak{m}_i \subseteq \mathfrak{m}_{n+1}$, which contradicts our assertion that $\mathfrak{m}_i$ are different maximal ideals. Hence Spec A is a finite set. □

The following theorem of Akizuki[3] gives the characterization of Artinian rings in terms of Noetherian rings.

Theorem 2.8 (Theorem of Akizuki) *The following conditions are equivalent:*

(1) *A is an Artinian ring.*
(2) *A is a Noetherian ring whose prime ideals are maximal.*

Proof (1) ⇒(2) First, we show that an Artin local ring $(A, \mathfrak{m})$ is Noetherian. To show this, first we show that the ideal $\mathfrak{m}$ is nilpotent. Since $\{\mathfrak{m}^n\}$ is a descending chain of ideals, there is an integer $n \geq 1$ such that $\mathfrak{m}^k = \mathfrak{m}^n$ for any $k \geq n$. If we put $I = \mathfrak{m}^n$, then we have $I^2 = I$ by definition. Assume that $I \neq 0$. Put

$$\mathcal{S} = \{J \mid J \text{ is an ideal of } A \text{ and } JI \neq 0\}.$$

Let J be a minimal element of $\mathcal{S}$. Since $xI \neq 0$ for some $0 \neq x \in J$, by the minimality of J, $J = (x)$. Since $0 \neq xI \subseteq (x) = J$ and $xI \cdot I = xI^2 = xI \neq 0$, we get $xI = (x)$ by minimality. Hence we can write $x = xi$ for some $i \in I$. Then we have $x(1 - i) = 0$ and then $x = 0$ since $1 - i \notin \mathfrak{m}$ is a unit. This contradicts our assumption that $x \neq 0$. Thus we conclude that $I = \mathfrak{m}^n = 0$.

[3] Yasuo Akizuki, 1902–1984.

Next, we show that A is Noetherian by induction on the *nilpotency index*

$$\ell = \min\{n \geq 0 \mid \mathfrak{m}^n = 0\}$$

of A. If $\ell = 0$, then A is a field and hence is Noetherian. We assume $\ell > 0$ and assume that A is Noetherian for rings with smaller nilpotency index. Then $A/\mathfrak{m}^{\ell-1}$ is an Artin local ring with smaller nilpotency index, and hence it is Noetherian. Since $\mathfrak{m}\cdot\mathfrak{m}^{\ell-1} = 0$, $\mathfrak{m}^{\ell-1}$ is a vector space over $A/\mathfrak{m}$, and is Artinian. So it does not have an infinite descending chain of subspaces, and hence is finite-dimensional. Hence $\mathfrak{m}^{\ell-1}$ is Noetherian as an A-module. In the exact sequence

$$0 \to \mathfrak{m}^{\ell-1} \to A \to A/\mathfrak{m}^{\ell-1} \to 0,$$

since every A-submodule of $A/\mathfrak{m}^{\ell-1}$ and $\mathfrak{m}^{\ell-1}$ is finitely generated and hence we obtain that every ideal of A is finitely generated. Hence A is a Noetherian ring.

Next, let A be an Artinian ring which is not necessarily a local ring. Let

$$I_1 \subseteq I_2 \subseteq \cdots \subseteq I_i \subseteq \cdots$$

be an ascending chain of ideals of A and $\mathfrak{m}$ be an maximal ideal of A. Since $A_{\mathfrak{m}}$ is an Artinian local ring, we have shown that $A_{\mathfrak{m}}$ is a Noetherian ring. Since $\operatorname{Max} A$ is a finite set, we can choose a number k such that for any $i \geq k$ and for any $\mathfrak{m} \in \operatorname{Max} A$, we have $I_k A_{\mathfrak{m}} = I_i A_{\mathfrak{m}}$. Hence by Proposition 1.142, we get $I_i = I_k$ for any $i \geq k$, and A is Noetherian.

(2) $\Rightarrow$(1) By our assumption (2), since every prime ideal of A is also a minimal prime ideal, $\operatorname{Spec} A$ is a finite set by Proposition 1.149 (see also Chap. 3). Hence by the same reasoning as in the proof of (1) $\Rightarrow$(2), we may assume that A is a local ring with the maximal ideal $\mathfrak{m}$. Since $\mathfrak{m}$ is the unique prime ideal of A and is finitely generated, $\mathfrak{m}$ is nilpotent by Exercise 1.17. Then we can conclude that A is Artinian by the following lemma (Lemma 2.9). □

Lemma 2.9 *If* $(A, \mathfrak{m})$ *is a Noetherian ring and if* $\mathfrak{m}$ *is nilpotent, then* A *is an Artinian ring.*

Proof We prove by the induction on the nilpotency index $\ell = \min\{n \geq 0 \mid \mathfrak{m}^n = 0\}$. If $\ell = 0$, then A is a field. Assume $\ell > 0$ and assume that our conclusion is true up to nilpotency index $\ell - 1$. Then $A/\mathfrak{m}^{\ell-1}$ is Artinian since $A/\mathfrak{m}^{\ell-1}$ is Noetherian local ring with smaller nilpotency index.

Also, $V = \mathfrak{m}^{\ell-1}$ is an ideal of A and a finite-dimensional vector space over $A/\mathfrak{m}$. Putting $B = A/V$, we have an exact sequence of A-modules,

$$0 \to V \to A \xrightarrow{f} B \to 0,$$

here $f : A \to B$ denotes the natural surjection.

Now, the submodules of V and B satisfy condition (DCC). Hence we can show that the ideals of A also satisfy (DCC). Indeed, if let $\{I_i\}_{i \geq 1}$ be a descending chain of ideals of A, $\{I_i \cap V\}_{i\geq 1}$ is a descending chain of submodules of V and $I_m \cap V = I_i \cap V$ for sufficiently large $m \geq 1$ and for every $i \geq m$. Also, $\{f(I_i) = (I_i + V)/V\}_{i\geq 1}$ is a descending chain of ideals of B. Since B is an Artinian ring, $f(I_n) = f(I_i)$ for sufficiently large n and for all $i \geq n$. If we put $k = \max\{m, n\}$, then for $i \geq k$, we have $I_k \cap V = I_i \cap V$ and $I_k + V = I_i + V$. Hence by Proposition 1.18, we have

$$I_k = (I_i + V) \cap I_k = I_i + (I_k \cap V) = I_i + (I_i \cap V) = I_i$$

and we show that A is an Artinian ring. □

Next, we introduce the concepts of Artinian modules and composition series.

Proposition–Definition 2.10 The following conditions on an A-module M are equivalent:

(1) If $N_1 \supseteq N_2 \supseteq \cdots \supseteq N_i \supseteq \cdots$ is a descending chain of A-submodules of M, then there exists $n \geq 1$ such that $N_n = N_i$ for every $i \geq n$.
(2) Any non-empty set $\mathcal{S}$ of A-submodules of M has a minimal element with respect to the inclusion relation.

We say that M is an Artinian module if these equivalent conditions are satisfied.

Remark 2.1 We have discussed above that an Artinian ring is a special Noetherian ring. But the concept of Artinian module is "dual" to the concept of Noetherian A-modules and *not* a subclass of Noetherian modules.

Example 2.11

(1) Let $\mathbb{Z}$ be the ring of integers and p be a prime number. Then $\mathbb{Q}/\mathbb{Z}_{(p)}$ is an Artinian $\mathbb{Z}$-module but is not a finitely generated $\mathbb{Z}$-module.
(2) Let k be a field and X be a variable on k. Then $M = k[X, X^{-1}]/k[X]$ is an Artinian $k[X]$-module but not finitely generated as a $k[X]$-module.

A finitely generated Artinian module is a Noetherian module.

Proposition 2.12 *If* M *is a* finitely generated *Artinian* A*-module, then* $A/\operatorname{Ann}_A(M)$ *is an Artinian ring and* M *is a Noetherian* A*-module.*

Proof For $x \in M, x \neq 0$, $A/\operatorname{Ann}_A(x) \cong Ax \subseteq M$ is an Artinian module. Also, submodules of Ax have one-to-one correspondence with ideals of $A/\operatorname{Ann}_A(x)$. Hence $A/\operatorname{Ann}_A(x)$ is an Artinian ring and is also a Noetherian ring. If M is generated by $x_1, \ldots, x_s$ as an A-module, then $\operatorname{Ann}_A(M) = \bigcap_{i=1}^{s} \operatorname{Ann}_A(x_i)$. Hence $A/\operatorname{Ann}_A(M)$ is a subring and a submodule of $\prod_{i=1}^{s} A/\operatorname{Ann}_A(x_i)$. This implies that $A/\operatorname{Ann}_A(M)$ is an Artinian ring and since M is a homomorphic image of $\bigoplus_{i=1}^{s} A/\operatorname{Ann}_A x_i$, M is also a Noetherian A-module. □

For a Noetherian and Artinian module, we can define the concept of composition series and length. An A-module M is said to be *simple* if $M \neq 0$ and there is no submodule of M other than M or 0. If $\mathfrak{m}$ is a maximal ideal of A, then the A-module $A/\mathfrak{m}$ is simple. Conversely, if M is a simple A-module, then for any $x \in M \setminus \{0\}$, $M \cong Ax \cong A/\operatorname{Ann}_A x$, and $\operatorname{Ann}_A x$ is maximal.

Proposition–Definition 2.13 Assume that M is a Noetherian and Artinian A-module.

(1) There is a sequence of submodules of M

$$0 = N_0 \subsetneq N_1 \subsetneq \cdots \subsetneq N_m = M$$

such that N_{i+1}/N_i is simple for $i = 1, \ldots, m$. We call such a sequence of A-modules a *composition series* of M. Conversely, if M has a composition series, then M is Noetherian and Artinian.

(2) We call the length m of the sequence the *length* of M, and denote it by $\ell_A(M)$. If $M = 0$, we define $\ell_A(M) = 0$. The length of M does not depend on the choice of composition series. On the other hand, if M does not have a composition series, then we define $\ell_A(M) = \infty$.

(3) If $0 \to L \xrightarrow{f} M \xrightarrow{g} N \to 0$ is an exact sequence of A-modules, then we have $\ell_A(M) = \ell_A(L) + \ell_A(N)$, where we consider that $\infty + m = m + \infty = \infty$ for $m \in \mathbb{Z}_{\geq 0} \cup \{\infty\}$.

Proof

(1) Assume that M is Noetherian and Artinian, and consider the set

$$\Gamma = \{N \subseteq M \mid N \text{ has a composition series } 0 = N_0 \subsetneq N_1 \subsetneq N_2 \subsetneq \cdots \subsetneq N_m = N\}.$$

Since $0 \in \Gamma$, we have $\Gamma \neq \emptyset$. So it has a maximal element N with respect to the inclusion relation, since M is Noetherian. If $N \neq M$, then the set of submodules

$$\Omega = \{L \subseteq M \mid N \subsetneq L\}$$

is non-empty, since $M \in \Omega$. Since M is Artinian, Ω has a minimal element L. Then the sequence

$$0 = N_0 \subsetneq N_1 \subsetneq N_2 \subsetneq \cdots \subsetneq N_m = N \subsetneq L$$

is a composition series of L, and hence $L \in \Gamma$. This contradicts the maximality of N. Hence $N = M$, and the assertion has been proved.

We prove the converse. Let

$$0 = N_0 \subsetneq N_1 \subsetneq \cdots \subsetneq N_m = M$$

be a composition series of M. Then

$$0 \to N_{i-1} \to N_i \to N_i/N_{i-1} \to 0$$

is exact. Since the simple module N_i/N_{i-1} is Noetherian and Artinian, it follows that N_i is Noetherian and Artinian by induction on i. So $M = N_m$ is Noetherian and Artinian.

Assuming that $\ell_A(M) < \infty$ is well-defined, we prove that $\ell_A(N)$ and $\ell_A(L)$ are also well-defined, and (3) holds. Note that M, N, and L are all Noetherian and Artinian by (1), and hence N and L also have composition series. If

$$0 = L_0 \subsetneq L_1 \subsetneq \cdots \subsetneq L_s = L$$

and

$$0 = N_0 \subsetneq N_1 \subsetneq \cdots \subsetneq N_t = N$$

are composition series of L and N, respectively, then

$$0 = L_0 \subsetneq L_1 \subsetneq \cdots \subsetneq L_s = M_0 \subsetneq M_1 \subsetneq \cdots \subsetneq M_t = M$$

is a composition series of M, where $M_i = g^{-1}(N_i)$. So it follows that $\ell_A(L)$ and $\ell_A(N)$ are also well-defined, and $\ell_A(M) = s + t = \ell_A(L) + \ell_A(N)$. Since $\ell_A(M) < \infty$ if and only if both $\ell_A(L)$ and $\ell_A(N)$ are finite, we are done.

(2) We need to prove that the length of a Noetherian and Artinian A-module M does not depend on its composition series. Take two composition series

$$0 = N_0 \subsetneq N_1 \subsetneq \ldots \subsetneq N_m = M; \quad 0 = L_0 \subsetneq L_1 \subsetneq \ldots \subsetneq L_t = M$$

and we prove that $t = m$ by induction on m. The proof of (3) above shows that if an A-module N has a composition series of length smaller than m, then the length of any subquotient of N is well-defined. Now let r be the smallest positive integer such that $N_1 \subseteq L_r$ but $N_1 \not\subseteq L_{r-1}$. Then the composite $N_1 \to L_r \to L_r/L_{r-1}$ is an isomorphism, and hence $L_r = N_1 + L_{r-1}$. So $L_{r-1} \cong L_r/N_1 \subseteq M/N_1$, and

$$\begin{aligned} m - 1 = \ell_A(M/N_1) &= \ell_A((M/N_1)/(L_r/N_1)) + \ell_A(L_r/N_1) \\ &= \ell_A(M/L_r) + \ell_A(L_{r-1}) = t - r + r - 1 = t - 1, \end{aligned}$$

and hence $m = t$, as desired, where all the ℓ_A that appear are well-defined by induction.

□

Exercise 2.1 Let $(A, \mathfrak{m}) \to (B, \mathfrak{n})$ be a flat ring homomorphism of Noetherian local rings which satisfies $\mathfrak{m}B \subseteq \mathfrak{n}$ and assume that $B/\mathfrak{m}B$ has finite length. If M is an A-module which has finite length, prove the following equality:

$$\ell_B(B \otimes_A M) = \ell_A(M) \cdot \ell_B(B/\mathfrak{m}B).$$

2.3 Associated Primes and Primary Decomposition

The concept of primary decomposition corresponds to prime factorization of integers. In this section, we explain primary decomposition of ideals from the viewpoint of associated prime ideals. The contents of this section are based on Bourbaki's book [Bo].

Definition 2.14 (Associated Prime Ideals) Let M be an A-module.

(1) If a prime ideal $\mathfrak{p}$ satisfies the following equivalent conditions, we say that $\mathfrak{p}$ is an *associated prime ideal* of M and we denote the set of associated prime ideals of M by $\operatorname{Ass}_A M$.

(a) $\mathfrak{p} = \operatorname{Ann}_A(x)$ holds for some $x \in M$.
(b) M has an A-submodule isomorphic to $A/\mathfrak{p}$.

Namely, we define

$$\operatorname{Ass}_A M = \{\mathfrak{p} \in \operatorname{Spec} A \mid \exists x \neq 0 \in M, \mathfrak{p} = \operatorname{Ann}_A(x)\}.$$

(2) For an ideal I of A, we call an element of $\operatorname{Ass}_A(A/I)$ an *associated prime ideal of* I.

Proposition 2.15 *Let M be an A-module.*

(1) *If $\mathfrak{a}$ is a maximal element of the set of ideals $\{\operatorname{Ann}_A(x) | x \in M \setminus \{0\}\}$, then $\mathfrak{a}$ is a prime ideal and hence $\mathfrak{a} \in \operatorname{Ass}_A M$.*
(2) *If A is a Noetherian ring, and $M \neq 0$, then $\operatorname{Ass}_A M \neq \emptyset$.*
(3) *Let A be a Noetherian ring and let $a \in A$. Then there exists $x \in M, x \neq 0$ such that $ax = 0$ if and only if $a \in \bigcup\{\mathfrak{p} \mid \mathfrak{p} \in \operatorname{Ass}_A M\}$.*

Proof (1) Assume that $\mathfrak{a} = \operatorname{Ann}_A(x)$ is a maximal element and not a prime ideal. Take $a, b \notin \mathfrak{a}, ab \in \mathfrak{a}$. Then by the maximality of $\mathfrak{a}$, $ax \neq 0$ and $\operatorname{Ann}_A(ax) \supseteq \mathfrak{a}$. Since $b \in \operatorname{Ann}_A(ax)$, this contradicts our assertion that $\mathfrak{a}$ is maximal.

(2) and (3) clearly follow from (1). □

Example 2.16

(1) If A is an integral d1omain, then $\operatorname{Ass}_A A = \{(0)\}$. Hence if $\mathfrak{p}$ is a prime ideal of A, then $\operatorname{Ass}_A(A/\mathfrak{p}) = \{\mathfrak{p}\}$.
(2) Let us give an example such that $M \neq 0$ and $\operatorname{Ass}_A M = \emptyset$ (A is *not* Noetherian). Let k be a field and let $A = k[X_1, X_2, \ldots, X_n, \ldots]/(X_1^2, \ldots, X_n^2, \ldots)$. A is the ring obtained by adjoining infinitely many x_i with $x_i^2 = 0$.

In this case, the unique prime ideal of A is $\mathfrak{m} = (x_1, x_2, \ldots)$. But since every $f \in A \setminus \{0\}$ involves only finitely many x_i, there is some x_j such that $x_j f \neq 0$. Namely, $0 :_A f \neq \mathfrak{m}$. Hence $\mathfrak{m} \notin \operatorname{Ass}_A A$, and we have $\operatorname{Ass}_A A = \emptyset$.

As we have seen above, the Noetherian property of A is very important in the theory of associated prime ideals. Hence *we always assume that all the rings are*

Noetherian in the rest of this section. Actually, we can construct some counterexamples to almost all statements if we do not assume the rings are Noetherian.

The following is the first basic theorem in this theory.

Theorem 2.17

(1) *If* $0 \to M' \to M \to M'' \to 0$ *is a short exact sequence of A-modules, then we have*

$$\mathrm{Ass}_A M' \subseteq \mathrm{Ass}_A M \subseteq \mathrm{Ass}_A M' \cup \mathrm{Ass}_A M''.$$

(2) *If M is a finitely generated A-module, then* $\mathrm{Ass}_A M$ *is a finite set.*

Proof (1) If $\mathfrak{p} \in \mathrm{Ass}_A M'$ then we have $A/\mathfrak{p} \subseteq M' \subseteq M$. Hence $\mathrm{Ass}_A M' \subseteq \mathrm{Ass}_A M$.

Let $\mathfrak{p} \in \mathrm{Ass}_A M$ and assume $\mathfrak{p} = \mathrm{Ann}_A(x)$ $(x \in M)$. If $0 \neq y \in Ax \cap M'$, then since $Ax \cong A/\mathfrak{p}$, we have $\mathrm{Ann}_A(y) = \mathfrak{p}$ and $\mathfrak{p} \in \mathrm{Ass}_A M'$. If we assume $Ax \cap M' = 0$, then let $\overline{x} \in M''$ be the image of x in M''. Then since $A\overline{x} \cong A/\mathfrak{p} \subseteq M''$, we have $\mathfrak{p} \in \mathrm{Ass}_A M''$ and (1) is proved.

To show (2), we prove the following lemma. From this lemma we show (2) by (1) and Example 2.16. □

Lemma 2.18 *If M is a finitely generated A-module, then there is a sequence* $0 = M_0 \subsetneq M_1 \subsetneq \cdots \subsetneq M_n = M$ *of A-submodules of M such that* $M_i/M_{i-1} \cong A/\mathfrak{p}_i (\forall i\ \mathfrak{p}_i \in \mathrm{Spec}\, A)$.

Proof Consider all the submodules of M satisfying the condition of the lemma and let N be a maximal element of this set. (Since M is a finitely generated A-module, M is a Noetherian module by Theorem 1.81, and there exists a maximal element.)

If $N = M$, there is nothing to prove. If $N \neq M$, since $\mathrm{Ass}_A(M/N) \neq \emptyset$, $\exists x \in M, x \notin N$, $\mathrm{Ann}_A((x+N)/N) = \mathfrak{q}$ for some prime ideal $\mathfrak{q}$. Put $N' = N + Ax \subseteq M$. Then since $N'/N \cong A/\mathfrak{q}$, N' satisfies the condition of the lemma, which contradicts the maximality of N. □

We call a filtration as in Lemma 2.18 a *Bourbaki filtration*.

The existence of a non-zerodivisor of a local ring $(A, \mathfrak{m})$ is described by $\mathrm{Ass}_A M$.

Proposition 2.19 *If* $(A, \mathfrak{m})$ *is a Noetherian local ring, the following conditions are equivalent:*

(1) $\mathfrak{m} \in \mathrm{Ass}\, A$.
(2) *Every element of* $\mathfrak{m}$ *is a zerodivisor of A.*

Proof (1) $\Rightarrow$(2) is clear by definition. If we assume (2), then we have $\mathfrak{m} \subseteq \bigcup_{P \in \mathrm{Ass}\, A} P$. Since $\mathrm{Ass}\, A$ is a finite set, by the prime avoidance lemma (Theorem 1.31) and Proposition 2.15, $\mathfrak{m} \subseteq P$ for some $P \in \mathrm{Ass}\, A$. Hence $P = \mathfrak{m} \in \mathrm{Ass}\, A$. □

Let us see the relation between $\mathrm{Ass}_A M$ and $\mathrm{Supp}_A M$.

Proposition 2.20 *Let M be an A-module and $S \subseteq A$ be a multiplicatively closed set.*

(1) *If we regard* $\mathrm{Spec}(S^{-1}A)$ *as a subset of* $\mathrm{Spec}\, A$, *we have*

$$\mathrm{Ass}_{S^{-1}A}(S^{-1}M) = \mathrm{Ass}_A(S^{-1}M) = \mathrm{Ass}_A M \cap \mathrm{Spec}(S^{-1}A).$$

(2) $\mathrm{Ass}_A M \subseteq \mathrm{Supp}_A M$.
(3) *A minimal element of* $\mathrm{Supp}_A M$ *is an element of* $\mathrm{Ass}_A M$. *Namely,* $\mathrm{Min}_A M \subseteq \mathrm{Ass}_A M$. *In particular, if M is finitely generated, then* $\mathrm{Min}_A M$ *is a finite set.*

Proof

(1) If $\mathfrak{p} \in \mathrm{Ass}_A M$, $\mathfrak{p} \cap S = \emptyset$, then we have $A/\mathfrak{p} \subseteq M$. Localizing by S, we have $S^{-1}A/S^{-1}\mathfrak{p} \subseteq S^{-1}M$. Hence we have $S^{-1}\mathfrak{p} \in \mathrm{Ass}_{S^{-1}A}(S^{-1}M)$.
Conversely, if $S^{-1}\mathfrak{p} \in \mathrm{Ass}_{S^{-1}A}(S^{-1}M)$ and if $\mathrm{Ann}_{S^{-1}A}(\frac{x}{s}) = S^{-1}\mathfrak{p}$, we have $\mathrm{Ann}_A(x) \subseteq \mathfrak{p}$ and since the image of $\mathfrak{p}x$ is 0 in $S^{-1}M$, $\exists s \in S$, $\mathrm{Ann}_A(sx) = \mathfrak{p}$.
(2) If $\mathfrak{p} \in \mathrm{Ass}_A M$, $\mathfrak{p} = \mathrm{Ann}_A(x)$ $(x \in M)$ then $\frac{x}{1}$ is not 0 in $M_\mathfrak{p}$. Hence $M_\mathfrak{p} \neq 0$.
(3) If $\mathfrak{p}$ is a minimal element of $\mathrm{Supp}_A M$, we have $\mathrm{Supp}_{A_\mathfrak{p}}(M_\mathfrak{p}) = \{\mathfrak{p}A_\mathfrak{p}\}$. Since $\mathrm{Ass}_A M_\mathfrak{p} \neq \emptyset$, we get $\mathfrak{p} \in \mathrm{Ass}_A M$ by (1). The last assertion follows from Theorem 2.17.

□

The following formula on $\mathrm{Ass}_A M$ is very useful.

Proposition 2.21 *Let M and N be A-modules, and assume that M is finitely generated. Then the following formula holds true:*

$$\mathrm{Ass}_A(\mathrm{Hom}_A(M, N)) = \mathrm{Supp}_A M \cap \mathrm{Ass}_A N.$$

Proof Since M is finitely generated, there is a surjective map $A^m \to M$ and then $\mathrm{Hom}_A(M, N) \subseteq \mathrm{Hom}_A(A^m, N) = N^m$. Hence $\mathrm{Ass}_A(\mathrm{Hom}_A(M, N)) \subseteq \mathrm{Ass}_A N^m$. Since $\mathrm{Ass}_A N^m = \mathrm{Ass}_A N$ by Theorem 2.17, we have $\mathrm{Ass}_A(\mathrm{Hom}_A(M, N)) \subseteq \mathrm{Ass}_A N$. It is clear that $\mathrm{Supp}(\mathrm{Hom}_A(M, N)) \subseteq \mathrm{Supp}_A M$. Hence we have

$$\mathrm{Ass}_A(\mathrm{Hom}_A(M, N)) \subseteq \mathrm{Supp}_A M \cap \mathrm{Ass}_A N.$$

Conversely, assume that $\mathfrak{p} \in \mathrm{Supp}_A M \cap \mathrm{Ass}_A N$. Since $A/\mathfrak{p} \subseteq N$, we have $\mathrm{Hom}_A(M, A/\mathfrak{p}) \subseteq \mathrm{Hom}_A(M, N)$. Again $\mathrm{Hom}_A(M, A/\mathfrak{p}) \subseteq \mathrm{Hom}_A(A^m, A/\mathfrak{p})$ and $\mathrm{Ass}_A(M, A/\mathfrak{p}) \subseteq \mathrm{Ass}_A(A/\mathfrak{p}) = \{\mathfrak{p}\}$. If we know $\mathrm{Hom}_A(M, A/\mathfrak{p}) \neq 0$, then $\mathfrak{p} \in \mathrm{Ass}_A(\mathrm{Hom}_A(M, N))$. To show that $\mathrm{Hom}_A(M, A/\mathfrak{p}) \neq 0$, it suffices to localize at $\mathfrak{p}$ since $\mathfrak{p} \in \mathrm{Supp}_A M$. Hence we may assume $(A, \mathfrak{p})$ is a local ring. By NAK, $M/\mathfrak{p}M \cong (A/\mathfrak{p})^r$ for some $r > 0$. So we can construct a nonzero map as the composite of the surjective maps $M \to M/\mathfrak{p}M \cong (A/\mathfrak{p})^r \to A/\mathfrak{p}$. □

We define the concept of primary ideals.

Proposition–Definition 2.22

(1) If an ideal $\mathfrak{q}$ of A satisfies the following equivalent conditions, we say that $\mathfrak{q}$ is a *primary ideal*, or, *$\mathfrak{q}$ is a $\mathfrak{p}$-primary ideal.*

 (a) $\mathrm{Ass}_A(A/\mathfrak{q}) = \{\mathfrak{p}\}$.
 (b) $\mathfrak{p} = \sqrt{\mathfrak{q}}$ is a prime ideal, and zerodivisors of $A/\mathfrak{q}$ are nilpotents.
 (c) $1 \notin \mathfrak{q}$, $\mathfrak{p} = \sqrt{\mathfrak{q}}$, and for $a, b \in A$, if $ab \in \mathfrak{q}$ then either $a \in \mathfrak{q}$ or $b \in \mathfrak{p}$ holds.
 (d) $\sqrt{\mathfrak{q}} = \mathfrak{p}$ is a prime ideal, and $i_{\mathfrak{p}}^{-1}(\mathfrak{q}A_{\mathfrak{p}}) = \mathfrak{q}$ holds.

(2) Let M be an A-module. If N is an A-submodule of M and if $\mathrm{Ass}_A(M/N) = \{\mathfrak{p}\}$, we say that N is a $\mathfrak{p}$-*primary submodule* of M.

Proof of the equivalence of (a)–(d) If we assume (a), $\mathfrak{p}$ is a prime ideal by the definition of Ass. The set of elements of A which are zerodivisors on $A/\mathfrak{q}$ is $\bigcup_{P \in \mathrm{Ass}_A(A/\mathfrak{q})} P = \mathfrak{p}$ by Proposition 2.15. Moreover, $\sqrt{\mathfrak{q}} = \bigcap_{P \in \mathrm{Min}_A A/\mathfrak{q}} P = \mathfrak{p}$, since $\emptyset \subsetneq \mathrm{Min}_A A/\mathfrak{q} \subseteq \mathrm{Ass}_A(A/\mathfrak{q}) = \{\mathfrak{p}\}$, and we have proved (b).

(b) $\Rightarrow$(c) is immediate.

(c) $\Rightarrow$(d). We first prove that $\mathfrak{p}$ is a prime ideal. If $1 \in \mathfrak{p} = \sqrt{\mathfrak{q}}$, then there exists some $n \geq 1$ such that $1 = 1^n \in \mathfrak{q}$, and this is a contradiction. Assume that $a, b \in A$, $ab \in \mathfrak{p}$, and $a \notin \mathfrak{p}$. Then there exists some $n \geq 1$ such that $a^n b^n \in \mathfrak{q}$. Since $a^n \notin \mathfrak{q}$, we have that $b^n \in \mathfrak{p}$. So $(b^n)^m \in \mathfrak{q}$ for some $m \geq 1$, and hence $b \in \mathfrak{p}$, and hence $\mathfrak{p}$ is a prime ideal. Also, if $a, s \in A$, $s \notin \mathfrak{p}$ and $as \in \mathfrak{q}$, we must have $a \in \mathfrak{q}$. Hence we have (d).

If we assume (d), then $\mathrm{Ass}_A(A/\mathfrak{q}) \subseteq \mathrm{Ass}_A(A_{\mathfrak{p}}/\mathfrak{q}A_{\mathfrak{p}}) = \{\mathfrak{p}\}$. □

Example 2.23

(1) If $\sqrt{I} = \mathfrak{m}$ is a maximal ideal of A, then since $\mathrm{Ass}_A(A/I) \subseteq \mathrm{Supp}_A(A/I) = \{\mathfrak{m}\}$, I is an $\mathfrak{m}$-primary ideal.
(2) When $\mathfrak{p}$ is a prime ideal and not a maximal ideal, then a power of $\mathfrak{p}$ is not necessarily a primary ideal. In fact, if $A = k[X, Y, Z]$ is a polynomial ring of 3 variables over a field k and if $\mathfrak{p} = (X^3 - YZ, Y^2 - XZ, Z^2 - X^2Y) = \mathrm{Ker}(A \to k[T^3, T^4, T^5])$ (see Example 1.46). Since we have

$$(X^3 - YZ)^2 - (Y^2 - XZ)(Z^2 - X^2Y) = X(X^5 - 3X^2YZ + XY^3 + Z^3) \in \mathfrak{p}^2$$

and $X \notin \mathfrak{p}$, $X^5 - 3X^2YZ + XY^3 + Z^3 \notin \mathfrak{p}^2$ we see that $\mathfrak{p}^2$ is *not* a primary ideal.

Primary ideals have the following properties.

Lemma 2.24 *If $\mathfrak{q}_i$ $(i = 1, \ldots, r)$ are $\mathfrak{p}$-primary ideals, then so is $\bigcap_{i=1}^r \mathfrak{q}_i$.*

Also, if Q_i $(i = 1, \ldots, r)$ are $\mathfrak{p}$-primary submodules of an A-module M, then so is $\bigcap_{i=1}^r Q_i$.

Proof Since the natural map $\pi : M/\bigcap_{i=1}^r Q_i \to \bigoplus_{i=1}^r M/Q_i$ is injective, we have

$$\mathrm{Ass}_A(M/\bigcap_{i=1}^r Q_i) \subseteq \mathrm{Ass}_A(\bigoplus_{i=1}^r M/Q_i) = \{\mathfrak{p}\}.$$

□

Definition 2.25 Let I be an ideal of A and N be an A-submodule of an A-module M.

(1) We call an expression $I = \bigcap_{i=1}^r \mathfrak{q}_i$, where $\mathfrak{q}_i$'s are primary ideals, a *primary decomposition* of I.
(2) Also we call $N = \bigcap_{i=1}^r Q_i$ a primary decomposition of N, if every Q_i is a primary submodule of M.
(3) Let $I = \bigcap_{i=1}^r \mathfrak{q}_i$ be a primary decomposition of I. Consider the conditions:

    (a) For every i, $\bigcap_{j \neq i} \mathfrak{q}_j$ is strictly bigger than I.
    (b) For every $i \neq j$, $\sqrt{\mathfrak{q}_i} \neq \sqrt{\mathfrak{q}_j}$.

    We say that the decomposition is *irredundant* if (a) holds. We say that it is *minimal* if both (a) and (b) holds.
    An irredundant and minimal primary decompositions of a submodule N of an A-module M is defined in the same manner.

If there exists a primary decomposition of an ideal I, then using Lemma 2.24, we get one that satisfies condition (b). Then removing redundancy if any, we get a minimal primary decomposition of I. *In the following, we always assume that every ring and every module is Noetherian.*

Theorem 2.26 (Lasker–Noether Decomposition Theorem) *For every A-module M and its submodule N, there is a primary decomposition of N. In particular, every ideal I of A has a primary decomposition.*

We need a lemma to prove the theorem.

Lemma 2.27 *Let M be an A-module and let T be any subset of* $\mathrm{Ass}_A M$. *If a submodule N of M is a maximal submodule of M satisfying the condition "*$\mathrm{Ass}_A N \subseteq T$,*" then* $\mathrm{Ass}_A N = T$ *and* $\mathrm{Ass}_A(M/N) = \mathrm{Ass}_A M \setminus T$.

Proof If $\mathfrak{p} \in \mathrm{Ass}_A(M/N)$, then there is a submodule N' such that $N \subsetneq N'$ and $N'/N \cong A/\mathfrak{p}$. Since $\mathrm{Ass}_A N \subseteq \mathrm{Ass}_A N' \subseteq \mathrm{Ass}_A N \cup \{\mathfrak{p}\}$ by Theorem 2.17, if $\mathfrak{p} \in T$, we have $\mathrm{Ass}_A N' \subseteq T$, contradicting the maximality of N. Hence we have $\mathfrak{p} \in \mathrm{Ass}_A M \setminus T$ and $\mathrm{Ass}_A(M/N) \subseteq \mathrm{Ass}_A M \setminus T$. Since $\mathrm{Ass}_A N \subseteq T$ and $\mathrm{Ass}_A M \subseteq \mathrm{Ass}_A N \cup \mathrm{Ass}_A(M/N)$ by Theorem 2.17, we must have $\mathrm{Ass}_A N = T$ and $\mathrm{Ass}_A(M/N) = \mathrm{Ass}_A M \setminus T$. □

Proof of Theorem 2.26 Considering M/N instead of M, we assume $N = 0$. For every $\mathfrak{p} \in \mathrm{Ass}_A M$, if we put $T = \mathrm{Ass}_A M \setminus \{\mathfrak{p}\}$ in Lemma 2.27, we can take a $\mathfrak{p}$-primary module $Q = Q[\mathfrak{p}]$ such that $\mathrm{Ass}_A(Q[\mathfrak{p}]) = \mathrm{Ass}_A M \setminus \{\mathfrak{p}\}$.

Next, put

$$Q = \bigcap_{\mathfrak{p}\in \mathrm{Ass}_A M} Q[\mathfrak{p}].$$

Since $Q \subseteq Q[\mathfrak{p}]$, $\mathrm{Ass}\, Q \subseteq \mathrm{Ass}_A\, Q[\mathfrak{p}]$. Hence

$$\mathrm{Ass}_A\, Q \subseteq \bigcap_{\mathfrak{p}\in \mathrm{Ass}_A M} (\mathrm{Ass}_A M \setminus \{\mathfrak{p}\}) = \emptyset.$$

This implies $Q = 0$, and we have proved the existence of a primary decomposition. □

In general, a primary decomposition $I = \bigcap_{i=1}^{r} \mathfrak{q}_i$ writing I as an intersection of primary ideals is not unique. But as we can see from the proof of Theorem 2.26, in an irredundant primary decomposition, the set $\{\sqrt{\mathfrak{q}_i}\}$ coincides with $\mathrm{Ass}_A(A/I)$. Hence for a prime ideal $\mathfrak{p}$, $\mathfrak{p}$ is an associated prime of I if and only if a $\mathfrak{p}$-primary ideal appears in an irredundant primary decomposition of I. If, moreover, the decomposition is minimal, there exists exactly one i such that $\mathfrak{q}_i = Q$ for each $Q \in \mathrm{Ass}_A(A/I)$.

In general, the choice of $\mathfrak{q}_i$ in a minimal primary decomposition is *not* uniquely determined only by I and $\sqrt{\mathfrak{q}_i} \in \mathrm{Ass}_A(A/I)$. But if $\sqrt{\mathfrak{q}_i}$ is a minimal prime ideal of I, then $\mathfrak{q}_i$ is uniquely determined.

Proposition–Definition 2.28 If $\mathfrak{p}$ is a minimal prime ideal of an ideal I, the $\mathfrak{p}$-primary ideal appearing in a minimal primary decomposition is $i_{\mathfrak{p}}^{-1}(I_{\mathfrak{p}})$.

We call $i_{\mathfrak{p}}^{-1}(I_{\mathfrak{p}})$ *the* $\mathfrak{p}$*-primary component* of I.

Proof If $I = \bigcap_{\mathfrak{p}'\in \mathrm{Ass}_A(A/I)} Q[\mathfrak{p}']$ is a minimal primary decomposition of I, where $\sqrt{Q[\mathfrak{p}']} = \mathfrak{p}'$, then by localizing at $\mathfrak{p}$ we get

$$IA_{\mathfrak{p}} = \bigcap_{\mathfrak{p}'\in \mathrm{Ass}_A(A/I)} Q[\mathfrak{p}']A_{\mathfrak{p}}.$$

Since $\mathfrak{p}$ is a minimal prime ideal of I, $Q[\mathfrak{p}']A_{\mathfrak{p}} = A_{\mathfrak{p}}$ for $\mathfrak{p}' \neq \mathfrak{p}$. Hence we have $IA_{\mathfrak{p}} = Q[\mathfrak{p}]A_{\mathfrak{p}}$. But since $Q[\mathfrak{p}]$ is a $\mathfrak{p}$-prime ideal, we have $Q_{\mathfrak{p}} = i_{\mathfrak{p}}^{-1}(Q_{\mathfrak{p}}A_{\mathfrak{p}}) = i_{\mathfrak{p}}^{-1}(IA_{\mathfrak{p}})$. □

Definition 2.29 Let $\mathfrak{p}$ be a prime ideal of A. We call the $\mathfrak{p}$-primary component of $\mathfrak{p}^n$ *the symbolic nth power of* $\mathfrak{p}$ and we denote it as

$$\mathfrak{p}^{(n)} = A \cap \mathfrak{p}^n A_{\mathfrak{p}}.$$

Exercise 2.2 Let $A = k[X, Y]$, where k is a field. Show that

$$(X^2, XY) = (X) \cap (X^2, Y) = (X) \cap (X^2, XY, Y^n)$$

for any $n \geq 1$. In particular, a $\mathfrak{p}$-primary component, where $\mathfrak{p} = \sqrt{\mathfrak{q}} = (X, Y)$, is *not* unique.

Exercise 2.3

(1) In Example 1.46, $\mathfrak{p} = (X^3 - YZ, Y^2 - XZ, Z^2 - X^2Y)$ and $\mathfrak{p}^{(2)} = (X^5 - 3X^3YZ + XY^3 + Z^3) + \mathfrak{p}^2$.
(2) Put $A = k[X, Y, Z]/(X^2 - YZ)$ and let x, y, z be the images of X, Y, Z in A, respectively. If $\mathfrak{p} = (x, y)$, show that $\mathfrak{p}^2 \neq \mathfrak{p}^{(2)} = (y)$.

The concept of *irreducible ideals* can be used to show the existence of primary decomposition. This concept is also important in the theory of Gorenstein rings.

Definition 2.30 Let I be an ideal of A. Then I is called an *irreducible ideal* if $I \neq A$, and I cannot be expressed as $I = J \cap K$, the intersection of two ideals properly containing I.

When A is Noetherian, then every ideal can be expressed as the intersection of finite number of irreducible ideals (Exercise 2.4). Hence the existence of primary decomposition for any ideal can be proved by the following proposition.

Proposition 2.31 *If* $\mathfrak{q}$ *is an* irreducible *ideal of* A*, then* $\mathfrak{q}$ *is a primary ideal.*

Proof Assume $\mathfrak{q}$ is an irreducible ideal of A. We will show that $\mathfrak{q}$ satisfies condition (c) of Proposition–Definition 2.22.

Take $a, b \in A$ and assume that $ab \in \mathfrak{q}, a \notin \mathfrak{q}$. Since the ascending chain

$$\mathfrak{q} : b \subseteq \mathfrak{q} : b^2 \subseteq \ldots \subseteq \mathfrak{q} : b^n \subseteq \ldots$$

terminates, there is a number n such that $\mathfrak{q} : b^n = \mathfrak{q} : b^{n+1}$. We will show that $b^n \in \mathfrak{q}$. For that purpose, we will show

$$\mathfrak{q} = (\mathfrak{q} : b) \cap (\mathfrak{q} + (b^n)).$$

Then since $\mathfrak{q}$ is irreducible, we have either $\mathfrak{q} = \mathfrak{q} : b$ or $\mathfrak{q} = \mathfrak{q} + (b^n)$. Since $a \in (\mathfrak{q} : b) \setminus \mathfrak{q}$, we have $b^n \in \mathfrak{q} + (b^n) = \mathfrak{q}$, and condition (c) of Proposition–Definition 2.22 is proved.

Take $x \in (\mathfrak{q} : b) \cap (b^n)$ and write $x = b^n y$ $(y \in A)$. Then since we have $bx = b^{n+1}y \in \mathfrak{q}$, we have $y \in \mathfrak{q} : b^{n+1} = \mathfrak{q} : b^n$ and then we have $x = b^n y \in \mathfrak{q}$. Since $(\mathfrak{q} : b) \cap (b^n) \subseteq \mathfrak{q}$, we have $\mathfrak{q} = (\mathfrak{q} : b) \cap (\mathfrak{q} + (b^n))$ as desired. □

Exercise 2.4 Show that any ideal in a Noetherian ring I is an intersection of finitely many irreducible ideals. (If there exists an ideal not satisfying this, take a maximal such ideal.)

We show that an Artinian ring is isomorphic to a product of Artinian local rings. For that purpose, we show the *"Chinese remainder theorem"*.

Lemma 2.32 (Chinese Remainder Theorem) *Let A be a ring.*

(1) *If I, J are ideals of A, satisfying the condition $I + J = A$, then the equality $IJ = I \cap J$ holds.*
(2) *Let $\{I_i\}_{1 \le i \le n}$ be ideals of A and assume that $I_i + I_j = A$ for every $i \ne j$. Then the natural ring homomorphism $\varphi : A \to \prod_{i=1}^{n} A/I_i$, $a \mapsto (\overline{a}, \overline{a}, \ldots, \overline{a})$ is a surjection and we have $\operatorname{Ker}\varphi = \bigcap_{i=1}^{n} I_i = \prod_{i=1}^{n} I_i$. In particular, we have an isomorphism*

$$A/\bigcap_{I=1}^{n} I_i \cong \prod_{i=1}^{n} A/I_i.$$

Proof

(1) Write $1 = i + j$ $(i \in I, j \in J)$. Then for every $x \in I \cap J$, we have $x = xi + xj \in IJ$.
(2) For $1 \le i \le n$, we have $I_i + \bigcap_{j \ne i} I_j = A$. Actually, if $I_i + \bigcap_{j \ne i} I_j \ne A$, we can take a prime ideal $\mathfrak{p}$ of A so that $I_i + \bigcap_{j \ne i} I_j \subseteq \mathfrak{p}$. But then for some $j \ne i$, $I_j \subseteq \mathfrak{p}$ and $I_i \subseteq \mathfrak{p}$, which contradicts our assertion $I_i + I_j = A$. Hence for every i, we can write $1 = x_i + y_i$ $(x_i \in \bigcap_{j \ne i} I_j, y_i \in I_i)$. Then we have

$$\varphi(x_i) = (\overline{x_i}, \overline{x_i}, \ldots, \overline{x_i}) = (0, \ldots, 0, 1, 0, \ldots, 0),$$

where 1 is in the ith place. Then for $a_1, a_2, \ldots, a_n \in A$, we have

$$\varphi(\sum_{i=1}^{n} a_i x_i) = (\overline{a_1}, \overline{a_2}, \ldots, \overline{a_n}),$$

where $\overline{a_i}$ means the image of a_i in A/I_i and this shows that $\varphi : A \to \prod_{i=1}^{n} A/I_i$ is surjective. We prove the equality $\bigcap_i I_i = \prod_i I_i$ by induction on n. If $n \le 2$, then the assertion is true by (1). Assume that $n \ge 3$. Then since $I_1 + \bigcap_{i \ge 2} I_i = A$, we have

$$\bigcap_{i} I_i = I_1 \cap (\bigcap_{i \ge 2} I_i) = I_1 \cdot (\bigcap_{i \ge 2} I_i) = I_1 \cdot (\prod_{i \ge 2} I_i)$$

by (1) and the induction assumption.

□

Theorem 2.33 (Structure Theorem of Artinian Rings) *Let A be an Artinian ring and let*

$$(0) = \bigcap_{i=1}^{n} Q_i$$

be a minimal primary decomposition of (0) *in* A. *If we put* $\mathfrak{m}_i = \sqrt{Q_i}$, *then the ring homomorphism*

$$\varphi : A \to \prod_{i=1}^{n} A/Q_i,$$

is a ring isomorphism and for every i, *we have* $A/Q_i \cong A_{\mathfrak{m}_i}$. *Hence we have*

$$A \cong \prod_{\mathfrak{m} \in \operatorname{Max} A} A_{\mathfrak{m}} \qquad (a \mapsto (\frac{a}{1} \in A_{\mathfrak{m}})_{\mathfrak{m} \in \operatorname{Max} A}).$$

Proof Since $\mathfrak{m}_i = \sqrt{Q_i}$, for every $i \neq j$, we have $Q_i + Q_j = A$. Hence from Lemma 2.32, the ring homomorphism $\Phi : A \to \prod_{i=1}^{n} A/Q_i, a \mapsto (\overline{a}, \overline{a}, \dots, \overline{a})$ is an isomorphism.

On the other hand, A/Q_i is an Artinian local ring whose maximal ideal is $\mathfrak{m}_i/Q_i$. Hence if $s \in A \setminus \mathfrak{m}_i$, the image $\overline{s}$ of s in A/Q_i is a unit of A/Q_i.

Hence the natural surjection $\varepsilon_i : A \to A/Q_i$ factors through the natural localization $h_i : A \to A_{\mathfrak{m}_i}$ by Proposition 1.126. On the other hand, we have $Q_i A_{\mathfrak{m}_i} = 0$, hence the surjection $A \to A/Q_i$ and the localization $h_i : A \to A_{\mathfrak{m}_i}$ coincides. Hence we have the desired result $A \cong \prod_{i=1}^{n} A_{\mathfrak{m}_i}$. □

Exercise 2.5 Let $(A, \mathfrak{m})$ be an Artinian local ring. Prove the following:

(0) is an irreducible ideal if and only if there exists some $z \in A \setminus \{0\}$ such that for every $x \in A \setminus \{0\}$ there exists some $y \in A$ such that $z = xy$.

We will see in Chap. 6 that (0) is irreducible in A if and only if A is Gorenstein.

2.4 Dimension

The dimension of a ring is defined by the length of chains of prime ideals. The notion of dimension is the most important invariant of a ring.

Definition 2.34

(1) Let A be a ring. The supremum of the lengths n of prime chains

$$\mathfrak{p}_n \supsetneqq \mathfrak{p}_{n-1} \supsetneqq \cdots \supsetneqq \mathfrak{p}_1 \supsetneqq \mathfrak{p}_0$$

is called the *Krull dimension* or the *dimension* of A and is denoted by

$$\dim A.$$

Let M be an A-module. We define

$$\dim_A M = \sup\{\dim(A/\mathfrak{p}) \mid \mathfrak{p} \in \operatorname{Supp}_A(M)\}.$$

(If $M = 0$, we define $\dim_A M = -\infty$.) If M is finitely generated, then we have $\dim_A M = \dim A/\mathrm{Ann}_A(M)$.

(2) For a prime ideal $\mathfrak{p}$ of A, the supremum of the lengths n of chains of prime ideals

$$\mathfrak{p} = \mathfrak{p}_n \supsetneq \mathfrak{p}_{n-1} \supsetneq \cdots \supsetneq \mathfrak{p}_1 \supsetneq \mathfrak{p}_0$$

is called the *height* of $\mathfrak{p}$ and is denoted by

$$\mathrm{ht}\,\mathfrak{p} \quad \text{or} \quad \mathrm{ht}_A\,\mathfrak{p}.$$

Thus $\dim A$ is the maximal value of $\mathrm{ht}\,\mathfrak{m}$ for all maximal ideals $\mathfrak{m}$ of A.

Remark 2.2 A prime chain of A/I or a localization $S^{-1}A$ can be considered as a prime chain in A. Hence we have $\dim A/I, \dim S^{-1}A \leq \dim A$. Considering the case that $S = A \setminus \mathfrak{p}$, we have that $\dim A_{\mathfrak{p}} = \mathrm{ht}\,\mathfrak{p} \leq \dim A$.

Example 2.35

(1) If A is an Artinian ring, then every prime ideal of A is a maximal ideal. Hence we have $\dim A = 0$. Conversely, a Noetherian ring with $\dim A = 0$ is an Artinian ring by Theorem 2.8.

(2) We have $\dim \mathbb{Z} = 1$. More generally, if A is a PID (and not a field), then $\dim A = 1$.

The following theorem is most fundamental in dimension theory of rings.

Theorem 2.36 (Krull's Altitude Theorem[4]) *Let A be a Noetherian ring. Assume an ideal I of A is generated by r elements $f_1, f_2, \cdots, f_r$. Then for each* minimal prime *ideal $\mathfrak{p}$ of I, we have $\mathrm{ht}_A\,\mathfrak{p} \leq r$.*

To prove this theorem, it suffices to localize at $\mathfrak{p}$ and consider in the local ring $A_{\mathfrak{p}}$. Hence in the following, we assume that $(A, \mathfrak{m})$ is a local ring and $\mathfrak{p} = \mathfrak{m}$.

First, we show the case $r = 1$.

Lemma 2.37 *Let $(A, \mathfrak{m})$ be a Noetherian local ring and assume that $\mathfrak{m} = \sqrt{(f)}$ for some $f \in \mathfrak{m}$. If $\mathfrak{p}$ is a prime ideal of A and $\mathfrak{p} \neq \mathfrak{m}$, then $\mathfrak{p}$ is a* minimal *prime ideal of A.*

Proof Let $\mathfrak{p}^{(\ell)}$ be the symbolic ℓth power of $\mathfrak{p}$. Then $\{\mathfrak{p}^{(\ell)}\}_{\ell \geq 1}$ is a descending sequence of $\mathfrak{p}$-primary ideals. Since

$$\{[\mathfrak{p}^{(\ell)} + (f)]/(f)\}_{\ell \geq 1}$$

is a descending chain of ideals in the Artinian ring $A/(f)$, we have that

$$\mathfrak{p}^{(\ell)} + (f) = \mathfrak{p}^{(\ell+1)} + (f)$$

[4] Wolfgang Krull, 1899–1971.

for some $\ell \geq 1$. Since $f \notin \mathfrak{p}$ and $\mathfrak{p}^{(\ell)}$ is a $\mathfrak{p}$-primary ideal, $\mathfrak{p}^{(\ell)} \cap (f) = f\mathfrak{p}^{(\ell)}$. Hence we have $\mathfrak{p}^{(\ell)} = \mathfrak{p}^{(\ell)} \cap [\mathfrak{p}^{(\ell+1)} + (f)] = \mathfrak{p}^{(\ell+1)} + \mathfrak{p}^{(\ell)} \cap (f) = \mathfrak{p}^{(\ell+1)} + f\mathfrak{p}^{(\ell)}$ and by (NAK), we have $\mathfrak{p}^{(\ell)} = \mathfrak{p}^{(\ell+1)}$. Then we have $\mathfrak{p}^{\ell} A_{\mathfrak{p}} = \mathfrak{p}^{\ell+1} A_{\mathfrak{p}}$ and then $(\mathfrak{p}A_{\mathfrak{p}})^{\ell} = 0$. Hence $A_{\mathfrak{p}}$ is an Artinian ring and $\mathfrak{p} \in \operatorname{Min} A$. □

Proof of Theorem 2.36 Let $(A, \mathfrak{m})$ be a Noetherian local ring and assume that $\mathfrak{m} = \sqrt{I}, I = (f_1, \ldots, f_n)$. Then we show that $\dim A \leq n$. Since we have proved the case $n = 1$, we assume that $n \geq 2$ and that our conclusion holds up to $n - 1$. Take a prime ideal $\mathfrak{p} \subsetneq \mathfrak{m}$ so that there is no prime ideal between $\mathfrak{p}$ and $\mathfrak{m}$. It suffices to show that $\operatorname{ht} \mathfrak{p} \leq n - 1$.

Since $\sqrt{I} = \mathfrak{m}$, $I \not\subseteq \mathfrak{p}$. Changing the numbers if necessary, we assume that $f_1 \notin \mathfrak{p}$. Since only $\mathfrak{m}$ is a prime ideal containing $\mathfrak{p}+(f_1)$, if we put $B = A/(\mathfrak{p}+(f_1))$, then B is an Artinian local ring and every element of $\mathfrak{m}$ is nilpotent in B. In particular, for $i = 2, \ldots, r$, there exists $n \geq 1$ such that

$$f_i^n = g_i + f_1 x_i \quad (g_i \in \mathfrak{p}, x_i \in A).$$

Let $J = (g_2, \ldots, g_n) \subseteq \mathfrak{p}$. If we can show that $\mathfrak{p}$ is a minimal prime ideal of J, then $\operatorname{ht} \mathfrak{p} \leq n - 1$ by induction, and the proof is complete.

Now take a prime ideal $\mathfrak{q} \subseteq \mathfrak{p}$ which is a minimal prime ideal of J. Then, since $f_i \in \sqrt{\mathfrak{q} + (f_1)}$ $(i = 2, \ldots, n)$, we have $\sqrt{\mathfrak{q} + (f_1)} \supseteq \sqrt{I} = \mathfrak{m}$. Since $\mathfrak{m}/\mathfrak{q}$ is a minimal prime ideal of $(\mathfrak{q} + (f_1))/\mathfrak{q}$, we have $\dim A/\mathfrak{q} = 1$. Hence we get $\mathfrak{p} = \mathfrak{q}$ and $\mathfrak{p}$ is a minimal prime of J. □

By this theorem, we can say that the height of every prime ideal of a Noetherian ring is finite. However, the dimension of the ring is not necessarily finite, see [Nag, (A1.1)]. If $(A, \mathfrak{m})$ is a Noetherian local ring, then $\dim A = \operatorname{ht}_A \mathfrak{m}$ and $d = \dim A$ is a non-negative integer. In particular, for a prime ideal $\mathfrak{p}$ of A, we have $\operatorname{ht}_A \mathfrak{p} = \operatorname{ht}_{A_{\mathfrak{p}}} \mathfrak{p}A_{\mathfrak{p}} = \dim A_{\mathfrak{p}}$ is finite.

Lemma 2.38 *Let A be a Noetherian ring, $P \in \operatorname{Spec} A$, and M an A-module. Then $P \in \operatorname{Min}_A M$ (that is, P is a minimal element of $\operatorname{Supp}_A M$) if and only if P is a minimal element of $\operatorname{Ass}_A M$.*

Proof The 'only if' part is clear from the fact that $\operatorname{Min}_A M \subseteq \operatorname{Ass}_A M \subseteq \operatorname{Supp}_A M$, see Proposition 2.20. We prove the 'if' direction. Take a prime ideal Q from the set $\Gamma = \{Q \in \operatorname{Supp}_A M \mid Q \subseteq P\}$ so that its height is minimal as possible. Then $Q \in \operatorname{Min}_A M$, and hence $Q \in \operatorname{Ass}_A M$. Since P is a minimal element of $\operatorname{Ass}_A M$ and $Q \subseteq P$, we have that $Q = P$. So $P \in \operatorname{Min}_A M$. □

Let A be a ring. For an ideal I of A, we define

$$\operatorname{ht}_A I = \inf_{\mathfrak{p} \in \mathrm{V}(I)} \operatorname{ht}_A \mathfrak{p}$$

and call it the *height of I*. In particular, we define $\operatorname{ht}_A A = \infty$. If A is Noetherian and $I \neq A$, then $\operatorname{ht}_A I$ is a non-negative integer.

Proposition 2.39 *Let I be an ideal of a Noetherian ring A of height n. Then we can take $f_1, f_2, \cdots, f_n \in I$ such that $\mathrm{ht}_A(f_1, f_2, \cdots, f_i) = i$ holds for every $0 \le i \le n$.*

Proof Assume that we have chosen $f_1, f_2, \cdots, f_i \in I$ $(i < n)$ so that for every $j \le i$ we have $\mathrm{ht}_A(f_1, f_2, \cdots, f_j) = j$. Then there are only finitely many prime ideals containing $J = (f_1, f_2, \cdots, f_i)$ and have height i, since they are minimal prime ideals of J. Also, if $\mathfrak{p}$ is such a prime ideal, then $I \not\subseteq \mathfrak{p}$ since $\mathrm{ht}_A \mathfrak{p} = i < \mathrm{ht}_A(I) = n$. Hence we can take $f_{i+1} \in I$ so that $f_{i+1} \notin \bigcup_{J \subseteq \mathfrak{p}, \mathrm{ht}_A \mathfrak{p} = i} \mathfrak{p}$ and then we have $\mathrm{ht}_A(f_1, f_2, \cdots, f_{i+1}) = i + 1$. □

Corollary 2.40 *Let $(A, \mathfrak{m})$ be a Noetherian local ring of dimension d. Then we can take $f_1, f_2, \cdots, f_d \in \mathfrak{m}$ so that $\mathfrak{m} = \sqrt{(f_1, f_2, \cdots, f_d)}$.*

Proposition 2.41 *If A is a Noetherian ring, then* $\dim A[X] = \dim A + 1$.[5]

Proof Put $B = A[X]$. For a chain $\mathfrak{p}_n \supsetneq \mathfrak{p}_{n-1} \supsetneq \cdots \supsetneq \mathfrak{p}_1 \supsetneq \mathfrak{p}_0$ of length n of prime ideals of A, we have a chain $(\mathfrak{p}_n, X)B \supsetneq \mathfrak{p}_n B \supsetneq \mathfrak{p}_{n-1} B \supsetneq \cdots \supsetneq \mathfrak{p}_1 B \supsetneq \mathfrak{p}_0 B$ of length $n+1$ of prime ideals of B. Hence we have $\dim B \ge \dim A + 1$.

We want to show the opposite inequality. It suffices to prove that for any maximal ideal $\mathfrak{n}$ of B, we have $\mathrm{ht}\, \mathfrak{n} \le \dim A + 1$. Let $\mathfrak{m} = \mathfrak{n} \cap A$. Replacing A by $A_\mathfrak{m}$ and B by $A_\mathfrak{m}[X]$, we may assume that $(A, \mathfrak{m})$ is local. We take $a_1, \ldots, a_d \in \mathfrak{m}$ such that $\mathfrak{m} = \sqrt{(a_1, \ldots, a_d)}$ and $d = \mathrm{ht}\, \mathfrak{m}$. Take $b \in \mathfrak{n}$ such that the image of b in $B/\mathfrak{m}B \cong (A/\mathfrak{m})[X]$ is nonzero. If P is a minimal prime of $(a_1, \ldots, a_d, b)B_\mathfrak{n}$, then $P \cap A = \mathfrak{m}$, and $P \supseteq \mathfrak{m}B_\mathfrak{n}$. Since $b \notin \mathfrak{m}B_\mathfrak{n}$ and any nonzero prime ideal of $B_\mathfrak{n}/\mathfrak{m}B_\mathfrak{n} \cong (A/\mathfrak{m})[X]_\mathfrak{n}$ is maximal by Proposition 1.52, we have that $P = \mathfrak{n}B_\mathfrak{n}$. Since $(a_1, \ldots, a_d, b)B_\mathfrak{n} \ne B_\mathfrak{n}$, we have that $(a_1, \ldots, a_d, b)B_\mathfrak{n}$ has at least one minimal prime, which must be $\mathfrak{n}B_\mathfrak{n}$. By Theorem 2.36, $\mathrm{ht}\, \mathfrak{n} = \mathrm{ht}\, \mathfrak{n}B_\mathfrak{n} \le d + 1 = \mathrm{ht}\, \mathfrak{m} + 1 \le \dim A + 1$, as desired. □

Definition 2.42 Let $(A, \mathfrak{m})$ be a Noetherian local ring of dimension d.

(1) A sequence $f_1, f_2, \ldots, f_d$ of elements in $\mathfrak{m}$ with $\mathfrak{m} = \sqrt{(f_1, f_2, \ldots, f_d)}$ is called a *system of parameters* (sometimes abbreviated as *s.o.p.* in some literature) of A (or, of $\mathfrak{m}$).

Let M be a finitely generated A-module and $f_1, f_2, \ldots, f_n$ be a sequence of elements of $\mathfrak{m}$. Then we say $(f_1, f_2, \ldots, f_n)$ is a system of parameters of M if $\dim_A M = n$ and $\dim_A M/((f_1, f_2, \ldots, f_n)M) = 0$

(2) We put $\mathrm{Assh}\, A = \{\mathfrak{p} \mid \mathfrak{p} \in \mathrm{Ass}_A A, \dim A/\mathfrak{p} = d\}$. $\mathrm{Assh}\, A$ is a subset of $\mathrm{Min}\, A$, and is not empty. If A is an integral domain, then $\mathrm{Assh}\, A = \{(0)\}$.

Proposition 2.43 *Let $(A, \mathfrak{m})$ be a local ring of dimension d. For $f \in \mathfrak{m}$, the following statements hold true:*

(1) $d - 1 \le \dim A/(f) \le d$.

[5] There is an example of non-Noetherian ring satisfying $\dim A[X] = 2 \dim A + 1$.

(2) $\dim A/(f) = d - 1$ *if and only if* $f \notin \bigcup_{\mathfrak{p} \in \operatorname{Assh} A} \mathfrak{p}$. *If this is the case, there is a system of parameters* $f_1, f_2, \ldots, f_d$ *of* A *where* $f = f_1$.

Proof

(1) Put $\overline{A} = A/(f)$. $\operatorname{Spec} \overline{A}$ is isomorphic to $\mathrm{V}(f) \subseteq \operatorname{Spec} A$ as an ordered set. So $\dim \overline{A} \leq \dim A = d$ is trivial. If $\dim \overline{A} \leq d - 2$, then we can take $g_1, g_2, \cdots, g_{d-2}$ so that $\mathfrak{m} = \sqrt{(f, g_1, g_2, \ldots, g_{d-2})}$, which contradicts $\dim A = d$ by Theorem 2.36. Hence we have $\dim \overline{A} \geq d - 1$.
(2) We have $\dim \overline{A} = \sup_{P \in \mathrm{V}(f)} \dim A/P$. If $f \notin \bigcup_{\mathfrak{p} \in \operatorname{Assh} A} \mathfrak{p}$, then $\mathrm{V}(f) \cap \operatorname{Assh} A = \emptyset$, and hence $\dim A/P < d$ for $P \in \mathrm{V}(f)$, and hence $\dim \overline{A} \leq d - 1$. By (1), we have $\dim \overline{A} = d - 1$ this case.

On the other hand, if $f \in \bigcup_{\mathfrak{p} \in \operatorname{Assh} A} \mathfrak{p}$, then taking $\mathfrak{p} \in \operatorname{Assh} A$ such that $f \in \mathfrak{p}$, we have that $d \geq \dim \overline{A} \geq \dim A/\mathfrak{p} = d$, and hence $\dim \overline{A} = d$ this case. □

Corollary 2.44 *If* $(A, \mathfrak{m})$ *is a local ring of dimension* d *and if* $f \in \mathfrak{m}$ *is a nonzerodivisor, then we have* $\dim A/(f) = d - 1$,

Proof Since $f \in \mathfrak{m} \setminus \bigcup_{P \in \operatorname{Ass} A} P$ and $\operatorname{Ass} A \supseteq \operatorname{Min} A \supseteq \operatorname{Assh} A$, we have that $f \in \mathfrak{m} \setminus \bigcup_{P \in \operatorname{Assh} A} P$, and we can apply Proposition 2.43. □

Chapter 3
Integral Extension and Integrally Closed Domains

In this chapter, we introduce the notion of integral dependence and integral closure. It is very important especially in connection with dimension theory.

In Sect. 3.1, we define the notion of integral dependence for a ring extension $A \subseteq B$ and prove the going-up theorem and the going-down theorem, showing that $\dim A = \dim B$ if $A \subseteq B$ is an integral extension.

In Sect. 3.2, we study Zariski's lemma, which plays a very important role in the theory of rings finitely generated over a field. The famous Nullstellensatz (zero-point theorem) naturally follows from this observation.

In Sect. 3.3, we introduce the notion of discrete valuation rings (DVR) and the concept of normal rings. After defining the depth of a finitely generated module over a Noetherian local ring, we introduce Serre's $(R_1) + (S_2)$ criterion of normality. Serre's (S_k) condition is also important in module theory.

In Sect. 3.4, we introduce the notion of reflexive modules. In a normal domain, a fractional ideal which is reflexive as a module is called a "divisorial ideal" and we come naturally to the notion of a "divisor class group" $\mathrm{Cl}(A)$ which generalizes the notion of "Unique Factorization" and a Noetherian domain A is a UFD if and only if A is normal and $\mathrm{Cl}(A) = 0$. Finally, we study Dedekind domains and prime factorizations of ideals over them.

3.1 Going-Up and Going-Down Theorems

In this section, we assume that B is an A-algebra.

Definition 3.1 Let $x \in B$. We say that x is *integral* over A if there exist $a_i \in A$ $(1 \leq i \leq n)$ such that the following equation in B holds true:

$$x^n + a_1 x^{n-1} + \cdots + a_n = 0.$$

S. Goto, K. Watanabe, *Commutative Algebra*, Graduate Texts in Mathematics 309,
https://doi.org/10.1007/978-981-95-4227-7_3

In other words, there exists a monic polynomial $f(T) \in A[T]$ such that $f(x) = 0$ in B.

Lemma 3.2 (Determinantal Trick) *Let A be a ring, M be a finitely generated A-module, I be an ideal of A, and $\phi : M \to M$ be an A-linear map such that $\phi(M) \subset IM$. Then there is a relation of the form*

$$\phi^n + a_1\phi^{n-1} + a_2\phi^{n-2} + \cdots + a_{n-1}\phi + a_n \operatorname{id}_M = 0$$

in $\operatorname{End}_A M$ *with $a_i \in I^i$ for each i with $1 \le i \le n$.*

Proof Assume that $m_1, \ldots, m_n \in M$ generates M as an A-module. Since we can write $\phi m_i = \sum c_{ij} m_j$ $(c_{ij} \in A,\ 1 \le i, j \le n)$, the $n \times n$ matrix $C = (c_{ij})$ satisfies

$$\phi \begin{pmatrix} m_1 \\ \vdots \\ m_n \end{pmatrix} = \begin{pmatrix} c_{11} & \cdots & c_{1n} \\ \cdots & \cdots & \cdots \\ c_{n1} & \cdots & c_{nn} \end{pmatrix} \begin{pmatrix} m_1 \\ \vdots \\ m_n \end{pmatrix}$$

or, putting $C = (c_{i,j})_{1 \le i,j \le n}$ and E_n be the identity matrix, we have

$$(\phi E_n - C) \begin{pmatrix} m_1 \\ \vdots \\ m_n \end{pmatrix} = \begin{pmatrix} 0 \\ \vdots \\ 0 \end{pmatrix}.$$

Note that the A-subalgebra $A[\phi]$ of $\operatorname{End}_A M$ generated by ϕ is commutative. Multiplying both sides of the equality by the adjoint matrix of $\phi E_n - C$, we get $\det(\phi E_n - C)m_i = 0$ $(i = 1, \ldots, n)$.

So we have $\det(\phi E_n - C) = 0$ in $A[\phi]$. Expanding the determinant, we get the equality of the form $\phi^n + a_1\phi^{n-1} + \cdots + a_n \operatorname{id}_M = 0$. □

Corollary 3.3 *Let A be a ring, M be a finitely generated A-module, and I be an ideal of A. If $M = IM$, then there exists some $a \in I$ such that $(1 - a)M = 0$.*

Proof Set $\phi = \operatorname{id}_M$ in Lemma 3.2. Then by the lemma, there exists some $a_1, \ldots, a_n \in I$ such that

$$\operatorname{id}_M^n + a_1 \operatorname{id}_M^{n-1} + a_2 \operatorname{id}_M^{n-2} + \cdots + a_{n-1}\phi + a_n \operatorname{id}_M = 0.$$

Letting $a = a_1 + \cdots + a_n$, we have that $1 - a$ is zero in $\operatorname{End}_A M$. □

Corollary 3.4 *Let A be a ring, M be a finitely generated A-module, and $f : M \to M$ be a surjective A-linear map. Then f is an isomorphism.*

Proof Let $B = A[f] \subseteq \operatorname{End}_A M$. Then B is commutative, M is a finitely generated B-module, and $M = fM$. By Corollary 3.3, there exists some $h \in fB$ such that $\operatorname{id}_M - h$ is zero on M. That is, $\operatorname{id}_M = h \in fB$. We can write $\operatorname{id}_M = fg = gf$ for some $g \in B \subseteq \operatorname{End}_A M$, since B is commutative and $\mathrm{id}_M \in fB$. This shows that f is an isomorphism with g its inverse. □

Proposition 3.5 *The following conditions on $x \in B$ are equivalent:*

(1) *x is integral over A.*
(2) *The subring $A[x]$ of B is a finitely generated A-module.*
(3) *There exists a subring C of B which contains $A[x]$ and is a finitely generated A-module.*
(4) *There exists a faithful $A[x]$-module M which is finitely generated as an A-module.*

Proof (1) ⇒(2) If $f(T)$ is a monic polynomial in $A[T]$ of degree n such that $f(x) = 0$, then $A[x] = \sum_{i=0}^{n-1} Ax^i$ by Proposition 1.40.

(2) ⇒(3) Enough to take $C = A[x]$.

(3) ⇒(4) $M = C$ satisfies the condition.

(4) ⇒(1) Let $\phi : M \to M$ be the homothety $m \mapsto xm$. Applying Lemma 3.2 to the A-linear map $\phi = (m \mapsto xm)$ and the ideal $I = A$, we have a relation of the form

$$\phi^n + a_1\phi^{n-1} + \cdots + a_{n-1}\phi + a_n \,\mathrm{id}_M = 0$$

in $A[\phi] \subseteq \operatorname{End}_A M$. Since M is faithful as an $A[x]$-module, the A-algebra map $\delta : A[x] \to \operatorname{End}_A M$ given by $\delta(f(x)) = f(\phi)$ is injective, and hence we get the relation

$$x^n + a_1x^{n-1} + \cdots + a_{n-1}x + a_n = 0$$

in $A[x]$, which shows that x is integral over A. □

A homomorphism of rings $f : A \to B$ is said to be *finite* if B is finitely generated as an A-module. We say that f is of *finite type* if B is finitely generated as an A-algebra. If B is an integral extension of the image $f(A)$ of f, we say that f is *integral*. We also say that B is finite (resp. of finite type, integral) over A, instead of saying that f is so. By definition, B is finite (resp. of finite type, integral) over A if and only if it is so over $f(A)$. Obviously, a surjective homomorphism is finite, integral, and of finite type.

Lemma 3.6 *Let $f : A \to B$ be a finite homomorphism of rings, and C be a finitely generated B-module. Then C is a finitely generated A-module. In particular, if C is a ring and $g : B \to C$ is finite, then gf is also finite.*

Proof Let $b_1, \ldots, b_m$ and $c_1, \ldots, c_n$ be generators of B over A and C over B as modules, respectively. It is easy to check that C is generated by $\{b_ic_j \mid 1 \le i \le m,\ 1 \le j \le n\}$ as an A-module. □

Lemma 3.7 *Let $f : A \to B$ be a ring homomorphism. Then the following are equivalent:*

(1) *f is finite.*
(2) *f is integral and of finite type.*
(3) *There exist some $x_1, \dots, x_n \in B$ such that $B = A[x_1, \dots, x_n]$ and x_i is integral over $B_{i-1} = A[x_1, \dots, x_{i-1}]$ for each i with $1 \le i \le n$, where $B_0 = f(A)$.*

Proof (1) ⇒(2) is obvious by Proposition 3.5.

(2) ⇒(3). Since f is of finite type, we can write $B = A[x_1, \dots, x_n]$. Since f is integral, then each x_i is integral over A, and hence over B_{i-1}, since $f(A) \subseteq B_{i-1}$.

(3) ⇒(1). By Proposition 3.5, each $B_i = B_{i-1}[x_i]$ is finite over B_{i-1}. By Lemma 3.6, $B = B_n$ is finite over $A = B_0$. □

Corollary 3.8 *If we put*

$$\overline{A}^B = \{x \in B \mid x \text{ is integral over } A\},$$

then $\overline{A}^B$ is an A-subalgebra of B.

Proof Take $x, y \in \overline{A}^B$. Then $A[x, y]$ is a finite A-algebra, and the subalgebras $A[x \pm y]$ and $A[xy]$ are contained in $A[x, y]$. Hence $x \pm y$, xy are integral over A. Moreover, for $a \in A$, the image of a in B is integral over A, since it is a root of the monic polynomial $f(X) = X - a \in A[X]$. So $\overline{A}^B$ contains $f(A)$, and the assertion follows. □

We call $\overline{A}^B$ the *integral closure* of A in B. If $\overline{A}^B = f(A)$, then we say that A is *integrally closed* in B. Note that B is integral over A if and only if $\overline{A}^B = B$. If, moreover, $A \to B$ is injective, we also say that the extension B/A is an integral extension.

The proofs of the following two propositions are very easy.

Proposition 3.9 *Let $f : A \to C$ and $g : C \to B$ be ring homomorphisms. If gf is integral (resp. of finite type, finite), then so is g. If f and g are integral, then so is gf. In particular, we have $\overline{\overline{A}^C}^B = \overline{A}^B$.*

Proof If $b \in B$, $\varphi(X) \in A[X]$ is a monic polynomial, and $\varphi(b) = 0$, then letting $\psi(X)$ be the image of $\varphi(X)$ by the map $A[X] \to C[X]$, we have that $\psi(X) \in C[X]$ is also monic, and $\psi(b) = \varphi(b) = 0$. So b is also integral over C, and the first assertion for integrity follows. Since $A[S] \subseteq C[S]$ for any subset S of B, the first assertion for the "finite-type" property is obvious. Now the first assertion for finiteness follows, since finiteness is equivalent to finite-type plus integrity by Lemma 3.7.

We prove the second assertion. Let $b \in B$. Since b is integral over C, there is a relation $b^n + c_1 b^{n-1} + \cdots + c_{n-1} b + c_n = 0$ in B with $c_1, \dots, c_n \in C$. Since C is integral over A, $C_0 = A[c_1, \dots, c_n]$ is finite over A by Lemma 3.7. Since b is integral over C_0, we have that $C_0[b]$ is finite over C_0. Hence $C_0[b]$ is finite over A by Lemma 3.6. Hence b is integral over A by Proposition 3.5, and B is integral over A.

We prove the third assertion. Note that $g(\overline{A}^C)$ is integral over A, and hence it is contained in $\overline{A}^B$. By the first assertion, $\overline{A}^B$ is integral over $\overline{A}^C$, and hence $\overline{A}^B \subseteq \overline{\overline{A}^C}^B$. On the other hand, by the second assertion, we have that $\overline{\overline{A}^C}^B \subseteq \overline{A}^B$, and the equality holds, as required. □

Proposition 3.10 *Let $f : A \to B$ be a homomorphism of rings.*

(1) *If f is integral (resp. of finite type, finite) and I is an ideal of B, then the ring extension $A/(I \cap A) \subseteq B/I$ is also integral (resp. of finite type, finite).*
(2) *Let A' be an A-algebra, and $B' = A' \otimes_A B$. Let $f' : A' \to B'$ be the ring homomorphism given by $f'(a') = a' \otimes 1$. If f is integral (resp. of finite type, finite), then so is f'. In particular, if f is integral (resp. of finite type, finite) and S is a multiplicatively closed subset of A, then $S^{-1}B$ is so over $S^{-1}A$.*

The proof is easy. The following lemma is simple but important.

Lemma 3.11 *Assume that the ring extension $A \subset B$ is an extension of integral domains. Then A is a field if and only if so is B.*

Proof If B is a field, for $x \in A, x \neq 0, x^{-1} \in B$ is integral over A. We show that $x^{-1} \in A$, which implies that A is a field.

Since x^{-1} is integral over A, there exists a monic polynomial $f(T) \in A[T]$ such that $f(x^{-1}) = 0$. If $f(T) = T^n + r_1T^{n-1} + \cdots + r_n$ with $r_1, \ldots r_n \in A$, expanding $x^n f(x^{-1})$ we get

$$1 = (-1)x\left(r_1 + \cdots + r_n x^{n-1}\right).$$

Hence x is a unit in A and $x^{-1} \in A$.

Conversely, assume that A is a field and take $b \in B, b \neq 0$. Since B is integral over A, we have an equation

$$b^n + r_1 b^{n-1} + \cdots + r_n = 0, \quad (r_1, \ldots, r_n \in A, r_n \neq 0).$$

Then we have $b(b^{n-1} + \ldots + r_{n-1}) = -r_n$ and since r_n is a unit in A, b is a unit in B. Thus B is a field, too. □

Corollary 3.12 *Let $A \to B$ be an integral ring homomorphism. If $P \in \operatorname{Spec} B$, then P is a maximal ideal of B if and only if $P \cap A$ is a maximal ideal in A.*

Proof Since $A/(P \cap A) \subseteq B/P$ is an integral ring extension, this corollary follows from Lemma 3.11. □

Corollary 3.13 *Assume that B is integral over A. If $P_1 \subseteq P_2$ are prime ideals of B and if $P_1 \cap A = P_2 \cap A$, then we must have $P_1 = P_2$.*

Proof Put $\mathfrak{p} = P_1 \cap A = P_2 \cap A$. Since $A_\mathfrak{p} \subseteq B_\mathfrak{p}$ is integral, $P_1 B_\mathfrak{p} \cap A_\mathfrak{p} = \mathfrak{p}A_\mathfrak{p}$ is a maximal ideal of $A_\mathfrak{p}$. So $P_1 B_\mathfrak{p}$ is also maximal by Corollary 3.12. Hence $P_1 = P_1 B_\mathfrak{p} \cap B = P_2 B_\mathfrak{p} \cap B = P_2$. □

Theorem 3.14 *Let $f : A \to B$ be an integral homomorphism. Then the following statements hold:*

(1) *(Lying-over theorem) Assume that $A \to B$ is injective. For every $\mathfrak{p} \in \operatorname{Spec} A$, there exists $P \in \operatorname{Spec} B$ such that $\mathfrak{p} = P \cap A$.*
(2) *(Going-up theorem) Let $P_1 \in \operatorname{Spec} B$ and put $\mathfrak{p}_1 = P_1 \cap A \in \operatorname{Spec} A$. If $\mathfrak{p}_2 \in \operatorname{Spec} A$ and $\mathfrak{p}_1 \subseteq \mathfrak{p}_2$, then there exists $P_2 \in \operatorname{Spec} B$ such that $P_1 \subseteq P_2$ and $P_2 \cap A = \mathfrak{p}_2$.*

Proof

(1) The ring extension $A_\mathfrak{p} \subseteq B_\mathfrak{p}$ is integral. Take a maximal ideal $\mathfrak{m}$ of $B_\mathfrak{p}$. Then since $\mathfrak{m} \cap A_\mathfrak{p}$ is a maximal ideal by Corollary 3.12, and we have $\mathfrak{m} \cap A_\mathfrak{p} = \mathfrak{p}A_\mathfrak{p}$. If we put $P = \mathfrak{m} \cap B$, then $P \cap A = \mathfrak{p}$.
(2) Since the ring extension $A/\mathfrak{p}_1 \subseteq B/P_1$ is integral, there is a prime ideal Q of B/P_1 which satisfies $Q \cap (A/\mathfrak{p}_1) = \mathfrak{p}_2/\mathfrak{p}_1$ by (1). Let P_2 be the inverse image of Q in B, then we have $P_2 \cap A = \mathfrak{p}_2$.

□

Corollary 3.15 *Let $f : A \to B$ be an integral ring homomorphism. Then* $\dim A \geq \dim B$. *If, moreover, f is injective, then* $\dim A = \dim B$.

Proof Let $P_n \supsetneq P_{n-1} \supsetneq \cdots \supsetneq P_1 \supsetneq P_0$ be a chain of prime ideals in B. Then by Corollary 3.13, we have a chain of prime ideals $P_n \cap A \supsetneq P_{n-1} \cap A \supsetneq \cdots \supsetneq P_1 \cap A \supsetneq P_0 \cap A$ in A. Hence we have $\dim A \geq \dim B$. Conversely, by Theorem 3.14, if $\mathfrak{p}_0 \subsetneq \mathfrak{p}_1 \subsetneq \cdots \subsetneq \mathfrak{p}_{n-1} \subsetneq \mathfrak{p}_n$ is a chain of prime ideal of A of length n, we can construct a chain of prime ideals $P_0 \subsetneq P_1 \subsetneq \cdots \subsetneq P_{n-1} \subsetneq P_n$ satisfying $P_i \cap A = \mathfrak{p}_i$ for $0 \leq i \leq n$ by Theorem 3.14. Indeed, we can find P_0 by Theorem 3.14, (1), and then we can construct P_i inductively, using Theorem 3.14, (2). Hence we have $\dim B \geq \dim A$. □

In the following, we denote the integral closure $\overline{A}^B$ of A in B simply by $\overline{A}$.

Proposition 3.16 *Let S be a multiplicatively closed subset of A. Then we have $S^{-1}(\overline{A}) = \overline{S^{-1}A}^{S^{-1}B}$, as subrings of $S^{-1}B$.*

Proof We put $\varphi : A \to S^{-1}A, a \mapsto a/1$. Since $\overline{A}$ is integral over A, $S^{-1}\overline{A} \subseteq S^{-1}B$ is integral over $S^{-1}A$. So $S^{-1}\overline{A} \subseteq \overline{S^{-1}A}^{S^{-1}B}$.

Conversely, if $\alpha = b/s \in S^{-1}B$ is integral over $S^{-1}A$, there is a monic polynomial $f(T) \in S^{-1}A$ such that $f(\alpha) = 0$, where $b \in B$ and $s \in S$. If $f(T) = T^n + (r_1/s_1)T^{n-1} + \cdots + r_n/s_n$ with $r_1, \ldots, r_n \in A$, $s_1, \ldots, s_n \in S$, putting $u = s_1 s_2 \cdots s_n \in S$, we can see that ub is integral over A and hence we have $\overline{S^{-1}A}^{S^{-1}B} \subseteq S^{-1}\overline{A}$. □

Definition 3.17 (Normal Domain) If an integral domain A is integrally closed in its field of fractions $K = \mathrm{Q}(A)$, we say that A is an *integrally closed domain*. We say that a ring A is *normal* if $A_\mathfrak{p}$ is an integrally closed domain for any $\mathfrak{p} \in \operatorname{Spec} A$.

3.18 Clearly, a localization of a normal ring is normal. A nonzero localization of an integrally closed domain is again an integrally closed domain. In particular, an integrally closed domain is normal. So A is normal if $A_\mathfrak{m}$ is integrally closed for each maximal ideal $\mathfrak{m}$ of A. Conversely, a normal domain is integrally closed. Let A be a normal domain, and $B = \overline{A}^{\mathrm{Q}(A)}$. Then $B_\mathfrak{p} = \overline{A_\mathfrak{p}}^{\mathrm{Q}(A)} = A_\mathfrak{p}$ for any $\mathfrak{p} \in \operatorname{Spec} A$, and hence $A = B$, as desired. In general, a normal ring need not be an integral domain. For example, $\mathbb{Q} \times \mathbb{Q}$ is a normal ring which is not a domain. A ring A is reduced if and only if $A_\mathfrak{p}$ is reduced for any $\mathfrak{p} \in \operatorname{Spec} A$. So a normal ring is reduced. The direct product of finitely many normal domains is normal. Conversely, if a normal ring A has only finitely many minimal primes $P_1, \ldots, P_r$, then $P_i + P_j = A$ for $i \neq j$ (otherwise $P_i + P_j \subseteq \mathfrak{m}$ for some maximal ideal, and this contradicts the assumption that $A_\mathfrak{m}$ is an integral domain), and hence $A \cong A/P_1 \times \cdots \times A/P_r$ by the Chinese remainder theorem. Being a localization of A, each A/P_i is a normal domain, and hence A is a finite direct product of integrally closed domains. In particular, a Noetherian normal ring is a finite direct product of Noetherian normal domains.

Exercise 3.1 Show that a UFD is a normal domain.

Exercise 3.2 Let K be a field, and A_λ ($\lambda \in \Lambda$) be normal subrings of K. Show that $\bigcap_{\lambda \in \Lambda} A_\lambda$ is normal.

Exercise 3.3 Let A be a normal domain and K be the field of fraction of A.

(1) Show that if L is a finite extension field of K and if $\alpha \in L$ is integral over A, then the minimal polynomial[1] $f(T)$ of α over K belongs to $A[T]$.
(2) Let k be a field and let $B = k[X^4, X^3Y, XY^3, Y^4]$. Show that $\alpha = XY$ is integral over B but does not satisfy (1).

The "going-down theorem" is more sensitive than the going-up theorem.

Theorem 3.19 (Going-Down Theorem) *Let B/A be an integral ring extension, B be an integral domain, and A be a normal domain. Assume that $P_1 \in \operatorname{Spec} B$ and $\mathfrak{p}_1, \mathfrak{p}_2 \in \operatorname{Spec} A$ such that $\mathfrak{p}_1 = P_1 \cap A$ and $\mathfrak{p}_1 \supseteq \mathfrak{p}_2$. Then there exists $P_2 \in \operatorname{Spec} B$ satisfying the conditions $P_1 \supseteq P_2$ and $\mathfrak{p}_2 = P_2 \cap A$.*

[1] We say that $f(T)$ is the *minimal polynomial* of α over K if $f(T) \in K[T] \setminus K$ is monic, $f(\alpha) = 0$, and has the smallest degree among such polynomials. Note that the degree of $f(T)$ equals the degree of the field extension $[K(\alpha) : K]$.

For the proof of this theorem, we need a lemma.

Lemma 3.20 *Let $A \subseteq B$ be an integral ring extension and A be a normal domain with field of fractions $K = Q(A)$. Then the following statements hold:*

(1) *Let $\alpha \in B$ and let $f(T)$ be the minimal polynomial of α over K. Then $f(T) \in A[T]$.*
(2) *If P is a prime ideal of A and if $\alpha \in PB$, then the coefficients of $f(T)$ above are elements of P except the leading coefficient.*

Proof

(1) Let $f(T)$ be the minimal polynomial of α and put $\deg f = n$. Then the roots of $f(T)$ are integral over A since they are conjugate with α. Hence the coefficients of $f(T)$ are integral over A since they are expressed by the elementary symmetric polynomials of roots of $f(T)$. Since A is a normal domain by our assumption, they are in A.
(2) Replacing B by the integral closure of B in the algebraic closure of $Q(B)$, we may assume that $f(T)$ decomposes as $f(T) = (T - \alpha_1)(T - \alpha_2) \cdots (T - \alpha_n)$ in $B[T]$ with $\alpha_1 = \alpha$. Since $f(T) \in K[T]$ is irreducible, for each j, there is a K-automorphism φ_j of $Q(B)$ such that $\varphi_j(\alpha) = \alpha_j$. Since $\varphi_j(B) \subseteq B$ and hence $\varphi_j(PB) \subseteq PB$, we have that $\alpha_j = \varphi_j(\alpha) \in PB$ for each j. So the coefficients of $f(T)$ lies in $PB \cap A$. By the lying-over theorem, there is a prime ideal Q of B such that $Q \cap A = P$. Hence
$$P \subseteq PB \cap A \subseteq Q \cap A = P,$$
and the assertion follows.

□

Proof of Theorem 3.19 Consider the ring extension B_{P_1}/A. If the equality $\mathfrak{p}_2 B_{P_1} \cap A = \mathfrak{p}_2$ holds, then $\kappa(\mathfrak{p}_2) = A_{\mathfrak{p}_2}/\mathfrak{p}_2 A_{\mathfrak{p}_2} \to (B_{P_1}/\mathfrak{p}_2 B_{P_1})_{\mathfrak{p}_2}$ is injective, and hence $(B_{P_1}/\mathfrak{p}_2 B_{P_1})_{\mathfrak{p}_2} \neq 0$. So we can take $\overline{Q} \in \operatorname{Spec}(B_{P_1}/\mathfrak{p}_2 B_{P_1})_{\mathfrak{p}_2}$. If we set $Q = \overline{Q} \cap B_{P_1}$, then $Q \cap A = \overline{Q} \cap A = (\overline{Q} \cap \kappa(\mathfrak{p}_2)) \cap A = \mathfrak{p}_2$. Then letting $P_2 = Q \cap B \in \operatorname{Spec} B$, $P_2 \subseteq P_1$ and $P_2 \cap A = \mathfrak{p}_2$ holds. Hence it suffices to show that $\mathfrak{p}_2 B_{P_1} \cap A = \mathfrak{p}_2$.

Take any element $0 \neq x \in \mathfrak{p}_2 B_{P_1} \cap A$. Since $x \in \mathfrak{p}_2 B_{P_1}$, we can take $a \in \mathfrak{p}_2 B$ and $s \in B \setminus P_1$ such that $x = a/s$. Let $f(T) = T^n + b_1 T^{n-1} + \cdots + b_n$ be the minimal polynomial of a over $K = Q(A)$. Since $s = a/x$, we have $K(a) = K(s)$, and hence $g(T) = x^{-n} f(xT) = T^n + (b_1/x)T^{n-1} + \cdots + (b_n/x^n)$ is the minimal polynomial of s over K. By Lemma 3.20, $b_i \in \mathfrak{p}_2$ and we have $b_i/x^i \in A$. If $x \notin \mathfrak{p}_2$, then since $b_i = (b_i/x^i) \cdot x^i \in \mathfrak{p}_2$, we have that $b_i/x^i \in \mathfrak{p}_2$. Then since $s^n \in \mathfrak{p}_2$, we get $s \in \mathfrak{p}_2$. So $s \in \mathfrak{p}_2 \subseteq \mathfrak{p}_1 \subseteq P_1$, and this contradicts our assumption $s \notin P_1$. Hence we get $x \in \mathfrak{p}_2$, and this completes the proof of the theorem. □

Exercise 3.4 Let $A \to B$ be a ring homomorphism and assume that B is flat over A. Then show that $A \to B$ satisfies the going-down theorem. Namely, if $P_1 \in$

Spec B and $\mathfrak{p}_1, \mathfrak{p}_2 \in \operatorname{Spec} A$ satisfy $\mathfrak{p}_1 = P_1 \cap A$ and $\mathfrak{p}_1 \supseteq \mathfrak{p}_2$, then there exists $P_2 \in \operatorname{Spec}(B)$ such that $P_1 \supseteq P_2$ and $\mathfrak{p}_2 = P_2 \cap A$.

Example 3.21 In Theorem 3.19, the assumption "A is normal" is essential. For example, let k be a field and $B = k[X, Y]$, a polynomial ring over k in 2 variables. Let $A = \{f \in B \mid f(0,0) = f(1,0)\}$. Since $X(X-1), Y \in A$, B is integral over A. Put $\mathfrak{p}_2 = XB \cap A \subseteq \mathfrak{p}_1 = (X-1, Y) \cap A$. If we put $P_1 = (X-1, Y)B$, there does not exist $P_2 \in \operatorname{Spec}(B)$ satisfying $P_2 \cap A = \mathfrak{p}_2$ and $P_2 \subseteq P_1$.

3.2 Zariski's Lemma and Noether Normalization

Proposition 3.22 (Artin–Tate Lemma) *Let $R \subseteq A \subseteq B$ be extensions of rings and assume that R is a Noetherian ring, B is a finitely generated R-algebra. If B/A is integral, then A is also a finitely generated R-algebra.*

Proof Put $B = R[x_1, x_2, \ldots, x_n]$ $(x_i \in B)$. Then since each x_i is integral over A, there exist monic polynomials $f_1, \ldots, f_n \in A[T]$ such that $f_i(x_i) = 0$. Let C be the (finitely generated) R-subalgebra of A generated by all the coefficients of $f_1, \ldots, f_n$. Then each x_i is integral over C. Since $B = C[x_1, x_2, \ldots, x_n]$, B is finitely generated as a C-module. Since C is Noetherian and A is a C-submodule of B, A is also a finitely generated C-module. Since C is a finitely generated R-algebra, so is A. □

Definition 3.23 We call a ring which is finitely generated over a field an *affine ring*. If an affine ring is an integral domain, we call it an *affine domain.*

Let us introduce fundamental properties of affine domains.

Theorem 3.24 (Zariski's Lemma) *Let $k \subseteq K$ be an extension of fields. If K is finitely generated as a k-algebra, then K is a finite extension of k. In other words, if an affine domain K over k is a field, then K is integral over k (in other words, K is an algebraic extension of k).*

Proof Put $K = k[x_1, x_2, \ldots, x_n]$. We order $x_1, x_2, \ldots, x_n$ so that $x_1, x_2, \ldots, x_r$ $(0 \le r \le n)$ are algebraically independent over k and $x_{r+1}, \ldots, x_n$ are algebraic over $k(x_1, x_2, \ldots, x_r)$. We will derive a contradiction, assuming that $r \ge 1$.

If we put $A = k[x_1, x_2, \ldots, x_r]$ and $F = \mathrm{Q}(A)$, then $K = F[x_{r+1}, \ldots, x_n]$ and since K/F is an integral extension, by Proposition 3.22 F is a finitely generated A-algebra. Let

$$F = A\left[g_i/f_i \mid i = 1, \ldots, s\right] \quad (f_i, g_i \in A,\ f_i \neq 0;\ 1 \le i \le s)$$

and put $f = f_1 \cdots f_s$. Then we get $F = A[1/f]$.

Being a polynomial ring in r variables, A is not a field, and hence $f \notin k$. On the other hand, $1/(f+1) \in F = A[1/f]$. So we have $f^m/(f+1) \in A$ for some m. But since f and $f+1$ are relatively prime in A, it is impossible. Hence we must have $r = 0$ and K/k is an algebraic extension. □

The *Noether normalization* is a fundamental property of affine rings. For a nonzero affine ring A over a field k, the transcendence degree $\text{tr.deg}_k A$ is the supremum of $r \geq 0$ such that there exists some algebraically independent elements $x_1, \ldots, x_r \in A$. We define $\text{tr.deg}_k 0 = -\infty$. If $A = k[z_1, \ldots, z_n]$ is an affine domain and $x_1, \ldots, x_r$ is any maximal algebraically independent subset of A, then each z_i is algebraic over $k(x_1, \ldots, x_r)$, and hence we have $r = \text{tr.deg}_k \mathrm{Q}(A) \geq \text{tr.deg}_k A \geq r$. Namely, $r = \text{tr.deg}_k \mathrm{Q}(A) = \text{tr.deg}_k A$.

Theorem 3.25 (Noether Normalization Theorem) *If A is a nonzero finitely generated k-algebra, then we can take $x_1 \ldots, x_r \in A$, which are algebraically independent over k and A is integral over $k[x_1, \ldots, x_r]$. If so, then $r = \dim A$ and is also equal to the transcendence degree $\text{tr.deg}_k A$ of A over k.*

Proof Let $A = k[z_1, \ldots, z_n]$. We prove the theorem by induction on n. If $n = 0$, then there is nothing to prove. So let $n > 0$, and the theorem is true for smaller n. If $z_1, \ldots, z_n$ are algebraically independent over k, then the theorem is obvious by Proposition 2.41.

If $z_1, \ldots, z_n$ are not algebraically independent, then for some nonzero polynomial $F(X_1, \ldots, X_n) \in k[X_1, \ldots, X_n]$ we have $F(z_1, \ldots, z_n) = 0$. Then by Lemma 3.26 below, we can take $A' = k[y_1, \ldots, y_{n-1}] \subseteq A$ so that A is integral over A'. Then we can take algebraically independent elements $x_1, \ldots, x_r \in A'$ such that A' is integral over $k[x_1, \ldots, x_r]$. Then $r = \dim A' = \dim A$ by Corollary 3.15, and A is also integral over $k[x_1, \ldots, x_r]$. It is obvious that $x_1, \ldots, x_r$ is a maximal algebraically independent subset of A. So if A is an affine domain, the theorem is obvious, since we have $r = \text{tr.deg}_k A$ then.

It remains to show that if $t_1, \ldots, t_u$ is an algebraically independent subset of A, then $u \leq r$ for general A. Taking any prime ideal Q of $S^{-1}A$, where $S = k[t_1, \ldots, t_u] \setminus \{0\}$ (it exists, since $0 \neq S^{-1}k[t_1, \ldots, t_u] = k(t_1, \ldots, t_u) \to S^{-1}A$ is injective), and letting $P = Q \cap A$, we have that $k[t_1, \ldots, t_u] \to A/P$ is still injective. So $u \leq \text{tr.deg}_k A/P = \dim A/P \leq \dim A = r$, as desired. □

Lemma 3.26 *Let $A = k[x_1, \ldots, x_n]$ be a finitely generated k-algebra, and assume that $x_1, \ldots, x_n$ satisfy $F(x_1, \ldots, x_n) = 0$ for a nonzero polynomial $F(X_1, \ldots, X_n) \in k[X_1, \ldots, X_n]$. Then the following statements hold:*

(1) *We can choose positive integers $r_1, \ldots, r_{n-1}$ so that the highest-degree term with respect to X_n of $F(X_1 + X_n^{r_1}, \ldots, X_{n-1} + X_n^{r_{n-1}}, X_n)$ is an element of k.*
(2) *If the positive integers r_i $(1 \leq i \leq n-1)$ satisfies the condition above, then putting $y_i = x_i - x_n^{r_i}$ $(i = 1, \ldots, n-1)$, x_n is integral over $k[y_1, \ldots, y_{n-1}]$.*

Proof (Masayoshi Nagata)

(1) Consider the subset

$$M := \{(m_1, \cdots, m_{n-1}, m_n) \mid \text{the coefficient of } X_1^{m_1} \cdots X_{n-1}^{m_{n-1}} X_n^{m_n} \text{ in } F(X_1, \ldots, X_n) \text{ is nonzero}\}$$

of $\mathbb{Z}^n_{\geq 0}$. To show (1), it suffices to show that we can choose $(r_1, \ldots, r_{n-1})$ so that there exists a unique $(m_1, \ldots, m_{n-1}, m_n) \in M$ such that

$$m_1 r_1 + \ldots + m_{n-1} r_{n-1} + m_n$$

takes the maximal value. We define $q_{i+1} : M \to \mathbb{Z}_{\geq 0}$ and r_i by reverse induction on $i = 1, \ldots, n-1$. If $i = n-1$, then we define $q_n(m_1, \ldots, m_n) = m_n$ and $r_{n-1} = 1 + \max q_n(M)$. If $i = n-2$, then we define $q_{n-1}(m_1, \ldots, m_n) = r_{n-1}m_{n-1} + m_n$ and $r_{n-2} = 1 + \max q_{n-1}(M)$, and so on. Then

$$q_1(m_1, \ldots, m_n) = r_1 m_1 + r_2 m_2 + \cdots + r_{n-1} m_{n-1} + m_n.$$

By construction, for $\boldsymbol{m} = (m_1, \ldots, m_n), \boldsymbol{m}' = (m'_1, \ldots, m'_n) \in M$, if $\boldsymbol{m}$ is larger than $\boldsymbol{m}'$ in the lexicographic order, then $q_1(\boldsymbol{m}) > q_1(\boldsymbol{m}')$. So the maximum value of q_1 is only taken by the maximum element of M with respect to the lexicographic order.

(2) If we think $0 = F(y_1 + x_n^{r_1}, \ldots, y_{n-1} + x_n^{r_{n-1}}, x_n) = G(y_1, \ldots, y_{n-1}, x_n)$ as an equation of x_n over $k[y_1, \ldots, y_{n-1}]$, this gives an integral equation of x_n over $k[y_1, \ldots, y_{n-1}]$, since the highest coefficient of x_n is a nonzero element of k.

□

Exercise 3.5 Give a polynomial ring over $\mathbb{C}$ which gives a Noether normalization for the following affine domains over $\mathbb{C}$:

(1) $\mathbb{C}[x^2, x^3]$;
(2) $\mathbb{C}[x^4, x^3y, xy^3, y^4]$;
(3) $\mathbb{C}[x, y, z]/(y^2z - x^2 - 1)$.

Exercise 3.6 Let $f : A \to B$ a ring homomorphism. Is it true that if $\mathfrak{m}$ is a maximal ideal of B, then $f^{-1}(\mathfrak{m})$ is a maximal ideal of A? What if A and B are affine rings over k and f is a homomorphism of k-algebras?

Exercise 3.7 Let A and B be affine domains over an *algebraically closed* field k. Show that $A \otimes_k B$ is also an affine domain over k.

The following theorem is called the *"Nullstellensatz" of Hilbert*[2] and is very famous in relation to algebraic geometry.

---

[2] Nullstellensatz is a German word that means "zero-point theorem."

Theorem 3.27 *Let k be an algebraically closed field and $A = k[X_1, \dots, X_n]$ a polynomial ring of n variables over k.*

(1) *(Weak Nullstellensatz) Every maximal ideal of A is of the form $(X_1 - a_1, \dots, X_n - a_n)$ for some $(a_1, \dots, a_n) \in k^n$.*
(2) *(Nullstellensatz) For every ideal I of A, we define*

$$V(I) = \{(a_1, \dots, a_n) \in k^n \mid \forall f \in I, f(a_1, \dots, a_n) = 0\}.$$

Then $g \in A$ satisfies the condition "$\forall (a_1, \dots, a_n) \in V(I), g(a_1, \dots, a_n) = 0$" if and only if $g \in \sqrt{I}$.

Proof

(1) Let $\mathfrak{m}$ be a maximal ideal of A. Then by Theorem 3.24, $A/\mathfrak{m}$ is a finite extension of k and hence the composition of canonical injection and canonical surjection $\phi : k \to A \to A/\mathfrak{m}$ is an isomorphism. If we put $a_i = \phi^{-1}(X_i + \mathfrak{m})$, then $X_i - a_i \in \mathfrak{m}$. Since $(X_1 - a_1, \dots, X_n - a_n)$ is a maximal ideal and contained in $\mathfrak{m}$, we must have $\mathfrak{m} = (X_1 - a_1, \dots, X_n - a_n)$.
(2) [ZS2] Let $B = A[Y]$ be a polynomial ring of $n + 1$ variables and let J be the ideal of B generated by I and $1 - gY$. For every $P = (a_1, \dots, a_n, b) \in k^{n+1}$, if $(a_1, \dots, a_n) \in V(I)$, then the value of $1 - gY$ at P is 1 by our definition and if $(a_1, \dots, a_n) \notin V(I)$, then by definition, $f(a_1, \dots, a_n) \neq 0$ for some $f \in I$. This implies that the common zero of the ideal J is empty. Then by (1), $(I, 1 - gY) = B$. This means that if $I = (f_1, \dots, f_r)$, we have

$$1 = \sum_{i=1}^{r} f_1 h_i + (1 - gY)h_{r+1} \qquad (*)$$

for some $h_1, \dots, h_r, h_{r+1} \in B$. Putting $Y = g^{-1}$ in (*), we have

$$1 = \sum_{i=1}^{r} f_i h_i(X_1, \dots, X_n, g^{-1}).$$

If g^{-s} is the highest power of g^{-1} appearing on the right hand side, multiplying both sides by g^s, we get $g^s \in I$.

□

Exercise 3.8 If k is not algebraically closed, describe the shape of a maximal ideal of A.

Exercise 3.9 Let A be a finitely generated $\mathbb{Z}$-algebra. Show that for a maximal ideal $\mathfrak{m}$ of A, $A/\mathfrak{m}$ is a *finite* field.

For an integral domain R, we define *absolute integral closure* as the integral closure $\overline{R}^{\Omega_R}$ of R in a fixed algebraic closure Ω_R of $\mathrm{Q}(R)$ and we denote it by R^+. This notion is very important in the latest developments of commutative ring theory.

Example 3.28 Let R be an integral domain. Prove the following statements concerning R^+. (See [Ar] for (1) and (2), and [Neu, (VI.7.5)] for (3).)

(1) For a prime ideal $\mathfrak{p}$ of R^+, we have $(R/(\mathfrak{p} \cap R))^+ \cong R^+/\mathfrak{p}$.
(2) For any prime ideals $\mathfrak{p}, \mathfrak{q}$ of R^+, $\mathfrak{p} + \mathfrak{q}$ is either the unit ideal (1) or a prime ideal.
(3) The absolute integral closure $\mathbb{Z}^+$ of the ring of integers $\mathbb{Z}$ is a *Bézout domain*. That is, any finitely generated ideal of $\mathbb{Z}^+$ is principal.

3.3 Discrete Valuation Rings, Dedekind Domains, and Serre's Criterion for Normality

Definition 3.29 Let $(A, \mathfrak{m})$ be a Noetherian local domain which is not a field. If $\mathfrak{m}$ is a principal ideal, A is called a *discrete valuation ring* (often called *DVR* for short).

By Krull's altitude theorem (Theorem 2.36), a DVR is one-dimensional. Let A be a DVR with $\mathfrak{m} = (a)$ and I be any nonzero ideal of A. Then A/I is an Artinian local ring, and hence $\mathfrak{m}^r = (a^r) \subseteq I$ for sufficiently large r. Letting $n = \min\{r \geq 0 \mid a^r \in I\}$, we have $I = (a^n)$. Indeed, if $I \supsetneq (a^n)$, then we can take $b \in I \setminus (a^n)$. Then we can take some $i \geq 0$ such that $a^i b \notin (a^n)$, but $a^{i+1}b \in (a^n)$. Since A is a domain, $a^i b \in (a^{n-1}) \setminus (a^n)$, and it is the basis of the one-dimensional space $(a^{n-1})/(a^n)$. Hence we have $(a^i b, a^n) = (a^{n-1})$. By NAK, $I \supseteq (a^i b) = (a^{n-1})$, and this contradicts the choice of n.

Hence the ideals of A are (a^n) $(n \geq 0)$ and 0. In particular, a DVR is a PID and is a normal domain.

Lemma 3.30 *If a Noetherian ring A has a principal prime ideal $\mathfrak{p} = (a)$ with* $\mathrm{ht}(\mathfrak{p}) > 0$, *then A is an integral domain, and* $\mathrm{ht}(\mathfrak{p}) = 1$.

Proof Since $\mathrm{ht}(\mathfrak{p}) > 0$, we can take a prime ideal $\mathfrak{q} \subsetneq \mathfrak{p} = (a)$. For any $x \in \mathfrak{q}$, we can write $x = ay$. Since $a \notin \mathfrak{q}$ and $\mathfrak{q}$ is a prime ideal, we have $y \in \mathfrak{q}$ and we have $\mathfrak{q} = a\mathfrak{q}$. So $\mathfrak{q}A_\mathfrak{p} = a\mathfrak{q}A_\mathfrak{p}$, and hence $\mathfrak{q}A_\mathfrak{p} = 0$ by NAK. Hence $\mathfrak{q} = \mathfrak{q}A_\mathfrak{p} \cap A = 0$, and A is an integral domain. $\mathrm{ht}(\mathfrak{p}) = 1$ is a consequence of Krull's altitude theorem. □

DVRs have the following characterization.

Proposition 3.31 *Let $(A, \mathfrak{m})$ be a Noetherian local domain.*

(1) *If $\mathfrak{m}$ is not a principal ideal and assume that for some $a \in \mathfrak{m}$, $a \neq 0$, we have $\mathfrak{m} \in \mathrm{Ass}_A(A/(a))$. Then A is not integrally closed.*
(2) *A is a DVR if and only if it is a Noetherian local normal domain with $\dim A = 1$.*
(3) *If every $\mathfrak{m}$-primary ideal of A is of the form $\mathfrak{m}^n$ $(n \geq 0)$, then A is a DVR or a field.*

Proof

(1) If $\mathfrak{m} \in \mathrm{Ass}_A(A/(a))$, there exists $b \notin (a)$ with $b\mathfrak{m} \subseteq (a)$ (cf. Definition 2.14). Since $\mathfrak{m}$ is not principal, $b\mathfrak{m} = (a)$ cannot happen and hence $b\mathfrak{m} \subseteq a\mathfrak{m}$. Hence we have $(b/a)\mathfrak{m} \subseteq \mathfrak{m}$, which implies b/a is integral over A. Since $b \notin (a)$, $b/a \notin A$ and thus A is not integrally closed.
(2) If A is a DVR, then it is a local PID which is not a field, and hence it is a one-dimensional Noetherian local normal domain. We prove the converse. Since $\dim A = 1$, for any $a \in \mathfrak{m}, a \neq 0$, we have $\mathfrak{m} \in \mathrm{Ass}_A(A/(a))$. Since A is integrally closed, $\mathfrak{m}$ must be principal by (1).
(3) If $a_1, \ldots, a_r$ with $r \geq 2$ is a minimal generator of $\mathfrak{m}$, then $\mathfrak{m}^2 \subsetneq \mathfrak{m}^2 + (a_1) \subsetneq \mathfrak{m}$ and $I = \mathfrak{m}^2 + (a_1)$ is not of the form $\mathfrak{m}^n$.

□

Lemma 3.32 *Let $(A, \mathfrak{m})$ be a local ring.*

(1) *If $\mathfrak{m} = (p)$ is a principal ideal and $\bigcap_{n\geq 1} \mathfrak{m}^n = 0$, then A is a DVR or an Artinian local ring.*
(2) *A one-dimensional local UFD is a DVR.*

Proof

(1) For each $a \in A \setminus \{0\}$, define $v(a) = \max\{n \in \mathbb{Z}_{\geq 0} \mid a \in \mathfrak{m}^n\}$. Then we can write $a = p^{v(a)}u$ with $u \in A \setminus pA$. Then $u \in \mathrm{U}(A)$, and $(a) = (p^{v(a)})$. For a nonzero ideal I of A, we define $v(I) = \min\{v(a) \mid a \in I \setminus \{0\}\}$. It is easy to see that $I = (p^{v(I)})$, and it is principal. So every ideal of A is principal, and hence A is Noetherian. So if A is not Artinian, then by Lemma 3.30, A is an integral domain, and hence is a PID. In this case, A is a local PID which is not a field, and hence is a DVR.
(2) Let $(A, \mathfrak{m})$ be a one-dimensional UFD. Let $a \in \mathfrak{m} \setminus \{0\}$. We can decompose as $a = p_1 \cdots p_r$ for some $r \geq 1$ with each p_i a prime element. In particular, A has a prime element. If p is a prime element, then $(p) \subseteq \mathfrak{m}$, and so $\mathfrak{m} = (p)$, since $\mathrm{ht}\,\mathfrak{m} = 1$. If $a \in A \setminus \{0\}$, then we can write $a = up_1p_2 \cdots p_r$ for some $u \in \mathrm{U}(A)$, $r \geq 0$, and prime elements p_i. Each p_i is a unit multiple of p, and hence we can write $a = vp^r$ for some $v \in \mathrm{U}(A)$. Then $a \notin \mathfrak{m}^{r+1} = (p^{r+1})$, otherwise we can write $a = wp^{r+1}$, and hence $v = wp \in \mathfrak{m}$ and this is a contradiction. So $\bigcap_{n\geq 1} \mathfrak{m}^n = 0$, and A is a DVR by (1).

□

If A is a DVR whose field of fractions is K, we say that A is a DVR of K.

Now we study Dedekind domains. The word *ideal* comes from "ideal numbers" and based on the fact that in some rings, the unique factorization does not hold for elements of the ring but does hold for "ideal numbers (= ideals)."

Such rings are called *Dedekind domains*.

Definition 3.33 (Dedekind Domains) We call a Noetherian normal domain of dimension 1 a *Dedekind domain*.

Example 3.34 Let $K \subseteq \mathbb{C}$ be a finite algebraic extension field of $\mathbb{Q}$. We say the integral closure A of $\mathbb{Z}$ in K the ring of *algebraic integers* of K. Since such rings of algebraic integers are Noetherian, they are Dedekind domains.

For example, let $K = \mathbb{Q}(\sqrt{d})$. Then $A = \mathbb{Z}[\sqrt{d}]$ if $d \equiv 2, 3 \pmod 4$ and $A = \mathbb{Z}[\frac{\sqrt{d}+1}{2}]$ if $d \equiv 1 \pmod 4$.

In the following theorem, we assume that the ring A is Noetherian for simplicity, although the theorem is still true without the assumption.

Theorem 3.35 (Prime Factorization of Ideals) *Let A be a Noetherian integral domain. Then the following conditions on A are equivalent.*

(1) *A is a Dedekind domain.*
(2) *A is not a field, and any ideal $I \neq 0$ of A is the product of a finite number of prime ideals.*

Proof (1) $\Rightarrow$(2). Since A is one-dimensional, it is not a field. Let $I \neq (0), (1)$ be any ideal of A and let $\mathfrak{q}_i$ $(1 \leq i \leq n)$ be primary components of I and assume $\mathfrak{q}_i$ is a $\mathfrak{p}_i$-primary ideal $(1 \leq i \leq n)$. Since every $A_{\mathfrak{p}_i}$ is a DVR, $\mathfrak{q}_i = \mathfrak{p}_i^{(a_i)} = \mathfrak{p}_i^{a_i}$ for some $a_i \geq 1$. Thus we have $I = \bigcap_{i=1}^n \mathfrak{p}_i^{a_i}$ and by the Chinese remainder theorem (Lemma 2.32), $I = \prod_{i=1}^n \mathfrak{p}_i^{a_i}$.

(2) $\Rightarrow$(1). Let $\mathfrak{m}$ be a maximal ideal of A. Then any $\mathfrak{m}$-primary ideal of $A_{\mathfrak{m}}$ is a power of $\mathfrak{m}$, and hence $A_{\mathfrak{m}}$ is a DVR by Proposition 3.31, (3). Hence $\dim A = 1$. Since each local ring $A_{\mathfrak{m}}$ is normal, A is normal. □

3.36 (Valuation Rings) Let G be an ordered group. That is, G is an additive group with a total order such that $a, b, c \in G$ and $a \leq b$ together imply $a + c \leq b + c$. A map $v : K \to G \cup \{\infty\}$ is called a *valuation* of the field K if $v^{-1}(\infty) = \{0\}$, $v : \mathrm{U}(K) \to G$ is a group homomorphism, that is, $v(ab) = v(a) + v(b)$ for $a, b \in \mathrm{U}(K)$, and $v(a + b) \geq \min(v(a), v(b))$ for $a, b \in K$.

A subring A of K is called a *valuation ring* of K if either $x \in A$ or $x^{-1} \in A$ holds for any $x \in K \setminus \{0\}$. If A is a valuation ring of K, then $G_A = \{\alpha A \mid \alpha \in K \setminus \{0\}\}$ is a totally ordered set. Indeed, for $\alpha, \beta \in K \setminus \{0\}$, we have either $\alpha\beta^{-1} \in A$ or $\beta\alpha^{-1} \in A$. So $\alpha + \beta$ is contained in αA or βA. In particular, if $\alpha, \beta \in A \setminus \mathrm{U}(A)$, then $\alpha + \beta \in A \setminus \mathrm{U}(A)$, and hence a valuation ring is a local domain. Since G_A is totally ordered, it is easy to see that A is a local Bézout domain (every finitely generated ideal of A is principal). Conversely, a local Bézout domain is a valuation ring (left to the reader). A Noetherian Bézout domain is nothing but a PID, and hence a Noetherian valuation ring is a DVR or a field. Conversely, DVRs and fields are Noetherian valuation rings.

If $v : K \to G \cup \{\infty\}$ is a valuation and put $G_{\geq 0} = \{g \in G | g \geq 0\}$, then $A_v = v^{-1}(G_{\geq 0} \cup \{\infty\})$ is a valuation ring of K whose unique maximal ideal is $\mathfrak{m}_v = v^{-1}(G_{>0} \cup \{\infty\})$. Conversely, if A is a valuation ring of K, then giving G_A the opposite of the inclusion relation, G_A is an ordered group. Then defining $v_A : K \to G_A \cup \{\infty\}$ by $v_A(\alpha) = \alpha A$ for $\alpha \in K \setminus \{0\}$ and $v_A(0) = \infty$, v_A is a valuation of K. It is easy to see that $A_{v_A} = A$, and v_{A_v} is equivalent to v if v is surjective, where we say that two valuations $v : K \to G \cup \{\infty\}$ and $v' : K \to G' \cup \{\infty\}$ are equivalent if there is an isomorphism $\psi : G \to G'$ of ordered groups such that $v' = \psi v$ on $K \setminus \{0\}$. For a valuation v of K, we call $v(K \setminus \{0\})$ the *value group* of v. Note that the value group of v is isomorphic to G_{A_v}, and is determined only by A_v.

A valuation of K is called *trivial* (resp. *discrete*) if $v(K) = 0$ (resp. $v(K) \cong \mathbb{Z}$). A discrete valuation $v : K \to \mathbb{Z} \cup \{\infty\}$ is said to be *normalized* if v is surjective. A valuation v of K is trivial (resp. discrete) if and only if $A_v = K$ (resp. A_v is a DVR). If A is a DVR and $\mathrm{Q}(A) = K$, then we say that A is a DVR of K. Thus a DVRs of K and normalized discrete valuations $v : K \to \mathbb{Z} \cup \{\infty\}$ are in one-to-one correspondence. If A is a DVR of A, then the corresponding $v : K \to \mathbb{Z} \cup \{\infty\}$ is given by $v(\alpha) = \sup\{n \in \mathbb{Z} \mid \alpha \in \pi^n A\}$, where π is any generator of the maximal ideal of A. This v is called the normalized discrete valuation of A, and we also denote it by v_A, identifying G_A by $\mathbb{Z}$.

Let us introduce the notion of regular sequences and depth, which will be very important in our further progress.

Definition 3.37 Let A be a ring and M be an A-module.

(1) We call a sequence $x_1, x_2, \dots, x_n$ of elements of A a *regular sequence* on M (resp. a *weak regular sequence* on M) or an M-sequence (resp. a weak M-sequence) if the following two conditions are (resp. condition (b) below is) satisfied:

 (a) $M/(x_1, x_2, \dots, x_n)M \neq 0$.
 (b) For every $1 \leq i \leq n$, x_i is a non-zerodivisor on $M/(x_1, \dots, x_{i-1})M$.

(2) Assume that $(A, \mathfrak{m})$ is a Noetherian local ring and M is a nonzero finitely generated A-module. We call the supremum of the lengths of M-regular sequences the *depth* of M and denote it by

$$\operatorname{depth} M \text{ or } \operatorname{depth}_A M.$$

For $M = 0$, we define $\operatorname{depth}_A M = \infty$.

Lemma 3.38 *Let $(A, \mathfrak{m})$ be a Noetherian local ring, and M be a finitely generated A-module. If $M \neq 0$, then* $\operatorname{depth}_A M \leq \dim_A M$. *In particular,* $\operatorname{depth}_A M$ *is finite.*

Proof If $x_1, x_2, \dots, x_n$ is an M-regular sequence,

$$\begin{aligned}\dim_A M = 1 + \dim_A M/x_1 M = 2 + \dim_A M/(x_1, x_2)M = \cdots \\ = n + \dim_A M/(x_1, \dots, x_n)M \geq n\end{aligned}$$

by Corollary 2.44. So $\operatorname{depth}_A M \leq \dim_A M < \infty$. □

Note that for a finitely generated module M over a Noetherian local ring $(A, \mathfrak{m})$, $\operatorname{depth}_A M = 0$ if and only if $\mathfrak{m} \in \operatorname{Ass}_A M$. This follows from the fact that the set of non-zerodivisors on M in $\mathfrak{m}$ is $\mathfrak{m} \setminus (\bigcup_{P \in \operatorname{Ass}_A M} P)$, and the prime avoidance.

The following three lemmas will be proved in much more general forms in Chap. 5, using homological methods, see Lemmas 5.9, 5.10, and 5.11, so we omit the proof (left to the readers as exercises).

Lemma 3.39 *Let* $(A, \mathfrak{m})$ *be a Noetherian local ring, and*

$$0 \to M \xrightarrow{\varphi} N \xrightarrow{\psi} L \to 0$$

be a short exact sequence of finitely generated A*-modules with* $\operatorname{depth}_A L = 0$ *and* $\operatorname{depth}_A N \geq 1$. *Then* $\operatorname{depth}_A M = 1$.

Lemma 3.40 *Let* $(A, \mathfrak{m})$ *be a Noetherian local ring,* M *be a finitely generated* A*-module, and* $x \in \mathfrak{m}$ *be a non-zerodivisor on* M. *Then* $\operatorname{depth}_A M \geq 2$ *if and only if* $\operatorname{depth}_A M/xM \geq 1$. *In particular, if* $\operatorname{depth}_A M/xM = 0$, *then* $\operatorname{depth}_A M = 1$.

Lemma 3.41 *Let* $(A, \mathfrak{m})$ *be a Noetherian local ring, and*

$$0 \to M \to N \to L \to 0$$

be a short exact sequence of finitely generated A*-modules. If* $\operatorname{depth}_A N \geq 2$ *and* $\operatorname{depth}_A L \geq 1$, *then* $\operatorname{depth}_A M \geq 2$.

The following Serre's conditions (R_k) and (S_k) are used frequently.

Definition 3.42 Let A be a Noetherian ring, M be a finitely generated A-module.

(1) We say that A satisfies the condition (R_k) if for every prime ideal $\mathfrak{p}$ with $\operatorname{ht}\mathfrak{p} \leq k$, $A_\mathfrak{p}$ is a regular local ring (cf. Chap. 5, Sect. 5.3). In this chapter, we only use (R_k) for $k \leq 1$. Note that a regular local ring of dimension 0 (resp. 1) is nothing but a field (resp. a DVR).
(2) An A-module M satisfies the condition (S_k) if for every prime ideal $\mathfrak{p}$, we have

$$\operatorname{depth} M_\mathfrak{p} \geq \min\{k, \operatorname{ht}\mathfrak{p}\}.$$

We say that the ring A satisfies (S_k) if A satisfies (S_k) as an A-module.

Note that the (S_0) condition is always satisfied. M satisfies the (S_1) condition if and only if $\operatorname{Ass}_A M \subseteq \operatorname{Min}_A A$. Indeed, to say that M satisfies (S_1) is nothing but to say that for any $P \in \operatorname{Spec} A$, $\operatorname{depth}_{A_P} M_P = 0$ implies $\operatorname{ht}_A P = 0$.

Now we prove Serre's $(R_1) + (S_2)$ normality criterion. We start by giving a characterization of the (S_2) condition for a Noetherian normal domain.

Lemma 3.43 *For a Noetherian domain* A, *the following are equivalent:*

(1) *For any* $a \in A \backslash (U(A) \cup \{0\})$, *the principal ideal* aA *does not have an embedded prime ideal.*
(2) $A = \bigcap_{\operatorname{ht}_A(P)=1} A_P$, *where the intersection is taken in the field of fractions* $Q(A)$ *(so the intersection is understood to be* $Q(A) = A$ *if* A *is a field).*
(3) A *satisfies the* (S_2) *condition.*

Proof (1) ⇒(2). It suffices to prove $\bigcap_{\mathrm{ht}_A(P)=1} A_P \subseteq A$. Let $a/b \in \bigcap_{\mathrm{ht}_A(P)=1} A_P$ with $a, b \in A$, $b \neq 0$. Let $bA = \mathfrak{q}_1 \cap \cdots \cap \mathfrak{q}_r$ be a minimal primary decomposition of bA. By assumption, each $\mathfrak{p}_i = \sqrt{\mathfrak{q}_i}$ is a minimal prime of bA. So $a \in \bigcap_i (bA_{\mathfrak{p}_i} \cap A) = \bigcap_i \mathfrak{q}_i = bA$, and hence $a/b \in A$, as desired.

(2) ⇒(1). If $a \in A \setminus (\mathrm{U}(A) \cup \{0\})$ and assume that there is an embedded prime of aA. Then we can take $b \in (\bigcap_{P \in \mathrm{Min}_A A/aA} (aA_P \cap A)) \setminus aA$. Then $b/a \in (\bigcap_{P \in \mathrm{ht}_A(P)=1} A_P) \setminus A$, as can be seen easily.

(1) ⇒(3). Let $P \in \mathrm{Spec}\, A$, and assume that $\mathrm{depth}\, A_P < 2$. If $P = 0$, then $\mathrm{depth}\, A_P \geq 0 = \mathrm{ht}\, P$. If $P \neq 0$, then we can take a nonzero element $x \in P$. Since $x \in PA_P$ is a non-zerodivisor on the local domain A_P, we have $\mathrm{depth}\, A_P = 1$. Hence $\mathrm{depth}\, A_P/xA_P = 0$. So $P \in \mathrm{Ass}_A A/xA = \mathrm{Min}_A A/xA$, and $\mathrm{depth}\, A_P = 1 = \mathrm{ht}\, P$ by Krull's altitude theorem. Hence $\mathrm{depth}\, A_P \geq \min(2, \mathrm{ht}\, P)$ for any $P \in \mathrm{Spec}\, A$.

(3) ⇒(1). If $a \in A \setminus (\mathrm{U}(A) \cup \{0\})$ and P is an embedded prime ideal of aA, then $\mathrm{ht}\, P \geq 2$ and $\mathrm{depth}_{A_P} A_P/aA_P = 0$. So $\mathrm{depth}\, A_P = 1 < \min(2, \mathrm{ht}\, P)$, and the (S_2) condition does not hold. □

Lemma 3.44 *Let A be a Noetherian ring. Then the following are equivalent:*

(1) *A is reduced.*
(2) *A satisfies Serre's conditions* (R_0) *and* (S_1).

Proof (1) ⇒(2). Since $(0) = \bigcap_{P \in \mathrm{Min}\, A} P$ is the minimal primary decomposition, A does not have an embedded prime, and (S_1) is satisfied. For each $P \in \mathrm{Min}\, A$, $PA_P = (0)A_P = 0$, and hence $A_P = A_P/PA_P$ is a field. So (R_0) is also satisfied.

(2) ⇒(1). Let $(0) = \mathfrak{q}_1 \cap \cdots \cap \mathfrak{q}_r$ be the minimal primary decomposition. Since A satisfies (S_1), we have that $\mathfrak{p}_i = \sqrt{\mathfrak{q}_i}$ is a minimal prime for each i. Since A satisfies (R_0), $A_{\mathfrak{p}_i}$ is a field, and hence the kernel $\mathfrak{q}_i$ of the localization $A \to A_{\mathfrak{p}_i}$ is a prime ideal. Hence $\mathfrak{q}_i = \mathfrak{p}_i$, and $\sqrt{(0)} = \mathfrak{p}_1 \cap \cdots \cap \mathfrak{p}_r = (0)$, as desired. □

The following criterion for normality is called "Serre's $(R_1) + (S_2)$ criterion," which was proved by Krull for integral domains, and by Serre[3] for the general case.

Theorem 3.45 (Serre's Criterion) *The following conditions for a Noetherian ring A are equivalent:*

(1) *A is normal.*
(2) *A satisfies Serre's conditions* (R_1) *and* (S_2).
(3) *For every prime ideal* $\mathfrak{p}$ *of height at most* 1, $A_{\mathfrak{p}}$ *is a DVR or a field, and* $A = A_1 \times \cdots \times A_r$ *is a finite direct product of integral domains such that*

$$A_i = \bigcap_{\mathfrak{p} \in \mathrm{Spec}\, A_i,\ \mathrm{ht}\, \mathfrak{p}=1} (A_i)_{\mathfrak{p}}. \qquad (\#)$$

[3] Jean-Pierre Serre, 1926–.

Proof (1) ⇒(2). Let $\mathfrak{p}$ be a prime ideal with $\operatorname{ht}\mathfrak{p} \leq 1$. Then $A_\mathfrak{p}$ is a DVR or a field, since it is a normal local domain of dimension at most one. Hence A satisfies the condition (R_1). To prove the (S_2) condition, we may assume that A is local. A Noetherian normal local ring is an integral domain, and hence Lemma 3.43, (1) ⇒(3) is applicable. Now the assertion follows from Proposition 3.31, (1) immediately.

(2) ⇒(3). The first assertion is clear from the (R_1) property. By Lemma 3.44, A is reduced. Let $P_1, \dots, P_r$ be the minimal prime ideals of A. We prove the second assertion by induction on r. If $r = 1$, then A is an integral domain, and the assertion is proved in Lemma 3.43. Assume that $r \geq 2$, and the assertion is already proved for smaller r. Assume that $I = (P_1 \cap \cdots \cap P_{r-1}) + P_r \neq A$. We can take $Q \in \operatorname{Ass}_A A/I$. The sequence

$$0 \to A \xrightarrow{\phi} A/(P_1 \cap \cdots \cap P_{r-1}) \oplus A/P_r \xrightarrow{\psi} A/I \to 0$$

is exact, where $\phi(a) = (a + (P_1 \cap \cdots \cap P_{r-1}), a + P_r)$ and

$$\psi(a_1 + (P_\cap \cdots \cap P_{r-1}), a_2 + P_r) = (a_1 - a_2) + I.$$

Since $Q \in \operatorname{Ass}_A A/I$, we have $\operatorname{depth}(A/I)_Q = 0$. Since $Q \supseteq P_r$ and $Q \supseteq P_i$ for some $i < r$, we have that $\operatorname{ht} Q \geq 1$. Since $(A/(P_1 \cap \cdots \cap P_{r-1}))_Q$ and $(A/P_r)_Q$ are reduced and hence satisfy (S_1), we have that

$$\operatorname{depth}_{A_Q}((A/(P_1 \cap \cdots \cap P_{r-1}))_Q \oplus (A/P_r)_Q) \geq 1.$$

It follows that $\operatorname{depth} A_Q = 1$ by the exact sequence and Lemma 3.39. Since A satisfies (S_2), we have that $\operatorname{ht} Q = 1$. Hence by the (R_1) condition, A_Q is a DVR. In particular, $\operatorname{Min} A_Q = \{(0)\}$. On the other hand, both $P_r A_Q$ and $P_i A_Q$ for some $i < r$ are minimal primes of A_Q, and $P_r A_Q \neq P_i A_Q$. This is a contradiction. So $I = A$, and we have $A \cong A/(P_1 \cap \cdots \cap P_{r-1}) \times A/P_r$ by the Chinese remainder theorem. It follows that both $A/(P_1 \cap \cdots \cap P_{r-1})$ and A/P_r satisfy the (R_1) and (S_2) conditions. Now the assertion follows immediately by the induction assumption.

(3) ⇒(1). It suffices to prove that each A_i is normal, and hence we may assume that $A = A_1$ is an integral domain. So A satisfies (S_2) by Lemma 3.43. The (R_1) condition is obvious. □

In the next section, the concept of reflexive modules will be important because of its relation to divisorial ideals.

Definition 3.46 (Reflexive Modules) Let M be an A-module. We write

$$M^* = \operatorname{Hom}_A(M, A)$$

and we call M^* the *dual module* of M. The dual module M^{**} of M^* is called the *double dual* of M.

We call the map

$$\phi_M : M \to M^{**} \text{ defined by } \phi_M(m)(f) = f(m) \ (m \in M, f \in M^*)$$

the *canonical map* and we define:

(1) If ϕ_M is injective, then we say M is a *torsionless* A-module.
(2) If ϕ_M is bijective, then we say M is a *reflexive* A-module.

Remark 3.1 Assume that A is Noetherian and M is finitely generated. Then the map ϕ_M is compatible with localization. Hence M is a reflexive A-module if and only if $\forall \mathfrak{p} \in \mathrm{Spec}(A)$, $M_\mathfrak{p}$ is a reflexive $A_\mathfrak{p}$-module.

Example 3.47 Let A be a Noetherian ring and M be a finitely generated A-module.

(1) A finitely generated free A-module is a reflexive A-module.
(2) A submodule of a torsionless module is torsionless. So a submodule of a finitely generated free A-module is torsionless. If $M = N^*$ for some finitely generated A-module N, then M is a submodule of a finitely generated A-free module, and is torsionless. Indeed, taking the dual of a surjective map $A^m \to N$, we get $M = N^* \hookrightarrow A^m$. Conversely, any torsionless module is a submodule of a finitely generated A-free module. Indeed, if M is torsionless, then M is embedded in M^{**}, and M^{**} is embedded in a finitely generated free module. If A is an integral domain, then M is torsionless if and only if it is torsion-free, as can be seen easily from the commutative diagram

$$\begin{array}{ccccc} M & \xrightarrow{\phi_M} & M^{**} & \hookrightarrow & A^m \\ \downarrow & & \downarrow & & \downarrow \\ M \otimes_A \mathrm{Q}(A) & \xrightarrow{\cong} & (M \otimes_A \mathrm{Q}(A))^{\star\star} & \hookrightarrow & \mathrm{Q}(A)^m, \end{array}$$

where $M^{**} \hookrightarrow A^m$ is an appropriate injection, $(-)^\star$ denotes the $\mathrm{Q}(A)$-dual, and the second row is obtained by the localization of the first. In particular, if A is an integral domain, then M is torsionless if and only if $\mathrm{Ass}_A M = \{(0)\}$.
(3) Hence if A is a PID, a finitely generated torsionless module is free.
(4) If $M \cong N^*$ for some A-module N, then M is a direct summand of $M^{**} = N^{***}$. Indeed, it is straightforward to check that the composite

$$M = N^* \xrightarrow{\phi_M} N^{***} = M^{**} \xrightarrow{\phi_N^*} N^* = M$$

is the identity map. If, moreover, A is a domain, then we can write $M^{**} \cong M \oplus L$ and $L \hookrightarrow L \otimes_A \mathrm{Q}(A) = 0$ by dimension counting. Hence M is reflexive this case.

(5) If $A = k[X, Y]/(X^2, XY, Y^2)$ and $\mathfrak{m} = (X, Y)$, then $(A/\mathfrak{m})^* \cong (A/\mathfrak{m})^2$ is not reflexive, since $(A/\mathfrak{m})^{***} \cong (A/\mathfrak{m})^8$.

Exercise 3.10 Let A be an integral domain, K be its field of fractions, and I be a nonzero A-submodule of K.

(1) Take $x \in I \setminus \{0\}$, and define $\rho : I^* \to K$ by $\phi \mapsto \phi(x)/x$. Show that ρ is A-linear, and is independent of the choice of x.
(2) Show that $\rho : I^* \to K$ maps I^* bijectively onto $I^{-1} := \{x \in K \mid xI \subseteq A\}$.
(3) Assuming that $I^* \neq 0$, prove that the double dual I^{**} is identified with an A-submodule of K containing I. Assuming moreover that I is an ideal of A, can we say that $(I^{-1})^{-1}$, which is identified with I^{**}, is an ideal of A?

Lemma 3.48 *Let A be a Noetherian ring, and $f : M \to N$ be an A-linear map between finite A-modules. Assume that M satisfies Serre's (S_1) condition. Then the following are equivalent:*

(1) *f is injective;*
(2) *f_P is injective for any $P \in \operatorname{Spec} A$ such that $\operatorname{ht} P = 0$.*

Proof We only have to prove (2) $\Rightarrow$(1). Assume the contrary, and assume that $K = \operatorname{Ker} f$ is nonzero. Then we can find $P \in \operatorname{Ass} K$, and we have $P \in \operatorname{Ass} K \subseteq \operatorname{Ass} M \subseteq \operatorname{Min} A$. Hence $K_P = 0$ by assumption, but this contradicts $P \in \operatorname{Ass} K \subseteq \operatorname{Supp} K$. □

Proposition 3.49 *Let A be a Noetherian ring, and $f : M \to N$ be an A-linear map between finitely generated A-modules. Assume that M satisfies Serre's (S_2) condition, and N satisfies (S_1) condition. Then the following are equivalent:*

(1) *f is bijective;*
(2) *f_P is bijective for any $P \in \operatorname{Spec} A$ such that $\operatorname{ht} P \leq 1$.*

Proof We only have to prove (2) $\Rightarrow$(1). By Lemma 3.48, we have that f is injective. So we want to show that f is surjective. Assume the contrary, and assume that $C = \operatorname{Coker} f$ is nonzero. Let $P \in \operatorname{Ass} C$. By assumption, we have that $\operatorname{ht} P \geq 2$. If $\operatorname{depth} N_P = 0$, then $\operatorname{ht} P = 0$ by (S_1) property of N, and this is a contradiction. So $\operatorname{depth} N_P \geq 1$. As $\operatorname{depth} C_P = 0$, we have that $\operatorname{depth} M_P = 1$. So $\operatorname{ht} P = 1$ by the (S_2) property of M, and this is also a contradiction. □

Lemma 3.50 *Let A be a reduced Noetherian ring, and M a finitely generated A-module. Then the following are equivalent:*

(1) *M is torsionless;*
(2) *M is a first syzygy, in the sense that there exists some injective map $M \to F_0$ such that F_0 is a finite free A-module;*
(3) *M satisfies Serre's (S_1) condition.*

Proof (1) $\Rightarrow$(2). There is a surjective A-linear map $\varphi : A^n \to M^*$, which induces an injective A-linear map $\varphi^* : M^{**} \to A^n$. So the composite $M \hookrightarrow M^{**} \xrightarrow{\varphi^*} A^n$ is also injective, and hence M is a first syzygy.

(2) $\Rightarrow$(3). Since A satisfies (S_1), we have that F_0 satisfies (S_1). The (S_1) condition, which says that any associated prime is of height zero, is inherited by a submodule, and hence M satisfies (S_1).

(3) $\Rightarrow$(1). By Lemma 3.48, we may assume that A is an Artinian local ring. Then A is a field, and the assertion is obvious, since any finite-dimensional vector space is reflexive. □

There is a criterion for reflexive property of modules over normal rings in terms of Serre's condition (S_2).

Proposition 3.51 *Let A be a Noetherian normal ring, and M a finitely generated A-module. Then the following are equivalent:*

(1) *M is reflexive;*
(2) *M is a second syzygy, in the sense that there exists some exact sequence $0 \to M \to F_0 \to F_1$ such that F_0 and F_1 are finite free A-modules.*
(3) *M satisfies Serre's (S_2) condition.*

Proof Let $A^n \xrightarrow{\varphi} A^m \xrightarrow{\psi} M^* \to 0$ be a presentation of M^*. Taking the dual,

$$0 \to M^{**} \xrightarrow{\psi^*} A^m \xrightarrow{\varphi^*} A^n$$

is exact, and (1) $\Rightarrow$(2) follows immediately.

(2) $\Rightarrow$(3). Assume the contrary, and let $P \in \operatorname{Spec} A$ such that $\operatorname{depth} M_P < 2$ and $\operatorname{depth} M_P < \operatorname{ht} P$. If $\operatorname{depth} M_P = 0$, then $P \in \operatorname{Ass} F_0 = \operatorname{Ass} A = \{(0)\}$, and hence $\operatorname{depth} M_P < \operatorname{ht} P = 0$. This is a contradiction. So $\operatorname{depth} M_P = 1$. Hence $\operatorname{ht} P \geq 2$. Since A satisfies the (S_2) condition, $\operatorname{depth} F_i \geq 2$. Then letting I be the image of $F_0 \to F_1$, $\operatorname{depth} I_P \geq 1$. So $\operatorname{depth} M_P \geq 2$ by Lemma 3.41, and this is a contradiction. This shows that M satisfies (S_2).

(3) $\Rightarrow$(1). By Proposition 3.49, we may assume that A is a local ring of dimension at most one. Then A is a DVR or a field, and M is a finitely generated free module. Hence M is reflexive. □

A slightly more general result will be treated in Chap. 6 (Proposition 6.53). See also Lemma 6.52 for the criterion for torsionless property.

3.4 Divisor Class Groups and UFD

3.52 (Assumption) *In this section, our ring A is always an integral domain and we denote by K the field of fractions of A.*

Definition 3.53

(1) For an A-submodule I of K, we define $I^{-1} = \{x \in K \mid xI \subseteq A\}$.
(2) An A-submodule I of K is called a *fractional ideal* of A if $I \neq 0$ and $I^{-1} \neq 0$.

(3) For a fractional ideal I of A, we set $\widetilde{I} = (I^{-1})^{-1}$, and call it the *divisorial hull* of I. Since $II^{-1} \subseteq A$, we have $I \subseteq \widetilde{I}$. It follows that I^{-1} is a fractional ideal again. We call I a *divisorial ideal* if I is a fractional ideal and $I = \widetilde{I}$ holds.

If I and J are two fractional ideals of A, then obviously $IJ \neq 0$. Moreover, since $IJI^{-1}J^{-1} \subseteq A$, we have $(IJ)^{-1} \supseteq I^{-1}J^{-1} \neq 0$. Hence the product IJ is also a fractional ideal again.

Exercise 3.11 Let k be a field.

(1) Letting $A = k[t^3, t^5]$ and $I = (t^3, t^5)$, compute I^{-1} and $\widetilde{I}$.
(2) Letting $A = k[X, Y]$ and $I = (X^3, Y^5)$, show that $I^{-1} = \widetilde{I} = A$.

Proposition 3.54

(1) *Let I be a fractional ideal of A. Then I is divisorial if and only if I is reflexive as an A-module. The fractional ideals I^{-1} and $\widetilde{I}$ are divisorial.*
(2) *If M is a reflexive A-module rank one, then M is isomorphic to some divisorial ideal.*
(3) *If A is Noetherian and normal, then an ideal I of A is divisorial if and only if every element of* $\mathrm{Ass}_A(A/I)$ *has height* 1.

Proof

(1) Let I be a fractional ideal of A. Then as in Exercise 3.10, $\rho : \mathrm{Hom}_A(I, A) \to I^{-1}$ given by $\rho(\varphi) = \varphi(x)/x$ is an isomorphism, where $x \in I \setminus \{0\}$, and the definition of ρ is independent of the choice of x. Hence we can identify the natural maps $I \to I^{**}$ and $I \to \widetilde{I}$, and the first assertion follows. Since A is an integral domain, the A-module I^* is reflexive, see Exercise 3.47, (3). So I^{-1} is divisorial. Applying this to the fractional ideal $J = I^{-1}$, we have that $J^{-1} = \widetilde{I}$ is also divisorial.
(2) Since M is torsion-free, $M \to M \otimes_A K \cong K$ is injective. So M can be identified with an A-submodule of K. Since $M \neq 0$ and $M^{-1} \cong M^* \neq 0$, M is a fractional ideal. By (1), M is a divisorial ideal.
(3) From the exact sequence $0 \to I \to A \to A/I \to 0$ and Proposition 3.51, we can see that I is reflexive if and only if for every prime ideal $\mathfrak{p}$ of A with $\mathrm{ht}\,\mathfrak{p} \geq 2$, $\mathrm{depth}(A/I)_{\mathfrak{p}} \geq 1$. Our conclusion follows from (1).

□

Now we introduce the notion of the *divisor class group* of A. We will explain it for *Krull domains*, which are slightly more general rings than Noetherian normal domains. For example, if A is a Noetherian integral domain, then the integral closure of A in a finite extension L of $K = \mathrm{Q}(A)$ is not always Noetherian but is a Krull domain (theorem of Yoshiro Mori).

Definition 3.55 We say that A is a *Krull domain* if it satisfies the following conditions:

(1) For every prime ideal $\mathfrak{p}$ of A with $\mathrm{ht}\,\mathfrak{p} = 1$, $A_{\mathfrak{p}}$ is a DVR.

(2) $A = \bigcap_{\mathrm{ht}\,\mathfrak{p}=1} A_{\mathfrak{p}}$.
(3) For any $a \in A \setminus \{0\}$, there are only finite number of prime ideals of height 1 containing a.

Being the intersection of normal domains, a Krull domain is normal. From Theorem 3.45, a Noetherian normal domain is Krull.

In the rest of this chapter, we assume that A is a Krull domain with $K = \mathrm{Q}(A)$, unless otherwise specified.

First, we define the divisor group of a Krull domain.

Definition 3.56

(1) Let $\mathrm{Ht}_1(A)$ be the set of prime ideal of A of height 1. For $\mathfrak{p} \in \mathrm{Ht}_1(A)$, we denote the normalized discrete valuation of $A_{\mathfrak{p}}$ by $v_{\mathfrak{p}}$.
(2) Let $\mathrm{Div}(A)$ be the free $\mathbb{Z}$-module with the free basis $\mathrm{Ht}_1(A)$. We call it the *divisor group* of A. If we consider $\mathfrak{p} \in \mathrm{Ht}_1(A)$ as an element of $\mathrm{Div}(A)$, we denote it by $[\mathfrak{p}]$ for clarity.
(3) We put an order $\geq$ in $\mathrm{Div}(A)$ by

$$\sum_{\mathfrak{p}\in\mathrm{Ht}_1(A)} n_{\mathfrak{p}}[\mathfrak{p}] \geq \sum_{\mathfrak{p}\in\mathrm{Ht}_1(A)} m_{\mathfrak{p}}[\mathfrak{p}] \iff n_{\mathfrak{p}} \geq m_{\mathfrak{p}} \ (\forall \mathfrak{p} \in \mathrm{Ht}_1(A)).$$

(4) For a fractional ideal $I \neq 0$ of A and for $\mathfrak{p} \in \mathrm{Ht}_1(A)$, we define $\mathrm{div}(I) \in \mathrm{Div}(A)$ by

$$\mathrm{div}(I) = \sum_{\mathfrak{p}\in\mathrm{Ht}_1(A)} v_{\mathfrak{p}}(I)[\mathfrak{p}],$$

where $v_{\mathfrak{p}}(I) = \min\{v_{\mathfrak{p}}(x) \mid x \in I\}$. Note that the sum is a finite sum. Namely, $v_{\mathfrak{p}}(I) \neq 0$ only for finitely many $\mathfrak{p}$. Indeed, if $x \in I \setminus \{0\}$ and $y \in I^{-1} \setminus \{0\}$, then $xA \subseteq I \subseteq y^{-1}A$. If $v_{\mathfrak{p}}(x) = 0$ and $v_{\mathfrak{p}}(y) = 0$, then $v_{\mathfrak{p}}(I) = 0$, since $A_{\mathfrak{p}} = (xA)_{\mathfrak{p}} \subseteq I_{\mathfrak{p}} \subseteq (y^{-1}A)_{\mathfrak{p}} = A_{\mathfrak{p}}$. When $f \in K \setminus \{0\}$ and $I = fA$, we denote $\mathrm{div}(I)$ simply by $\mathrm{div}(f)$.
(5) For $D \in \mathrm{Div}(A)$, we define $A(D) = \{f \in K \setminus \{0\} \mid \mathrm{div}(f) + D \geq 0\} \cup \{0\}$. If $D = \sum_{\mathfrak{p}\in\mathrm{Ht}_1(A)} n_{\mathfrak{p}}[\mathfrak{p}]$, then $A(D) = \bigcap_{\mathfrak{p}\in\mathrm{Ht}_1(A)} \mathfrak{p}^{-n_{\mathfrak{p}}} A_{\mathfrak{p}}$ by definition.

Proposition 3.57 *Let I, I' be nonzero fractional ideals of A.*

(1) *For any $\mathfrak{p} \in \mathrm{Ht}_1(A)$, we have $I_{\mathfrak{p}} = \pi^{v_{\mathfrak{p}}(I)} A_{\mathfrak{p}}$, where π is any prime element of the DVR $A_{\mathfrak{p}}$. In particular, if $I \subseteq I'$, then* $\mathrm{div}(I) \geq \mathrm{div}(I')$.
(2) $\mathrm{div}(II') = \mathrm{div}(I) + \mathrm{div}(I')$.
(3) *If $D = \sum_{\mathfrak{p}\in\mathrm{Ht}_1(A)} n_{\mathfrak{p}}[\mathfrak{p}] \in \mathrm{Div}(A)$, then* $\mathrm{div}(A(D)) = -D$.
(4) $I^{-1} = A(\mathrm{div}\, I) = \bigcap_{a\in I\setminus\{0\}} a^{-1}A$. *In particular, we have* $\mathrm{div}(I^{-1}) = -\mathrm{div}\, I$ *and* $\mathrm{div}(\widetilde{I}) = \mathrm{div}(I)$.
(5) *If I' is divisorial, then $I \subseteq I'$ if and only if* $\mathrm{div}(I) \geq \mathrm{div}(I')$.

Proof

(1) This is easy.
(2) follows from (1).
(3) Let $\mathfrak{p} \in \mathrm{Ht}_1(A)$. If $\alpha \in A(D) \subseteq \mathfrak{p}^{-n_\mathfrak{p}} A_\mathfrak{p}$, then $v_\mathfrak{p}(\alpha) \geq -n_\mathfrak{p}$. So $v_\mathfrak{p}(I) \geq -n_\mathfrak{p}$. It follows that $\operatorname{div}(A(D)) \geq -D$. To prove the equality, it suffices to prove that for each $\mathfrak{p}$, there exists some $\alpha \in A(D)$ such that $v_\mathfrak{p}(\alpha) = -n_\mathfrak{p}$. Take $a \in v_\mathfrak{p}^{-1}(-n_\mathfrak{p})$. Take $\Gamma = \{\mathfrak{q} \in \mathrm{Ht}_1(A) \mid v_\mathfrak{q}(a) + n_\mathfrak{q} < 0\}$. Then Γ is finite and $\mathfrak{p} \notin \Gamma$. Take $b \in (\bigcap_{\mathfrak{q}\in\Gamma} \mathfrak{q}) \setminus \mathfrak{p}$, and set $r = -\min\{v_\mathfrak{q}(a) + n_\mathfrak{q} \mid \mathfrak{q} \in \Gamma\}$. Then $v_\mathfrak{q}(ab^r) + n_\mathfrak{q} \geq 0$ for any $\mathfrak{q} \in \mathrm{Ht}_1(A)$, and we have $ab^r \in A(D)$. Since $v_\mathfrak{p}(ab^r) = v_\mathfrak{p}(a) = -n_\mathfrak{p}$, we have proved that $\operatorname{div}(A(D)) = -D$.
(4) We prove $I^{-1} = A(\operatorname{div} I)$. For each $\mathfrak{p} \in \mathrm{Ht}_1(A)$, we have

$$A(\operatorname{div} I) \cdot I \subseteq (\mathfrak{p}^{-v_\mathfrak{p}(I)} A_\mathfrak{p})(v_\mathfrak{p}(I) A_\mathfrak{p}) \subseteq A_\mathfrak{p}.$$

So $A(\operatorname{div} I)I \subseteq \bigcap_{\mathfrak{p}\in\mathrm{Ht}_1(A)} A_\mathfrak{p} = A$, and $A(\operatorname{div} I) \subseteq I^{-1}$. Let $x \in K \setminus A(\operatorname{div} I)$. Then there exists some $\mathfrak{p} \in \mathrm{Ht}_1(A)$ such that $v_\mathfrak{p}(x) < -v_\mathfrak{p}(I)$. When we take $y \in I$ such that $v_\mathfrak{p}(y) = v_\mathfrak{p}(I)$, then $v_\mathfrak{p}(xy) < 0$, and hence $xy \notin A$. So $x \notin I^{-1}$, and this shows $I^{-1} \subseteq A(\operatorname{div} I)$. Hence we have $I^{-1} = A(\operatorname{div} I)$. So $\operatorname{div} I^{-1} = -\operatorname{div} I$ by (3). Hence $\operatorname{div} \widetilde{I} = \operatorname{div}(I^{-1})^{-1} = -\operatorname{div} I^{-1} = \operatorname{div} I$. Finally, $I^{-1} = A :_K I = \bigcap_{a\in I} A :_K aA = \bigcap_{a\in I\setminus\{0\}} a^{-1}A$.
(5) We only need to prove the 'if' part. Let $J = (I')^{-1}$. Then $\operatorname{div} J = -\operatorname{div}(I') \geq -\operatorname{div}(I)$. So $\operatorname{div}(IJ) \geq 0$, and hence $IJ \subseteq \bigcap_{\mathfrak{p}\in\mathrm{Ht}_1(A)} A_\mathfrak{p} = A$. This shows that $I \subseteq J^{-1} = \widetilde{I'} = I'$, as desired.

□

We denote the set of divisorial ideals by $\mathrm{DI}(A)$, and its subset consisting of principal fractional ideals by $\mathrm{PI}(A)$.

Corollary 3.58 *Let $D \in \mathrm{Div}(A)$. Then $A(D) \in \mathrm{DI}(A)$. The map* $\operatorname{div} : \mathrm{DI}(A) \to \mathrm{Div}(A)$ *is an isomorphism of abelian groups with $D \mapsto A(-D)$ its inverse, where* $\mathrm{DI}(A)$ *is considered as a multiplicative abelian group such that the product of I and I' is $\widetilde{II'}$.*

Proof Let $D \in \mathrm{Div}(A)$. By Proposition 3.57, we have $\operatorname{div} A(-D) = -(-D) = D$. Hence $A(D) = A(\operatorname{div} A(-D)) = \widetilde{A(-D)^{-1}}$. So $A(-D) = A(D)^{-1}$. Hence $A(D) = A(\widetilde{-D})^{-1} = (A(D)^{-1})^{-1} = \widetilde{A(D)}$, and $A(D)$ is divisorial. For $I, I' \in \mathrm{DI}(A)$, $\operatorname{div}(\widetilde{II'}) = \operatorname{div}(II') = \operatorname{div}(I) + \operatorname{div}(I')$, and this shows that $\operatorname{div} : \mathrm{DI}(A) \to \mathrm{Div}(A)$ is a homomorphism of abelian groups. Since $\operatorname{div} A(-D) = -(-D) = D$ and $A(-\operatorname{div} I) = A(\operatorname{div} I^{-1}) = (I^{-1})^{-1} = I$ for $I \in \mathrm{DI}(A)$, we have that $\operatorname{div} : \mathrm{DI}(A) \to \mathrm{Div}(A)$ is an isomorphism with $D \mapsto A(-D)$ its inverse. □

Next, we define the divisor class group of A.

Definition 3.59

(1) We define

$$\mathrm{P}(A) = \{\mathrm{div}(f) \mid f \in K,\ f \neq 0\},$$

and call the element of $\mathrm{P}(A)$ a *principal divisor*. $\mathrm{P}(A)$ is a subgroup of $\mathrm{Div}(A)$.

(2) We define

$$\mathrm{Cl}(A) = \mathrm{Div}(A)/\mathrm{P}(A)$$

and call it the *divisor class group* of A. We denote by $\mathrm{cl}(D)$ the image of $D \in \mathrm{Div}(A)$ in $\mathrm{Cl}(A)$.

The following properties of $\mathrm{Cl}(A)$ easily follow from the definition.

Proposition 3.60 *Let $a \in K, a \neq 0$.*

(1) $\mathrm{div}(a) = 0$ *if and only if* $a \in \mathrm{U}(A)$ ($\mathrm{U}(A)$ *denotes the unit group of* A).
(2) *There is an exact sequence*

$$1 \to \mathrm{U}(A) \to \mathrm{U}(K) \xrightarrow{\mathrm{div}} \mathrm{Div}(A) \to \mathrm{Cl}(A) \to 0,$$

where $\mathrm{U}(A)$ *and* $\mathrm{U}(K) = K \setminus \{0\}$ *are groups under the multiplication, and* $\mathrm{Div}(A)$ *and* $\mathrm{Cl}(A)$ *are groups under addition. The map* $\mathrm{U}(A) \to \mathrm{U}(K)$ *is the inclusion, and* $\mathrm{Div}(A) \to \mathrm{Cl}(A)$ *is the canonical surjective map.*
(3) *The subset* $\mathrm{PI}(A)$ *of* $\mathrm{DI}(A)$ *is a subgroup, and the isomorphism* $\mathrm{div} : \mathrm{DI}(A) \to \mathrm{Div}(A)$ *maps* $\mathrm{PI}(A)$ *isomorphically onto* $\mathrm{P}(A)$. *So* div *induces an isomorphism between* $\mathrm{DI}(A)/\mathrm{PI}(A)$ *and* $\mathrm{Cl}(A)$.

The proof is straightforward.
In the next proposition, we do not assume that A is a Krull domain.

Proposition 3.61 *Let A be an integral domain.*

(1) *A is a UFD if and only if A is a Krull domain and* $\mathrm{Cl}(A) = 0$.
(2) *If A is Noetherian, A is a UFD if and only if every prime ideal* $\mathfrak{p}$ *with* $\mathrm{ht}\mathfrak{p} = 1$ *is principal.*

Proof

(1) Let A be a UFD, and P be its prime ideal of height one. Then A_P is a one-dimensional UFD by Lemma 1.63, and hence it is a DVR by Lemma 3.32. If $a \in A \setminus \{0\}$ and $a = up_1 \cdots p_r$ with $u \in \mathrm{U}(A)$, $r \geq 0$, and each p_i a prime element, then $a \notin P$ for any height-one prime P such that $P \notin \{(p_1), (p_2), \ldots, (p_r)\}$. Hence A is Krull. Since each P is principal, we have that $\mathrm{Div}(A) = \mathrm{P}(A)$, and we have $\mathrm{Cl}(A) = 0$.

Conversely, for any $a \in A \setminus \{0\}$, put $\mathrm{div}(a) = \sum n_{\mathfrak{p}}[\mathfrak{p}]$ and we will show that a is either a unit or a product of prime elements by induction on $n = \sum_{\mathfrak{p}} n_{\mathfrak{p}}$. If

$n = 0$, then $\operatorname{div} a = 0$, and a is a unit. Assume that $n > 0$, and the assertion is true for smaller n. Then $n_{\mathfrak{p}} > 0$ for some $\mathfrak{p} \in \mathrm{Ht}_1(A)$. Since $\mathrm{Cl}(A) = 0$, $\mathfrak{p} = (p)$ is principal. Since $a \in \mathfrak{p}$, we have $a = pb$ for some $b \in A$. Then by the induction hypothesis, b is a product of prime elements or a unit, and so a is a product of prime elements.

(2) It suffices to show the 'if' part. Assume that A is not a UFD. Then there exists some $a \in A \setminus (\mathrm{U}(A) \cup \{0\})$ such that a is not a product of prime elements. We take such an element a so that (a) is maximal among such. When we take $P \in \mathrm{Min}_A A/aA$, then $\operatorname{ht} P = 1$. By assumption, we can write $\mathfrak{p} = (p)$ and so $a = a_1 p$ for some $a_1 \neq 0$. If $a_1 \in \mathrm{U}(A)$, then a is a prime element, and this contradicts the choice of a. So $a_1 \in A \setminus (\mathrm{U}(A) \cup \{0\})$. Then $(a_1) \supsetneq (a)$, and a_1 is a product of prime elements by the maximality of (a). Then $a = a_1 p$ is also a product of prime elements, and this is also a contradiction.

□

We have shown the following theorem in Chap. 1 (Theorem 1.61).

Theorem 3.62 *A is a UFD if and only if $A[X]$ is a UFD.*

Remark 3.2 If we change $A[X]$ to $A[[X]]$, the same result is not true. For example, let k be a field and $a, b, c \geq 2$ be positive integers relatively coprime to each other. Then if we put $A = k[X, Y, Z]/(X^a + Y^b + Z^c)$ then A is a UFD but if $1/a + 1/b + 1/c \leq 1$, $A[[X]]$ is not a UFD (cf. [Sam] or [Fo]).

The following theorem, which is due to Masayoshi Nagata, is also useful to compute divisor class groups. For $\mathfrak{p} \in \mathrm{Ht}_1(A)$, we denote the class of $[\mathfrak{p}]$ in $\mathrm{Cl}(A)$ by $\mathrm{cl}(\mathfrak{p})$.

Theorem 3.63 *Let $S \subseteq A$ be a multiplicatively closed set. Then we have*

$$\mathrm{Cl}(S^{-1}A) \cong \mathrm{Cl}(A)/H,$$

where H is the subgroup of $\mathrm{Cl}(A)$ generated by $\{\mathrm{cl}(\mathfrak{p}) \mid \mathfrak{p} \in \mathrm{Ht}_1(A), \mathfrak{p} \cap S \neq \emptyset\}$.

Hence if A is a UFD, so is $S^{-1}A$. Also, if S is generated by prime elements and if $S^{-1}A$ is a UFD, then so is A.

Proof It is clear that $\mathrm{Ht}_1(S^{-1}A) = \{\mathfrak{p} \in \mathrm{Ht}_1(A) \mid \mathfrak{p} \cap S = \emptyset\}$. Hence it is natural to regard $\mathrm{Div}(S^{-1}A)$ as a factor group of $\mathrm{Div}(A)$ and we define $\phi : \mathrm{Div}(A) \to \mathrm{Div}(S^{-1}A)$ by

$$\phi([\mathfrak{p}]) = \begin{cases} [\mathfrak{p}(S^{-1}A)] & (\mathfrak{p} \cap S = \emptyset) \\ 0 & (\text{otherwise}). \end{cases}$$

Then $\mathrm{P}(S^{-1}A)$ is also a factor group of $\mathrm{P}(A)$ and we get the commutative diagram

$$\begin{array}{ccccccccc}
 & & 0 & & 0 & & 0 & & \\
 & & \downarrow & & \downarrow & & \downarrow & & \\
0 & \longrightarrow & \mathrm{Div}(S, A) \cap \mathrm{P}(A) & \longrightarrow & \mathrm{P}(A) & \longrightarrow & \mathrm{P}(S^{-1}A) & \longrightarrow & 0 \\
 & & \downarrow & & \downarrow & & \downarrow & & \\
0 & \longrightarrow & \mathrm{Div}(S, A) & \longrightarrow & \mathrm{Div}(A) & \xrightarrow{\phi} & \mathrm{Div}(S^{-1}A) & \longrightarrow & 0 \\
 & & \downarrow & & \downarrow & & \downarrow & & \\
0 & \longrightarrow & H' & \longrightarrow & \mathrm{Cl}(A) & \longrightarrow & \mathrm{Cl}(S^{-1}A) & \longrightarrow & 0 \\
 & & \downarrow & & \downarrow & & \downarrow & & \\
 & & 0 & & 0 & & 0 & &
\end{array}$$

with exact rows, where $\mathrm{Div}(S, A)$ is the free $\mathbb{Z}$-submodule of $\mathrm{Div}(A)$ with the basis $\{[\mathfrak{p}] \mid \mathfrak{p} \in \mathrm{Ht}_1(A),\ \ \mathfrak{p} \cap S \neq \emptyset\}$, and H' is the kernel of the map $\overline{\phi} : \mathrm{Cl}(A) \to \mathrm{Cl}(S^{-1}A)$ induced by ϕ. The second and the third columns are also exact, and hence so is the first by the snake lemma. Hence $H' = H$, and the assertion follows. □

Let us compute some examples using this theorem.

Example 3.64 (**[Sam, Fo]**)

(1) Assume that k is a field and $A = k[X, Y, Z]/(X^2 - YZ)$. Then we have $\mathrm{Cl}(A) \cong \mathbb{Z}/2\mathbb{Z}$.
(2) $B = \mathbb{R}[X, Y, Z]/(X^2 + Y^2 + Z^2)$ is a UFD ($\mathbb{R}$ is the field of real numbers).

Proof In the following, we write x, y, z the image of X, Y, Z, respectively, in A or B.

(1) Since $A_x \cong k[X, X^{-1}, Y, Z]/(X^2 - YZ)$ and $yz = x^2$ Cy, z are units of A_x since so is x^2. Thus we have $A_x \cong k[X, X^{-1}, Y, Y^{-1}]$. Being a localization of the polynomial ring $k[X, Y]$, it is a UFD. Hence by Theorem 3.63, $\mathrm{Cl}(A) \cong H$, where H is the subgroup of $\mathrm{Cl}(A)$ generated by $\{[\mathfrak{p}] \mid x \in \mathfrak{p} \in \mathrm{Ht}_1(A)\}$. The prime ideals of A of height one containing x are $\mathfrak{p}_1 = (x, y)$ and $\mathfrak{p}_2 = (x, z)$. Since $\mathfrak{p}_1^2 = (x^2, xy, y^2) = (yz, y^2, xy) = y(x, y, z)$, we have $\widetilde{\mathfrak{p}_1^2} = (y)$, which implies $2\,\mathrm{cl}(\mathfrak{p}_1) = 0$. Also, since $\mathfrak{p}_1\mathfrak{p}_2 = (x^2, xy, xz, yz) = x(x, y, z)$, we have $\mathrm{cl}(\mathfrak{p}_1) + \mathrm{cl}(\mathfrak{p}_2) = 0$ and we get $\mathrm{Cl}(A) \cong H \cong \mathbb{Z}/2\mathbb{Z}$. (It is straightforward to check that $\mathfrak{p}_1$ is not a principal ideal, and $\mathrm{Cl}(A)$ is nontrivial.)
(2) Since $B/xB = \mathbb{R}[y, z]/(y^2 + z^2)$ is an integral domain, we have $\mathrm{Cl}(B) \cong \mathrm{Cl}(B_x)$ by Theorem 3.63. So it suffices to show that

$$B_x = (\mathbb{R}[X, X^{-1}])[Y, Z]/(Y^2 + Z^2 + X^2) \cong (\mathbb{R}[\eta, \zeta]/(\eta^2 + \zeta^2 + 1))[X, X^{-1}]$$

is a UFD, where $\eta = y/x$ and $\zeta = z/x$. So it suffices to show that the integral domain $C := \mathbb{R}[\eta, \zeta]/(\eta^2+\zeta^2+1)$ is a UFD. If $(a, b) \in \mathbb{C}^2$ and $a^2+b^2+1 = 0$, then there exists some $(s, t, u) \in \mathbb{R}^3$ such that $sa + tb + u = 0$ and $(s, t) \neq (0, 0)$. Indeed, since $(a, b) \notin \mathbb{R}^2$, we have that either $a \notin \mathbb{R}$ or $b \notin \mathbb{R}$. If $b \notin \mathbb{R}$, then we can write $a = -tb - u$, since $1, b$ is an $\mathbb{R}$-basis of $\mathbb{C}$. If $a \notin \mathbb{R}$, then we can write $b = -sa - u$. Note that $(s\eta + t\zeta + u) \subseteq C$ is a maximal ideal with $C/(s\eta + t\zeta + u) \cong \mathbb{C}$ for $(s, t, u) \in \mathbb{R}^3$ with $(s, t) \neq (0, 0)$. For any height one prime ideal (or equivalently, a maximal ideal) P of C, we can take a maximal ideal $Q \in \operatorname{Spec} \mathbb{C}[\eta, \zeta]/(\eta^2 + \zeta^2 + 1)$ lying over P, and we can write $Q = (\eta - a, \zeta - b)$ for some $(a, b) \in \mathbb{C}^2$ with $a^2 + b^2 + 1 = 0$ by the Nullstellensatz, Theorem 3.27.

Then as we have shown, there exists some $(s, t, u) \in \mathbb{R}^3$ such that $s\eta + t\zeta + u \in Q \cap C = P$. Since $(s\eta + t\zeta + u)$ is maximal, we have that $P = (s\eta + t\zeta + u)$ is principal, and hence C is a UFD. Thus B is a UFD. This fact also follows from a result of Salmon [Salm]. □

The unique factorization property is very delicate and many properties that usually hold are not true. For example, the property is not preserved by extensions or contractions of the base fields.

Example 3.65 Let $\mathbb{R}$ and $\mathbb{C}$ be the fields of real numbers and complex numbers, respectively.

(1) $A = \mathbb{C}[X, Y]/(X^2 + Y^2 - 1)$ and $A' = \mathbb{R}[X, Y]/(X^2 + Y^2 + 1)$ are UFDs, but $A'' = \mathbb{R}[X, Y]/(X^2 + Y^2 - 1)$ is not.
(2) $B' = \mathbb{R}[X, Y, Z]/(X^2 + Y^2 + Z^2 - 1)$ is a UFD, but $B = \mathbb{C}[X, Y, Z]/(X^2 + Y^2 + Z^2 - 1)$ is not.

Proof

(1) Over $\mathbb{C}$, $X^2 + Y^2$ factors as $X^2 + Y^2 = (X + iY)(X - iY)$. If we change by the coordinates by $X' = X + iY$ and $Y' = X - iY$, then since $A \cong \mathbb{C}[X', Y']/(X'Y' - 1) \cong \mathbb{C}[X', X'^{-1}]$, A is a UFD.

    In A'', put $t = (y - 1)^{-1}$, then from $x^2 = (1 + t^{-1})^2 - 1 = 2t^{-1} + t^{-2}$ we get $2t = (tx)^2 - 1$. Hence we have $t \in \mathbb{R}[tx]$, and it follows that $A''[t] = \mathbb{R}[tx, t^{-1}]$. Being the localization of the polynomial ring $\mathbb{R}[tx]$, we have that $A''[t]$ is a UFD. The only prime ideal of A'' containing $t^{-1} = y - 1$ is $\mathfrak{p} = (x, y - 1)$ and $\mathfrak{p}$ is not principal (the ideal $(X, Y - 1)$ of $\mathbb{R}[X, Y]$ is not generated by $(X^2 + Y^2 - 1)$ and one more element). Thus we know that A'' is not a UFD. We have $\mathfrak{p}^2 = (y - 1)$, which shows that $\operatorname{Cl}(A'') \cong \mathbb{Z}/(2)$. We have already proved that A' is a UFD in the proof of Example 3.64.
(2) Like the latter half of (1), putting $t = (z - 1)^{-1}$ in B', we have $B'[t] = \mathbb{R}[tx, ty, t^{-1}]$. Being a localization of the polynomial ring $\mathbb{R}[tx, ty]$, the right-hand side is a UFD. Since $B'/(z - 1) \cong \mathbb{R}[X, Y]/(X^2 + Y^2)$ is a domain, $z - 1$ a prime element of B'. Hence B' is a UFD.

Putting $X' = X + iY$, $Y' = X - iY$ in B, we have $B \cong \mathbb{C}[X', Y', Z]/(X'Y' + Z^2 - 1)$. Putting $t = (z-1)^{-1}$ again, we have $B[t] = \mathbb{C}[tx', ty', t^{-1}]$, which shows that $B[t]$ is a UFD. The prime ideals of height 1 containing $z - 1$ are $\mathfrak{p}_1 = (x', z-1)$ and $\mathfrak{p}_2 = (y', z-1)$. Hence by Theorem 3.63, $\mathrm{Cl}(B)$ is generated by $\mathrm{cl}(\mathfrak{p}_1), \mathrm{cl}(\mathfrak{p}_2)$. Since $\widetilde{\mathfrak{p}_1\mathfrak{p}_2} = (z-1)$, $\mathrm{cl}(\mathfrak{p}_1) + \mathrm{cl}(\mathfrak{p}_2) = 0$ and $\widetilde{\mathfrak{p}_1^n}$ is not principal for all n, the order of $\mathrm{cl}(\mathfrak{p}_1)$ is infinite and we have $\mathrm{Cl}(B) \cong \mathbb{Z}$.

□

Chapter 4
Completions and Properties of Complete Local Rings

In this chapter, we introduce "topology" and "completion" on a ring A using a filtration on A. Most important is the "$\mathfrak{a}$-adic topology" defined by the filtration $\{\mathfrak{a}^n\}_{n\geq 1}$, where $\mathfrak{a}$ is an ideal of A. Also the $\mathfrak{a}$-adic topology on an A-module M defined by $\{\mathfrak{a}^n M\}_{n\geq 1}$ is very important.

Given a topology on A or M, the notion of "completion" is defined by using Cauchy sequences with respect to the topology. Especially, for a Noetherian local ring $(A, \mathfrak{m})$, the $\mathfrak{m}$-adic completion $\widehat{A}$ of A is very important in the theory of commutative rings, because complete local rings have very nice properties and we can simplify many arguments considering $\widehat{A}$ instead of A.

In Sect. 4.1, we explain generality of completion. The Artin–Rees lemma is very important and the concepts of Rees algebra $R_A(\mathfrak{a})$ and the associated graded ring $\mathrm{G}_{\mathfrak{a}}(A)$ will be very important in many aspects of commutative algebra. Also, the formal power series rings appear as the completion of polynomial rings.

In Sect. 4.2, we study Henselian pairs and their relationship to the lifting of idempotents. We prove Hensel's lemma, which asserts that a complete local ring $(A, \mathfrak{m})$ is Henselian. As a consequence, we prove that if A is a complete local ring and $A \subseteq B$ is a finite ring extension, then B is a direct product of finitely many complete local rings.

In Sect. 4.3, we discuss the structure theorem for complete local rings. As a result, we can assert that *a Noetherian complete local ring is a homomorphic image of a complete regular local ring.*

4.1 $\mathfrak{a}$-Adic Completions and Formal Power Series Rings

To define a topology on a ring and construct a theory through completion is a very important technique in commutative algebra.

S. Goto, K. Watanabe, *Commutative Algebra*, Graduate Texts in Mathematics 309,
https://doi.org/10.1007/978-981-95-4227-7_4

There are several ways to define a topology on a ring. In this section, we restrict ourselves to the typical one used in commutative algebra.

Definition 4.1 Let A be a ring and M be an A-module.

(1) A *filtration* of A is a family $\{I_n\}_{n\geq 0}$ of ideals of A satisfying the following three conditions:

  (a) $I_0 = A$;
  (b) $I_n \supseteq I_{n+1}, \forall n \geq 0$;
  (c) $I_m I_n \subseteq I_{m+n}, \forall m, n \geq 0$.

(2) Also, a *filtration on M* is a family $\{M_n\}_{n\geq 0}$ of A-submodules of M satisfying the property $M = M_0$, $M_n \supseteq M_{n+1}, \forall n \geq 0$.

Definition 4.2 Given a filtration $\{M_n\}_{n\geq 0}$ on M, we can define a *topology* on M letting $\{x + M_n\}_{n\geq 0}$ as a system of fundamental neighborhoods of each $x \in M$. This topology makes M a topological additive group. We say that a sequence $\{x_n\}_{n\geq 1}$ of elements of M forms a *Cauchy sequence* if for any integer $k \geq 1$, there exists some $N(k)$ such that for any $m, n \geq N(k)$, we have $x_m - x_n \in M_k$.

We say that a sequence $\{x_n\}_{n\geq 1}$ of elements of M *converges* to $x \in M$ if for every k, there exists $N(k)$ such that for all $n \geq N(k)$, $x - x_n \in M_k$. A convergent sequence is a Cauchy sequence.

The following $\mathfrak{a}$-adic filtration is the most typical one in our theory.

Example 4.3 Let A be a ring and $\mathfrak{a}$ be an ideal of A.

(1) For an A-module M, the filtration $\{\mathfrak{a}^n M\}_{n\geq 0}$ is called the *$\mathfrak{a}$-adic filtration* of M. The topology defined by the $\mathfrak{a}$-adic filtration on M is called the *$\mathfrak{a}$-adic topology* on M.
(2) Let N be an A-submodule of M. The filtration $\{N_n\}_{n\geq 0}$ defined by $N_n = \mathfrak{a}^n M \cap N$ is called the *induced filtration on N*. The topology defined by the induced filtration from M is the induced topology on N from M.

The $\mathfrak{a}$-adic filtration $\{\mathfrak{a}^n\}$ of A makes A a topological ring. That is, A is a topological additive group, and the product $A \times A \to A$ is continuous. Moreover, the $\mathfrak{a}$-adic filtration $\{\mathfrak{a}^n M\}$ makes M a topological A-module. That is, M is a topological additive group, and the action $A \times M \to M$ is continuous.

4.4 The filtrations $\{M_n\}_{n\geq 0}$ and $\{M'_n\}_{n\geq 0}$ define the same topology on M if and only if for any $n \geq 0$, there exist positive integers $k = k(n)$ and $l = l(n)$ such that $M'_k \subseteq M_n$ and $M_l \subseteq M'_n$.

The following theorem shows that if A is Noetherian, M is a finitely generated A-module and if N is a submodule of M, the $\mathfrak{a}$-adic topology on N defines the same topology as the induced topology on N from the $\mathfrak{a}$-adic topology on M.

Theorem 4.5 (Artin–Rees Lemma) *Let A be a Noetherian ring, $\mathfrak{a}$ be an ideal of A, M be a finitely generated A-module and N be an A-submodule of M.*

Then there exists an integer $k \geq 0$ such that

$$\mathfrak{a}^n M \cap N = \mathfrak{a}^{n-k}(\mathfrak{a}^k M \cap N)$$

for every $n \geq k$.

To prove this theorem, it is effective to use the notion of *Rees algebras* introduced by D. Rees.[1]

Definition 4.6 Let A be a ring, $\mathfrak{a}$ be an ideal of A and M be an A-module. We define

$$R_A(\mathfrak{a}) = A[\mathfrak{a}T] = \bigoplus_{n\geq 0} \mathfrak{a}^n T^n, \quad \mathrm{R}_M(\mathfrak{a}) = \bigoplus_{n\geq 0} (\mathfrak{a}^n M)T^n$$

and call them the *Rees algebra* of $\mathfrak{a}$ and the *Rees module* of M with respect to $\mathfrak{a}$. Here T is an indeterminate, which distinguishes the degree of an element.

We will study the theory of graded rings and graded modules in Chap. 7. For each $n \geq 0$, we call

$$R_A(\mathfrak{a})_n = \mathfrak{a}^n T^n \quad \text{and} \quad \mathrm{R}_M(\mathfrak{a})_n = \mathfrak{a}^n M T^n$$

the homogeneous components of degree n. For every m and n, we have

$$\mathrm{R}_A(\mathfrak{a})_m \mathrm{R}_M(\mathfrak{a})_n \subseteq \mathrm{R}_M(\mathfrak{a})_{m+n}.$$

Lemma 4.7 *If A is a Noetherian ring, then so is $\mathrm{R}_A(\mathfrak{a})$ for any ideal $\mathfrak{a}$. If M is a finitely generated A-module, then $\mathrm{R}_M(\mathfrak{a})$ is a finitely generated $\mathrm{R}_A(\mathfrak{a})$-module.*

Proof If $a_1, a_2, \ldots, a_r$ generates the ideal $\mathfrak{a}$, then $\mathrm{R}_A(\mathfrak{a})$ is generated by $a_1 T, a_2 T, \ldots, a_r T$ as an A-algebra. Hence $\mathrm{R}_A(\mathfrak{a})$ is a Noetherian ring by Theorem 1.43.

Also, a set of generators of M, considered as elements of $M = \mathfrak{a}^0 M T^0 = \mathrm{R}_M(\mathfrak{a})_0 \subseteq \mathrm{R}_M(\mathfrak{a})$, is also a set of generators of $\mathrm{R}_M(\mathfrak{a})$ as an $\mathrm{R}_A(\mathfrak{a})$-module. Hence if M is a finitely generated A-module, then $\mathrm{R}_M(\mathfrak{a})$ is a finitely generated $\mathrm{R}_A(\mathfrak{a})$-module. □

Proof of Theorem 4.5 Put $L = \bigoplus_{n\geq 0}(N \cap \mathfrak{a}^n M)T^n$. Then L is an $\mathrm{R}_A(\mathfrak{a})$-submodule of $\mathrm{R}_M(\mathfrak{a})$. Since $\mathrm{R}_M(\mathfrak{a})$ is a finitely generated $\mathrm{R}_A(\mathfrak{a})$-module, so is L.

We can take a finite number of generators of L over $\mathrm{R}_A(\mathfrak{a})$ in the form $f_i T^{d_i}$ $(f_i \in N \cap \mathfrak{a}^{d_i} M)$. Then if $n \geq k = \max\{d_i\}$, then comparing the homogeneous components, we have $L_n = [\mathrm{R}_A(\mathfrak{a})]_{n-k} L_k$, which gives the desired result. □

4.8 The topology on M, defined by a filtration $\{M_n\}_{n\geq 0}$ is said to be *separated* if $\bigcap_{n\geq 0} M_n = 0$. If M is separated, then M is a metric space with the distance d given by $d(x, x') = 1/2^n$ for $x \neq x'$, where $n = \max\{i \mid x - x' \in M_i\}$. If so,

[1] David Rees, 1918–2013.

M is Hausdorff. Conversely, if M is Hausdorff, then it is a T_1-space, and $\{0\}$ is a closed subset. So for each $x \in M \setminus \{0\}$, there exists some n such that $0 \notin M_n + x$, or equivalently, $x \notin M_n$. This is equivalent to saying that $\bigcap_{n\geq 0} M_n = 0$. If M is separated, then the concepts of Cauchy and convergent sequences agree with those of a metric space.

Theorem 4.9 *Let A be a Noetherian ring, M a finitely generated A-module, and $\mathfrak{a}$ an ideal of A. If we put*

$$L = \bigcap_{n\geq 0} \mathfrak{a}^n M,$$

there exists $a \in \mathfrak{a}$ such that $(1-a)L = 0$. Hence for $x \in M$, we have

$$x \in \bigcap_{n\geq 0} \mathfrak{a}^n M \iff \exists a \in \mathfrak{a}, (1-a)x = 0.$$

Proof From Theorem 4.5, we can take $k \geq 0$ such that for every $n \geq k$, we have $\mathfrak{a}^n M \cap L = \mathfrak{a}^{n-k}(\mathfrak{a}^k M \cap L)$. Since $\mathfrak{a}^n M \cap L = L$ from the definition of L, we have $L = \mathfrak{a}L$. Using the determinantal trick (Lemma 3.3), we can choose $a \in \mathfrak{a}$ so that $(1-a)L = 0$. By this choice, $\Rightarrow$ of the last equivalence is obvious. Since $x = ax = a^2x = \cdots$ and $a \in \mathfrak{a}$, the direction $\Leftarrow$ also holds. □

Corollary 4.10 *Let A be a Noetherian ring, $\mathfrak{a}$ be an ideal of A and M be a finitely generated A-module. If for every $\mathfrak{p} \in \mathrm{Ass}_A M$ we have $\mathfrak{p} + \mathfrak{a} \neq A$, then $\bigcap_{n\geq 0} \mathfrak{a}^n M = 0$.*

Proof If $x \in \bigcap_{n\geq 0} \mathfrak{a}^n M$, then there exists some $a \in \mathfrak{a}$ such that $(1-a)x = 0$. If $x \neq 0$, then there exists some $\mathfrak{p} \in \mathrm{Ass}_A(M)$ such that $1 - a \in \mathfrak{p}$. Since $1 = (1-a) + a \in \mathfrak{p} + \mathfrak{a}$, we have $\mathfrak{p} + \mathfrak{a} = A$. □

Corollary 4.11 (Krull's Intersection Theorem) *Assume that A is a Noetherian ring, $\mathfrak{a} \neq A$ is an ideal of A and M is a finitely generated A-module. Then we have $\bigcap_{n\geq 0} \mathfrak{a}^n M = 0$ in the following cases:*

(1) *A is a local ring.*
(2) *A is an integral domain and M is a torsion-free A-module.*

Proof

(1) If $(A, \mathfrak{m})$ is local and $\mathfrak{a} \neq A$ and $\mathfrak{p} \in \mathrm{Ass}_A M$, then $\mathfrak{a} + \mathfrak{p} \subseteq \mathfrak{m} \neq A$, and the assertion follows from Corollary 4.10.
(2) Since M is torsion-free, we have $\mathrm{Ass}_A M = \{(0)\}$. The assertion follows from Corollary 4.10.

□

Exercise 4.1 In Corollary 4.10, the Noetherian property of A is necessary. In fact, let R be the integral closure of $\mathbb{Z}$ in complex number field $\mathbb{C}$ (namely, R is the set of all algebraic integers). Then show that $\dim R = 1$ and for every positive integer n and for any prime ideal P of R, we have $P = P^n$.

Definition 4.12 Let M be an A-module and $\{M_n\}_{n\geq 0}$ be a filtration on M. We say that the topology with respect to this filtration is *complete* if it is separated and any Cauchy sequence in M has a limit in M (converges to some element of M).

We define

$$\widehat{M} = \varprojlim M/M_n$$

and we call the map $M \to \widehat{M}$ the *completion* of M with respect to this topology. We also call the A-module $\widehat{M}$ the completion of M.

Proposition 4.13 *The module M is complete if and only if the natural map $M \to \widehat{M}$ is bijective.*

Proof Since $\widehat{M} = \varprojlim M/M_n$ is embedded in $\prod M/M_n$, the kernel of the completion agrees with the kernel of the map $M \to \prod M/M_n$, and it is $\bigcap_n M_n$. So the completion $M \to \widehat{M}$ is injective if and only if the topology of M is separated. Let $(m_n + M_n) \in \varprojlim M/M_n$, where $m_n \in M$ for each $n \geq 0$. That is, $(m_n + M_n) \in \prod_n M/M_n$ with $m_i + M_n = m_n + M_n$ for every integers i, n with $i \geq n \geq 0$. Then the sequence (m_n) is a Cauchy sequence. If $(m_n + M_n)$ is in the image of the completion $M \to \widehat{M}$, then $(m + M_n) = (m_n + M_n)$. Namely, $m - m_n \in M_n$ for each n. This shows that $m = \lim m_n$. Conversely, if $m = \lim m_n$, then for each n, there exists some $k(n)$ such that $m - m_i \in M_n$ for $i \geq k(n)$. Letting $i = \max(k(n), n)$, we have that $m - m_n = m - m_i + m_i - m_n \in M_n$, and we have that the image of m by the completion $M \to \widehat{M}$ is (m_n). So every Cauchy sequence in M has a limit in M if and only if the completion $M \to \widehat{M}$ is surjective. □

The filtration $\{\widehat{M_n}\}_{n\geq 0}$, on $\widehat{M}$ defined by $\widehat{M_n} = \operatorname{Ker}(\widehat{M} = \varprojlim M/M_n \to M/M_n)$ defines a filtration on $\widehat{M}$ as an A-module. Since $M \to M/M_n$ is surjective and factors through $\widehat{M}$, the projection $\widehat{M} \to M/M_n$ is also surjective. It follows that the induced map $\widehat{M}/\widehat{M_n} \to M/M_n$ is an isomorphism for every n. Hence $\widehat{M}$ is complete with respect to the topology defined by the filtration $\{\widehat{M_n}\}_{n\geq 0}$.

Let $\mathfrak{a}$ be an ideal of a ring A and $\widehat{A}$ be the completion of A with respect to the $\mathfrak{a}$-adic topology. Then $\widehat{A}$ is a ring and for an A-module M, the completion $\widehat{M}$ with respect to the $\mathfrak{a}$-adic topology on M is an $\widehat{A}$-module. Indeed, $\widehat{A}$ acts on the $A/\mathfrak{a}^n$-module $M/\mathfrak{a}^n M$ through the projection $\widehat{A} \to A/\mathfrak{a}^n$, and this makes the projective system $\{M/\mathfrak{a}^n M\}$ that of $\widehat{A}$-modules, and thus $\widehat{M} = \varprojlim M/\mathfrak{a}^n M$ is an $\widehat{A}$-module. If M, M' are A-modules and $u : M \to M'$ is a homomorphism of A-modules, then a homomorphism $\widehat{u} : \widehat{M} \to \widehat{M'}$ is induced from u naturally as an induced homomorphism of projective limits.

Lemma 4.14 *Let A be a ring and $\mathfrak{a}$ be its ideal. If A is complete with respect to the $\mathfrak{a}$-adic topology, then $\mathfrak{a} \subseteq \mathrm{J}(A)$.*

Proof Let $x \in \mathfrak{a}$. Then for each $y \in A$, the infinite sum $1 + xy + x^2y^2 + \cdots$ exists in A, since

$$1, 1 + xy, 1 + xy + x^2y^2, \ldots$$

is a Cauchy sequence, and A is complete. Since $(1 - xy)(1 + xy + x^2y^2 + \cdots) = 1$, $1 - xy \in \mathrm{U}(A)$. This shows that $x \in \mathrm{J}(A)$, and thus $\mathfrak{a} \subseteq \mathrm{J}(A)$. □

The ring of formal power series is the completion of polynomial rings and very important in commutative algebra.

4.15 Let $A = R[X]$ be a polynomial ring over a commutative ring R. The completion of A with respect to $\mathfrak{a} = (X)$ is called the *formal power series ring* over R and denoted by $R[[X]]$. Note that $A = \bigoplus_{i \geq 0} A_i$, where $A_i = A \cdot X^i$. For $i \geq 0$, set $B_i = \bigoplus_{0 \leq j \leq i} A_i$. Then $B_i \cong A/\mathfrak{a}^{i+1}$ as an R-module, and the projective system $\{A/\mathfrak{a}^{i+1}\}$ is isomorphic to $\{(B_i), (f_{ij})\}$, where $f_{ij} : B_j \to B_i$ is given by $f_{ij}(a_0, a_1, \ldots, a_j) = (a_0, a_1, \ldots, a_i)$ for $0 \leq i \leq j$. By Exercise B.4, we have that $B = \prod_{i \geq 0} A_i \to R[[X]]$ is an isomorphism of R-modules. The element in $R[[X]]$ corresponding to $(a_i X^i)_{i \geq 0} \in B$ is nothing but the limit of the Cauchy sequence $(a_0, a_0 + a_1X, a_0 + a_1X + a_2X^2, \ldots)$, which we denote by $a_0 + a_1X + a_2X^2 + a_3X^3 + \cdots$, and we call an element of $R[[X]]$ a *formal power series*. So we have that $R[[X]]$ is the ring of formal power series:

$$R[[X]] = \{a_0 + a_1X + a_2X^2 + \cdots \mid (a_0, a_1, a_2, \ldots) \in R^{\mathbb{Z}_{\geq 0}}\}.$$

The addition and the multiplication is given by

$$\begin{aligned} f + g &= (a_0 + b_0) + (a_1 + b_1)X + (a_2 + b_2)X^2 + \cdots \text{ and} \\ fg &= a_0b_0 + (a_1b_0 + a_0b_1)X + (a_2b_0 + a_1b_1 + a_0b_2)X^2 + \cdots \end{aligned}$$

for $f = a_0 + a_1X + a_2X^2 + \cdots$ and $g = b_0 + b_1X + b_2X^2 + \cdots$.

For example, if we put $f_n = 1 + X + \ldots + X^n (n \geq 0)$, the sequence $\{f_n\}_{n \geq 0}$ forms a Cauchy sequence in A. But this sequence does not have a limit in A. In $R[[X]]$, the limit of this Cauchy sequence $\{f_n\}_{n \geq 0}$ is the power series

$$\sum_{n=0}^{\infty} X^n.$$

More generally, the completion of the polynomial ring $R[X_1, X_2, \ldots, X_n]$ with respect to the $(X_1, X_2, \ldots, X_n)$-adic topology is also called the formal power series ring, and is denoted by $R[[X_1, X_2, \ldots, X_n]]$.

Let us study the properties of formal power series rings.

Proposition 4.16 *Let $A = R[[X]]$ be a formal power series ring over R.*

(1) $f = \sum_{n\geq 0} a_n X^n \in \mathrm{U}(A) \Leftrightarrow a_0 \in \mathrm{U}(R)$. *Hence* $X \in \mathrm{J}(A)$.
(2) *For every* $\mathfrak{m} \in \operatorname{Max} R$, *we have* $\mathfrak{m}A + XA \in \operatorname{Max} A$. *By this correspondence, there is a one-to-one correspondence between* $\operatorname{Max} R$ *and* $\operatorname{Max} A$. *Hence if* R *is a local ring, so is* A.

Proof

(1) Let $g = \sum_{n\geq 0} b_n X^n \in A$. We have $fg = 1$ if and only if

$$a_0b_0 = 1,\ a_1b_0 + a_0b_1 = 0,\ a_2b_0 + a_1b_1 + a_0b_2 = 0, \ldots.$$

So if $f \in \mathrm{U}(A)$, then $a_0 \in \mathrm{U}(A)$. Conversely, if $a_0 \in \mathrm{U}(A)$, then defining b_i inductively:

$$b_0 = a_0^{-1},\ b_1 = -a_0^{-1}a_1b_0,\ b_2 = -a_0^{-1}(a_2b_0 + a_1b_1), \ldots,$$

and $fg = 1$ is satisfied, and the equivalence has been established.
So $1 - fX \in \mathrm{U}(A)$ for any $f \in A$. By Exercise 1.15, we have $X \in \mathrm{J}(A)$.
(2) Since $(X) \subseteq \mathfrak{n}$ for any maximal ideal $\mathfrak{n}$ of A by (1), the correspondence

$$\mathfrak{m} \in \operatorname{Max} R \cong \operatorname{Max} A/(X) \mapsto \pi^{-1}(\mathfrak{m}) \in \{\mathfrak{n} \in \operatorname{Max} A \mid \mathfrak{n} \supseteq (X)\} = \operatorname{Max} A$$

is a bijection, where $\pi : A \to A/(X) = R$ is the canonical projection, see Exercise 1.12. Since $\pi^{-1}(\mathfrak{m}) = \mathfrak{m} + (X) = \mathfrak{m}A + (X)$, we are done.

□

If A is Noetherian, the completion of A with respect to the $\mathfrak{a}$-adic topology is Noetherian. Also, if M is a finitely generated A-module, we have $\widehat{M} \cong \widehat{A} \otimes_A M$ and the ring extension $A \to \widehat{A}$ is flat (cf. Theorem 1.115).

Theorem 4.17 *Let* A *be a Noetherian ring,* $\mathfrak{a}$ *be an ideal of* A. *Then the following statements hold for the* $\mathfrak{a}$*-adic topology on* A:

(1) *If* $0 \to N \xrightarrow{u} M \xrightarrow{v} L \to 0$ *is an exact sequence of finitely generated* A*-modules, then the induced sequence*

$$0 \to \widehat{N} \xrightarrow{\widehat{u}} \widehat{M} \xrightarrow{\widehat{v}} \widehat{L} \to 0$$

of $\widehat{A}$*-modules is also exact.*
(2) *For an* A*-module* M, *there is a natural homomorphism* $\widehat{A} \otimes_A M \to \widehat{M}$ *of* $\widehat{A}$*-modules, which is an isomorphism if* M *is finitely generated.*
(3) *The natural map* $A \to \widehat{A}$ *is a flat ring homomorphism.*
(4) *We have* $\widehat{\mathfrak{a}^n} = \mathfrak{a}^n\widehat{A} = \widehat{\mathfrak{a}}^n$ $(n \geq 0)$. *Hence the topology of* $\widehat{A}$ *induced from the* $\mathfrak{a}$*-adic topology on* A *is equal to the* $\widehat{\mathfrak{a}}$*-adic topology on* $\widehat{A}$.

Proof (1) The sequence

$$0 \to N/(\mathfrak{a}^n M \cap N) \to M/\mathfrak{a}^n M \to L/\mathfrak{a}^n L \to 0$$

is exact for any n, and hence we get an exact sequence

$$0 \to \varprojlim N/(\mathfrak{a}^n M \cap N) \to \widehat{M} \to \widehat{L} \to 0$$

by Lemma B.10. By Theorem 4.5, the $\mathfrak{a}$-adic topology on N is equal to the topology induced from the $\mathfrak{a}$-adic topology on M. Since the completion is determined only by the topology, (1) is proved.

(2), (3) The map is obtained as the composite $\widehat{A} \otimes_A M \to \widehat{A} \otimes_A \widehat{M} \to \widehat{M}$, where the second map is the action of $\widehat{A}$. Assume that M is finitely generated, and take a presentation

$$A^m \to A^n \to M \to 0$$

of M. Then we have the following commutative diagram with exact rows:

$$\begin{array}{ccccccc}
\widehat{A} \otimes_A A^m & \longrightarrow & \widehat{A} \otimes_A A^n & \longrightarrow & \widehat{A} \otimes_A M & \longrightarrow & 0 \\
\downarrow & & \downarrow & & \downarrow & & \\
\widehat{A^m} & \longrightarrow & \widehat{A^n} & \longrightarrow & \widehat{M} & \longrightarrow & 0\,.
\end{array}$$

Since the morphism $\widehat{A} \otimes_A A \to \widehat{A}$ is isomorphic, for every $n \geq 1$, we can also show that $\widehat{A} \otimes_A A^n \to \widehat{A^n}$ is an isomorphism for all $n \geq 1$. Hence $\widehat{A} \otimes_A M \to \widehat{M}$ is an isomorphism by the five lemma. From (1), $A \to \widehat{A}$ is flat, since $\widehat{A} \otimes_A I \to \widehat{A}$ is injective for any ideal I of A.

(4) From (2) and (3), we have $\mathfrak{a}^n \widehat{A} = \widehat{A} \otimes_A \mathfrak{a}^n = \widehat{\mathfrak{a}^n}$. Since $\mathfrak{a}\widehat{A} = \widehat{\mathfrak{a}}$, we have $\widehat{\mathfrak{a}}^n = \mathfrak{a}^n \widehat{A} = \widehat{\mathfrak{a}^n}$. □

Corollary 4.18 *If A is a Noetherian local ring with maximal ideal $\mathfrak{m}$, the completion $\widehat{A}$ of A with respect to the $\mathfrak{m}$-adic filtration is a faithfully flat A-module (cf. Definition 5.76).*

Theorem 4.19 *Let A be a Noetherian ring and $\mathfrak{a}$ be an ideal of A. Then the completion $\widehat{A}$ of A with respect to the $\mathfrak{a}$-adic topology is also a Noetherian ring.*

4.20 Before proving the theorem, let us take a detour. Let $\mathfrak{a}$ be an ideal of a ring A. Also, let us assume that a filtration $\{M_n\}_{n \geq 0}$ on an A-module M satisfies the condition

$$\mathfrak{a} M_n \subseteq M_{n+1} \qquad (\forall n \geq 0).$$

Then we define

$$G_{\mathfrak{a}}(A) = \bigoplus_{n \geq 0} \mathfrak{a}^n/\mathfrak{a}^{n+1}, \quad G(M) = \bigoplus_{n \geq 0} M_n/M_{n+1}$$

and call them the *associated graded ring* of A with respect to $\mathfrak{a}$ and the *associated graded module* of M with respect to $\mathfrak{a}$.

$G(M)$ is a graded module over a graded ring $G_{\mathfrak{a}}(A)$ (cf. Definition 7.1). The homogeneous components of $G_{\mathfrak{a}}(A)$ and $G(M)$ are $\mathfrak{a}^n/\mathfrak{a}^{n+1}$ and M_n/M_{n+1}, respectively. We denote $G(M)$ by $G_{\mathfrak{a}}(M)$, if the filtration of M is $\mathfrak{a}$-adic, that is, $M_n = \mathfrak{a}^n M$.

For each $n \geq 0$, we define $\phi_n : \mathfrak{a}^n/\mathfrak{a}^{n+1} \otimes_{A/\mathfrak{a}} M/\mathfrak{a}M \to \mathfrak{a}^n M/\mathfrak{a}^{n+1}M$ by $\phi_n((a + \mathfrak{a}^{n+1}) \otimes (x + \mathfrak{a}M)) = ax + \mathfrak{a}^{n+1}M$. We denote the sum of ϕ_n by

$$\phi : G_{\mathfrak{a}}(A) \otimes_{A/\mathfrak{a}} M/\mathfrak{a}M \to G_{\mathfrak{a}}(M).$$

Since each ϕ_n is surjective, the map ϕ is surjective. Note also that ϕ is the action of $G_{\mathfrak{a}}(A)$ on $G_{\mathfrak{a}}(M)$. Namely, $\phi(\overline{f} \otimes \overline{m}) = \overline{fm}$ for $\overline{f} \in G_{\mathfrak{a}}(A)$ and $\overline{m} \in M/\mathfrak{a}M$. In other words, the $G_{\mathfrak{a}}(A)$-module $G_{\mathfrak{a}}(M)$ is generated by $M/\mathfrak{a}M$.

The following proposition is the key of our proof.

Proposition 4.21 *Let* $\mathfrak{a}$ *be an ideal of a ring* A *and* $\{M_n\}_{n \geq 1}$ *a filtration on* M *satisfying the condition*

$$\mathfrak{a}M_n \subseteq M_{n+1} \qquad (\forall n \geq 0).$$

Assume that A *is* $\mathfrak{a}$*-adically complete and* M *is separated with respect to the topology defined by this filtration. If* $G(M)$ *is a finitely generated* $G_{\mathfrak{a}}(A)$*-module, then* M *is a finitely generated* A*-module.*

Proof Since $G(M)$ is a finitely generated $G_{\mathfrak{a}}(A)$-module, we can choose $\{y_i \in M_{d_i}\}_{1 \leq i \leq r}$ ($d_i \geq 0$), so that the images $f_i = \overline{y_i} \in M_{d_i}/M_{d_i+1} = [G(M)]_{d_i}$ ($1 \leq i \leq r$) generate $G(M)$ as a $G_{\mathfrak{a}}(A)$-module. Then we can show that M is generated by $\{y_i\}_{1 \leq i \leq r}$.

Actually, suppose that there exists $x \in M \setminus \sum_{i=1}^r Ay_i$. Put $x_0 = x$. Since $x_0 \neq 0$, there exists some $k_0 \geq 0$ such that $x_0 \in M_{k_0} \setminus M_{k_0+1}$. Then $\overline{x_0} \in [G(M)]_{k_0}$ is expressed as

$$\overline{x_0} = \sum_{i=1}^{r} \alpha_{0i} f_i$$

with $\alpha_{0i} \in G_{\mathfrak{a}}(A)$. Comparing the components in $[G_{\mathfrak{a}}(A)]_{k_0}$, we may assume that $\alpha_{0i} \in [G_{\mathfrak{a}}(A)]_{k_0-d_i}$ for each i. So we may assume that $\alpha_{0i} = \overline{a}_{0i}$ for some $a_{0i} \in \mathfrak{a}^{k_0-d_i}$, where $\overline{a_{i0}}$ is the image of a_{0i} in $\mathfrak{a}^{k_0-d_i}/\mathfrak{a}^{k_0-d_i+1} = [G_{\mathfrak{a}}(A)]_{k_0-d_i}$. Put $x_1 := x_0 - \sum_{i=1}^r a_{0i} y_i \in M_{k_0+1}$. Then $x_1 \notin \sum_{i=1}^r Ay_i$. Repeating the same process for x_1, we can construct a sequence of elements $\{x_i\}_{i \geq 0} \in M$, sequence of positive integers

$\{k_n\}_{n\geq 0}$ and elements $\{a_{ni}\}_{n\geq 0, 1\leq i\leq r}$ of A, satisfying the following four conditions for every $n \geq 0$:

(1) $k_n < k_{n+1}$;
(2) $a_{ni} \in \mathfrak{a}^{k_n - d_i}$;
(3) $x_n \in M_{k_n}$;
(4) $x_n - x_{n+1} = \sum_{i=1}^r a_{ni} y_i$.

Then we have $x = x_{n+1} + \sum_{i=1}^r (\sum_{m=0}^n a_{mi}) y_i$ for every $\forall n \geq 0$. Since A is complete, the unique infinite sum $a_i = \sum_{n=0}^\infty a_{ni}$ exists in A for every i. Note that $\{x_n\}$ converges to 0. Since M is separated, we can take the unique limit, and we have

$$x = \lim_{n\to\infty} x_{n+1} + \sum_{i=1}^r \left(\sum_{m=0}^\infty a_{mi}\right) y_i = \sum_{i=1}^r a_i y_i.$$

Since $x \notin \sum_{i=1}^r Ay_i$, this is a contradiction. □

Corollary 4.22 *Let A be a ring and $\mathfrak{a}$ be an ideal of A. Let M be an A-module and $\{M_n\}_{n\geq 0}$ be a filtration on M satisfying $\mathfrak{a}M_n \subseteq M_{n+1}$, $\forall n \geq 0$. Assume that A is complete with respect to the $\mathfrak{a}$-adic topology and M is separated with respect to this filtration. If $\mathrm{G}(M)$ is a Noetherian $\mathrm{G}_\mathfrak{a}(A)$-module, then M is a Noetherian A-module.*

Proof Let N be an A-submodule of M. Define a filtration $\{N_n\}_{n\geq 0}$ on N by $N_n = N \cap M_n$, $\mathrm{G}(N)$ is a $\mathrm{G}_\mathfrak{a}(A)$-submodule of $\mathrm{G}(M)$ and hence is finitely generated. Since the topology of N is induced from M, N is also separated. Then by Proposition 4.21, N is a finitely generated A-module. □

Proof of Theorem 4.17 By Theorem 4.17, (4), the topology of $\widehat{A}$ is $\mathfrak{a}\widehat{A}$-adic topology. On the other hand, by Proposition 4.13, we have $\mathrm{G}_\mathfrak{a}(A) = \mathrm{G}_{\mathfrak{a}\widehat{A}}(\widehat{A})$. Since $\mathrm{G}_\mathfrak{a}(A)$ is a Noetherian ring, so is $\mathrm{G}_{\mathfrak{a}\widehat{A}}(\widehat{A})$. Hence $\widehat{A}$ is a Noetherian ring by Corollary 4.22. □

Corollary 4.23 *If R is a Noetherian ring, then so is the formal power series ring $R[[X_1, X_2, \ldots, X_n]]$, and the natural injection*

$$R[X_1, X_2, \ldots, X_n] \to R[[X_1, X_2, \ldots, X_n]]$$

is flat.

The following proposition is very useful.

Lemma 4.24 *Let A be a ring, $\mathfrak{a}$ be an ideal of A, and M be an A-module. Assume that A is complete with respect to the $\mathfrak{a}$-adic topology, and M is separated with respect to the $\mathfrak{a}$-adic topology. If $y_1, \ldots, y_n \in M$ and their images $\overline{y}_1, \ldots, \overline{y}_n \in M/\mathfrak{a}M$ generate $M/\mathfrak{a}M$ as an $A/\mathfrak{a}$-module, then $y_1, \ldots, y_n$ generate M as an A-module.*

Proof Note that $G_{\mathfrak{a}}(M)$ is generated by $G_{\mathfrak{a}}(M)_0 = M/\mathfrak{a}M$ as a $G_{\mathfrak{a}}(A)$-module. Indeed, the action $\phi : G_{\mathfrak{a}}(A) \otimes_{A/\mathfrak{a}} M/\mathfrak{a}M \to G_{\mathfrak{a}}(M)$ is surjective. So by Proposition 4.21, M is finitely generated as an A-module. Since $\mathfrak{a} \subseteq J(A)$ by Lemma 4.14, the assertion follows from NAK. □

Let $(A, \mathfrak{m})$ and $(B, \mathfrak{n})$ be local rings, and $f : A \to B$ be a homomorphism of rings. We say that f is *local* if $f(\mathfrak{m}) \subseteq \mathfrak{n}$. This is equivalent to saying that $f^{-1}(\mathfrak{n}) = \mathfrak{m}$.

Proposition 4.25 *Let $f : (S, \mathfrak{n}) \to (A, \mathfrak{m})$ be a local homomorphism of Noetherian local rings and assume that S is complete with respect to $\mathfrak{n}$-adic topology. If $A/\mathfrak{n}A$ is a finitely generated $S/\mathfrak{n}$-module, then A is a finitely generated S-module. In fact, if we take $a_1, a_2, \ldots, a_n \in A$ so that the images of $a_1, a_2, \ldots, a_n$ in $A/\mathfrak{n}A$ generate $A/\mathfrak{n}A$ as an $S/\mathfrak{n}$-module, then $a_1, a_2, \ldots, a_n$ generate A as an S-module.*

Proof In view of Lemma 4.24, it suffices to show that A is separated. That is to say, $\bigcap_{n\geq 0} \mathfrak{n}^n A = 0$. This is a consequence of Corollary 4.11. □

Exercise 4.2 Let I be an ideal of a ring R and let $A = R[[X_1, X_2, \ldots, X_n]]$ be a formal power series ring over R. We put $J = IA + (X_1, X_2, \ldots, X_n)A$. Show that if R is complete with respect to the I-adic topology, then A is complete with respect to the J-adic topology.

Exercise 4.3 Let k be a field. Show that $k[x][[t]]$ and $k[[t]][x]$ are not isomorphic. In which direction is there a natural inclusion morphism?

Exercise 4.4 Let k be a field and put $B = k[X, Y]/(Y^2 - X^2(X - 1))$.

(1) Show that B is an integral domain.
(2) Let A be the localization of B at the maximal ideal $\mathfrak{m} = (X, Y)B$. Is the $\mathfrak{m}$-adic completion of A an integral domain?

Exercise 4.5 Let k be a field. Show that the ring $A = k[[x]] \otimes_k k[[x]]$ is *not* a Noetherian ring. (Hint: Use the fact that the transcendence degree of the field $k((x))$ over $k(x)$ is infinite, where $k((x))$ denotes the field of fractions of $k[[x]]$.)

4.2 Henselian Pairs and Hensel's Lemma

4.26 Let $(A, \mathfrak{m})$ be a Noetherian local ring. We say that A is a *complete local ring* if A is complete with respect to the $\mathfrak{m}$-adic topology.

Complete local rings have many nice properties, and many problems concerning local rings get easier over complete local rings. One of the reasons is the theorem called "Hensel's lemma" below (Corollary 4.30) holds true for complete local rings.

4.27 Let A be a ring and I an ideal of A. Until the end of this chapter, a pair is a pair (A, I) where A is a ring and I is an ideal of A. For a pair (A, I) and an object ξ

over A, we denote the object over A/I corresponding to ξ by $\overline{\xi}$, provided it makes sense. Also, let $\pi : A \to \overline{A} = A/I$ denote the canonical projection.

We say that (A, I) is a *Henselian pair* if: $I \subseteq \mathrm{J}(A)$, where $\mathrm{J}(A)$ is the Jacobson radical of A; and for any monic polynomial $f \in A[X]$ and a factorization $\overline{f} = g_0h_0$ in $\overline{A}[X]$ with g_0 and h_0 monic and $(g_0, h_0) = \overline{A}[X]$, there exists a factorization $f = gh$ in $A[X]$ with g, h monic and $\overline{g} = g_0$ and $\overline{h} = h_0$. If so, then $(g, h)M + IM = M$, where M is the finitely generated free A-module $A[X]/(f)$. Since $I \subseteq \mathrm{J}(A)$, $(g, h)M = M$ by NAK. Since $f \in (g, h)$, we have $(g, h) = A[X]$. A local ring $(A, \mathfrak{m})$ is said to be a *Henselian local ring* if $(A, \mathfrak{m})$ is a Henselian pair.

Lemma 4.28

(1) *If (A, I) is a Henselian pair and $(\overline{A}, J)$ is a Henselian pair, then $(A, \pi^{-1}(J))$ is also a Henselian pair.*
(2) *If I is a nilpotent ideal, then (A, I) is a Henselian pair.*

Proof

(1) is easy, and is left to the reader.
(2) There exists some $n \geq 1$ such that $I^n = 0$. Obviously, $I \subseteq \mathrm{J}(A)$, and so we are to prove the lifting of the factorization. We prove it by induction on n. If $n = 1$, then the assertion is trivial. So assume that $n \geq 2$, and the assertion is true for smaller n. Since $(I^2)^{n-1} = I^{n+(n-2)} = 0$, (A, I^2) is a Henselian pair by induction. So it suffices to show that $(A/I^2, I/I^2)$ is a Henselian pair by (1). So we may assume that $n = 2$. Let $f \in A[X]$ be a monic polynomial, and $\overline{f} = g_0h_0$ be a factorization in $\overline{A}[X]$ with g_0, h_0 monic and $(g_0, h_0) = \overline{A}[X]$. Let $g_1, h_1 \in A[X]$ be monic polynomials such that $\overline{g}_1 = g_0$ and $\overline{h}_1 = h_0$. We can write $f - g_1h_1 = \sum_{i=0}^{d-1} a_iX^i$ with $d = \deg f$ and $a_i \in I$ for $0 \leq i \leq d-1$. Since $(g_0, h_0) = \overline{A}[X]$, we can write $X^i = u_ig_0 + v_ih_0$ with $u_i, v_i \in \overline{A}[X]$. Since h_0 is monic, we can write $u_i = q_ih_0 + r_i$ with $\deg r_i < \deg h_0$. Replacing u_i by $r_i = u_i - q_ih_0$ and v_i by $v_i + q_ig_0$, we may further assume that $\deg u_i < \deg h_0$. Then $\deg v_i = \deg(v_ih_0) - \deg h_0 = \deg(X^i - u_ig_0) - \deg h_0 < d - \deg h_0 = \deg g_0$. Taking $U_i, V_i \in A[X]$ so that $\deg U_i = \deg u_i$, $\deg V_i = \deg v_i$, $\overline{U}_i = u_i$, and $\overline{V}_i = v_i$, we define $g = g_1 + \sum_{i=0}^{d-1} a_iV_i$ and $h = h_1 + \sum_{i=0}^{d-1} a_iU_i$. Since g_1 is monic and $\deg V_i < \deg g_0 = \deg g_1$, we have that g is monic. Similarly, h is monic. Since $a_i \in I$ and $I^2 = 0$, $f - gh = (f - g_1h_1) + \sum_i a_i(U_ig_1 + V_ih_1) = \sum_i a_i(X^i - (U_ig_1 + V_ih_1)) = 0$, as desired.

□

Theorem 4.29 *Let (A, I) be a pair such that A is complete with respect to the I-adic topology. Then (A, I) is a Henselian pair.*

Proof $I \subseteq \mathrm{J}(A)$ is proved in Lemma 4.14. Let $f \in A[X]$ be a monic polynomial, $\overline{f} = g_0h_0$ be a factorization in $\overline{A}[X]$ such that $(g_0, h_0) = (1)$ and g_0, h_0 are monic. We set $A_n = A/I^{n+1}$ for $n \geq 0$. Then $(A_n, I^n/I^{n+1})$ is a Henselian pair by Lemma 4.28. So by induction on n, we can construct monic polynomials

$g_n, h_n \in A_n[X]$ such that $g_n + I^n A[X] = g_{n-1}, h_n + I^n A[X] = h_{n-1}$ for $n \geq 1$, and $f + I^{n+1}A[X] = g_n h_n$ in $A_n[X]$ for $n \geq 0$. Write $g_n = X^d + \sum_{i=0}^{d-1}(a_{ni} + I^{n+1})X^i$ and $h_n = X^e + \sum_{j=0}^{e-1}(b_{nj} + I^{n+1})X^j$ with $a_{ni}, b_{nj} \in A$, where $d = \deg g_0$ and $e = \deg h_0$. Then (a_{ni}) and (b_{nj}) are Cauchy sequences for each i and j. Since A is complete, $a_i = \lim a_{ni}$ and $b_j = \lim b_{nj}$ exist. Letting $g = X^d + \sum_i a_i X^i$ and $h = X^e + \sum_j b_j X^j$, g and h are monic, and we have that $g + I^{n+1}A[X] = g_n$ and $h + I^{n+1}A[X] = h_n$ for each $n \geq 0$. In particular, $\overline{g} = g_0$ and $\overline{h} = h_0$. Since A is separated, $f - gh \in \bigcap_{n\geq 0} I^{n+1}A[X] = 0$, as desired. □

The following important property of a complete local ring follows immediately.

Corollary 4.30 (Hensel's Lemma[2]) *A complete local ring is a Henselian local ring.*

Corollary 4.31 *Assume that a complete local ring $(A, \mathfrak{m}, K)$ contains a field k. We regard the composition map $k \xrightarrow{\iota} A \xrightarrow{\varepsilon} K$ (where ι is the embedding and ε is the natural surjection) as a field extension. If the extension K/k is separably generated (see* (E.13) *for the definition), then A contains a field $k_0 \supseteq k$ such that $\varepsilon(k_0) = K$.*

Proof Let B be a separating transcendence basis of K over k, and let X be a subset of A which is mapped bijectively onto B by the natural surjection $\pi : A \to A/\mathfrak{m} = K$. Since B is algebraically independent over k, so is X, and the map $\pi : k[X] \to k[B]$ is an isomorphism. So we have $k[X] \cap \mathfrak{m} = 0$. It follows that $k(X)$ is contained in A. Replacing k by $k(X)$, we may assume that the field extension K/k is separable algebraic.

Let Ω be the set of subfields of A containing k. Applying Zorn's lemma, A has a maximal subfield k_0 containing k. It remains to show that $\varepsilon(k_0) = K$. Assume the contrary, and take $x \in A$ such that $\varepsilon(x) \notin \varepsilon(k_0)$. Let $f(X) = \varepsilon(F(X)) \in \varepsilon(k_0)[X]$ be the minimal polynomial of $\varepsilon(x)$ over $\varepsilon(k_0)$. Here we can assume that $F(X)$ is a monic polynomial in $k_0[X]$. Since the field extension $k \subseteq K$ is separable algebraic, $f(X)$ does not have multiple roots. Since $\varepsilon(x)$ is a root of $f(X)$, we have a decomposition $f(X) = (X - \varepsilon(x))g(X)$ and $X - \varepsilon(x)$ and $g(X)$ are coprime in $K[X]$. Then by Theorem 4.30, $F(X)$ factors as $F(X) = (X - y)G(X)$ ($y \in A, \varepsilon(y) = \varepsilon(x)$). Since y is a root of $F(X) \in k_0[X]$, $k_0(y)$ is an extension field, properly containing k_0. Indeed, since $f(X) \in \varepsilon(k_0)[X]$ is irreducible, $k_0[X]/(F) \cong \varepsilon(k_0)[X]/(f)$ is a field, and hence the map $k_0[X]/(F) \to k_0[y]$ given by $X \mapsto y$ is an isomorphism. However, this contradicts our assumption that k_0 is a maximal subfield of A. Hence we have proved that $\varepsilon(k_0) = K$. □

One of the important properties of Henselian local rings is lifting of idempotents. As an application of this property, we can show that "a ring finite over a complete local ring is isomorphic to a product of complete local rings."

Let us begin with the definition of idempotents.

---

[2] Kurt Wilhelm Sebastian Hensel, 1861–1941.

Definition 4.32 Let A be a ring. If $e \in A$ satisfies

$$e^2 = e,$$

we say that e is an *idempotent*. If this is the case, we have $(1-e)e = 0$ and since $(1-e)^2 = (1-e) - e(1-e) = 1-e$, $1-e$ is also an idempotent. Note that 0 and 1 are idempotents, which we call the *trivial idempotents*.

4.33 The following conditions on a ring A are equivalent:

(1) A decomposes as $A = A_1 \times A_2$, where neither A_1 nor A_2 is a zero ring.
(2) A has a nontrivial idempotent.

Indeed, if (1) holds, then $e = (1, 0)$ is a nontrivial idempotent of A. Conversely, if e is a nontrivial idempotent of A, then $A_1 = Ae$ and $A_2 = A(1-e)$ are nonzero rings with the addition and the multiplication of A, and $\phi : A \to A_1 \times A_2$ given by $\phi(x) = (ex, (1-e)x)$ is an isomorphism of rings. Note that the map $A \to Ae$ ($x \mapsto xe$) is a surjective ring homomorphism, which induces an isomorphism $A/(1-e)A \cong Ae$. Similarly, we have $A/eA \cong A(1-e)$.

4.34 Let P be a finitely generated projective A-module with constant rank r (for the definition, see D.8). The characteristic polynomial $\chi(\phi)$ is defined to be $\det(X\,\mathrm{id} - \phi \otimes 1) \in A[X]$, where $X\,\mathrm{id} - \phi \otimes 1 \in \mathrm{End}_{A[X]}(P \otimes_A A[X])$ is given by $(X\,\mathrm{id} - \phi \otimes 1)(p \otimes f) = p \otimes fX - \phi(p) \otimes f$ for $p \in P$ and $f \in A[X]$, and det is the determinant (for the definition, see D.9). Note that $\chi(\phi)$ is a monic polynomial of degree r. By the Caylay–Hamilton theorem, we have that $(\chi(\phi))(\phi)$ as an element of $A[\phi] \subseteq \mathrm{End}_A P$ is zero. Note also that if $f \in A[X]$ is a monic polynomial of degree r, then $B = A[X]/(f)$ is a free A-module of rank r. When we set $\phi : B \to B$ to be the A-linear map $b \to Xb$, then $\chi(\phi) = f$, since $\chi(\phi)$ is divisible by f and both f and $\chi(\phi)$ are monic of degree r.

Theorem 4.35 *Let (A, I) be a pair. The following are equivalent:*

(1) $I \subseteq \mathrm{J}(A)$; *and for any A-algebra Λ, not necessarily commutative, if Λ is finitely generated as an A-module, then any idempotent e_0 of $\overline{\Lambda} = \Lambda/I\Lambda$ lifts to an idempotent e of Λ.*
(2) *For any finite (commutative) A-algebra B, the map $e \mapsto \overline{e}$ gives a bijection from the set of idempotents of B to the set of idempotents of $\overline{B} = B/IB$.*
(3) (A, I) *is Henselian.*

Proof (1) $\Rightarrow$(2). The map $e \mapsto \overline{e}$ is surjective by assumption. We prove that the map is injective by induction on $r = r(B) = \sup\{\dim_{A/\mathfrak{m}} B/\mathfrak{m}B \mid \mathfrak{m} \in \mathrm{Max}\, A\}$. If $r = 0$, then $B = 0$, and there is nothing to prove. Assume that $r > 0$, and the assertion is true for smaller r. Let e_0 be an idempotent of $\overline{B}$, and e and e' be idempotents of B that lift e_0, that is, $\overline{e} = \overline{e'} = e_0$. Then $r(\overline{B}(1-e_0)) + r(\overline{B}e_0) = r$. If $r(\overline{B}e_0) < r$, then since $I \subseteq \mathrm{J}(A)$, we have that $r(Be) = r(\overline{B}e_0) = r(Be') < r$. Since the idempotents e and ee' of eB lift the idempotent e_0 of $e_0\overline{B} = \overline{eB}$, we have $e = ee'$. Since the idempotents e' and ee' of $e'B$ lift the idempotent e_0 of

$e_0\overline{B} = \overline{e'B}$, we have $e' = ee'$. Hence $e = e'$. Similarly, if $r(\overline{B}(1 - e_0)) < r$, then $1 - e = (1 - e)(1 - e') = 1 - e'$, and hence $e = e'$.

(2) $\Rightarrow$(3). Assume that $I \not\subseteq \mathrm{J}(A)$. Then there exists some maximal ideal $\mathfrak{m}$ of A such that $I + \mathfrak{m} = A$. Then letting B be the finite A-algebra $B = A/(I \cap \mathfrak{m})$, we see that $B \cong A/I \times A/\mathfrak{m}$, and $B/IB \cong A/I$. Then the idempotents $(0, 1)$ and 0 of B are mapped to $0 \in B/IB$, and the map between the set of idempotents is not injective, and this contradicts the assumption. Hence $I \subseteq \mathrm{J}(A)$.

Let $f \in A[X]$ be a monic polynomial, and $g_0, h_0 \in \overline{A}[X]$ monic polynomials such that $(g_0, h_0) = \overline{A}[X]$ and $\overline{f} = g_0 h_0$. Set $B = A[X]/(f)$. Then B is finite over A, and $\overline{B} = \overline{A}[X]/(\overline{f}) \cong \overline{A}[X]/(g_0) \times \overline{A}[X]/(h_0)$ by the Chinese remainder theorem. Let $\overline{e} \in \overline{B}$ be the idempotent corresponding to $(1, 0) \in \overline{A}[X]/(g_0) \times \overline{A}[X]/(h_0)$. By assumption, $\overline{e}$ lifts to an idempotent $e \in B = A[X]/(f)$. Then we get a decomposition $B = Be \times B(1 - e)$. Note that B is finitely generated free as an A-module, and hence its direct summands Be and $B(1 - e)$ are finitely generated projective as A-modules. Note that for each maximal ideal $\mathfrak{m}$ of A, we have $\operatorname{rank}_{A_\mathfrak{m}}(Be)_{A_\mathfrak{m}} = \operatorname{rank}_{\overline{A_\mathfrak{m}}} \overline{A_\mathfrak{m}}[X]/(g_0) = \deg g_0$ and $\operatorname{rank}_{A_\mathfrak{m}}(B(1 - e))_{A_\mathfrak{m}} = \deg h_0$. Hence Be and $B(1 - e)$ have constant ranks, and $\operatorname{rank}_A Be = \deg g_0$ and $\operatorname{rank}_A B(1 - e) = \deg h_0$. Letting $\phi : B \to B$ be the multiplication by X on B, we have that $f(X) = \chi(\phi) = \chi(\phi|_{Be})\chi(\phi|_{B(1-e)})$ in $A[X]$. When we set $g = \chi(\phi|_{Be})$ and $h = \chi(\phi|_{B(1-e)})$, then $f = gh$, g and h are monic, and $\overline{g} = \chi(\overline{\phi}|_{\overline{A}[X]/(g_0)}) = g_0$ and $\overline{h} = \chi(\overline{\phi}|_{\overline{A}[X]/(h_0)}) = h_0$, as desired.

(3) $\Rightarrow$(1). We have $I \subseteq \mathrm{J}(A)$ by the definition of Henselian pairs.

Let e_0 be an idempotent of $\overline{\Lambda}$. Then we can take some $b \in \Lambda$ such that $\overline{b} = e_0$. When we set $c = b(1 - b)$, then $\overline{c} = e_0(1 - e_0) = 0$, and hence $c \in I\Lambda$. By the determinantal trick (Lemma 3.2), there exists some monic polynomial

$$\varphi(X) = X^n + c_1X^{n-1} + c_2X^{n-2} + \cdots + c_n$$

such that $c_1, c_2, \ldots, c_n \in I$ and $\varphi(c) = 0$. So letting $f(X) = \varphi(X(X - 1))$, we have that $f(b) = 0$. Since $\overline{f}(X) = \overline{\varphi}(X(X - 1)) = X^n(X - 1)^n$, there is a factorization $f = gh$ such that g and h are monic, and $\overline{g} = X^n$ and $\overline{h} = (X - 1)^n$. Since $(g, h) = A[X]$ (see (4.27)), we have a direct decomposition $A[X]/(f) \cong A[X]/(g) \times A[X]/(h)$. We have a commutative diagram

$$\begin{array}{ccccccc}
A[X]/(g) \times A[X]/(h) & \xrightarrow{\cong} & A[X]/(f) & \xrightarrow{X \mapsto b} & A[b] & \longrightarrow & \Lambda \\
\downarrow & & \downarrow & & \downarrow{\scriptstyle b \mapsto e_0} & & \downarrow \\
\overline{A}[X]/(X^n) \times \overline{A}[X]/((X-1)^n) & \xrightarrow{\cong} & \overline{A}[X]/X^n(X-1)^n & \xrightarrow{X \mapsto e_0} & \overline{A}[e_0] & \longrightarrow & \overline{\Lambda}\,.
\end{array}$$

So when we set $e \in A[b] \subseteq \Lambda$ to be the image of $(0, 1) \in A[X]/(g) \times A[X]/(h)$, then $e = e^2$, and $e_0 = \overline{e} \in \overline{\Lambda}$, as desired. □

Remark 4.1 Even if (A, I) is a Henselian pair and Λ is a non-commutative A-algebra which is finitely generated as an A-module, the correspondence $e \mapsto \overline{e}$ from the set of idempotents of Λ to the set of idempotents of $\overline{\Lambda}$ need not be injective. For example, if k is a field, $A = k[x]/(x^2)$, $I = (x) \subseteq A$, and $\Lambda = M_2(A)$, the A-algebra of the 2×2 matrices with coefficients in A, then both

$$e = \begin{pmatrix} 1 & 0 \\ 0 & 0 \end{pmatrix} \quad \text{and} \quad e' = \begin{pmatrix} 1 & x \\ x & 0 \end{pmatrix}$$

are idempotents that lift the same idempotent in $\overline{\Lambda} = M_2(k)$, but $e \neq e'$.

Corollary 4.36 *If (A, I) is a Henselian pair and B is a finite A-algebra, then (B, IB) is also a Henselian pair.*

Corollary 4.37 *Let $(A, \mathfrak{m})$ be a Henselian local ring and B be a finite A-algebra. Then B is the direct product of finitely many local rings.*

Proof We prove this corollary by induction on the number of maximal ideals n of the semi-local ring B. If $n \leq 1$, then there is nothing to prove. Assume that $n \geq 2$. Then since each maximal ideal of B contains $\mathfrak{m}B$, $B/\mathfrak{m}B$ is a finite-dimensional $A/\mathfrak{m}$-algebra with n maximal ideals. Since $n \geq 2$, the ring $B/\mathfrak{m}B$ has a non-trivial idempotent $\overline{e}$. By Theorem 4.35, $\overline{e}$ is lift to an idempotent e of B, and we have a decomposition $B \cong B/(e) \times B/(1 - e)$. By the induction assumption, each of $B/(e)$ and $B/(1 - e)$ are the direct product of finitely many local rings, and hence so is B. □

Remark 4.2 We can prove Corollary 4.37 for the case that A is a complete local ring, not using Theorem 4.35. Let $\mathrm{J}(B) = \bigcap_{M \in \mathrm{Max}\, B} M$ be the Jacobson radical of B. Then by the Chinese remainder theorem, for every $n \geq 1$,

$$B/\mathrm{J}(B)^n = \prod_{M \in \mathrm{Max}\, B} B/M^n$$

and then we have $\widehat{B} = \prod_{M \in \mathrm{Max}\, B} \widehat{B_M}$. Since B is $\mathfrak{m}$-adically complete and there exists some r such that $\mathrm{J}(B)^r \subseteq \mathfrak{m}B \subseteq \mathrm{J}(B)$, it is also $\mathrm{J}(B)$-adically complete, and we are done.

The lifting of idempotents as in Theorem 4.35 is important in the representation theory of (non-commutative) algebras. See [LW].

4.3 Structure Theorem of Complete Local Rings

4.38 Let $(A, \mathfrak{m}, k)$ be a local ring. If the characteristic of the residue field k is 0, then A contains the rational number field $\mathbb{Q}$ since $\mathfrak{m} \cap \mathbb{Z} = \{0\}$. In this case, we say that A is of *equicharacteristic* (or *equal characteristic*) 0. If $\mathrm{char}(A) = p$ is a prime number, then k has the same characteristic p. In this case, we say that A is of equicharacteristic p. A contains the field $\mathbb{F}_p$ this case. In these two cases, we say that

A is of equicharacteristic. A is of equicharacteristic if and only if $\mathrm{char}(A) = \mathrm{char}(k)$ if and only if A contains a field.

Otherwise k has a prime characteristic p, and $\mathrm{char}(A)$ is either 0 or a power p^e of p with $e \geq 2$. In this case, we say that A is of *unequal characteristic*. If, moreover, $\mathrm{char}(A) = 0$, then we say that A has a *mixed characteristic*. A local domain with an unequal characteristic has a mixed characteristic.

Until the end of this chapter, we assume that $(A, \mathfrak{m}, k)$ *is a Noetherian local ring.*

Definition 4.39 Assume that A is complcte and that the characteristic of k is $p \geq 0$. If a subring R of A satisfies the following conditions, we say that R is a *coefficient ring* of A:

(1) R is a complete local ring whose maximal ideal is pR.
(2) $A = R + \mathfrak{m}$.

If A contains a field, then the coefficient ring R is a field isomorphic with k. In this case R is called the *coefficient field* of A. If R is not Artinian, then A is of mixed characteristic, and R is a complete DVR by Lemma 3.32.

The following theorem is due to I. S. Cohen.[3]

Theorem 4.40 (Structure Theorem of Complete Local Rings [Coh]) *A complete local ring* $(A, \mathfrak{m}, k)$ *has a coefficient ring* R *and hence there is a surjective homomorphism from* $R[[X_1, \dots, X_n]]$ *to* A.

This theorem is one of the most important theorems in commutative ring theory. Many theorems are proved by reducing to the case of complete local rings. This is because of the fact that any complete local ring is a homomorphic image of a complete regular local ring. (For the definition and properties of regular local rings, cf. Chap. 5, Sect. 5.4. To read this section, it is enough to know that a formal power series ring over a field or a DVR is a regular local ring.) Also the fact that a complete local ring is an *excellent ring* is due to this structure theorem.

The most important part of Cohen's theorem is to show the existence of a coefficient ring. Here we prove only the case when A contains a field. The proof of the mixed characteristic case, we refer to other literature (for example, [Ma3, §29]).

Proof of Theorem 4.40 First, we prove that if A contains a coefficient ring R, then A is a homomorphic image of $S = R[[X_1, X_2, \dots, X_n]]$ for some n.

Let $x_1, x_2, \dots, x_n$ be a generator of $\mathfrak{m}$. Then we define

$$\varphi : S = R[[X_1, \dots, X_n]] \to A$$

by $\varphi(X_i) = x_i$ $(1 \leq \forall i \leq n)$, and we extend it continuously, with respect to the $(X_1, \dots, X_n)$-adic topology of S and the $\mathfrak{m}$-adic topology of A. Then φ

[3] Irvine Sol Cohen, 1917–1955.

is a local homomorphism of local rings and since the maximal ideal of S is $\mathfrak{n} = (p, X_1, X_2, \ldots, X_n)S$, we have $\mathfrak{n}A = \mathfrak{m}$ by our definition. Since $R \hookrightarrow S \to A/\mathfrak{n}A = A/\mathfrak{m} = k$ is surjective, A is generated by 1 as an S-module by Proposition 4.25. In other words, $\varphi : S \to A$ is surjective. □

Next, we show that if A contains a field k_0, then A has a coefficient field. We have shown it if $k_0 \subseteq k$ is separably generated in Corollary 4.31. In particular, the assertion is true for the case that k_0 is of characteristic zero. Hence in the following, we assume that A contains a field of characteristic $p > 0$.

Claim 4.1 If $\mathfrak{m}^p = 0$, then A contains a subfield L such that $A = L + \mathfrak{m}$.

Proof of Claim 4.1 By our assumption $\mathfrak{m}^p = 0$, $F = \{a^p \mid a \in A\}$ is a subring of A and is a field. Let us take a maximal subfield L of A such that $F \subseteq L$. Take $\alpha = \overline{a} \in A/\mathfrak{m}$, where $\overline{a}$ denotes the image of $a \in A$ in $A/\mathfrak{m}$. We show that $\alpha \in [L + \mathfrak{m}]/\mathfrak{m}$.

Assume that $\alpha \notin [L + \mathfrak{m}]/\mathfrak{m}$. Since $\alpha^p = \overline{a^p}$, $\alpha^p \in [F + \mathfrak{m}]/\mathfrak{m} \subseteq [L + \mathfrak{m}]/\mathfrak{m}$. Hence if we put $b = a^p$, $b \notin L^p$ (if $b = a^p = c^p$ for some $c \in L$, since $a^p - c^p = (a - c)^p = 0$, we have $a = c \in L$, contradicting our assumption $\alpha \notin [L + \mathfrak{m}]/\mathfrak{m}$). Hence the polynomial $f(X) = X^p - b \in L[X]$ is irreducible and since $f(a) = 0$, $L[a]$ is a subfield of A, which contradicts our assumption on the maximality of L. Hence we have $A = L + \mathfrak{m}$. □

For every $n \geq 1$, we let $A_n = A/\mathfrak{m}^n$ and $\varphi_n : A_{n+1} \to A_n$ be the natural surjection.

Claim 4.2 We can take a subfield L_n of A_n for each n so that $L_1 = A_1 = k$ and for each $n \geq 1$, we have $\varphi_n(L_{n+1}) = L_n$.

Proof of Claim 4.2 Put $L_1 = A_1$. Assume that subfields $L_i \subseteq A_i$ have already been defined for $1 \leq i \leq n$, and satisfy $\varphi_i(L_{i+1}) = L_i$ for $1 \leq i \leq n-1$. Now, $R = \varphi_n^{-1}(L_n)$ is a subring of A_{n+1} and the morphism $\varepsilon : R \to L_n$, $a \mapsto \varphi_n(a)$ is a surjection. If we put $J = \operatorname{Ker} \varepsilon$, then $J \in \operatorname{Max} R$ and $J^2 = 0$ since $J \subseteq \mathfrak{m}^n/\mathfrak{m}^{n+1}$. Hence (R, J) is a local ring. By Claim 1, the ring R contains a subfield L_{n+1} so that we have $R = L_{n+1} + J$. From the definition of R, we have $\varepsilon(L_{n+1}) = L_n$. □

Proof Now define $L = \varprojlim L_n$. Not only that each $\varphi_n : L_{n+1} \to L_n$ is surjective, it is also injective, since L_{n+1} and L_n are fields. So L is isomorphic to L_n for every n, and hence L is a field. We consider the map between the projective limits

$$L = \varprojlim L_n \to \varprojlim A_n = \widehat{A} = A.$$

Since L is a field and $A \neq 0$, we have that this map is also injective, and L is a subfield of A. Finally, the map $L \to A \to A/\mathfrak{m} = A_1$ agrees with $L \cong L_1 \cong A_1$ by construction. Hence $L + \mathfrak{m} = A$, as desired. □

We note here three important theorems deduced from Theorem 4.40. For the proofs, we refer the readers to [Bo, Ma1], and [Nag].

Theorem 4.41 *If A is a Noetherian complete local domain, then A contains a subring R, which is a complete* regular *local ring and A is a finitely generated R-module.*

Theorem 4.42 *Let A be a Noetherian complete local domain and L be a finite extension field of the field of fractions of A. Then the integral closure $\overline{A}^L$ of A in L is a finitely generated A-module and is a Noetherian complete* local *domain.*

Corollary 4.43 *Let A be a Noetherian local ring and assume that the completion $\widehat{A}$ of A is* reduced. *Then the integral closure $\overline{A}$ of A in the total quotient ring* $\mathrm{Q}(A)$ *of A is a finitely generated A-module.*

Proof We have $\widehat{A} \otimes_A \overline{A} \subseteq \overline{\widehat{A}}$. Since $\overline{\widehat{A}}$ is a finitely generated $\widehat{A}$-module by Theorem 4.41, so is $\widehat{A} \otimes_A \overline{A}$. Since $\widehat{A}$ is a faithfully flat A-module (Corollary 4.18), $\overline{A}$ is a finitely generated A-module. □

Chapter 5
Homological Methods in Commutative Algebra

In this chapter, we introduce homological methods in commutative algebra. Serre's criterion of regularity of local rings, which characterizes them as local rings of finite global dimension, was revolutionary. For example, this criterion made the question "Is a localization of a regular local ring again regular?" almost trivial. Also, many notions in commutative algebra have transparent descriptions because of homological methods. We assume that the reader is familiar with basics of homological algebra such as the functors Ext, Tor, and so on. Please refer to the Appendices if necessary.

In this chapter and the next, we study Cohen–Macaulay and Gorenstein properties, which are the most investigated concepts in modern commutative algebra.

The name "Cohen–Macaulay" comes from the research on "unmixedness theorem" by F. S. Macaulay for polynomial rings, and by I. S. Cohen for regular local rings.

In commutative ring theory, when a new ring appears, we are interested to know, first if it is Cohen–Macaulay or not and then when it is Gorenstein. Such characterizations very frequently reflect essential properties of the ring.

5.1 Projective Dimension and Cohen–Macaulay Rings and Modules

Definition 5.1 (Projective Dimension) Let A be a commutative ring, and M be an A-module. The infimum of $n \in \mathbb{Z}$ such that there exists a projective resolution (see Appendix C, Sect. C.1) of M

$$\cdots \to P_n \to \cdots \to P_1 \to P_0 \to M \to 0$$

S. Goto, K. Watanabe, *Commutative Algebra*, Graduate Texts in Mathematics 309,
https://doi.org/10.1007/978-981-95-4227-7_5

such that $P_i = 0$ for $i > n$ is denoted by $\mathrm{hd}_A M$, and we call it the *projective dimension* of M. As we consider that $\inf \emptyset = \infty$ and $\inf \mathbb{Z} = -\infty$, we consider that $\mathrm{hd}_A M = \infty$ if M does not have a projective resolution of finite length, and $\mathrm{hd}_A M = -\infty$ if $M = 0$. In particular, $\mathrm{hd}_A M \leq 0$ if and only if M is a projective module.

We will see that for a general (non-regular) ring, modules with finite projective dimensions are special, and a "generic" module has infinite projective dimension.

Example 5.2

(1) Let $x \in A$ be a non-zerodivisor. When we set $M = A/xA$, then M has a projective resolution $0 \to A \xrightarrow{x} A \to M \to 0$, and hence we have $\mathrm{hd}_A M \leq 1$. If, moreover, $(A, \mathfrak{m})$ is local and $x \in \mathfrak{m}$, then $\mathrm{hd}_A M = 1$, since $\mathrm{Ext}^1_A(M, A) \cong M \neq 0$, see Lemma 5.3.
(2) Let A be a commutative ring, and set $R = A[X]/(X^2)$. We denote the image of X in R by x. When we set $M = R/xR$, since we have $\mathrm{Ker}[R \xrightarrow{x} R] = xR \cong M$, the projective resolution of M as an R-module is

$$\cdots \xrightarrow{x} R \xrightarrow{x} R \xrightarrow{x} R \to M \to 0$$

and repeats infinitely. So it is easy to see that $\mathrm{Ext}^i_R(M, M) = M$ using this projective resolution. Applying Lemma 5.3 below, we have that $\mathrm{hd}_R M = \infty$.

There is a characterization of $\mathrm{hd}_A M$ using the vanishing of Ext modules.

Lemma 5.3 *Let M be an A-module and $n \geq 0$. The following are equivalent:*

(1) $\mathrm{hd}_A M \leq n$.
(2) *For any A-module X, we have* $\mathrm{Ext}^{n+1}_A(M, X) = 0$.
(3) *For any A-module X and integer $i > n$, we have* $\mathrm{Ext}^i_A(M, X) = 0$.

If, moreover, A is Noetherian and M is finitely generated, then since we have $\mathrm{Ext}^i_A(M, \varinjlim X_\alpha) \cong \varinjlim \mathrm{Ext}^i_A(M, X_\alpha)$ *(Proposition C.8), we can restrict X to be finitely generated A-modules in* (2) *and* (3).

Considering the long exact sequence of Ext functors, we get the following two corollaries.

Corollary 5.4 *Let $0 \to X \to Y \to Z \to 0$ be a short exact sequence of A-modules. If two modules out of the three modules X, Y, and Z have finite projective dimension, then so does the third.*

Corollary 5.5 *Let $0 \to L \to M \to N \to 0$ be a short exact sequence of A-modules, and assume that $\mathrm{hd}_A M$ is finite. Then the following hold true:*

(1) *If* $\mathrm{hd}_A L > \mathrm{hd}_A M$, *then* $\mathrm{hd}_A N = \mathrm{hd}_A L + 1$.
(2) *If* $\mathrm{hd}_A L = \mathrm{hd}_A M$, *then* $\mathrm{hd}_A N \leq \mathrm{hd}_A L + 1 = \mathrm{hd}_A M + 1$.
(3) *If* $\mathrm{hd}_A L < \mathrm{hd}_A M$, *then* $\mathrm{hd}_A N = \mathrm{hd}_A M$.
(4) *If* $\mathrm{hd}_A N > \mathrm{hd}_A M$, *then* $\mathrm{hd}_A L = \mathrm{hd}_A N - 1$.

(5) *If* $\mathrm{hd}_A N = \mathrm{hd}_A M$, *then* $\mathrm{hd}_A L \le \mathrm{hd}_A M = \mathrm{hd}_A N$.
(6) *If* $\mathrm{hd}_A N < \mathrm{hd}_A M$, *then* $\mathrm{hd}_A L = \mathrm{hd}_A M$.

If A is local or the polynomial ring $k[X_1, X_2, \ldots, X_n]$ over a field k, then it is known that any projective A-module is free [Qu, Su, Ba]. We only prove the following here.

Proposition 5.6 *A finitely presented flat module P over a local ring $(A, \mathfrak{m})$ is free. In particular, a finitely generated projective A-module is free.*

Proof Letting $n = \dim_{A/\mathfrak{m}} P/\mathfrak{m}P$, we consider a short exact sequence

$$0 \to K \xrightarrow{f} A^n \to P \to 0.$$

Since P is flat, the exact sequence remains exact after we apply $\otimes_A A/\mathfrak{m}$, and hence we get an exact sequence $0 \to K/\mathfrak{m}K \to (A/\mathfrak{m})^n \to P/\mathfrak{m}P \to 0$. As $n = \dim_{A/\mathfrak{m}} P/\mathfrak{m}P$, we have that $K/\mathfrak{m}K = 0$ by dimension counting. Since K is finitely generated, we have that $K = 0$ by NAK, Theorem 1.86, and hence $P \cong A^n$. Hence P is free. □

In general, a projective module need not be free. For example, if A is a Dedekind domain, then any nonzero ideal I of A is invertible, and hence is projective as an A-module. Indeed, when we set $I^{-1} = \{x \in \mathrm{Q}(A) \mid xI \subseteq A\}$, since we have $II^{-1} = A$, we can express $1 = \sum_{\alpha=1}^{n} i_\alpha j_\alpha$, where $i_\alpha \in I$ and $j_\alpha \in I^{-1}$. Then we can define the A-linear maps $\varphi : A^n \to I$ and $\psi : I \to A^n$, by $(a_\alpha)_{1 \le \alpha \le n} \mapsto \sum_{\alpha=1}^{n} a_\alpha i_\alpha$ and $i \mapsto (ij_\alpha)_{1 \le \alpha \le n}$, respectively, and it holds that $\varphi\psi = 1_I$. So ψ is a split monomorphism, and hence I is projective. If I is free, then we have $I \cong A$ is principal. So if A is not a PID, then A has at least one projective ideal which is not free.

Generalizing the above, we get:

Proposition–Definition 5.7 (Invertible Ideal) Let A be an integral domain whose field of fractions is K. For an A-submodule I of K, we set $I^{-1} = \{x \in K \mid xI \subseteq A\}$. We say that I is an invertible fractional ideal if $II^{-1} = A$. Note that an invertible fractional ideal is finitely generated projective as an A-module.

It is easy to see that a nonzero finitely generated submodule I of the field of fractions $K = \mathrm{Q}(A)$ of an integral domain A is a fractional ideal. That is, $I \neq 0$ and $I^{-1} \neq 0$. A finitely generated projective module is reflexive. So an invertible fractional ideal is a divisorial ideal by Proposition 3.54.

Let A be a Noetherian ring, I be an ideal of $A (I \neq A)$, and M be a finitely generated A-module. We define $\mathrm{depth}(I, M) = \inf\{i \in \mathbb{Z} \mid \mathrm{Ext}_A^i(A/I, M) \neq 0\}$, and call it the *I-depth* of M (or the *M-grade* of I), which we denote $\mathrm{depth}(I, M)$ or $\mathrm{depth}_A(I, M)$. If $M/IM \neq 0$, then $\mathrm{depth}(I, M)$ is the supremum of the lengths of the M-sequences in I, by the following theorem.

Theorem 5.8 *Let A be a Noetherian ring, and I an ideal of A. For a finitely generated A-module M with $M/IM \neq 0$ and a positive integer r, the following are equivalent:*

(1) *There exists an M-sequence of length r consisting of elements of I.*
(2) $\mathrm{Ext}^i_A(A/I, M) = 0$ *for any $i < r$.*
(3) *For any finitely generated A-module N such that* $\mathrm{Supp}_A(N) \subseteq \mathrm{V}(I)$, *we have* $\mathrm{Ext}^i_A(N, M) = 0$ *for any $i < r$.*

Proof Induction on r. Assume that $r = 1$. We can take an M-regular element from I if and only if $I \not\subseteq \mathfrak{p}$ for any $\mathfrak{p} \in \mathrm{Ass}_A(M)$. This is equivalent to saying that

$$\mathrm{Ass}_A(\mathrm{Hom}_A(A/I, M)) = \mathrm{Supp}_A(A/I) \cap \mathrm{Ass}_A(M) = \emptyset.$$

That is, $\mathrm{Hom}_A(A/I, M) = 0$. If so, $\mathrm{Ass}_A(\mathrm{Hom}_A(N, M)) = \emptyset$ for any N such that $\mathrm{Supp}_A(N) \subseteq \mathrm{V}(I)$, and the theorem is true for this case.

Next, assume that $r \geq 2$, and the theorem is true for smaller r. In particular, there exists an M-sequence of length $r - 1$ in I, and $\mathrm{Ext}^i_A(N, M) = 0$ for any finitely generated A-module N such that $\mathrm{Supp}_A(N) \subseteq \mathrm{V}(I)$ and $i < r - 1$.

We prove (1) $\Rightarrow$(3). Let $x_1, \ldots, x_r$ be an M-sequence in I, and set $x = x_1$. Applying the functor $\mathrm{Ext}^i_A(N, *)$ to the short exact sequence

$$0 \to M \xrightarrow{x} M \to M/xM \to 0, \tag{$*$}$$

we get a long exact sequence

$$\cdots \to \mathrm{Ext}^i_A(N, M/xM) \to \mathrm{Ext}^{i+1}_A(N, M) \xrightarrow{x} \mathrm{Ext}^{i+1}_A(N, M) \to \cdots . \tag{$**$}$$

As $x^n \in \mathrm{Ann}_A(N)$ for some $n \geq 1$, the action of x on $\mathrm{Ext}^i_A(N, M)$ is nilpotent for all i.

As $x_2, \ldots, x_n$ is an M/xM-sequence, we have $\mathrm{Ext}^i_A(N, M/xM) = 0$ for $i < r - 1$ by induction. So we have that $\mathrm{Ext}^i_A(N, M) \xrightarrow{x} \mathrm{Ext}^i_A(N, M)$ is injective for $i < r$ by the exact sequence ($**$). As this map is also nilpotent, we have that $\mathrm{Ext}^i_A(N, M) = 0$ for $i < r$.

(3) $\Rightarrow$(2) is trivial.

We show (2) $\Rightarrow$(1). As $r - 1 \geq 1$, we can take an M-regular element $x \in I$ by the induction assumption. As there is an exact sequence

$$\mathrm{Ext}^i_A(N, M) \to \mathrm{Ext}^i_A(N, M/xM) \to \mathrm{Ext}^{i+1}_A(N, M),$$

$\mathrm{Ext}^i_A(N, M/xM) = 0$ for $i < r - 1$. By the induction assumption, I contains an M/xM-sequence $x_2, \ldots, x_r$, and hence $x, x_2, \ldots, x_r$ is an M-sequence of length r contained in I. □

Using the long exact sequence of Ext, we get the following *depth lemma*.

Lemma 5.9 (Depth Lemma) *Let* $0 \to L \to M \to N \to 0$ *be a short exact sequence of finitely generated modules over a Noetherian ring* A. *Then the following hold:*

(1) $\operatorname{depth}_A(I, M) \leq \max\{\operatorname{depth}_A(I, L), \operatorname{depth}_A(I, N)\}$.
(2) *If* $\operatorname{depth}_A(I, M) > \operatorname{depth}_A(I, N)$, *then we have* $N/IN \neq 0 \neq L/IL$, *and* $\operatorname{depth}_A(I, L) = \operatorname{depth}_A(I, N) + 1$.
(3) *If* $\operatorname{depth}_A(I, L) = \operatorname{depth}_A(I, M)$, *then we have* $\operatorname{depth}_A(I, N) \geq \operatorname{depth}_A(I, L) - 1$.

Lemma 5.10 *Let* A *be a Noetherian ring,* I *an ideal of* A, *and* M *a finitely generated* A*-module such that* $M \neq IM$. *Let* $x \in I$ *be* M*-regular. Then we have* $\operatorname{depth}(I, M/xM) = \operatorname{depth}(I, M) - 1$.

Proof If $r = \operatorname{depth}(I, M/xM)$ and $a_1, \dots, a_r$ is an M/xM-sequence in I, then $x, a_1, \dots, a_r$ is an M-sequence in I, and hence $\operatorname{depth}(I, M) \geq \operatorname{depth}(I, M/xM) + 1$. On the other hand, as $0 \to M \xrightarrow{x} M \to M/xM \to 0$ is exact, so $\operatorname{depth}(I, M/xM) \geq \operatorname{depth}(I, M) - 1$ by Lemma 5.9, and the result follows. □

Let A be a Noetherian ring, I an ideal of A, and M a finitely generated A-module. An M-sequence $a_1, \dots, a_s$ in I is called *maximal* if there is no $x \in I$ such that $a_1, \dots, a_s, x$ is an M-sequence.

Lemma 5.11 *Let* A *be a Noetherian ring,* I *its ideal, and* M *a finitely generated* A*-module. Let* $a_1, \dots, a_s$ *be an* M*-sequence in* I.

(1) *The following are equivalent:*
    (a) $a_1, \dots, a_s$ *is maximal.*
    (b) $\operatorname{depth}(I, M/(a_1, \dots, a_s)M) = 0$.
    (c) $s = \operatorname{depth}(I, M) < \infty$.
(2) *If* $M/IM \neq 0$, *then* $s \leq \operatorname{depth}(I, M) < \infty$. *If so, there exists some* $a_{s+1}, \dots, a_r \in I$ *such that* $a_1, \dots, a_r$ *is a maximal* M*-sequence in* I, *where* $r = \operatorname{depth}(I, M)$.
(3) *If* $M/IM = 0$, *then* $\operatorname{depth}(I, M) = \infty$.
(4) $\operatorname{depth}(I, M)$ *depends only on* $\sqrt{I}$.

Proof

(1) (a) ⇔(b) is trivial. As $\operatorname{depth}(I, M) = s + \operatorname{depth}(I, M/(a_1, \dots, a_s)M)$ by Lemma 5.10, (b) ⇔(c) is also clear.
(2) Assume that $M/IM \neq 0$ and $\operatorname{depth}(I, M) = \infty$. Then we can find an infinite sequence $a_{s+1}, \dots, a_r, \dots$ of elements of I such that $a_1, \dots, a_r$ is an M-sequence for $r \geq s + 1$, since $a_1, \dots, a_r$ cannot be maximal by (1). But this is absurd, since $0 \subsetneq (a_1) \subsetneq (a_1, a_2) \subsetneq \cdots$ is an infinite ascending chain of ideals in A then. The second assertion is obvious from $\operatorname{depth}(I, M/(a_1, \dots, a_s)M) = r - s$ and Theorem 5.8.

(3) If $M/IM = 0$, then $\mathrm{Ext}^i_A(A/I, M) = 0$ for any $i \in \mathbb{Z}$ (because its support is contained in $\mathrm{Supp}\, M \cap \mathrm{Supp}\, A/I = \emptyset$), and we have $\mathrm{depth}(I, M) = \inf \emptyset = \infty$.
(4) This follows from (1) $\Leftrightarrow$(3) of Theorem 5.8.

□

For finitely generated modules of finite projective dimension over a Noetherian local ring $(A, \mathfrak{m})$, the following beautiful theorem holds.

Theorem 5.12 (Auslander[1]–Buchsbaum[2] Theorem 1) *Let $(A, \mathfrak{m})$ be a Noetherian local ring, and M be a nonzero finitely generated A-module of finite projective dimension. Then we have*

$$\mathrm{hd}_A M + \mathrm{depth}_A M = \mathrm{depth}\, A.$$

Proof We set $r = \mathrm{depth}\, A$ and $s = \mathrm{hd}_A M$. We prove the theorem by induction on s. If $s = 0$, then M is free, and hence the theorem is true. If $s = 1$, then we can take the minimal free resolution (see (C.14) for the definition)

$$0 \to G \to F \to M \to 0$$

of M, where F and G are finite free, and $G \subseteq \mathfrak{m}F$. From this exact sequence, we get the exact sequence of Ext modules

$$\to \mathrm{Ext}^i_A(A/\mathfrak{m}, F) \to \mathrm{Ext}^i_A(A/\mathfrak{m}, M) \to \mathrm{Ext}^{i+1}_A(A/\mathfrak{m}, G) \to \mathrm{Ext}^{i+1}_A(A/\mathfrak{m}, F) \to .$$

Since we have $G \subseteq \mathfrak{m}F$, The map $\mathrm{Ext}^i_A(A/\mathfrak{m}, G) \to \mathrm{Ext}^i_A(A/\mathfrak{m}, F)$ is zero for all i. From this and Theorem 5.8, $\mathrm{depth}_A M = r - 1$ follows easily.

Next, assume that $s > 1$ and the theorem is true for smaller s. If the number of generators of M is n, then we can take a surjective map $\phi : F = A^n \to M$. When we set $\mathrm{Ker}\, \phi = K$, then we have $\mathrm{hd}_A K = s - 1$ and $K \subseteq \mathfrak{m}A^n$. By the induction assumption, we have $\mathrm{depth}_A K = r - s + 1$. Hence by Theorem 5.8, $\mathrm{Ext}^i_A(A/\mathfrak{m}, K) = 0$ for $i \leq r - s$, and $\mathrm{Ext}^{r-s+1}_A(A/\mathfrak{m}, K) \neq 0$. Hence by the exact sequence

$$\to \mathrm{Ext}^i_A(A/\mathfrak{m}, F) \to \mathrm{Ext}^i_A(A/\mathfrak{m}, M) \to \mathrm{Ext}^{i+1}_A(A/\mathfrak{m}, K) \to \mathrm{Ext}^{i+1}_A(A/\mathfrak{m}, F) \to,$$

we have that $\mathrm{Ext}^i_A(A/\mathfrak{m}, M) = 0$ for $i < r - s$ and $\mathrm{Ext}^{r-s}_A(A/\mathfrak{m}, M) \neq 0$. □

Now we define Cohen–Macaulay rings. We could say that this notion is perhaps the most important in commutative algebra. Indeed, when a new class of rings

[1] Maurice Auslander, 1926–1994.
[2] David Alvin Buchsbaum, 1929–2021.

appears, what is considered as a problem is, "When is a ring in this class Cohen–Macaulay?"

Lemma 5.13 *Let* $(A, \mathfrak{m})$ *be a Noetherian local ring, and* M *be a nonzero finitely generated* A*-module. Then we have*

$$\dim_A M \geq \operatorname{depth}_A M.$$

Proof See Lemma 3.38 □

Definition 5.14 Let A be a Noetherian ring, and M be a finitely generated A-module. We say that M is *Cohen–Macaulay* if either $\dim_{A_\mathfrak{m}} M_\mathfrak{m} = \operatorname{depth}_{A_\mathfrak{m}} M_\mathfrak{m}$ or $M_\mathfrak{m} = 0$ for any $\mathfrak{m} \in \operatorname{Max} A$. We say that the ring A is Cohen–Macaulay if it is so as an A-module.

If $(A, \mathfrak{m})$ is local, and M is a nonzero finitely generated A-module such that $\dim M \leq \operatorname{depth} M$, then M is Cohen–Macaulay by Lemma 5.13. In particular, a Noetherian local ring A is a Cohen–Macaulay ring if and only if $\dim A \leq \operatorname{depth}_A A$.

Theorem 5.15 *Let* $(A, \mathfrak{m})$ *be a Noetherian local ring,* M *a nonzero finitely generated* A*-module, and* $x \in \mathfrak{m}$ *an* M*-regular element. Then we have that* M *is Cohen–Macaulay if and only if so is* M/xM*.*

Proof This is because $\dim_A M/xM = \dim_A M - 1$ and $\operatorname{depth}_A M/xM = \operatorname{depth}_A M - 1$. □

Lemma 5.16 *Let* A *be a Noetherian ring,* I *be an ideal of* A*, and* M *be a finitely generated* A*-module. Then* $\operatorname{depth}(I, M) = \inf_{P \in \mathrm{V}(I)} \operatorname{depth}_{A_P} M_P$.

Proof We set $r = \operatorname{depth}(I, M)$ and $s = \inf_{P \in \mathrm{V}(I)} \operatorname{depth}_{A_P} M_P$. If $M = IM$, then $\mathrm{V}(I) \cap \operatorname{Supp} M = \operatorname{Supp} M/IM = \emptyset$, and the both r and s are ∞. Assuming that $M \neq IM$, we prove $r = s$ by induction on r. If $r = 0$, then $\operatorname{Hom}_A(A/I, M) \neq 0$, and hence $\mathrm{V}(I) \cap \operatorname{Ass} M = \operatorname{Ass} \operatorname{Hom}_A(A/I, M) \neq \emptyset$. If $P \in \mathrm{V}(I) \cap \operatorname{Ass} M$, then $\operatorname{depth}_{A_P} M_P = 0$, and hence $s = 0$. Assume that $r > 0$. Then we have an M-regular element $a \in I$. If $P \in \mathrm{V}(I) \cap \operatorname{Supp} M$, then $a \in PA_P$ is M_P-regular, and hence $\operatorname{depth}_{A_P}(M/aM)_P = \operatorname{depth}_{A_P} M_P - 1$. So $\inf_{P \in \mathrm{V}(I)} \operatorname{depth}_{A_P}(M/aM)_P = s - 1$. On the other hand, $\operatorname{depth}(I, M/aM) = r - 1$. By the induction assumption, we get $r - 1 = s - 1$, and hence $r = s$. □

Corollary 5.17 *Let* A *be a Noetherian ring, and* I *be an ideal of* A*. Then we have* $\operatorname{depth}(I, A) \leq \operatorname{ht} I$.

Proof $\operatorname{depth}(I, A) = \inf_{P \in \mathrm{V}(I)} \operatorname{depth} A_P \leq \inf_{P \in \mathrm{V}(I)} \dim A_P = \operatorname{ht} I$. □

Lemma 5.18 *Let* $(A, \mathfrak{m})$ *be a Noetherian local ring, and* M *and* N *be finitely generated* A*-modules. Then* $\operatorname{Ext}^i_A(M, N) = 0$ *if* $i < \operatorname{depth} N - \dim M$.

Proof Induction on $d = \dim M$. Take a filtration $0 = M_0 \subseteq M_1 \subseteq M_2 \subseteq \cdots \subseteq M_r = M$ such that for each i with $1 \leq i \leq r$, there exists some $P_i \in \operatorname{Spec} A_i$ such that $M_i/M_{i-1} \cong A/P_i$. It suffices to show that $\operatorname{Ext}^i_A(A/P_i, N) = 0$ for $i <$

depth $N - d$. Take $x \in A \setminus P_i$. Since $0 \to A/P_i \xrightarrow{x} A/P_i \to A/(P_i + Ax) \to 0$ is exact,

$$\operatorname{Ext}_A^i(A/P_i, N) \xrightarrow{x} \operatorname{Ext}_A^i(A/P_i, N) \to \operatorname{Ext}_A^{i+1}(A/(P_i + Ax), N)$$

is exact. Since $i + 1 < \operatorname{depth} N - (d - 1)$ and $\dim A/(P_i + Ax) \leq \dim A/P_i - 1 \leq d - 1$, we have that $\operatorname{Ext}_A^{i+1}(A/(P_i + Ax), N) = 0$ by the induction assumption, and hence $\operatorname{Ext}_A^i(A/P_i, N) = 0$ by NAK, Theorem 1.86, as desired. □

Lemma 5.19 *If* $(A, \mathfrak{m})$ *is a Noetherian local ring, and* M *a Cohen–Macaulay* A*-module, then*

$$\operatorname{Ass}_A M = \operatorname{Assh}_A M := \{P \in \operatorname{Supp} M \mid \dim A/P = \dim_A M\}.$$

In particular, for $x \in \mathfrak{m}$, $\dim M/xM = \dim M - 1$ *if and only if* $\operatorname{ht} x(A/\operatorname{Ann}_A M) = 1$ *if and only if* x *is* M*-regular.*

Proof If $P \in \operatorname{Ass}_A M$, then $\operatorname{Ext}_A^0(A/P, M) \neq 0$. So $\dim A/P \geq \operatorname{depth} M = \dim_A M$ by Lemma 5.18. The last assertion is obvious, since $\operatorname{Assh}_A M \subseteq \operatorname{Min}_A M \subseteq \operatorname{Ass}_A M$ in general, and

$$\{x \in \mathfrak{m} \mid \dim M/xM = \dim M\} = \bigcup_{P \in \operatorname{Assh}_A M} P,$$

$$\{x \in \mathfrak{m} \mid \operatorname{ht} x(A/\operatorname{Ann}_A M) = 0\} = \bigcup_{P \in \operatorname{Min}_A M} P, \qquad \text{and}$$

$$\{x \in \mathfrak{m} \mid x \text{ is not } M\text{-regular}\} = \bigcup_{P \in \operatorname{Ass}_A M} P.$$

□

Corollary 5.20 *Let* $(A, \mathfrak{m})$ *be a Noetherian local ring, and* M *a finitely generated* A*-module. Then the following are equivalent:*

(1) *M is Cohen–Macaulay.*
(2) *There is a system of parameters of M that is an M-sequence.*
(3) *Any system of parameters of M is an M-sequence.*

Proof (3) ⇒(2) ⇒(1) is obvious.

(1) ⇒(3). Induction on $n = \dim M$. Let $x_1, \ldots, x_n$ be a system of parameters on M. Then $\dim M/x_1M = n - 1$. So x_1 is M-regular by Lemma 5.19, and hence M/x_1M is Cohen–Macaulay. Since $x_2, \ldots, x_d$ is a system of parameters of M/x_1M, it is an M/x_1M-sequence. Hence, $x_1, \ldots, x_d$ is an M-sequence, as desired. □

Lemma 5.21 *Let A be a Noetherian local ring, M be a Cohen–Macaulay finitely generated A-module, and $P \in \operatorname{Supp}_A M$. Then* $\operatorname{depth}(P, M) = \operatorname{depth} M_P = \dim M_P$.

Proof We have $\operatorname{depth}(P, M) \le \operatorname{depth} M_P \le \dim M_P < \infty$. We prove the lemma by induction on $\operatorname{depth}(P, M)$. If $\operatorname{depth}(P, M) = 0$, then $\operatorname{Hom}_A(A/P, M) \neq 0$, and hence $\dim A/P \ge \operatorname{depth} M = \dim M$ by Lemma 5.18. Since $P \in \operatorname{Supp} M$, we have that $P \in \operatorname{Assh}_A M$, and hence $\dim M_P = 0$. If $\operatorname{depth}(P, M) > 0$, then we can take an M-regular element a in P. Then $M_P/aM_P \neq 0$ by NAK, Theorem 1.86, and hence

$$\operatorname{depth}(P, M) = \operatorname{depth}(P, M/aM) + 1 = \dim(M/aM)_P + 1 = \dim M_P$$

by the induction assumption. □

Corollary 5.22 *Let A be a Noetherian ring, and M be a Cohen–Macaulay finitely generated A-module. Then for $P \in \operatorname{Spec} A$, M_P is a Cohen–Macaulay A_P-module. If S is a multiplicatively closed subset of A, then $S^{-1}M$ is a Cohen–Macaulay $S^{-1}A$-module.*

Proof This is immediate by Lemma 5.21. □

Proposition 5.23 *Let A be a Cohen–Macaulay ring, and $I \neq A$ be an ideal. Then we have* $\operatorname{depth}(I, A) = \operatorname{ht} I$.

Proof $\operatorname{depth}(I, A) = \inf_{P \in V(I)} \operatorname{depth} A_P = \inf_{P \in V(I)} \dim A_P = \operatorname{ht} I$. □

Lemma 5.24 *Let $(A, \mathfrak{m})$ be a Cohen–Macaulay local ring, and I be a proper ideal. Then we have* $\dim A = \dim A/I + \operatorname{ht} I$.

Proof Induction on $h = \operatorname{ht} I = \operatorname{depth}(I, A)$. If $h = 0$, then we can take $P \in V(I) \cap \operatorname{Min}_A A$. As $\operatorname{Min}_A A = \operatorname{Assh}_A A$, $\dim A = \dim A/P \le \dim A/I \le \dim A$, and the assertion holds. If $h > 0$, then we take an A-regular element $a \in I$. We set $\overline{A} = A/aA$, and $\overline{I} = I/aA$. Then $\overline{A}$ is a Cohen–Macaulay local ring. We have $\dim \overline{A} = \dim A - 1$, $\dim \overline{A}/\overline{I} = \dim A/I$, and $\operatorname{ht} \overline{I} = \operatorname{depth}(\overline{I}, \overline{A}) = \operatorname{depth}(I, \overline{A}) = \operatorname{depth}(I, A) - 1 = \operatorname{ht} I - 1$. Combining these, we get the assertion. □

Let $(A, \mathfrak{m})$ be a Noetherian local ring, and $a_1, \dots, a_r \in \mathfrak{m}$. We say that $a_1, \dots, a_r$ is a *subsystem of parameters* of A, if $\dim A - \dim A/(a_1, \dots, a_r) = r$. This is equivalent to saying that there exist some $a_{r+1}, \dots, a_d \in \mathfrak{m}$ such that $a_1, \dots, a_r, a_{r+1}, \dots, a_d$ is a system of parameters of A.

Let A be a Noetherian ring and I a proper ideal of A generated by $\operatorname{ht} I$ elements. Then by Theorem 2.36, for each minimal prime ideal P of I, we have $\operatorname{ht} P = \operatorname{ht} I$. It is easy to see that the height of $I = (a_1, \dots, a_r)$ is r if and only if $I \neq A$, and for each $P \in V(I)$, $a_1, \dots, a_r$ is a subsystem of parameters of A_P.

Theorem 5.25 (Unmixedness Theorem) *Let A be a Noetherian ring. Then the following are equivalent:*

(1) *A is Cohen–Macaulay.*
(2) *For each ideal $I = (a_1, \ldots, a_r)$ of A, if $r = \operatorname{ht} I$ and $P \in \operatorname{Ass}_A A/I$, we have $\operatorname{ht} P = r$.*

Proof (1) $\Rightarrow$(2). Since A_P is Cohen–Macaulay and $a_1, \ldots, a_r$ is its subsystem of parameters of A_P, $a_1, \ldots, a_r$ is an A_P-sequence. So A_P/IA_P is also Cohen–Macaulay. Since $P \in \operatorname{Ass}_A A/I$, we have that $\operatorname{depth}_{A_P} A_P/IA_P = 0$. So $\operatorname{ht} P = \operatorname{depth}(P, A) = \operatorname{depth} A_P = r + \operatorname{depth}_{A_P} A_P/IA_P = r = \operatorname{ht} I$, as desired.

(2) $\Rightarrow$(1). Let $\mathfrak{m}$ be a maximal ideal of A of height d. If $d > 0$, we can take $a_1 \in \mathfrak{m} \setminus \bigcup_{P \in \operatorname{Ass}_A A} P$, by assumption (2) applied to the ideal $I = 0$ and $r = 0$ ($P \in \operatorname{Ass}_A A$ has height zero), and a_1 is A-regular. If $d > 1$, we can take $a_2 \in \mathfrak{m} \setminus \bigcup_{P \in \operatorname{Ass}_A(A/(a_1))} P$. Continuing like this, we get an A-sequence $a_1, \ldots, a_d \in \mathfrak{m}$. So $\operatorname{depth} A_{\mathfrak{m}} \geq \operatorname{depth}(\mathfrak{m}, A) \geq d = \dim A_{\mathfrak{m}}$, and hence A is Cohen–Macaulay. □

5.2 Injective Dimension and Global Dimension

The notion of injective modules is dual to that of projective modules. It appears in many places in mathematics in abstract ways, but is not so often treated concretely. In this section, we treat injective modules concretely as long as possible so that they can be visualized.

Definition 5.26 (Injective Modules) Let A be a commutative ring. We say that an A-module I is *injective* or an injective A-module if the functor $\operatorname{Hom}_A(-, I)$ is exact. In other words, I is called injective if for any A-module Y, any submodule $X \subseteq Y$, and any A-linear map $f : X \to I$, there exists an A-linear map $g : Y \to I$ which agrees with f on X.

In the definition above, as the pair $X \subseteq Y$, we only need to consider the ring A and its ideal $\mathfrak{a} \subseteq A$. Namely:

Lemma 5.27 *For an A-module I, the following are equivalent:*

(1) *The A-module I is injective.*
(2) *For any ideal $\mathfrak{a}$ of A, $\operatorname{Ext}^1_A(A/\mathfrak{a}, I) = 0$.*
(2)′ *For any ideal $\mathfrak{a}$ of A and any A-linear map $f : \mathfrak{a} \to I$, f can be extended to $A \to I$.*
(3) *For any A-module M and any integer $i > 0$, we have $\operatorname{Ext}^i_A(M, I) = 0$.*

Proof (2) $\Rightarrow$(1). Let Y be an A-module, $X \subseteq Y$ be its submodule, and $f : X \to I$ be an A-linear map. Then we define $\mathcal{S}$ to be the set of pairs (Z, g) with Z an A-submodule of Y containing X and $g : Z \to I$ an A-linear map such that $g|_X = f$. In particular, $(X, f) \in \mathcal{S}$. For two elements $(Z, g), (Z', g')$ of $\mathcal{S}$, we say that $(Z, g) \leq (Z', g')$ if $Z \subseteq Z'$ and $g'|_Z = g$. Then it is easy to see that $\mathcal{S}$ is an ordered set with this relation, and any chain of $\mathcal{S}$ has an upper bound. By Zorn's lemma there exists a maximal element (Z, g) of $\mathcal{S}$. We want to show that $Z = Y$.

If $Z \subsetneq Y$, then when we take $y \in Y \setminus Z$ and set $Z' = Z + Ay$, then Z'/Z is generated by the image of y, and letting $\mathfrak{a}$ be the ideal $Z :_A y$ of A, we have $Z'/Z \cong A/\mathfrak{a}$. Taking the long exact sequence of Ext associated with the short exact sequence $0 \to Z \to Z' \to A/\mathfrak{a} \to 0$, we get $\mathrm{Hom}_A(Z', I) \to \mathrm{Hom}_A(Z, I) \to \mathrm{Ext}^1_A(A/\mathfrak{a}, I)$. Since we have $\mathrm{Ext}^1_A(A/\mathfrak{a}, I) = 0$ by assumption, the map $\mathrm{Hom}_A(Z', I) \to \mathrm{Hom}_A(Z, I)$ is surjective, and thus g is extended to some $g' : Z' \to I$. This contradicts the maximality of $(Z, g) \in \mathcal{S}$. Hence we have $Z = Y$. Namely, f can be extended to $g : Y \to X$. So I is an injective module.

(1) $\Rightarrow$(3). Let $P_\bullet \to M$ be a projective resolution of an A-module M. As the functor $\mathrm{Hom}_A(*, I)$ is exact, the cochain complex $\mathrm{Hom}_A(P_., I)$ is acyclic, and hence $\mathrm{Ext}^i_A(M, I) = 0$ for $i > 0$.

(3) $\Rightarrow$(2) is trivial.

We prove (2) $\Leftrightarrow$(2)$'$. As $0 \to \mathfrak{a} \to A \to A/\mathfrak{a} \to 0$ is a short exact sequence, we get an exact sequence

$$\mathrm{Hom}_A(A, I) \to \mathrm{Hom}_A(\mathfrak{a}, I) \to \mathrm{Ext}^1_A(A/\mathfrak{a}, I) \to 0 = \mathrm{Ext}^1_A(A, I).$$

So $\mathrm{Hom}_A(A, I) \to \mathrm{Hom}_A(\mathfrak{a}, I)$ is surjective (namely, (2)$'$ holds) if and only if $\mathrm{Ext}^1_A(A/\mathfrak{a}, I) = 0$ (that is, (2) holds), as desired. □

Corollary 5.28 *If A is a Noetherian ring, S is its multiplicatively closed subset, and I is an injective A-module, then $S^{-1}I$ is an injective $S^{-1}A$-module.*

Proof Note that every ideal of $S^{-1}A$ is of the form $\mathfrak{a}S^{-1}A$ with $\mathfrak{a}$ an ideal of A, and $\mathrm{Ext}^1_{S^{-1}A}(S^{-1}A/\mathfrak{a}S^{-1}A, S^{-1}I) \cong S^{-1}\mathrm{Ext}^1_A(A/\mathfrak{a}A, I)$, if A is Noetherian. The assertion follows immediately from Lemma 5.27. □

We see some examples.

Example 5.29

(1) Let A be an integral domain, and K be the field of fractions of A. Then any K-vector space V is injective as an A-module.
(2) Let A be a PID. Then an A-module I is injective if and only if I is *divisible*, that is, for any $x \in I$ and $a \in A \setminus \{0\}$, there exists some $y \in I$ such that $ay = x$.[3] If M is divisible, then M/N is also divisible for any A-submodule N of M.

Proof (1) This is because $\mathrm{Hom}_A(-, V) = \mathrm{Hom}_K(- \otimes_A K, V)$, and both $- \otimes_A K$ and $\mathrm{Hom}_K(-, V)$ are exact. (2) As A is a PID, any ideal $\mathfrak{a}$ of A is expressed as $\mathfrak{a} = (a)$. Assume that M is a divisible module. Then we can take $y \in M$ such that $ay = x$, and let $f : \mathfrak{a} \to I$ be the A-linear map determined by $f(a) = x$. Then $g : A \to I$ given by $g(r) = ry$ is the lift of f to A. Conversely, assume that I is injective, $x \in I$, and $a \in A \setminus \{0\}$. Then define $f : \mathfrak{a} = (a) \to I$ by

[3] Roughly speaking, $y = a^{-1}x$, and x is divisible by a.

$f(a) = x$, and take a lift $g : A \to I$ of f to A. Then defining $y = f(1)$, we have that $ay = f(a) = x$, and I is divisible. □

Now we are going to prove that any A-module is (isomorphic to) a submodule of some injective module. We start with the ring $\mathbb{Z}$ of integers.

Lemma 5.30 *We have the following:*

(1) *$\mathbb{Q}$ and $\mathbb{Q}/\mathbb{Z}$ are injective $\mathbb{Z}$-modules.*
(2) *Any $\mathbb{Z}$-module M is (isomorphic to) a submodule of some injective $\mathbb{Z}$-module.*

Proof

(1) follows immediately from Example 5.29.
(2) Let M be a $\mathbb{Z}$-module. Then we can write $M \cong F/X$ with F a free $\mathbb{Z}$-module, and X its submodule. Then $F \to F \otimes_{\mathbb{Z}} \mathbb{Q}$ is injective, and hence $M \cong F/X \to (F \otimes_{\mathbb{Z}} \mathbb{Q})/X$ is also injective. As $F \otimes_{\mathbb{Z}} \mathbb{Q}$ is a $\mathbb{Q}$-vector space, it is an injective $\mathbb{Z}$-module. As $\mathbb{Z}$ is a PID, its quotient $(F \otimes_{\mathbb{Z}} \mathbb{Q})/X$ is also an injective $\mathbb{Z}$-module by Example 5.29.

□

Before we move on, we remark:

Lemma 5.31 *Let $\varphi : A \to B$ be a ring homomorphism, M be a B-module, and N be an A-module. Then the A-module* $\mathrm{Hom}_A(M, N)$ *has also the structure of a B-module given by $(sf)(t) = f(st)$ ($f \in$* $\mathrm{Hom}_A(M, N), s \in B$*). Moreover, we have a natural isomorphism of B-modules*

$$\mathrm{Hom}_B(M, \mathrm{Hom}_A(B, N)) \cong \mathrm{Hom}_A(M, N).$$

Therefore, if N is an injective A-module, then $\mathrm{Hom}_A(B, N)$ *is an injective B-module.*

Proof For $\psi \in \mathrm{Hom}_B(M, \mathrm{Hom}_A(B, N))$ and $m \in M$, when we define $\alpha(\psi)(m) = \psi(m)(1)$, then $\alpha(\psi) \in \mathrm{Hom}_A(M, N)$, and it is easy to see that

$$\alpha : \mathrm{Hom}_B(M, \mathrm{Hom}_A(B, N)) \to \mathrm{Hom}_A(M, N)$$

is a B-linear map. When we define $\beta : \mathrm{Hom}_A(M, N) \to \mathrm{Hom}_B(M, \mathrm{Hom}_A(B, N))$ by $(\beta(f)(m))(s) = f(sm)$ for $f \in \mathrm{Hom}_A(M, N)$, $m \in M$ and $s \in B$, it is straightforward to check that β is the inverse of α. □

Proposition 5.32 *Let A be a ring. Then any A-module M is isomorphic to a submodule of an injective A-module.*

Proof We can consider that M is a $\mathbb{Z}$-module. Choose an injective $\mathbb{Z}$-module X such that M is its $\mathbb{Z}$-submodule, and set $I = \mathrm{Hom}_{\mathbb{Z}}(A, X)$. Then composing the A-linear injective maps

$$M \cong \mathrm{Hom}_A(A, M) \subseteq \mathrm{Hom}_{\mathbb{Z}}(A, M) \hookrightarrow \mathrm{Hom}_{\mathbb{Z}}(A, X) = I,$$

M is isomorphic to an A-submodule of an injective A-module, since I is an injective A-module by Lemma 5.31. □

Definition 5.33 An A-homomorphism $f : M \to N$ is said to be *essential* if for any submodule X of N, $f(M) \cap X = 0$ implies $X = 0$. If M is a submodule of N and the inclusion $i : M \to N$ is essential (this condition is equivalent to saying that if $g : N \to X$ is a homomorphism such that gi is injective, then g is injective), then we say that M is an *essential submodule* of N, or N is an *essential extension* of M.

Remark 5.1

(1) An extension $M \subseteq N$ is essential if and only if for any $n \in N \setminus \{0\}$, there exists some $a \in A$ such that $an \in M \setminus \{0\}$.
(2) Assume that A is Noetherian. If $M \subseteq N$ is an essential extension, then for any multiplicatively closed subset S of A, $S^{-1}M \subseteq S^{-1}N$ is also essential. Indeed, if $n/s \in S^{-1}N \setminus \{0\}$, then $tn \in N \setminus \{0\}$ for any $t \in S$. Let $I = 0 :_A tn$ be the maximal element of $\{0 :_A tn \mid t \in S\}$. By assumption, $Atn \cap M \neq 0$, and we can find some $a \in A$ such that $tan \in M \setminus \{0\}$. If $u \in S$ and $utan = 0$, then $a \in 0 :_A utn = 0 :_A tn$, and hence $tan = 0$, and this is a contradiction. Hence $utan \neq 0$ for any $u \in S$. This shows that $an/s \in S^{-1}M \setminus \{0\}$, and $S^{-1}M \subseteq S^{-1}N$ is essential, as desired.

Theorem 5.34 (Injective Envelope) *Let M be an A-module. Then there exists an injective A-module E and an essential monomorphism $f : M \to E$. Such a pair (E, f) is unique up to isomorphisms over M.*

We call the injective A-module M the *injective envelope* (or *injective hull*) of M, and denote it by $\mathrm{E}_A(M)$.

Proof Take an injective A-module I containing M. We set

$$\mathcal{S}_1 = \{J \mid J \text{ is an } A\text{-submodule of } I,\ M \subseteq J, \text{ and } J \text{ is an essential extension of } M\}.$$

As $M \in \mathcal{S}_1$, we have that $\mathcal{S}_1$ is nonempty. The set $\mathcal{S}_1$ is ordered by incidence relation, and any nonempty chain of $\mathcal{S}_1$ has an upper bound. So $\mathcal{S}_1$ has a maximal element J. Set

$$\mathcal{S}_2 = \{L \mid L \text{ is a submodule of } I \text{ and } L \cap J = 0\}.$$

Then $0 \in \mathcal{S}_2$, and $\mathcal{S}_2$ also satisfies the assumption of Zorn's lemma. Let $L \in \mathcal{S}_2$ be its maximal element, and consider the inclusion $i : J \to I$ and the canonical surjection $\varepsilon : I \to I/L$. Then the composition εi is injective, and by the maximality of L, it is essential. As εi is an injective map and I is an injective module, there exists an injective A-linear map $\alpha : I/L \to I$ such that $\alpha\varepsilon i = i$. Looking at the sequence of extensions $M \subseteq J \subseteq \operatorname{Im}\alpha \subseteq I$, as $M \subseteq J$ and $J \subseteq \operatorname{Im}\alpha$ are essential, the extension $M \subseteq \operatorname{Im}\alpha$ is also essential. By the maximality of J, we have that $J = \operatorname{Im}\alpha$. As εi is an isomorphism, the embedding $i : J \to I$ is a split monomorphism, and hence J is injective. □

Definition 5.35 (Injective Resolution) For an A-module M, take an injective A-module I^0 containing M as its submodule, and define X^0 by the short exact sequence $0 \to M \to I^0 \to X^0 \to 0$. Then do the same thing to X^0 and so on, and we get an exact sequence of A-modules

$$0 \to M \xrightarrow{\varepsilon} I^0 \xrightarrow{\partial^0} I^1 \xrightarrow{\partial^1} \cdots \xrightarrow{\partial^{i-1}} I^i \xrightarrow{\partial^i} I^{i+1} \xrightarrow{\partial^{i+1}} \cdots \qquad (\sharp)$$

in which each I^i is an injective A-module. Such an exact sequence is called an *injective resolution* of M.

By Theorem 5.34, the injective resolution ($\sharp$) can be constructed so that ε and each ∂^i $(i \geq 0)$ are essential. We call such an injective resolution the *minimal injective resolution* of M.

Note that the minimal injective resolution of M is unique up to isomorphisms. For example, $\mathbb{Q}$ is an injective $\mathbb{Z}$-module, and $\mathbb{Z}$ is an essential submodule of $\mathbb{Q}$. Hence, the short exact sequence of $\mathbb{Z}$-modules

$$0 \to \mathbb{Z} \to \mathbb{Q} \to \mathbb{Q}/\mathbb{Z} \to 0$$

is the minimal injective resolution of the $\mathbb{Z}$-module $\mathbb{Z}$.

Definition 5.36 (Injective Dimension) For an A-module M, we set

$$\begin{aligned} \mathcal{S} = \{m \in \mathbb{Z} \mid {} & \text{There exists an injective resolution} \\ & 0 \to M \to I^0 \to I^1 \to \cdots \to I^n \to \cdots \text{ of } M \text{ with } I^i = 0 \text{ for } i > m\}, \end{aligned}$$

and we define $\operatorname{id}_A M = \inf \mathcal{S}$, where $\inf \emptyset = \infty$ and $\inf \mathbb{Z} = -\infty$. We call $\operatorname{id}_A M$ the *injective dimension* of M.

Thus the injective dimension of the zero module is $-\infty$.

Lemma 5.37 *If* $0 \to M \to I^0 \to I^1 \to \cdots \to I^i \to \cdots$ *is the minimal injective resolution of an A-module M, then we have* $\operatorname{id}_A M = \inf\{i \in \mathbb{Z} \mid I^j = 0 \textit{ for all } j > i\}$.

Proof In fact, if $J^\cdot$ is an injective resolution of M, then there is a direct decomposition $J^\cdot \cong I^\cdot \oplus K^\cdot$, where $K^\cdot$ is an exact sequence whose terms are injective. We set $d_I^i : I^i \to I^{i+1}$ and $d_J^i : J^i \to J^{i+1}$ are the differential maps, and define $M_I^i = \operatorname{Ker} d_I^i$ and $M_J^i = \operatorname{Ker} d_J^i$. By induction on i, we define $\varphi^i : M_I^i \to M_J^i$, $\rho^i : M_J^i \to M_I^i$, $\psi^i : I^i \to J^i$, and $\pi^i : J^i \to I^i$ such that $\psi^\cdot : I^\cdot \to J^\cdot$ and $\pi^\cdot : J^\cdot \to I^\cdot$ are cochain maps such that $\pi^\cdot \psi^\cdot = \operatorname{id}$, and φ^i and ρ^i are the restrictions of ψ^i and π^i, respectively. When $i = 0$, then $M_I^0 = M_J^0 = M$. We define both φ^0 and ρ^0 to be id_M. If $i > 0$ and the maps are defined for smaller i, then φ^i and ρ^i are defined to be the unique maps which make the diagram (with exact rows)

$$
\begin{array}{ccccccc}
M_I^{i-1} & \longrightarrow & I^{i-1} & \xrightarrow{d_I^{i-1}} & M_I^i & \longrightarrow & 0 \\
\downarrow{\scriptstyle \varphi^{i-1}} & & \downarrow{\scriptstyle \psi^{i-1}} & & \downarrow{\scriptstyle \varphi^i} & & \\
M_J^{i-1} & \longrightarrow & J^{i-1} & \xrightarrow{d_J^{i-1}} & M_J^i & \longrightarrow & 0 \\
\downarrow{\scriptstyle \rho^{i-1}} & & \downarrow{\scriptstyle \pi^{i-1}} & & \downarrow{\scriptstyle \rho^i} & & \\
M_I^{i-1} & \longrightarrow & I^{i-1} & \xrightarrow{d_I^{i-1}} & M_I^i & \longrightarrow & 0
\end{array}
$$

commutative. As $\pi^{i-1}\psi^{i-1} = \mathrm{id}$, we have that $\rho^i\varphi^i = \mathrm{id}$. As J^i is injective, we can extend the composite of $M_I^i \xrightarrow{\varphi^i} M_J^i \xrightarrow{\upsilon^i} J^i$ to some $\psi^i : I^i \to J^i$, where υ^i is the inclusion. We have that ψ^i is injective, because $\upsilon^i\varphi^i$ is injective and $M_I^i \to I^i$ is an essential monomorphism. Then letting L be the push-out of the diagram $I^i \leftarrow M_I^i \xrightarrow{\varphi^i} M_J^i$, there is an induced injective map $h : L \to J^i$. Let $q : L \to I^i$ induced by $\mathrm{id} : I^i \to I^i$ and $M_J^i \xrightarrow{\rho^i} M_I^i \to I^i$. Since I^i is injective, q can be extended to $\pi^i : J^i \to I^i$ via h. It is easy to see that $\pi^i\psi^i = \mathrm{id}$, and we are done.

Letting $K^\bullet = \mathrm{Coker}\,\psi^\bullet$, we have the decomposition $J^\bullet = I^\bullet \oplus K^\bullet$. As the decomposition shows, $I^\bullet$ is the shortest injective resolution, and the assertion of the lemma follows. □

Proposition 5.38 *Let M be an A-module, and $n \geq 0$ be a non-negative integer. Then the following are equivalent:*

(1) $\mathrm{id}_A M \leq n$.
(2) *For any ideal $\mathfrak{a}$ of A, $\mathrm{Ext}_A^{n+1}(A/\mathfrak{a}, M) = 0$.*
(3) *For any A-module X and any integer $i > n$, we have $\mathrm{Ext}_A^i(X, M) = 0$.*

Proof We prove the proposition by induction on n. If $n = 0$, this is nothing but Lemma 5.27. Let $n > 0$, and assume that the lemma is true for smaller n.

(1) ⇒(3) ⇒(2) is trivial. We prove (2) ⇒(1). There exists an exact sequence

$$0 \to M \to I^0 \xrightarrow{d^0} I^1 \xrightarrow{d^1} I^2 \to \cdots \to I^{n-1} \to M^n \to 0 \qquad (\#)$$

with I^i is injective for $i = 0, 1, \ldots, n-1$. Letting $M^i = \mathrm{Im}\, d^{i-1}$ for $i = 1, \ldots, n-1$ and $M^0 = M$, we have short exact sequences

$$0 \to M^{i-1} \to I^{i-1} \to M^i \to 0$$

for $i = 1, 2, \ldots, n$. Then we have an exact sequence

$$0 = \mathrm{Ext}_A^{n+1-i}(A/\mathfrak{a}, I^{i-1}) \to \mathrm{Ext}_A^{n+1-i}(A/\mathfrak{a}, M^i) \to \mathrm{Ext}_A^{n+2-i}(A/\mathfrak{a}, M^{i-1}).$$

By induction on i, we have that $\operatorname{Ext}_A^{n+1-i}(A/\mathfrak{a}, M^i) = 0$ for $i = 1, 2, \dots, n$. By Lemma 5.27, M^n is injective, and hence the sequence (#) is an injective resolution of M of length $\leq n$. Hence $\operatorname{id}_A M \leq n$. □

The *global dimension* of a ring is defined as the maximum value of $\operatorname{hd}_A M$.

Definition 5.39 (Global Dimension) Let A be a ring. We set

$$\operatorname{gl.dim} A = \sup\{\operatorname{hd}_A M \mid M \text{ is an } A\text{-module}\},$$

and we call $\operatorname{gl.dim} A$ the *global dimension* of A.

Example 5.40 Any module over a field is free. Hence the global dimension of a field is 0. A Dedekind domain has global dimension 1.

The global dimension of a ring can be computed only using the projective dimensions of cyclic modules.

Lemma 5.41 (M. Auslander) $\operatorname{gl.dim} A = \sup\{\operatorname{hd}_A A/\mathfrak{a} \mid \mathfrak{a}$ *is an ideal of* $A\}$.

Proof Set $n = \sup\{\operatorname{hd}_A A/\mathfrak{a} \mid \mathfrak{a}$ is an ideal of $A\}$. It suffices to show that $\operatorname{gl.dim} A \leq n$. We may assume that $n < \infty$. Let M be any A-module. For any A-module X and any integer $i > n$, we have $\operatorname{Ext}_A^i(A/\mathfrak{a}, X) = 0$. So $\operatorname{id}_A X \leq n$, and hence for any $i > n$, we have that $\operatorname{Ext}_A^i(M, X) = 0$. Hence $\operatorname{hd}_A M \leq n$, as desired. □

5.3 Structure Theorem of Injective Modules, Macaulay's Inverse System, and Matlis Duality

At a first glance, injective modules are more difficult to understand than projective modules, but we can express one of them using the inverse system of Macaulay. Moreover, injective modules over Noetherian rings have a beautiful structure theorem. In addition, Noetherian and Artinian modules over a complete Noetherian local rings have a very beautiful duality theorem of Matlis. In this section, we will see these facts.

Theorem 5.42 *For a ring A, the following are equivalent:*

(1) *A is a Noetherian ring.*
(2) *For any family of injective A-modules* $\{I_\lambda\}_{\lambda\in\Lambda}$, *the direct sum* $\bigoplus_{\lambda\in\Lambda} I_\lambda$ *is again A-injective.*

Proof (1) ⇒(2). Let $\{I_\lambda\}_{\lambda\in\Lambda}$ be a family of A-modules, and M be a finitely generated A-module. We set $I = \bigoplus_{\lambda\in\Lambda} I_\lambda$. Then the map

$$h : \bigoplus_{\lambda \in \Lambda} \mathrm{Hom}_A(M, I_\lambda) \to \mathrm{Hom}_A(M, I)$$

given by $h(\sum_\lambda \psi_\lambda)(m) = \sum_\lambda \psi_\lambda(m)$ is an isomorphism, see Corollary B.14. As the ring A is Noetherian, M has a free resolution $\mathbb{P}.$ such that each $\mathbb{P}_i$ is finite free for $i \geq 0$. Then we have

$$\mathrm{Ext}^i_A(M, I) \cong \mathrm{H}^i(\mathrm{Hom}_A(\mathbb{P}., I)) \cong \mathrm{H}^i(\bigoplus_{\lambda \in \Lambda} \mathrm{Hom}_A(\mathbb{P}., I_\lambda)) \cong \bigoplus_{\lambda \in \Lambda} \mathrm{Ext}^i_A(M, I_\lambda).$$

So if each I_λ is injective, then for each ideal $\mathfrak{a}$ of A, we have

$$\mathrm{Ext}^1_A(A/\mathfrak{a}, I) \cong \bigoplus_{\lambda \in \Lambda} \mathrm{Ext}^1_A(A/\mathfrak{a}, I_\lambda) = 0,$$

and hence the direct sum $I = \bigoplus_{\lambda \in \Lambda} I_\lambda$ is an injective module, as desired.

(2) $\Rightarrow$(1) Let $\mathfrak{a}_1 \subseteq \mathfrak{a}_2 \subseteq \cdots \subseteq \mathfrak{a}_n \subseteq \cdots$ be an ascending chain of ideals of A, and set $\mathfrak{a} = \bigcup_{n \geq 1} \mathfrak{a}_n$. For each $n \geq 1$, we take an injective module E_n so that $A/\mathfrak{a}_n \subseteq E_n$. Then for $a \in \mathfrak{a}$, we can take n such that $a \in \mathfrak{a}_n$. Then for all $\ell \geq n$, we have $a \in \mathfrak{a}_\ell$, and hence we get

$$\varphi : \mathfrak{a} \to \bigoplus_{n \geq 1} E_n, \quad \varphi(a) = (a + \mathfrak{a}_1, a + \mathfrak{a}_2, \ldots, a + \mathfrak{a}_n, \ldots) \in \bigoplus_{n \geq 1} \mathfrak{a}/\mathfrak{a}_n \subseteq \bigoplus_{n \geq 1} E_n.$$

By assumption (2), $E = \bigoplus_{n \geq 1} E_n$ is injective. So letting $i : \mathfrak{a} \to A$ the inclusion, there exists an A-linear map $g : A \to E$ such that $gi = \varphi$. Then we have $\varphi(\mathfrak{a}) \subseteq A \cdot g(1)$. If $g(1) = (x_n)_{n \geq 1}$ ($x_n \in E_n$), then we have $\mathfrak{a}/\mathfrak{a}_n \subseteq E_n$ is contained in Ax_n. By the definition of the direct sum, $x_n = 0$ for $n \gg 0$. So $\mathfrak{a} = \mathfrak{a}_n$ for $n \gg 0$, and hence A is Noetherian. □

Exercise 5.1 Using the proposition above, prove the following *Eakin–Nagata theorem.*

Theorem *Let A be a subring of a Noetherian ring B. If B is a finite module over A, then A is Noetherian.*

We show a structure theorem on injective modules over Noetherian rings. We say that an A-module M is said to be *indecomposable* if $M \neq 0$, and M cannot be decomposed as $M \cong M_1 \oplus M_2$ with both M_1 and M_2 nonzero.

Lemma 5.43 *Let A be a Noetherian ring, and M be an A-module.*

(1) $\mathrm{Ass}_A(\mathrm{E}_A(M)) = \mathrm{Ass}_A(M)$.

(2) *If E is indecomposable and injective, then there exists a unique prime ideal $\mathfrak{p}$ such that $E \cong \mathrm{E}_A(A/\mathfrak{p})$, and $\mathrm{Ass}_A E = \{\mathfrak{p}\}$. Conversely, for any prime ideal $\mathfrak{p}$, $\mathrm{E}_A(A/\mathfrak{p})$ is an indecomposable injective A-module.*
(3) *If $x \in \mathrm{E}_A(A/\mathfrak{p}) \setminus \{0\}$, then $\mathrm{Ann}_A(x)$ is a $\mathfrak{p}$-primary irreducible ideal. Conversely, if $\mathfrak{q}$ is a $\mathfrak{p}$-primary irreducible ideal, then $\mathrm{E}_A(A/\mathfrak{q}) \cong \mathrm{E}_A(A/\mathfrak{p})$, and hence there exists some $x \in \mathrm{E}_A(A/\mathfrak{p})$ such that $\mathrm{Ann}_A(x) = \mathfrak{q}$.*
(4) *The action of an element of $A \setminus \mathfrak{p}$ on $\mathrm{E}_A(A/\mathfrak{p})$ is bijective. So $\mathrm{E}_A(A/\mathfrak{p})$ is an $A_\mathfrak{p}$-module, and $\mathrm{E}_A(A/\mathfrak{p}) \cong \mathrm{E}_{A_\mathfrak{p}}(\kappa(\mathfrak{p}))$, where $\kappa(\mathfrak{p}) = A_\mathfrak{p}/\mathfrak{p}A_\mathfrak{p}$.*

Proof

(1) As $M \subseteq \mathrm{E}_A(M)$, we have $\mathrm{Ass}_A(M) \subseteq \mathrm{Ass}_A(\mathrm{E}_A(M))$. Conversely, if $\mathfrak{p} \in \mathrm{Ass}_A(\mathrm{E}_A(M))$, then $A/\mathfrak{p} \subseteq \mathrm{E}_A(M)$. As $M \subseteq \mathrm{E}_A(M)$ is essential, $M \cap A/\mathfrak{p} \neq 0$, and hence $\mathfrak{p} \in \mathrm{Ass}_A(M)$.
(2) Take $\mathfrak{p} \in \mathrm{Ass}_A(E)$. Then $M \cong A/\mathfrak{p} \subseteq E$. As E is injective, the embedding $i : M \subseteq E$ extends to $\phi : \mathrm{E}_A(M) \to E$. As $M \subseteq \mathrm{E}_A(M)$ is essential, ϕ is injective. As $\mathrm{E}_A(M)$ is injective, ϕ splits, and $\mathrm{E}_A(M)$ is a direct summand of E. As E is indecomposable and $\mathrm{E}_A(M) \neq 0$, we have that $\mathrm{E}_A(M) \cong E$.

    If $E \cong \mathrm{E}_A(A/\mathfrak{p})$, then $\mathrm{Ass}_A E = \mathrm{Ass}_A \mathrm{E}_A(A/\mathfrak{p}) = \mathrm{Ass}_A A/\mathfrak{p} = \{\mathfrak{p}\}$, and the uniqueness of $\mathfrak{p}$ is also obvious.

    Conversely, consider the injective envelope $E = \mathrm{E}_A(A/\mathfrak{p})$ of $M \cong A/\mathfrak{p}$. Obviously $E \neq 0$. If $E = E_1 \oplus E_2$ with $E_1 \neq 0$ and $E_2 \neq 0$, letting $N_i = M \cap E_i$ $(i = 1, 2)$, N_1 and N_2 are nonzero ideals of $A/\mathfrak{p} = M$. So $E_1 \cap E_2 \supseteq N_1 \cap N_2 \neq 0$, and this is absurd.
(3) Set $E = \mathrm{E}_A(A/\mathfrak{p})$, and assume that $A/\mathfrak{q} \cong Ax \subseteq E$. By (1), $\emptyset \neq \mathrm{Ass}_A(A/\mathfrak{q}) \subseteq \mathrm{Ass}_A(E) = \{\mathfrak{p}\}$, and hence $\mathrm{Ass}_A(A/\mathfrak{q}) = \{\mathfrak{p}\}$, and $\mathfrak{q}$ is $\mathfrak{p}$-primary, see Proposition–Definition 2.22. So there is an injection $A/\mathfrak{p} \to A/\mathfrak{q} \cong Ax \subseteq \mathrm{E}_A(A/\mathfrak{p}) = E$. So $Ax \subseteq E$ is also essential, and hence $\mathrm{E}_A(A/\mathfrak{q}) \cong \mathrm{E}_A(A/\mathfrak{p})$. Obviously, $\mathfrak{q} \neq A$. So if $\mathfrak{q}$ is not irreducible, then we can write $\mathfrak{q} = \mathfrak{a} \cap \mathfrak{b}$ where $\mathfrak{a} \supsetneq \mathfrak{q}$ and $\mathfrak{b} \supsetneq \mathfrak{q}$. As $\mathfrak{a}x \neq 0$ and $\mathfrak{b}x \neq 0$, we have that $\mathfrak{a}x \cap A/\mathfrak{p}$ and $\mathfrak{b}x \cap A/\mathfrak{p}$ are nonzero ideals of $A/\mathfrak{p}$. So

$$0 \subsetneq (\mathfrak{a}x \cap A/\mathfrak{p}) \cap (\mathfrak{b}x \cap A/\mathfrak{p}) \subseteq \mathfrak{a}x \cap \mathfrak{b}x \cong (\mathfrak{a}/(\mathfrak{a} \cap \mathfrak{b})) \cap (\mathfrak{b}/(\mathfrak{a} \cap \mathfrak{b})) = 0,$$

    and this is absurd. So $\mathfrak{q}$ is irreducible.

    Conversely, let $\mathfrak{q}$ be an irreducible $\mathfrak{p}$-primary ideal. We show that $\mathrm{E}_A(A/\mathfrak{q}) \cong \mathrm{E}_A(A/\mathfrak{p})$. By Lemma 5.44 below, The 0 ideal of $A/\mathfrak{q}$ is irreducible. Take $M \cong A/\mathfrak{p}$, which is a submodule of $A/\mathfrak{q}$. The essential extension $M \to \mathrm{E}_A(M)$ factors through $\phi : A/\mathfrak{q} \to \mathrm{E}_A(M)$ by the injectivity of $\mathrm{E}_A(M)$. As $\mathrm{Ker}(\phi) \cap M = 0$ and 0 of $A/\mathfrak{q}$ is irreducible, we have that $\mathrm{Ker}(\phi) = 0$. As $M \subseteq A/\mathfrak{q} \subseteq \mathrm{E}_A(M)$ and $\mathrm{E}_A(M)$ is an essential extension of M, we have that $\phi : A/\mathfrak{q} \to \mathrm{E}_A(M)$ is also essential. So $\mathrm{E}_A(A/\mathfrak{q}) = \mathrm{E}_A(M) \cong \mathrm{E}_A(A/\mathfrak{p})$. In particular, $\mathrm{E}_A(A/\mathfrak{p})$ contains an A-submodule which is isomorphic to $A/\mathfrak{q}$, and the last assertion follows.
(4) Set $E = \mathrm{E}_A(A/\mathfrak{p})$. As $\mathrm{Ass}_A(E) = \{\mathfrak{p}\}$, for each $s \in A \setminus \mathfrak{p}$, $s_E : E \to E$ is injective, since the annihilator of a nonzero element of E is $\mathfrak{p}$-primary. As E

is injective, s_E is split injective. By the indecomposability of E, s_E must be an isomorphism. Letting s^{-1} acts on E by $(s_E)^{-1}$, E is an $A_\mathfrak{p}$-module in a natural way. Localizing $A/\mathfrak{p} \to E$ by $\mathfrak{p}$, we get an injective map $\kappa(\mathfrak{p}) = (A/\mathfrak{p})_\mathfrak{p} \to E_\mathfrak{p} = E$. As $E = E_\mathfrak{p}$ is an injective $A_\mathfrak{p}$-module and the extension $\kappa(\mathfrak{p}) \to E$ is essential, $E = \mathrm{E}_{A_\mathfrak{p}}(\kappa(\mathfrak{p}))$.

□

We used the following lemma in the proof of the lemma above.

Lemma 5.44 *Let $\mathfrak{q}$ be an ideal of A. Then $\mathfrak{q}$ is an irreducible ideal of A if and only if 0 is an irreducible ideal of $A/\mathfrak{q}$.*

Proof This follows from the correspondence theorem (Exercise 1.5), since the irreducibility of $\mathfrak{q}$ is a property of the set of ideals of A containing $\mathfrak{q}$ as an ordered set (the smallest element cannot be expressed as the infimum of two larger elements).

□

In Matlis duality, the injective envelope of the residue field plays a very important role.

Lemma 5.45 *Let $(A, \mathfrak{m})$ be a Noetherian local ring, and $E = \mathrm{E}_A(A/\mathfrak{m})$ the injective envelope of the residue field $A/\mathfrak{m}$. Then the following hold:*

(1) $\mathrm{Hom}_A(A/\mathfrak{m}, E) = [0 :_E \mathfrak{m}] \cong A/\mathfrak{m}$.
(2) *If $\mathfrak{q}$ is an $\mathfrak{m}$-primary ideal, then* $\ell_A(A/\mathfrak{q}) = \ell_A([0 :_E \mathfrak{q}])$.
(3) *If $\mathfrak{q}$ is an irreducible $\mathfrak{m}$-primary ideal, then $[0 :_E \mathfrak{q}]$ is cyclic, and $[0 :_E \mathfrak{q}] \cong A/\mathfrak{q}$.*

Proof

(1) By definition, E is the injective envelope of $V_0 \cong A/\mathfrak{m}$. The $A/\mathfrak{m}$-vector space $V := [0 :_E \mathfrak{m}]$ contains V_0. If $V \supsetneq V_0$, then we can take some $A/\mathfrak{m}$-subspace W of V such that $W \cap V_0 = 0$, and this contradicts the fact that E is an essential extension of V_0.
(2) Let $\mathfrak{q} = I_0 \subsetneq I_1 \subsetneq \ldots \subsetneq I_{l-1} = \mathfrak{m} \subsetneq I_l = A$ be a sequence of ideals of A such that $I_{i+1}/I_i \cong A/\mathfrak{m}$. Since we have that $\mathrm{Hom}_A(-, E)$ is an exact functor and $\mathrm{Hom}_A(A/\mathfrak{m}, E) \cong A/\mathfrak{m}$, we have that $\ell_A(A/\mathfrak{q}) = \ell_A([0 :_E \mathfrak{q}]) = l$.
(3) By Lemma 5.43, (3), we have that $\mathrm{E}_A(A/\mathfrak{q}) = E$, and E contains an A-submodule M which is isomorphic to $A/\mathfrak{q}$. Since $\mathrm{Ann}_A(M) = \mathfrak{q}$, we have $M \subseteq [0 :_E \mathrm{Ann}_A M] = [0 :_E \mathfrak{q}]$. Since we have $\ell_A(A/\mathfrak{q}) = \ell_A([0 :_E \mathfrak{q}])$ by (2), if follows that $M = [0 :_E \mathfrak{q}]$, and hence $[0 :_E \mathfrak{q}]$ is a cyclic module.

□

Theorem 5.46 *An injective module over a Noetherian ring A is isomorphic to the direct sum of $\mathrm{E}_A(A/\mathfrak{p})$ for various $\mathfrak{p} \in \mathrm{Spec}(A)$. If M is a finitely generated A-module, then we have*

$$\mathrm{E}_A(M) \cong \bigoplus_{\mathfrak{p} \in \mathrm{Ass}(A)} \mathrm{E}_A(A/\mathfrak{p})^{\oplus n(\mathfrak{p})},$$

where $n(\mathfrak{p}) = \dim_{\kappa(\mathfrak{p})} \mathrm{Hom}_{A_{\mathfrak{p}}}(\kappa(\mathfrak{p}), M_{\mathfrak{p}})$, *and the sum is a finite sum, since* $\mathrm{Ass}_A(M)$ *is a finite set.*

Proof Since we have shown in Lemma 5.43 that any indecomposable injective module is of the form $\mathrm{E}_A(A/\mathfrak{p})$, it suffices to show that any injective A-module E is the direct sum of indecomposable injective modules. Let S be the set of injective submodules of E which are direct sums of indecomposable injective modules. When we define the relation $\leq$ of S by defining $E_1 \leq E_2$ if and only if E_1 is a direct summand of E_2 for $E_1, E_2 \in S$, S has a maximal element E_0. As E_0 is injective, we can write $E = E_0 \oplus E'$. If $E' \neq 0$, then we have an injection $A/\mathfrak{q} \hookrightarrow E'$ for some $\mathfrak{q} \in \mathrm{Ass}_A E' \neq \emptyset$. Then we have a split injection $\mathrm{E}_A(A/\mathfrak{q}) \hookrightarrow E'$, and we can write $E' = \mathrm{E}_A(A/\mathfrak{q}) \oplus E''$. Then we have $E = E_0 \oplus \mathrm{E}_A(A/\mathfrak{q}) \oplus E''$, and this contradicts the maximality of E_0. Hence we have $E' = 0$, and $E = E_0$. So E is the direct sum of indecomposable injective modules. So the first half of the theorem has been proved.

Next, we prove the second half. First, we prove that $\mathrm{Hom}_{A_{\mathfrak{p}}}(\kappa(\mathfrak{p}), \mathrm{E}_A(A/\mathfrak{q})_{\mathfrak{p}}) \cong \kappa(\mathfrak{q})$ if $\mathfrak{p} = \mathfrak{q}$ and 0 otherwise. Indeed, if this Hom is nonzero, then $\mathfrak{p}A_{\mathfrak{p}} \in \mathrm{Ass}_{A_{\mathfrak{p}}} \mathrm{E}_A(A/\mathfrak{q})_{\mathfrak{p}} = \{\mathfrak{q}A_{\mathfrak{p}}\}$, and $\mathfrak{q} = \mathfrak{p}$. If $\mathfrak{q} = \mathfrak{p}$, then $\kappa(\mathfrak{p}) \to \mathrm{E}_A(A/\mathfrak{p})_{\mathfrak{p}}$ is the injective envelope, and hence the image must be the whole socle of $\mathrm{E}_A(A/\mathfrak{p})_{\mathfrak{p}}$, and the dimension of the Hom in problem is one. By Lemma 5.43, (1), we can write $\mathrm{E}_A(M) = \bigoplus_{\mathfrak{p}\in \mathrm{Spec}\, A} \mathrm{E}_A(A/\mathfrak{p})^{\oplus n(\mathfrak{p})}$, where $n(\mathfrak{p})$ is some cardinal number. Let $\mathfrak{p} \in \mathrm{Spec}\, A$. Then since $M_{\mathfrak{p}} \to \mathrm{E}_A(M)_{\mathfrak{p}}$ is the injective envelope as an $A_{\mathfrak{p}}$-module, we have

$$\begin{aligned} n(\mathfrak{p}) &= \dim_{\kappa(\mathfrak{p})} \mathrm{Hom}_{A_{\mathfrak{p}}}(\kappa(\mathfrak{p}), \mathrm{E}_A(A/\mathfrak{p})^{\oplus n(\mathfrak{p})}) \\ &= \dim_{\kappa(\mathfrak{p})} \mathrm{Hom}_{A_{\mathfrak{p}}}(\kappa(\mathfrak{p}), \mathrm{E}_A(M)_{\mathfrak{p}}) = \dim_{\kappa(\mathfrak{p})} \mathrm{Hom}_{A_{\mathfrak{p}}}(\kappa(\mathfrak{p}), M_{\mathfrak{p}}). \end{aligned}$$

Indeed, if the last equality is not true, then there is a map $\psi \in \mathrm{Hom}_{A_{\mathfrak{p}}}(\kappa(\mathfrak{p}), \mathrm{E}_A(M)_{\mathfrak{p}})$ such that $\mathrm{Im}\, \psi \not\subseteq M_{\mathfrak{p}}$. As $\mathrm{Im}\, \psi$ is at most one-dimensional, we have $\mathrm{Im}\, \psi \cap M_{\mathfrak{p}} = 0$, and this contradicts the fact that $M_{\mathfrak{p}} \to \mathrm{E}_A(M)_{\mathfrak{p}}$ is essential. □

Macaulay's inverse system is a good way to make injective modules easy to understand.[4]

Definition 5.47 (Macaulay's Inverse System) Let $A = k[X_1, \ldots, X_d]$ be the polynomial ring. We define

$$I = k[X_1^{\pm 1}, \ldots, X_d^{\pm 1}] / \sum_{i=1}^{d} X_i k[X_i, X_1^{\pm 1}, \ldots, X_{i-1}^{\pm 1}, X_{i+1}^{\pm 1}, \ldots, X_d^{\pm 1}].$$

Being a quotient of an A-module by its A-submodule, I is an A-module. Note that the canonical composite map

[4] Francis Sowerby Macaulay, 1862–1937 [Mac] is one of the major sources of commutative algebra.

$$k[X_1^{-1}, \ldots, X_n^{-1}] \hookrightarrow k[X_1^{\pm 1}, \ldots, X_n^{\pm n}] \to I$$

is an isomorphism of k-vector spaces, and it is often written as $I = k[X_1^{-1}, \ldots, X_n^{-1}]$. The action of A on the right-hand side is given by

$$(X_1^{a_1} X_2^{a_2} \cdots X_n^{a_n}) \cdot (X_1^{-b_1} X_2^{-b_2} \cdots X_n^{-b_n}) = \begin{cases} X_1^{a_1-b_1} X_2^{a_2-b_2} \cdots X_n^{a_n-b_n} & (\forall i\ a_i \le b_i) \\ 0 & (\exists i\ a_i > b_i) \end{cases},$$

and the equality $I = k[X_1^{-1}, \ldots, X_n^{-1}]$ is justified.

When we set I_n to be the k-span of

$$\{X_1^{-b_1} \cdots X_d^{-b_d} \mid (b_1, \ldots, b_d) \in \mathbb{Z}_{\ge 0}^d, \sum b_i \le n\},$$

then $I_n = [0 :_I \mathfrak{m}^{n+1}]$, and is an A-submodule of I. Since $I = \varinjlim I_n$, we have that each I_n is an $\widehat{A} = k[[X_1, \ldots, X_n]]$-module, and hence so is I.

Example 5.48 Let $d = 2$, $A = k[X, Y]$, and $I = k[X^{-1}, Y^{-1}]$. Then we have $X^2(X^{-2} + Y^{-2}) = X^2 \cdot X^{-2} + X^2 \cdot Y^{-2}$. The first term is 1, while the second term is zero, since the exponent of X in the product is positive. Therefore, $X^2(X^{-2} + Y^{-2}) = 1$. Note that $\mathrm{Ann}_A(X^{-2} + Y^{-2}) = (XY, X^2 - Y^2)$. Similarly, $\mathrm{Ann}_A(X^{-1}Y^{-1}) = (X^2, Y^2)$.

Theorem 5.49 *If $A = k[X_1, \ldots, X_d]$ and $\mathfrak{m} = (X_1, \ldots, X_d)$, then we have $\mathrm{E}_A(A/\mathfrak{m}) = I$. Also, when we set $\widehat{A} = k[[X_1, \ldots, X_d]]$, $\widehat{\mathfrak{m}} = \widehat{A}(X_1, \ldots, X_d)$, then $\mathrm{E}_{\widehat{A}}(\widehat{A}/\widehat{\mathfrak{m}}) = I$.*

Proof For any nonzero polynomial $f(X_1^{-1}, \ldots, X_d^{-1})$ of degree n, there exists a monomial $M = X_1^{a_1} \cdots X_d^{a_d}$ of degree n such that $Mf \in I_0 = k$. So the extension $I_0 = 0 :_I \mathfrak{m} \subseteq I$ is essential. So we get an injective A-linear map $I \to E = E_A(I_0) = E_A(A/\mathfrak{m})$. We set $I_n = [0 :_I \mathfrak{m}^{n+1}]$ and $E_n = [0 :_E \mathfrak{m}^{n+1}]$. By Lemma 5.45, $\dim_k E_n = \dim_k A/\mathfrak{m}^{n+1} = \binom{n+d}{d} = \dim_k I_n$. Thus $I_n \to E_n$ is an isomorphism. As $I = \varinjlim I_n$ and $E = \varinjlim E_n$, we have that $I \to E$ is also an isomorphism, and the first assertion has been proved. As $A/\mathfrak{m} \to E = \mathrm{E}_A(A/\mathfrak{m})$ is the injective hull, $A_\mathfrak{m}/\mathfrak{m}A_\mathfrak{m} \to \mathrm{E}_{A_\mathfrak{m}}(A_\mathfrak{m}/\mathfrak{m}A_\mathfrak{m})$ is also the injective hull. So the proof of the second assertion is reduced to the following lemma. □

Lemma 5.50 *Let $(A, \mathfrak{m})$ be a Noetherian local ring, and $\widehat{A}$ be its completion. Then $E = \mathrm{E}_A(A/\mathfrak{m})$ has a structure of an $\widehat{A}$-module in a natural way, and we have $E = \mathrm{E}_{\widehat{A}}(\widehat{A}/\widehat{\mathfrak{m}})$.*

Proof For any $x \neq 0 \in E$, $\mathfrak{q} = \mathrm{Ann}_A(x)$ is an $\mathfrak{m}$-primary ideal, and hence $A/\mathfrak{q} \cong \widehat{A}/\mathfrak{q}\widehat{A}$. So E is an $\widehat{A}$-module in a natural way. As $A/\mathfrak{m} \to E$ is also essential as an extension of $\widehat{A}$-modules, we get an injective $\widehat{A}$-linear map

$E \to \widehat{E}$, where $\widehat{E} = \mathrm{E}_{\widehat{A}}(\widehat{A}/\widehat{m})$. We set $E_n = \mathrm{Hom}_A(A/\mathfrak{m}^{n+1}, E)$ and $\widehat{E}_n = \mathrm{Hom}_A(A/\mathfrak{m}^{n+1}, \widehat{E}) = \mathrm{Hom}_{\widehat{A}}(\widehat{A}/\widehat{\mathfrak{m}}^n, \widehat{E})$. Then the canonical map $E_n \to \widehat{E}_n$ is also injective, and $\ell_A(E_n) = \ell_A(A/\mathfrak{m}^{n+1}) = \ell_{\widehat{A}}(\widehat{A}/\widehat{\mathfrak{m}}^{n+1}) = \ell_{\widehat{A}}(\widehat{E}_n) = \ell_A(\widehat{E}_n)$ by Lemma 5.45. So $E_n \to \widehat{E}_n$ must be bijective. Taking the inductive limit, we have that $E \to \widehat{E}$ is an isomorphism, as desired. □

Remark 5.1 By the structure theorem of complete local rings, if A is a complete local ring containing a field, then we can write $A \cong k[[X_1, \ldots, X_n]]/\mathfrak{a}$, where $k \cong A/\mathfrak{m}$. Then we have

$$\mathrm{E}_A(A/\mathfrak{m}) = [0 :_I \mathfrak{a}]; \quad I = k[X_1^{-1}, \ldots, X_d^{-1}],$$

and hence we may well say that we can describe $\mathrm{E}_A(A/\mathfrak{m})$ using Macaulay's inverse system for such an A.

Example 5.51 Let $A = k[X_1, \ldots, X_n]$, and $\mathfrak{m} = (X_1, \ldots, X_n)$. Using Macaulay's inverse system, we can describe the irreducible $\mathfrak{m}$-primary ideals of A. (If $\mathfrak{q}$ is an $\mathfrak{m}$-primary ideal, then $\mathfrak{q}$ is irreducible if and only if $A/\mathfrak{q}$ is Gorenstein, see Chap. 6. So this is also the structure theorem of Artinian Gorenstein quotients of A.) As $\mathrm{E}_A(A/\mathfrak{q}) = I$, we have that $M = A/\mathfrak{q} \subseteq I$. If $\varphi \in I$ is a generator of M, then $\mathfrak{q} = \mathrm{Ann}_A(\varphi)$. So any irreducible $\mathfrak{m}$-primary ideal is of the form $\mathrm{Ann}_A(\varphi)$ with $\varphi \in I$.

Exercise 5.2 Let $A = k[X, Y, Z]$. Compute the generators of the ideal $\mathrm{Ann}_A(\varphi)$ for each of the following $\varphi \in I$ (as $A = k[X, Y, Z]$ is of dimension three, the number of generators of an irreducible $\mathfrak{m}$-primary ideal is an odd integer which is greater than or equal to 3 (Junzo Watanabe [WJ] and Buchsbaum–Eisenbud [BE2])).

(1) $X^{-a}Y^{-b}Z^{-c}$ (a, b, c are positive integers);
(2) $X^{-3} + Y^{-3} + Z^{-3}$;
(3) $X^{-4} + Y^{-3}Z^{-1} + X^{-2}Y^{-1}Z^{-1} + Y^{-1}Z^{-3}$.

We may consider that Artinian modules are dual to Noetherian modules. We introduce the notion which is dual to the *number of minimal generators* $\mu_A(M) = \dim_{A/\mathfrak{m}} M/\mathfrak{m}M$ of Noetherian modules over a Noetherian local ring $(A, \mathfrak{m})$.

Definition 5.52 (The Type of Artinian Modules, Socles) Let $(A, \mathfrak{m})$ be a local ring, and M an Artinian A-module.

(1) The A-submodule $[0 :_M \mathfrak{m}]$ of M is called the *socle* of M, and is denoted by $\mathrm{Soc}_A M$. The socle $\mathrm{Soc}_A M \cong \mathrm{Hom}_A(A/\mathfrak{m}, M)$ is a left exact functor on M. A nonzero module of finite length has a nontrivial socle. As a nonzero Artinian A-module M contains a nonzero finitely generated Artinian submodule, which is nonzero and of finite length, M has a nontrivial socle. It follows easily that $\mathrm{Soc}_A M \subseteq M$ is the smallest essential submodule of M.
(2) Note that $\mathrm{Soc}_A M$ is a finite-dimensional (or equivalently, Artinian) vector space. We define $\mathrm{r}(M) = \dim_{A/\mathfrak{m}} \mathrm{Soc}_A M$ and call it the *type* of M. By Theorem 5.46, we have $\mathrm{E}_A(M) \cong \mathrm{E}_A(A/\mathfrak{m})^{\mathrm{r}(M)}$.

Next, we will state and prove Matlis duality. From the theorem, we will see that if $(A, \mathfrak{m})$ is a Noetherian complete local ring, then the category of Noetherian A-modules and the category of Artinian A-modules are contravariantly equivalent to each other under the functor $\mathrm{Hom}_A(-, \mathrm{E}_A(A/\mathfrak{m}))$.

Theorem 5.53 (Matlis Duality[5]) *Let $(A, \mathfrak{m})$ be a Noetherian local ring, M an A-module, and set $E = \mathrm{E}_A(A/\mathfrak{m}) = \mathrm{E}_{\widehat{A}}(\widehat{A}/\widehat{\mathfrak{m}})$. Then we have the following:*

(1) *We have* $\mathrm{Hom}_A(E, E) \cong \mathrm{Hom}_{\widehat{A}}(E, E) \cong \widehat{A}$.
(2) *An Artinian A-module has a structure of an $\widehat{A}$-module, and it is Artinian as an $\widehat{A}$-module. Conversely, an Artinian $\widehat{A}$-module is an Artinian A-module. We set $(-)^\vee = \mathrm{Hom}_A(-, E)$, and $(-)^\dagger = \mathrm{Hom}_{\widehat{A}}(-, E)$. Then $(-)^\vee$ and $(-)^\dagger$ are contravariant exact functors. If M is a Noetherian A-module, then $M^\vee$ is an Artinian A-module, and if M is an Artinian A-module, then $M^\vee = M^\dagger$ is a Noetherian $\widehat{A}$-module.*
(3) *If M is Noetherian, then the map $\delta_M : M \to M^{\vee\vee}$ given by $m \mapsto (\varphi \mapsto \varphi m)$ is identified with the completion $M \to \widehat{M}$. If M is Artinian, then $\widehat{\delta}_M : M \to M^{\dagger\dagger}$ is an isomorphism. Thus if $A = \widehat{A}$ is complete, then $(-)^\dagger$ gives an anti-equivalence between the categories of Noetherian A-modules and that of Artinian A-modules.*
(4) *If M is Noetherian, then $\mathrm{r}(M^\vee) = \mu_A(M)$. If M is Artinian, then $\mathrm{r}(M) = \mu_{\widehat{A}}(M^\vee)$.*
(5) *If M is of finite length, then $\ell_A(M) = \ell_A(M^\vee)$.*

Proof First, note that for an A-module N with $\mathrm{Supp}\, N \subseteq \{\mathfrak{m}\}$, we have $N = \bigcup W$ where W is an A-submodule of finite length. As each W is an $\widehat{A}$-module, so N has a structure of an $\widehat{A}$-module. Note that $W^\vee = W^\dagger$ for each W, and hence $N^\vee = \varprojlim W^\vee = \varprojlim W^\dagger = N^\dagger$. We set $A_n = A/\mathfrak{m}^{n+1}$, and $E_n = \mathrm{Hom}_A(A_n, E) = [0 :_E \mathfrak{m}^{n+1}]$.

Note that for a finite-dimensional $A/\mathfrak{m}$-vector space V, we have

$$V^\vee = \mathrm{Hom}_A(V, E) = \mathrm{Hom}_{A/\mathfrak{m}}(V, A/\mathfrak{m}) = V^*,$$

where $(-)^* = \mathrm{Hom}_{A/\mathfrak{m}}(-, A/\mathfrak{m})$ is the dual as the vector space. So (5) is true for vector spaces. Since $(-)^\vee$ is an exact functor and the length is additive with respect to short exact sequences, (5) follows easily using by the induction on length.

Similarly, $V \to V^{\vee\vee}$ in (3) is identified with the canonical map $V \to V^{**}$ for finite-dimensional $A/\mathfrak{m}$-vector spaces V, and is an isomorphism. Since $(-)^{\vee\vee}$ is exact, using the induction on the length and the five lemma, (3) is true for arbitrary M of finite length.

Now let N be an A-module with $\mathrm{Supp}\, N \subseteq \{\mathfrak{m}\}$. Then letting $N_n = [0 :_N \mathfrak{m}^n] \subseteq N$, we have that $N = \varinjlim N_n$, as Av is of finite length for any $v \in N$. So $N^\vee = \mathrm{Hom}_A(N, E) = \varprojlim \mathrm{Hom}_A(N_n, E) = \varprojlim \mathrm{Hom}_{A_n}(N_n, E_n)$. If M is a

[5] Eben Matlis, 1923–2015.

Noetherian A-module, then there is an exact sequence $A^r \to M \to 0$ for some r. Then $0 \to M^\vee \to E^r$ is exact, and $\operatorname{Supp} M^\vee \subseteq \{\mathfrak{m}\}$. So

$$M^{\vee\vee} = \varprojlim \operatorname{Hom}_{A_n}((M^\vee)_n, E_n) = \varprojlim \operatorname{Hom}_{A_n}(\operatorname{Hom}_A(A_n, \operatorname{Hom}_A(M, E)), E_n)$$
$$= \varprojlim \operatorname{Hom}_{A_n}(\operatorname{Hom}_A(M_n, E), E_n) = \varprojlim \operatorname{Hom}_{A_n}(\operatorname{Hom}_{A_n}(M_n, E_n), E_n),$$

where $M_n = A_n \otimes_A M$. Since $E_n = \mathrm{E}_{A_n}(A/\mathfrak{m})$, the last module is $\varprojlim M_n = \widehat{M}$ by (3) for finite length modules, which has already been proved. It is straightforward to check that the canonical map $M \to M^{\vee\vee}$ corresponds to the completion $M \to \widehat{M}$ with this isomorphism. Letting $M = A$, we have proved $\operatorname{Hom}_A(E, E) = \widehat{A}$. Since $E = \mathrm{E}_{\widehat{A}}(\widehat{A}/\widehat{\mathfrak{m}})$, we get $\operatorname{Hom}_{\widehat{A}}(E, E) = \widehat{A}$, replacing the ring A by $\widehat{A}$. So we have proved (1). □

Before we continue the proof of Theorem 5.53, we prove two lemmas.

Lemma 5.54 *Let A be a ring, and set* $\operatorname{Mod} A$ *to be the category of A-modules. Let $\mathcal{A}$ be an abelian category, and $F : \operatorname{Mod} A \to \mathcal{A}$ be an exact functor. Then the following conditions are equivalent:*

(1) *For any maximal ideal $\mathfrak{m}$ of A, $F(A/\mathfrak{m}) \neq 0$.*
(2) *For any A-module M, if $M \neq 0$, then $F(M) \neq 0$.*
(3) *F is faithful. That is, if $f : M \to N$ is a nonzero A-linear map, then $F(f) : F(M) \to F(N)$ is nonzero.*

Proof (1) $\Rightarrow$(2). Assume the contrary, and let M be a nonzero A-module such that $F(M) = 0$. Taking $x \in M \setminus \{0\}$ and considering $Ax \subseteq M$, we have $F(Ax) \subseteq F(M) = 0$, and hence we may assume that $M = A/I$ is a cyclic module. Taking a maximal ideal $\mathfrak{m}$ that contains I, we have an epimorphism $0 = F(M) = F(A/I) \to F(A/\mathfrak{m})$, and hence $F(A/\mathfrak{m}) = 0$. This is a contradiction.

(2) $\Rightarrow$(3). Assume the contrary, and let $f : M \to N$ be a nonzero A-linear map such that $F(f) : F(M) \to F(N)$ is zero. Then letting $I = \operatorname{Im} f$, we have that $I \neq 0$, but $F(I) = 0$.

(3) $\Rightarrow$(1). Let $\mathfrak{m}$ be a maximal ideal of A. Then the identity map $1_{A/\mathfrak{m}} : A/\mathfrak{m} \to A/\mathfrak{m}$ is nonzero. So the identity morphism $1_{F(A/\mathfrak{m})} : F(A/\mathfrak{m}) \to F(A/\mathfrak{m})$ is nonzero. In particular, $F(A/\mathfrak{m})$ is nonzero. □

Lemma 5.55 *Let $(A, \mathfrak{m})$ be a Noetherian local ring, and we set $E = \mathrm{E}_A(A/\mathfrak{m})$. For an A-module N, the following are equivalent:*

(1) *N is an Artinian module.*
(2) $\operatorname{Supp}_A N \subseteq \{\mathfrak{m}\}$, *and* $[0 :_N \mathfrak{m}]$ *is finite-dimensional.*
(3) $\mathrm{E}_A(N) \cong E^n$ *for some nonnegative integer n.*
(4) *N is a submodule of a module of the form E^n.*
(5) *$N^\vee$ is a Noetherian $\widehat{A}$-module.*
(6) *There exists a Noetherian $\widehat{A}$-module M such that $N \cong M^\dagger$ as an A-module.*

If these conditions are satisfied, then $N^\vee = N^\dagger$ is a Noetherian $\widehat{A}$-module, and $N \to N^{\dagger\dagger}$ is an isomorphism.

Proof (1) ⇒(2). For each $\nu \in N$, the submodule $A\nu$ is Artinian and Noetherian, and is of finite length. It follows that the annihilator of ν is $\mathfrak{m}$-primary. So $\operatorname{Supp}_A N \subseteq \{\mathfrak{m}\}$. Since $\operatorname{Soc}_A N \subseteq N$ is an Artinian vector space, it is finite-dimensional.

(2) ⇒(3). By Theorem 5.46, this is trivial.

(3) ⇒(4). This is trivial.

(4) ⇒(5). If $0 \to N \to E^n$ is exact, then $N^\vee = N^\dagger$, and $\widehat{A}^n \to N^\vee \to 0$ is an exact sequence of $\widehat{A}$-modules. So the implication is clear.

(5) ⇒(1). Letting C be the category of A-modules, The functor

$$(-)^\vee = \operatorname{Hom}_A(-, E) : C \to (C)^{\mathrm{op}}$$

is exact, since E is an injective module. Since $(A/\mathfrak{m})^\vee \cong A/\mathfrak{m} \neq 0$, $(-)^\vee$ is faithful by Lemma 5.54.

Now assume that $N^\vee$is a Noetherian $\widehat{A}$-module, but N is not Artinian. Then there is an infinite descending chain of A-submodules

$$E^n \supseteq N = N_0 \supsetneq N_1 \supsetneq N_2 \supsetneq \cdots .$$

Then we get an infinite sequence of surjective homomorphisms

$$\widehat{A}^n = (E^n)^\vee \to N^\vee = N_0^\vee \to N_1^\vee \to N_2^\vee \to \cdots$$

Letting $K_i = \operatorname{Ker}(\widehat{A}^n = N^\vee \to N_i^\vee) \cong (N/N_i)^\vee$, we have an infinite ascending chain

$$0 = K_0 \subsetneq K_1 \subsetneq K_2 \subsetneq \cdots$$

of submodules of $\widehat{A}^n$ (note that $K_i/K_{i-1} \cong (N_{i-1}/N_i)^\vee \neq 0$ by faithfulness). This contradicts the fact that $\widehat{A}$ is a Noetherian ring.

Now we know that (1)–(5) are equivalent.

Now consider the map $\widehat{\delta} : E \to E^{\dagger\dagger} = \operatorname{Hom}_{\widehat{A}}(\operatorname{Hom}_{\widehat{A}}(E, E), E) = \operatorname{Hom}_{\widehat{A}}(\widehat{A}, E) = E$. It is easy to see that this is nothing but the identity map, and the canonical map $E \to E^{\dagger\dagger}$ is an isomorphism. If N is an A-module which satisfies conditions (1)–(5), then $N \subseteq E^n$ for some n. Then E^n/N is also Artinian, and there is an exact sequence of the form

$$0 \to N \to E^n \to E^m.$$

As $\widehat{\delta} : E^n \to (E^n)^{\dagger\dagger}$ and $\widehat{\delta} : E^m \to (E^m)^{\dagger\dagger}$ are isomorphisms, we have that $\delta_N : N \to N^{\dagger\dagger}$ is also an isomorphism by the five lemma. So letting $M = N^\dagger$, (1) implies (6).

(6) ⇒(5). If $N \cong M^\dagger$ as A-modules for some finitely generated $\widehat{A}$-module M, then the isomorphism is also that of $\widehat{A}$-modules, since $M^\dagger$ is an Artinian A-module. So $N^\vee \cong N^\dagger \cong M^{\dagger\dagger} \cong \widehat{M} = M$.

The last assertions have already been proved above. □

Proof of Theorem 5.53 (Continued) We continue the proof of Theorem 5.53. (2) is now obvious from Lemma 5.55. The first part of (3) has already been proved above. The second part of (3) is proved in Lemma 5.55. So the rest of the assertions of (3) are now clear. To prove (4), we may assume that A is complete. For an Artinian module N, $\mathrm{r}(N)$ is the smallest n such that there is an injection $N \to E^n$, and for a Noetherian module M, $\mu(M)$ is the smallest n such that there is a surjection $A^n \to M$. Now the first assertion follows from the fact that there is a surjection $A^n \to M$ if and only if there is an injection $M^\vee \to E^n$. The second assertion is proved similarly. □

5.4 Regular Local Rings

The earliest and the most impactful event about the introduction of homological algebra into commutative algebra would be that the following definition is made possible.

Definition 5.56 (Regular Local Ring) Let $(A, \mathfrak{m})$ be a Noetherian local ring of dimension d. We say that A is a *regular local ring* if $\mathrm{gl.dim}\, A < \infty$.

In what follows, let $(A, \mathfrak{m})$ be Noetherian local ring of dimension d.

Example 5.57

(1) A is a regular local ring of dimension 0 if and only if A is a field.
(2) A is a regular local ring of dimension 1 if and only if A is a DVR.

Proof Assume that $\mathrm{gl.dim}\, A < \infty$ and $d = 0$. By Theorem 5.12, we have $\mathrm{gl.dim}\, A \leq \mathrm{depth}\, A \leq d$. So $d = 0$ implies that $\mathrm{gl.dim}\, A = 0$, and any A-module is free. In particular, $A/\mathfrak{m}$ is free, and hence $\mathfrak{m} = 0$. This shows that A is a field. Conversely, if A is a field, then any A-module is free, and so $\mathrm{gl.dim}\, A = 0$. Also, since 0 is maximal, A is a Noetherian local ring with $d = 0$.

Assume that $\mathrm{gl.dim}\, A < \infty$ and $d = 1$. Since $\mathrm{hd}_A A/\mathfrak{m} \leq 1$ and $0 \to \mathfrak{m} \to A \to A/\mathfrak{m} \to 0$ is exact, $\mathfrak{m}$ is free of rank one by Corollary 5.5, (4). So $\mathfrak{m}$ is a principal ideal, and hence A is a DVR by Definition 3.29. Conversely, assume that A is a DVR. Then clearly, $d = 1$. If $\mathfrak{m} = (x)$, then

$$0 \to A \xrightarrow{x} A \to A/\mathfrak{m} \to 0$$

is a free resolution of $A/\mathfrak{m}$, and $\mathrm{gl.dim}\, A < \infty$ by Theorem C.15 and Proposition C.10. □

The following theorem gives a characterization of regular local rings.

Theorem 5.58 *Let $(A, \mathfrak{m})$ be a d-dimensional Noetherian local ring. Then the following are equivalent:*

(1) *A is a regular local ring.*
(2) $\mathrm{hd}_A A/\mathfrak{m} \le d$.
(3) *The number of generators $\mu(\mathfrak{m})$ of $\mathfrak{m}$ is d.*
(4) *$\mathfrak{m}$ is generated by a regular sequence.*

If these conditions are satisfied, then A is a normal domain, and A is Cohen–Macaulay.

Proof (1) ⇒(2) follows from Theorem 5.12. (2) ⇒(1) follows from Theorem C.15.

(1) ⇒(3). Induction on d. The case that $d = 0$ is proved in Example 5.57. So we may assume that $d > 0$. Let $\mathrm{hd}_A A/\mathfrak{m} = n$, and

$$0 \to F_n \xrightarrow{\partial} F_{n-1} \to \cdots \to F_0 \to A/\mathfrak{m} \to 0$$

be the minimal free resolution of $A/\mathfrak{m}$. Since $d > 0$, we have $n > 0$. By Theorem 5.12, we have $\operatorname{depth} A \ge n > 0$. So $\mathfrak{m} \notin \mathrm{Ass}_A(A)$. Hence $\mathfrak{m} \not\subseteq \bigcup_{\mathfrak{p}\in \mathrm{Ass}_A(A)} \mathfrak{p} \cup \mathfrak{m}^2$. In other words, we can take a minimal set of generators $x_1, \ldots, x_d$ of $\mathfrak{m}$ such that x_1 is A-regular. We set $x = x_1$.

We prove that A/xA is a regular local ring. If so, as A/xA is a regular local ring of dimension $d-1$, $\mathfrak{m}/xA$ is generated by $d-1$ elements, and hence $\mathfrak{m}$ is generated by d elements. Now looking at the minimal free resolution

$$(\sharp) \quad 0 \to F_n \to \cdots \to F_2 \to F_1 \to \mathfrak{m} \to 0$$

of $\mathfrak{m}$, the element x acts on F_i $(1 \le i \le n)$ and $\mathfrak{m}$ as a non-zerodivisor. Applying the functor $A/xA \otimes_A -$ to ($\sharp$), we get a minimal free resolution

$$0 \to F_n/xF_n \to \cdots \to F_2/xF_2 \to F_1/xF_1 \to \mathfrak{m}/x\mathfrak{m} \to 0$$

of the A/xA-module $\mathfrak{m}/x\mathfrak{m}$. Hence $\mathrm{hd}_{A/xA}\mathfrak{m}/x\mathfrak{m} = n - 1$.

Letting $x_1 = x, x_2, \ldots, x_m$ be a minimal set of generators of $\mathfrak{m}$, and set $J = (x_2, x_3, \ldots, x_m)$. When we set $\overline{x_i} = x_i \mod x\mathfrak{m}$, we have

$$\mathfrak{m}/x\mathfrak{m} = A\overline{x_1} \oplus \sum_{i=2}^{m} A\overline{x_i}, \qquad A\overline{x_1} \cong A/\mathfrak{m}.$$

Indeed, for $a_i \in A$ $(1 \le i \le m)$, if $a_1x_1 + \sum_{i=2}^m a_ix_i \in x\mathfrak{m}$, then we have $a_1x_1 + \sum_{i=2}^m a_ix_i \in \mathfrak{m}^2$. As $x_1, x_2, \ldots, x_m$ is a minimal set of generators of $\mathfrak{m}$, it follows that $a_i \in \mathfrak{m}$, and hence $a_1x_1 = xa_1 \in x\mathfrak{m}$. So $A/\mathfrak{m}$ is a direct summand of $\mathfrak{m}/x\mathfrak{m}$ as an A/xA-module, and hence $\mathrm{hd}_{A/xA}A/\mathfrak{m} \le \mathrm{hd}_{A/xA}\mathfrak{m}/x\mathfrak{m} \le n-1$. As $A/\mathfrak{m}$ is also the residue field of A/xA, we have that $\mathrm{gl.dim}\, A/xA < \infty$ by Theorem C.15, and we have shown that A/xA is a regular local ring.

(3) ⇒(4). We first prove that A is an integral domain by induction on d. If $d = 0$, then $\mathfrak{m} = 0$, and $\mathfrak{m}$ is generated by the empty sequence, which is a regular sequence. So let $d > 0$. Let $x_1, \ldots, x_d$ be a minimal set of generators of $\mathfrak{m}$. Then the maximal

ideal $\mathfrak{m}/x_1A$ of A/x_1A is generated by $d-1$ elements. Since $\dim A/x_1A = d-1$, the ring A/x_1A also satisfies assumption (3), and it is an integral domain by induction. So x_1A is a prime ideal of height one. By Lemma 3.30, A is an integral domain, as desired.

Now we prove that $(x_1, \ldots, x_d)$ is a regular sequence by induction on d. Since we know that A is an integral domain, x_1 is a regular element. As A/x_1A satisfies (3), $(x_2, \ldots, x_d)$ is A/x_1A-regular, and this completes the proof. Now the Cohen–Macaulay property of A is clear.

(4) $\Rightarrow$(2). For any A-sequence $(x_1, \ldots, x_r)$, we have $\mathrm{hd}_A A/(x_1, \ldots, x_r) = r$. This can be shown by induction on r, using the short exact sequence

$$0 \to A/(x_1, \ldots, x_i) \to A/(x_1, \ldots, x_i) \to A/(x_1, \ldots, x_{i+1}) \to 0.$$

So if $\mathfrak{m}$ is generated by a regular sequence, then we have $\mathrm{hd}_A A/\mathfrak{m} < \infty$, and hence (2) follows.

The fact that every regular local ring is integrally closed follows from Theorem 3.45 above and Theorem 5.59, which follows. □

Traditionally, regular local rings had been defined by condition (3) in Theorem 5.58. In the era before homological algebra, the following theorem was a difficult open problem. However, it became almost trivial, thanks to the characterization using the global dimension.

Theorem 5.59 (Serre) *If A is a regular local ring, then for any $\mathfrak{p} \in \mathrm{Spec}\, A$, the localization $A_\mathfrak{p}$ is also regular.*

Proof If we localize the free resolution of $A/\mathfrak{p}$ of finite length at the prime $\mathfrak{p}$, then we get a free resolution of finite length of $\kappa(\mathfrak{p})$ as an $A_\mathfrak{p}$-module, and we have that $\mathrm{gl.dim}\, A_\mathfrak{p} \le \mathrm{hd}_{A_\mathfrak{p}} \kappa(\mathfrak{p}) < \infty$, as desired. □

As another important fact on regular local rings, M. Auslander and D. A. Buchsbaum showed the following theorem.

Theorem 5.60 (Auslander–Buchsbaum 2) *A regular local ring is a UFD.*

Proof Let $(A, \mathfrak{m})$ be a regular local ring of dimension d, and let $x = x_1, x_2, \ldots, x_d$ be a minimal set of generators of $\mathfrak{m}$. As we have seen in the proof of Theorem 5.58, x is a prime element. By Theorem 3.63, it suffices to show that $R = A[x^{-1}]$ is a UFD. By Proposition 3.61, it suffices to show that any height-one prime ideal of A is principal. Let P be a prime ideal of R with $\mathrm{ht}_R\, P = 1$. Letting $\mathfrak{p} = P \cap A$, we have $\mathfrak{p} \in \mathrm{Spec}\, A$, and $\mathrm{ht}_A\, \mathfrak{p} = 1$. We take a minimal free resolution

$$0 \to L_n \to L_{n-1} \to \cdots \to L_0 \to \mathfrak{p} \to 0$$

of the A-module $\mathfrak{p}$, and by localization, we get the finite free resolution

$$0 \to F_n \xrightarrow{d_n} F_{n-1} \to \cdots \to F_1 \xrightarrow{d_1} F_0 \to P \to 0 \qquad (*)$$

of the R-module P. Take a maximal ideal M of R, and set $Q = M \cap A$. Then Q is a prime ideal of A not containing x. Hence we have $Q \subsetneq \mathfrak{m}$, and thus $\dim A_Q = \dim R_M < d$, and by the induction assumption, R_M is a UFD, and the ideal PR_M of R_M is principal if $P \subseteq M$, and $PR_M = R_M$ otherwise. In either case, $PR_M \cong R_M$ for any $M \in \operatorname{Max} R$. So the R-module P is a projective R-module of rank one. Since each F_i is a finitely generated free R-module, the resolution $(*)$ splits. Letting $M_i = \operatorname{Im} d_i$ and $M_0 = P$, we have that $F_i = M_i \oplus M_{i+1}$ for $i = 0, 1, \ldots, n-1$. So $F_0 = M_0 \oplus M_1$ implies $F_0 \oplus M_2 = M_0 \oplus F_1$. This implies $F_0 \oplus F_2 = M_0 \oplus F_1 \oplus M_3$, and so on, and we have $F_0 \oplus F_2 \oplus F_4 \oplus \cdots = M_0 \oplus F_1 \oplus F_3 \oplus \cdots$. Namely, there exist some finite free R-modules F and G such that $P \oplus F \cong G$. Since $\operatorname{rank}_A P = 1$, letting $r = \operatorname{rank}_A F$, we have $\operatorname{rank}_A G = r + 1$, and hence using the isomorphism of exterior powers $R \cong \bigwedge^{r+1} G = \bigwedge^{r+1}(P \oplus F) \cong P \otimes_A \bigwedge^r F \cong P$, we have that $P \cong R$. Hence P is principal, as desired. □

5.5 Flat Modules and Flat Homomorphisms of Rings

The tensor product is an important device in commutative algebra. Associated with it, the notion of flat modules (see Theorem 1.115 for the definition) appears. Especially, flatness of ring homomorphisms is an important and convenient tool. In this section, we study the properties of flat modules more deeply.

Proposition 5.61 *For an A-module N, the following are equivalent:*

(1) *N is a flat A-module.*
(2) *For any A-module M and an integer $i \geq 1$, we have* $\operatorname{Tor}_i^A(M, N) = 0$.
(3) *For any finitely generated A-module M, we have* $\operatorname{Tor}_1^A(M, N) = 0$.
(4) *For any finitely generated ideal I of A, we have* $\operatorname{Tor}_1^A(A/I, N) = 0$.
(5) *For any finitely generated ideal I of A, we have that $I \otimes_A N \to N$ $(a \otimes n \mapsto an)$ is injective.*

Proof See Proposition C.9. □

Exercise 5.3 Let $A \to B$ be a flat morphism of rings, and M be a finitely presented A-module. Show that there is an isomorphism $\operatorname{Hom}_A(M, N) \otimes_A B \cong \operatorname{Hom}_B(M \otimes_A B, N \otimes_A B)$ for any A-module N.

Lemma 5.62 *Let M and N be A-modules, and B a flat A-algebra. Then the following hold:*

(1) *We have an isomorphism*

$$B \otimes_A \operatorname{Tor}_i^A(M, N) \cong \operatorname{Tor}_i^B(B \otimes_A M, B \otimes_A N).$$

for $i \geq 0$.
(2) *If A is Noetherian and M is finitely generated, then we have an isomorphism*

$$B \otimes_A \mathrm{Ext}^i_A(M, N) \cong \mathrm{Ext}^i_B(B \otimes_A M, B \otimes_A N)$$

for $i \geq 0$.

Proof Let $\mathbb{P}$ be a free resolution of M. In (2), we can take $\mathbb{P}$ so that each P_i is finitely generated. By the flatness, $B \otimes_A \mathbb{P}$ is a free resolution of $B \otimes_A M$.

(1)

$$B \otimes_A \mathrm{Tor}^A_i(M, N) = B \otimes_A \mathrm{H}_i(\mathbb{P} \otimes_A N) \cong \mathrm{H}_i(B \otimes_A (\mathbb{P} \otimes_A N)) \cong$$
$$\mathrm{H}_i((B \otimes_A \mathbb{P}) \otimes_B (B \otimes_A N)) \cong \mathrm{Tor}^B_i(B \otimes_A M, B \otimes_A N).$$

(2) Since each term of $\mathbb{P}$ is finitely presented, the canonical map

$$B \otimes \mathrm{Hom}_A(\mathbb{P}, N) \to \mathrm{Hom}_B(B \otimes_A \mathbb{P}, B \otimes_A N)$$

is an isomorphism, see Exercise 5.3. Hence we have

$$B \otimes_A \mathrm{Ext}^i_A(M, N) \cong B \otimes_A \mathrm{H}^i(\mathrm{Hom}_A(\mathbb{P}, N)) \cong \mathrm{H}^i(B \otimes_A \mathrm{Hom}_A(\mathbb{P}, N)) \cong$$
$$\mathrm{H}^i(\mathrm{Hom}_B(B \otimes_A \mathbb{P}, B \otimes_A N)) \cong \mathrm{Ext}^i_B(B \otimes_A M, B \otimes_A N),$$

as desired.

□

We list some basic properties of flat modules.

Proposition 5.63 *The following hold:*

(1) *If $A \to A'$ is a ring homomorphism, and M is a flat A-module, then $M \otimes_A A'$ is A'-flat.*
(2) *M is A-flat if and only if for any prime ideal $\mathfrak{p}$ of A, $M_\mathfrak{p}$ is $A_\mathfrak{p}$-flat (instead of "any prime ideal," we can say "any maximal ideal").*
(3) *Let $0 \to M' \to M \to M'' \to 0$ be a short exact sequence of A-modules, and assume that M'' is A-flat. Then M is A-flat if and only if M' is A-flat.*

Proof

(1) For an A'-module N, we have an isomorphism $(M \otimes_A A') \otimes_{A'} N \cong M \otimes_A N$, and hence this functor on N is exact.
(2) By Lemma 5.62, we get $\mathrm{Tor}^A_i(M, A/\mathfrak{a})_\mathfrak{p} \cong \mathrm{Tor}^{A_\mathfrak{p}}_i(M_\mathfrak{p}, (A/\mathfrak{a})_\mathfrak{p})$. Hence if $\mathrm{Tor}^A_i(M, A/\mathfrak{a}) \neq 0$ for some $\mathfrak{a}$ and $i > 0$, then for any $\mathfrak{p} \in \mathrm{Supp}(\mathrm{Tor}^A_i(M, A/\mathfrak{a}))$, $\mathrm{Tor}^{A_\mathfrak{p}}_i(M_\mathfrak{p}, (A/\mathfrak{a})_\mathfrak{p}) \neq 0$.
(3) For any A-module N, there is an exact sequence

$$0 = \mathrm{Tor}^A_2(M'', N) \to \mathrm{Tor}^A_1(M', N) \to \mathrm{Tor}^A_1(M, N) \to \mathrm{Tor}^A_1(M'', N) = 0$$

of Tor, and the assertion follows immediately.

□

Example 5.64

(1) Being a direct summand of a free module, a projective A-module is A-flat. If A is Noetherian, a finitely generated flat A-module is projective, see Lemma 5.109 below. In addition, if $(A, \mathfrak{m})$ is a Noetherian local ring, then a finitely generated flat A-module is A-free. So for a finitely generated module over a Noetherian local ring, the three notions "free," "projective," and "flat" are the same.
(2) The localization $S^{-1}A$ as an A-module is A-flat.
(3) If $(A, \mathfrak{m})$ is a Noetherian local ring and $\widehat{A}$ is its completion, then $\widehat{A}$ is a flat A-module. Since $\widehat{A}/\mathfrak{m}\widehat{A} \cong A/\mathfrak{m} \neq 0$, it is a faithfully flat A-module, see Definition 5.76 below.

By Proposition 5.63, flatness can be checked locally. For modules over a Noetherian local ring, there is a local criterion of flatness. It utilizes the associated graded rings and modules defined in Chap. 4.

Let $\mathfrak{a}$ be an ideal of a ring A, and M be an A-module. As in (4.20), we denote the associated graded ring $\bigoplus_{n\geq 0} \mathfrak{a}^n/\mathfrak{a}^{n+1}$ by $\mathrm{G}_{\mathfrak{a}}(A)$, and the associated graded module $\bigoplus_{n\geq 0} \mathfrak{a}^n M/\mathfrak{a}^{n+1}M$ by $\mathrm{G}_{\mathfrak{a}}(M)$. As pointed out in (4.20), the action

$$\phi_n : \mathfrak{a}^n/\mathfrak{a}^{n+1} \otimes_{A/\mathfrak{a}} M/\mathfrak{a}M \to \mathfrak{a}^n M/\mathfrak{a}^{n+1}M$$

is surjective, and hence the action $\phi : \mathrm{G}_{\mathfrak{a}}(A) \otimes_{A/\mathfrak{a}} M/\mathfrak{a}M \to \mathrm{G}_{\mathfrak{a}}(M)$ is also surjective, see (4.20).

Using these notions, we state the local criterion of flatness.

Theorem 5.65 (Local Criterion of Flatness) *For a ring A and an ideal $\mathfrak{a}$, we consider the following four conditions:*

(1) *M is a flat A-module.*
(2) *$M/\mathfrak{a}M$ is a flat $A/\mathfrak{a}$-module, and $\mathrm{Tor}_1^A(A/\mathfrak{a}, M) = 0$.*
(3) *$M/\mathfrak{a}M$ is a flat $A/\mathfrak{a}$-module, and the map ϕ in (4.20) is bijective.*
(4) *For any $n \geq 1$, $M/\mathfrak{a}^n M$ is a flat $A/\mathfrak{a}^n$-module.*

Then we have (1) ⇒(2) ⇒(3) ⇒(4). If $\mathfrak{a}$ is nilpotent, or A is Noetherian and the A-module $I \otimes_A M$ is separated with respect to the $\mathfrak{a}$-adic topology for any ideal I of A, that is, $\bigcap_{n\geq 1} \mathfrak{a}^n(I \otimes_A M) = 0$, then the four conditions are equivalent.

Proof (1) ⇒(2) ⇒(3) is clear from Proposition 5.61.

We prove (3) ⇒(4). Let $n \geq 1$. By the short exact sequence $0 \to \mathfrak{a}/\mathfrak{a}^n \to A/\mathfrak{a}^n \to A/\mathfrak{a} \to 0$, we get an exact sequence

$$0 \to \mathrm{Tor}_1^{A/\mathfrak{a}^n}(A/\mathfrak{a}, M/\mathfrak{a}^n M) \to \mathfrak{a}/\mathfrak{a}^n \otimes_A M \to A/\mathfrak{a}^n \otimes_A M \to A/\mathfrak{a} \otimes_A M \to 0.$$

So we get a commutative diagram

$$
\begin{array}{ccccccc}
 & & 0 & & 0 & & \\
 & & \downarrow & & \downarrow & & \\
 & & \mathrm{Tor}_1^{A/\mathfrak{a}^n}(A/\mathfrak{a}, M/\mathfrak{a}^n M) & & \mathrm{Tor}_1^{A/\mathfrak{a}^{n-1}}(A/\mathfrak{a}, M/\mathfrak{a}^{n-1} M) & & \\
 & & \downarrow & & \downarrow & & \\
\mathfrak{a}^{n-1}/\mathfrak{a}^n \otimes_A M & \longrightarrow & \mathfrak{a}/\mathfrak{a}^n \otimes_A M & \longrightarrow & \mathfrak{a}/\mathfrak{a}^{n-1} \otimes_A M & \longrightarrow & 0 \\
\phi_{n-1} \downarrow \cong & & \downarrow & & \downarrow & & \\
0 \to \mathfrak{a}^{n-1} M/\mathfrak{a}^n M & \longrightarrow & M/\mathfrak{a}^n M & \longrightarrow & M/\mathfrak{a}^{n-1} M & \longrightarrow & 0 \\
 & & \downarrow & & \downarrow & & \\
 & & M/\mathfrak{a}M & & M/\mathfrak{a}M & & \\
 & & \downarrow & & \downarrow & & \\
 & & 0 & & 0 & . &
\end{array}
$$

with exact rows and columns. Since ϕ_{n-1} is an isomorphism, using the induction on n and the snake lemma, we get

$$\mathrm{Tor}_1^{A/\mathfrak{a}^n}(A/\mathfrak{a}, M/\mathfrak{a}^n M) = 0 \qquad (\forall n \geq 1).$$

In order to show that $M/\mathfrak{a}^n M$ is $A/\mathfrak{a}^n$-flat to complete the proof of (3) $\Rightarrow$(4), we only need to prove that the implication (2) $\Rightarrow$(1) is true for the case that $\mathfrak{a}$ is nilpotent. Let X be any A-module. By nilpotency, $\mathfrak{a}^\ell X = 0$ for some $\ell \geq 1$. In order to show that $\mathrm{Tor}_1^A(X, M) = 0$, we may assume that $\mathfrak{a}X = 0$, using the induction on ℓ and the short exact sequence $0 \to \mathfrak{a}X \to X \to X/\mathfrak{a}X \to 0$. Take a presentation $0 \to N \to F \to M \to 0$ of the A-module M, where F is free. Since we have $\mathrm{Tor}_1^A(A/\mathfrak{a}, M) = 0$, the sequence $0 \to N/\mathfrak{a}N \to F/\mathfrak{a}F \to M/\mathfrak{a}M \to 0$ is exact. Since $M/\mathfrak{a}M$ is $A/\mathfrak{a}$-flat, the sequence

$$0 \to X \otimes_{A/\mathfrak{a}} N/\mathfrak{a}N \to X \otimes_{A/\mathfrak{a}} F/\mathfrak{a}F \to X \otimes_{A/\mathfrak{a}} M/\mathfrak{a}M \to 0 \qquad (*)$$

is exact. As $X \otimes_{A/\mathfrak{a}} L/\mathfrak{a}L = X \otimes_A L$ for any A-module L, the exact sequence $(*)$ is nothing but

$$0 \to X \otimes_A N \to X \otimes_A F \to X \otimes_A M \to 0.$$

Therefore, by the long exact sequence of Tor

$$\ldots \to \mathrm{Tor}_1^A(X, F) \to \mathrm{Tor}_1^A(X, M) \to X \otimes_A N \to X \otimes_A F \to X \otimes_A M \to 0,$$

it follows that $\mathrm{Tor}_1^A(X, M) = 0$, and hence M is A-flat.

(4) $\Rightarrow$(1). The case that $\mathfrak{a}$ is nilpotent is trivial. Assuming that that A is Noetherian, and that for any ideal I of A, $\bigcap_{n\geq 0} \mathfrak{a}^n(I \otimes_A M) = 0$, we want to prove that M is A-flat. It suffices to prove that for an ideal I of A, $I \otimes_A M \overset{\psi}{\to} M$ is injective (Proposition 5.61). Assume the contrary, and let $0 \neq x \in \mathrm{Ker}\,\psi$. We can take $n \geq 1$ so that $x \notin \mathfrak{a}^n\cdot(I \otimes_A M)$ by assumption. By Theorem 4.5, there exists some $k \gg 0$ such that

$$\mathfrak{a}^k \cap I \subseteq \mathfrak{a}^n I.$$

From the exact sequence

$$0 \to \mathfrak{a}^k \cap I \overset{\iota}{\to} I \overset{\varepsilon}{\to} I/(\mathfrak{a}^k \cap I) \to 0,$$

we get a commutative diagram with exact rows

$$\begin{array}{ccccccc}
(\mathfrak{a}^k \cap I) \otimes_A M & \xrightarrow{\iota\otimes 1_M} & I \otimes_A M & \xrightarrow{\varepsilon\otimes 1_M} & (I/(\mathfrak{a}^k \cap I)) \otimes_A M & \longrightarrow & 0\ . \\
 & & \Big\downarrow{\scriptstyle\psi} & & \Big\downarrow{\scriptstyle\rho} & & \\
 & & M & \longrightarrow & A/\mathfrak{a}^k \otimes_A M & \longrightarrow & 0
\end{array}$$

By our assumption (4), ρ is injective. Since $\psi(x) = 0$, we have that $x \in \mathrm{Ker}(\varepsilon \otimes 1_M) = \mathrm{Im}(\iota \otimes 1_M)$. On the other hand, since $\mathfrak{a}^k \cap I \subseteq \mathfrak{a}^n I$, we have $x \notin \mathfrak{a}^n(I \otimes_A M) \supseteq \mathrm{Im}(\iota \otimes 1_M) \ni x$, and this is a contradiction. Hence M is A-flat. □

From the theorem, we can show the following.

Corollary 5.66 *Let $(A, \mathfrak{m}, k)$ and $(A', \mathfrak{m}', k')$ be Noetherian local rings, and $f : A \to A'$ be a local homomorphism. Let M and N be A'-modules. Then the following hold:*

(1) *If N is a finitely generated A'-module, then N is A-flat if and only if* $\mathrm{Tor}_1^A(N, k) = 0$.
(2) *Assume that both M and N are finitely generated A'-modules, and N is A-flat. Let $u \in$* $\mathrm{Hom}_{A'}(M, N)$. *Then the following two conditions are equivalent:*
    (a) *u is injective and* $\mathrm{Coker}(u)$ *is A-flat.*
    (b) *$u \otimes 1 : M \otimes_A k \to N \otimes_A k$ is injective.*

Proof

(1) Since $I \otimes_A N$ is a finitely generated A'-module for any ideal I of A and $\mathfrak{m}A' \subseteq \mathfrak{m}'$, we have that $I \otimes_A N$ is separated with respect to the $\mathfrak{m}$-adic topology. Since $A/\mathfrak{m}$ is a field, $N/\mathfrak{m}N$ is $A/\mathfrak{m}$-flat. The assertion follows immediately from Theorem 5.65.

(2) (a) $\Rightarrow$(b) is obvious. Assuming condition (b), we apply the functor $-\otimes_A k$ to the two exact sequences

$$0 \to X \to M \to Y = \mathrm{Im}(u) \to 0, \quad 0 \to Y \to N \to Z = \mathrm{Coker}(u) \to 0.$$

As the composition of $M \otimes_A k \to Y \otimes_A k \to N \otimes_A k$ is injective, the map $M \otimes_A k \to Y \otimes_A k$ is injective, and hence is bijective. So $Y \otimes_A k \to N \otimes_A k$ is also injective. Since N is A-flat, we have $\mathrm{Tor}_1^A(Z, k) = 0$. Since $Z = \mathrm{Coker}(u)$ is a finitely generated A'-module, it is A-flat by (1). By Proposition 5.61, (3), Y is also A-flat. Since $M \otimes_A k \to Y \otimes_A k$ is injective and $\mathrm{Tor}_1^A(Y, k) = 0$, we have that $X \otimes_A k = 0$. As X is a finitely generated A'-module, $X = 0$ by NAK, Theorem 1.86. Hence u is injective. □

For a ring homomorphism $f : A \to A'$, we say that f is *flat* if A' as an A-module through f is A-flat.

Corollary 5.67 *Let* $(A, \mathfrak{m}) \xrightarrow{f} (B, \mathfrak{n}) \xrightarrow{g} (C, \mathfrak{l})$ *be a sequence of local homomorphisms of Noetherian local rings. Assume that* f *and* gf *are flat. Then* g *is flat if and only if* $\overline{g} : B/\mathfrak{m}B \to C/\mathfrak{m}C$ *induced by* g *is flat.*

Proof The 'only if' part follows from Proposition 5.63. We prove the 'if' part. Since $I \otimes_B C$ is separated with respect to the $\mathfrak{m}B$-adic topology for every ideal I of B and $C/\mathfrak{m}C$ is $B/\mathfrak{m}B$-flat, it suffices to show that $\mathrm{Tor}_1^B(B/\mathfrak{m}B, C) = 0$ by Theorem 5.65. Let $\mathbb{P}$ be a projective resolution of the A-module $A/\mathfrak{m}$. Since B is A-flat, $B \otimes_A \mathbb{P}$ is a projective resolution of $B/\mathfrak{m}B$. So

$$\mathrm{Tor}_1^B(B/\mathfrak{m}B, C) = \mathrm{H}_1(B/\mathfrak{m}B \otimes_B C) \cong \mathrm{H}_1(A/\mathfrak{m} \otimes_A C) \cong \mathrm{Tor}_1^A(A/\mathfrak{m}, C) = 0,$$

since C is also A-flat. □

We will see that flat ring homomorphisms preserve various ring theoretic properties. We start with the behaviors of associated primes.

Theorem 5.68 *Let* A *be a Noetherian ring, and* $f : A \to A'$ *be a ring homomorphism. Let* M *be an* A*-module, and* N *be an* A*-flat* A'*-module. Then we have*

$$\mathrm{Ass}_{A'}(M \otimes_A N) = \bigcup_{\mathfrak{p} \in \mathrm{Ass}_A(M)} \mathrm{Ass}_{A'}(A/\mathfrak{p} \otimes_A N).$$

Proof We prove the direction $\supseteq$. If $\mathfrak{p} \in \mathrm{Ass}_A(M)$, then we may consider that $A/\mathfrak{p} \subseteq M$. Since N is A-flat, we get an injective map $A/\mathfrak{p} \otimes_A N \to M \otimes_A N$. So we have $\mathrm{Ass}_{A'}(A/\mathfrak{p} \otimes_A N) \subseteq \mathrm{Ass}_{A'}(M \otimes_A N)$ (as the proof shows, we do not need the Noetherian property of A for this direction).

For the direction $\subseteq$, we prove a lemma. □

Lemma 5.69 *Let A be a Noetherian ring, $f : A \to A'$ be a ring homomorphism, $\mathfrak{p} \in \operatorname{Spec} A$, and M be an A-module such that $\operatorname{Ass}_A(M) = \{\mathfrak{p}\}$. Assume that $\mathfrak{p}^n M = 0$ for some $n \geq 1$. Let N be an A-flat A'-module. Then for any $\mathfrak{q} \in \operatorname{Ass}_{A'}(M \otimes_A N)$, $f^{-1}(\mathfrak{q}) = \mathfrak{p}$.*

Proof $\mathfrak{q} \in \operatorname{Ass}_{A'}(M \otimes_A N)$ is a prime ideal of A' that contains $\operatorname{Ann}_{A'}(M \otimes_A N)$. As $\mathfrak{p}^n A' \subseteq \operatorname{Ann}_{A'}(M \otimes_A N) \subseteq \mathfrak{q}$, we have that $f^{-1}(\mathfrak{q}) \supseteq \mathfrak{p}$. Conversely, if $x \in A \setminus \mathfrak{p}$, then x is M-regular. So it is also $M \otimes_A N$-regular by the flatness of N. So $x \notin f^{-1}(\mathfrak{q})$, and this completes the proof of $f^{-1}(\mathfrak{q}) = \mathfrak{p}$.

We continue the proof of Theorem 5.68, ($\subseteq$). As $\mathfrak{q} \in \operatorname{Ass}_{A'}(M \otimes_A N)$ is the annihilator of some element of $M \otimes_A N$, we may assume that M is finitely generated. From a primary decomposition of M, we get a finite set of finitely generated quotient modules M_i ($i = 1, \ldots, n$) of M such that $\operatorname{Ass}_A(M_i) = \{\mathfrak{p}_i\} \subseteq \operatorname{Ass}_A M$ for each i, and $M \subseteq \bigoplus_{i=1}^n M_i$. Since N is A-flat, the injectivity is preserved by $- \otimes_A N$, and hence we may assume that $\operatorname{Ass}_A(M) = \{\mathfrak{p}\}$.

By Lemma 5.69 and Proposition 2.20, we may localize at $\mathfrak{p}$, and hence we may assume that $(A, \mathfrak{p})$ is a Noetherian local ring. Then M has a composition series $0 = M_0 \subseteq M_1 \subseteq \cdots \subseteq M_r = M$, where $M_i / M_{i-1} \cong A/\mathfrak{p}$. Then

$$0 = M_0 \otimes_A N \subseteq M_1 \otimes_A N \subseteq \cdots \subseteq M_r \otimes_A N = M \otimes_A N$$

and $(M_i \otimes_A N)/(M_{i-1} \otimes_A N) \cong A/\mathfrak{p} \otimes_A N$ by flatness. So $\operatorname{Ass}_{A'}(M_i \otimes_A N) \subseteq \operatorname{Ass}_{A'}(A/\mathfrak{p} \otimes_A N)$ for any i, as can be seen easily from the induction on i. Letting $i = r$, we get $\operatorname{Ass}_{A'}(M \otimes_A N) \subseteq \operatorname{Ass}_{A'}(A/\mathfrak{p} \otimes_A N)$, as desired. □

From now, until the end of this section, let $(A, \mathfrak{m}, k)$ be a Noetherian local ring.

For a homomorphism $f : A \to A'$, we call $A'/\mathfrak{m}A' = A' \otimes_A k$ the *fiber* or the *closed fiber* of f. Geometrically, $A'/\mathfrak{m}A'$ represents the inverse image ${}^a f^{-1}(\mathfrak{m})$ of the closed point $\mathfrak{m}$ of $\operatorname{Spec} A$ by the morphism ${}^a f : \operatorname{Spec} A' \to \operatorname{Spec} A$. If f is flat, then properties of A' can often be expressed in terms of those of A and $A'/\mathfrak{m}A'$.

Theorem 5.70 *Let $f : (A, \mathfrak{m}, k) \to (A', \mathfrak{m}', k')$ be a local homomorphism between Noetherian local rings. Then we have*

$$\dim A' \leq \dim A + \dim A'/\mathfrak{m}A'.$$

If, moreover, the property

$$f^{-1}(\mathfrak{q}) \neq \mathfrak{m} \text{ for any } \mathfrak{p} \in \operatorname{Spec} A \setminus \{\mathfrak{m}\} \text{ and any minimal prime } \mathfrak{q} \text{ of } A'/\mathfrak{p}A' \qquad (*)$$

holds, then $\dim A' = \dim A + \dim A'/\mathfrak{m}A'$. *In particular, if f is a flat local homomorphism, then* $\dim A' = \dim A + \dim A'/\mathfrak{m}A'$.

Proof Induction on $n = \dim A$. If $n = 0$, then $\mathfrak{m}$ is nilpotent, and hence $\mathfrak{m}A'$ is contained in any minimal prime of A', and hence $\dim A'/\mathfrak{m}A' = \dim A'$.

Assume that $n > 0$, and the theorem is true for smaller n. We can take $a \in \mathfrak{m} \setminus \bigcup_{\mathfrak{p} \in \mathrm{Min}\, A} \mathfrak{p}$ by prime avoidance. Then $\dim A/aA = n - 1$. By the induction assumption,

$$\dim A' \leq \dim A'/aA' + 1 \leq \dim A/aA + \dim A'/\mathfrak{m}A' + 1 = \dim A + \dim A'/\mathfrak{m}A',$$

and the first assertion has been proved.

Now assume $(*)$. For any minimal prime ideal $\mathfrak{q}$ of A', $f^{-1}(\mathfrak{q}) \neq \mathfrak{m}$, otherwise when we take a minimal prime $\mathfrak{p}$ of A, then since $\mathfrak{p} \subseteq \mathfrak{m} = f^{-1}(\mathfrak{q})$, and so $\mathfrak{q}$ is a minimal prime of $A'/\mathfrak{p}A'$, and this contradicts $(*)$. So by prime avoidance, we can take $a \in \mathfrak{m} \setminus \left(\left(\bigcup_{\mathfrak{p} \in \mathrm{Min}\, A} \mathfrak{p}\right) \cup \left(\bigcup_{\mathfrak{q} \in \mathrm{Min}\, A'} \mathfrak{q} \cap A\right)\right)$. Then $\dim A/aA = \dim A - 1$ and $\dim A'/aA' = \dim A' - 1$. As the condition $(*)$ holds for $f \otimes 1 : A/aA \to A'/aA' = A' \otimes_A A/aA$, $\dim A'/aA' = \dim A/aA + \dim A'/\mathfrak{m}A'$ by the induction assumption, and the second assertion has been proved.

If f is flat local, then for any prime ideal $\mathfrak{p} \neq \mathfrak{m}$ of A, any minimal prime $\mathfrak{q}$ of $\mathfrak{p}A'$ is an element of $\mathrm{Ass}_{A'}(A/\mathfrak{p} \otimes_A A')$. By Lemma 5.69, we have $f^{-1}(\mathfrak{q}) = \mathfrak{p} \neq \mathfrak{m}$, and the condition $(*)$ is satisfied, and the last assertion follows. □

For a flat local homomorphism $(A, \mathfrak{m}) \to (B, \mathfrak{n})$, a similar additivity formula holds for depth. Namely:

Theorem 5.71 *Let $f : (A, \mathfrak{m}) \to (B, \mathfrak{n})$ be a flat local homomorphism between Noetherian local rings. Then*

$$\operatorname{depth} B = \operatorname{depth} A + \operatorname{depth} B/\mathfrak{m}B$$

holds. In particular, B is Cohen–Macaulay if and only if A and $B/\mathfrak{m}B$ are Cohen–Macaulay.

Proof Set $\operatorname{depth} A = r$, and let $(x_1, \ldots, x_r) \subseteq \mathfrak{m}$ be an A-sequence. Since B is A-flat, $(x_1, \ldots, x_r)$ is also B-regular. As $\operatorname{depth} B/(x_1, \ldots, x_r)B = \operatorname{depth} B - r$, we may assume that $\operatorname{depth} A = 0$ and hence $\mathfrak{m} \in \mathrm{Ass}_A(A)$. Set $\operatorname{depth} B/\mathfrak{m}B = s$, and we take $(y_1, \ldots, y_s) \subseteq \mathfrak{n}$ so that they form a $B/\mathfrak{m}B$-sequence. We apply Corollary 5.66, (2) to $y_1 : B \to B$. As this map satisfies condition (b) there, we have that $y_1 : B \to B$ is injective and B/y_1B is A-flat. Next, we apply the same corollary to $y_2 : B/y_1B \to B/y_1B$ and so on, and conclude that $(y_1, \ldots y_s)$ is a B-sequence and $B/(y_1, \ldots y_s)B$ is A-flat. From an injective map $A/\mathfrak{m} \hookrightarrow A$, we get an injective map $B/(\mathfrak{m}B + (y_1, \ldots y_s)B) \hookrightarrow B/(y_1, \ldots, y_s)B$ by flatness. By the choice of s, $\mathfrak{n} \in \mathrm{Ass}(B/((\mathfrak{m} + (y_1, \ldots, y_s)B))) \subseteq \mathrm{Ass}(B/(y_1, \ldots, y_s)B)$, and $\operatorname{depth} B/(y_1, \ldots y_s)B = 0$. So $\operatorname{depth} B = s$, as desired.

The second assertion is clear from the first assertion and $\dim B = \dim A + \dim B/\mathfrak{m}B$. □

Corollary 5.72 *Let $(A, \mathfrak{m})$ be a Noetherian local ring. Then* $\operatorname{depth} \widehat{A} = \operatorname{depth} A$ *and* $\dim \widehat{A} = \dim A$, *where $\widehat{A}$ is the completion of A. In particular, $\widehat{A}$ is Cohen–Macaulay if and only if A is so.*

Proof Note that $\widehat{A}$ is A-flat. So $\operatorname{depth} \widehat{A} = \operatorname{depth} A + \operatorname{depth} \widehat{A}/\mathfrak{m}\widehat{A} = \operatorname{depth} A$. We also have $\dim \widehat{A} = \dim A + \dim \widehat{A}/\mathfrak{m}\widehat{A} = \dim A$. The last assertion is clear. □

Corollary 5.73 *Let A be a Cohen–Macaulay ring, B be a Noetherian ring, and $f : A \to B$ be a flat homomorphism such that $B \otimes_A \kappa(P)$ is Cohen–Macaulay for any $P \in {}^a f(\operatorname{Max}(B))$, where $\kappa(P) = A_P/PA_P$ is the residue field of the local ring A_P. Then B is Cohen–Macaulay. In particular, the polynomial ring $A[x_1, \dots, x_n]$ and the formal power series ring $A[[x_1, \dots, x_n]]$ are Cohen–Macaulay.*

Proof Let $\mathfrak{n} \in \operatorname{Max}(B)$, and set $P = {}^a f(\mathfrak{n}) = A \cap \mathfrak{n} \in \operatorname{Spec} A$. Then $A_P \to B_\mathfrak{n}$ is flat local, A_P is Cohen–Macaulay, and $B_\mathfrak{n}/PB_\mathfrak{n} \cong (B \otimes_A \kappa(P))_\mathfrak{n}$ is also Cohen–Macaulay. By Theorem 5.71, we have that $B_\mathfrak{n}$ is Cohen–Macaulay. Hence B is Cohen–Macaulay, and the first assertion has been proved.

Now we want to prove that $A[x_1, \dots, x_n]$ and $A[[x_1, \dots, x_n]]$ are Cohen–Macaulay, if A is so. In order to do so, we may assume that $n = 1$. Let $B = A[x]$. Then B is Noetherian by the Hilbert basis theorem. It is an A-free module with the set of monomials as its A-basis, so B is A-flat. For $P \in \operatorname{Spec} A$, $B \otimes_A \kappa(P) \cong \kappa_P[x]$ is a polynomial ring over a field, and is regular. So B is Cohen–Macaulay.

Let $B = A[[x]]$ now. As $B = \prod_{r \in \mathbb{Z}} Ax^r$ is a direct product of A-flat modules and A is Noetherian, B is A-flat by Theorem C.13. (This is also clear from the fact that $A[x]$ is A-flat and $A[[x]]$ is the completion of $A[x]$ with respect to the (x)-adic topology.) Let $\mathfrak{n} \in \operatorname{Max}(B)$, and $P = A \cap \mathfrak{n}$. Since $x \in \mathrm{J}(B)$, the Jacobson radical of B, we have that $P \in \operatorname{Max} A$ and $\mathfrak{n} = P + (x)$. So $B \otimes_A \kappa(P) \cong B/PB \cong A/P[[x]]$ is a formal power series ring over the field A/P, which is a DVR. So B is Cohen–Macaulay. □

Let X be a (partially) ordered set. A subset C of X is called a *chain* of X if C is totally ordered (with respect to the order of X). The *length* of a (finite) chain C is $|C| - 1$. A chain C of X is said to be *maximal* if there is no chain C' of X such that $C' \supsetneq C$. For $x, y \in X$ with $x \leq y$, a chain from x to y in X is a chain of $[x, y] = \{z \in X \mid x \leq z \leq y\}$ by definition. A maximal chain of $[x, y]$ is called a *saturated chain* from x to y in X. We say that X is *pure* if maximal chains of X have the same finite length. We say that X is *catenary* if $[x, y]$ is pure for any $x, y \in X$ with $x \leq y$. We define $\dim X$ to be the supremum of the lengths of chains of X.

For a ring A, a chain of prime ideals of A is a chain of $\operatorname{Spec} A$ by definition. So $\dim A = \dim \operatorname{Spec} A$. We say that A is *catenary* if $\operatorname{Spec} A$ is so as an ordered set. We say that A is *universally catenary* if the polynomial ring $A[x_1, \dots, x_n]$ is catenary for any $n \geq 0$.

Lemma 5.74 *Let A be a catenary ring, S a multiplicatively closed subset, and I an ideal of A. Then $S^{-1}A$ and A/I are catenary.*

Proof It is easy to see that if X is a catenary ordered set and Y its *order ideal* (that is, Y is a subset of X such that $y \in Y$, $x \in X$, and $x \leq y$ imply $x \in Y$), then both Y and $X \setminus Y$ are catenary. Since $\operatorname{Spec} A/I$ is isomorphic to $\mathrm{V}(I) \subseteq \operatorname{Spec} A$ as an ordered set and $\operatorname{Spec} A \setminus \mathrm{V}(I)$ is an order ideal, A/I is catenary. Similarly,

Spec $S^{-1}A$ is isomorphic to the order ideal $\{P \in \operatorname{Spec} A \mid P \cap S = \emptyset\}$ of Spec A, and hence $S^{-1}A$ is also catenary. □

Theorem 5.75 *A homomorphic image of a Cohen–Macaulay ring is universally catenary.*

Proof Let A be a Cohen–Macaulay ring, and we want to show that $A/I[x_1, \ldots, x_n]$ is catenary for any ideal I of A and $n \geq 0$. By Lemma 5.74, it suffices to show that $B = A[x_1, \ldots, x_n]$ is catenary. Note that B is Cohen–Macaulay. Let $P, Q \in \operatorname{Spec} B$ with $P \subseteq Q$. Then clearly, any saturated chain of prime ideals from P to Q has length at most ht Q, and is finite. Let $P = P_0 \subsetneq P_1 \subsetneq P_2 \subsetneq \cdots \subsetneq P_r = Q$ be a saturated chain. Let i be an integer with $1 \leq i \leq r$. Since B_{P_i} is Cohen–Macaulay, we have $\operatorname{ht} P_i = \dim B_{P_i} = \dim B_{P_i}/P_{i-1}B_{P_i} + \operatorname{ht} P_{i-1}B_{P_i} = \operatorname{ht} P_{i-1} + 1$. So $r = \operatorname{ht} Q - \operatorname{ht} P$ is finite and is independent of the saturated chain. □

Next, we overview the properties of the notion of faithful flatness. In order to prove certain property of A, we sometimes take a faithfully flat extension B of A, prove the property for B, and we recover the result of A using it. This technique is called (faithfully) *flat descent*.

Definition 5.76 Let F be a flat A-module. Then the following conditions are equivalent:

(1) For any maximal ideal $\mathfrak{m}$ of A, we have $\mathfrak{m}F \neq F$.
(2) For any nonzero A-module M, we have $M \otimes_A F \neq 0$.
(3) The functor $- \otimes_A F$ is faithful (see Lemma 5.54).

The equivalence follows from Lemma 5.54. If these conditions are satisfied, F is called *faithfully flat*. A ring homomorphism $A \to B$ is said to be faithfully flat if B is so as an A-module.

The following is obvious from the definition.

Proposition 5.77 *If $A \to B$ is a faithfully flat ring homomorphism, then for an A-linear map $f : M \to N$, we have that f is injective (resp. surjective) if and only if $f \otimes 1 : M \otimes_A B \to N \otimes_A B$ is injective (resp. surjective).*

Example 5.78

(1) If $(A, \mathfrak{m}) \to (B, \mathfrak{n})$ is a flat local homomorphism between local rings, it is faithfully flat.
(2) If $A \to B$ is faithfully flat, then A is a pure subring of B (like $A \to S^{-1}A$, there are many flat homomorphism that is not pure), where we say that A is a *pure subring* of B if for any A-module M, the morphism $M \to M \otimes_A B$ given by $m \mapsto 1 \otimes m$ is injective (Definition 7.5).

Proof (1) By the definition of a local homomorphism, we have $\mathfrak{m}B \subseteq \mathfrak{n} \neq B$.
(2) We want to show that for any A-module M, the map $M = M \otimes_A A \to M \otimes_A B$ is injective. Let $m : B \otimes_A B \to B$ be the map given by $m(b \otimes b') = bb'$, and $i : B \to B \otimes_A B$ be the map given by $i(b) = 1 \otimes b$. Then letting B act on $B \otimes_A B$

by $b(b' \otimes b'') = b' \otimes bb''$, both m and i are B-linear, and obviously $mi = 1_B$. Hence B is a pure subring of $B \otimes_A B$ through $i : B \to B \otimes_A B$. So

$$1_M \otimes i : M \otimes_A B \to M \otimes_A (B \otimes_A B) \qquad ((1_M \otimes i)(m \otimes b) = m \otimes (1 \otimes b)$$

is injective. As this is a result of applying the functor $- \otimes_A B$ to $M \to M \otimes_A B$, we have that $M \to M \otimes_A B$ is also injective, by the faithful flatness of B. □

The following proposition is useful.

Proposition 5.79 *Let* $(A, \mathfrak{m})$ *be a Noetherian local ring, and* K *an arbitrary extension field of* $A/\mathfrak{m}$. *Then there is a complete Noetherian local ring* $(B, \mathfrak{n})$ *which is a faithfully flat algebra over* A *such that:*

(1) $\mathfrak{m}B = \mathfrak{n}$;
(2) $B/\mathfrak{n} \cong K$.

Proof Similarly to the structure theorem of complete local rings, we prove this only for the case that A contains a field. For the general case, we refer the reader to [EGAIII-1, $(0_{\text{III}}.10.3.1)$]. As $\mathfrak{m}\widehat{A}$ is the maximal ideal of $\widehat{A}$ and $\widehat{A}$ is complete local, we may assume that A is complete. Then by the structure theorem of complete local rings (Theorem 4.40), we may write $A = k[[X_1, \ldots, X_n]]/\mathfrak{a}$, where $k \cong A/\mathfrak{m}$ is the coefficient field. We set $B = K[[X_1, \ldots, X_n]]/\mathfrak{a}K[[X_1, \ldots, X_n]]$. The canonical map $A \to B$ is a local homomorphism, $\mathfrak{m}B = \mathfrak{n}$, and $B/\mathfrak{n} \cong K$, where $\mathfrak{m} = (X_1, \ldots, X_n)A$ and $\mathfrak{n} = (X_1, \ldots, X_n)B$. To prove that B is A-flat, replacing $A \to B$ by $A_n \to B_n$ for various $n \geq 1$ by the local criterion (Theorem 5.65), where $A_n = A/\mathfrak{m}^{n+1}$ and $B_n = B/\mathfrak{m}^{n+1}B$, we may assume that $\mathfrak{a}$ is $\mathfrak{m}$-primary. This case is trivial, since we have $B = K \otimes_k A$ then. □

5.6 Koszul Complexes and Čech Complexes

The Koszul complex,[6] which was first introduced in differential geometry, is also an important tool in commutative algebra.

Consider a free module $E \cong A^n$ over a ring A. Let $\{T_i\}_{1 \leq i \leq n}$ be a free basis of E over A. Let $\bigwedge E$ denote the exterior algebra of E over the ring A (see Appendix D for definition). Note that $\bigwedge E = \bigoplus_{r=0}^{n} \bigwedge^r E$ is a finite-dimensional graded A-algebra, not necessarily commutative, and $\bigwedge^r E$ is an A-free module with the basis $\{T_I \mid I \subseteq \{1, 2, \ldots, n\},\ \sharp I = r\}$, where for a sequence $J = (j_1, \ldots, j_r) \in \{1, \ldots, n\}^r$, we denote $T_{j_1} \wedge \cdots \wedge T_{j_r} \in \bigwedge^r E$ by T_J. For $I = \{i_1, i_2, \ldots, i_r\}$ $(1 \leq i_1 < i_2 < \ldots < i_r \leq n)$, we set $T_I = T_J$, where J is the ordered sequence $(i_1, \ldots, i_r)$. Note that for $r = 0$, we have that $\bigwedge^0 E = A$ with its basis $T_\emptyset = 1$.

[6] Jean-Louis Koszul (1921–2018).

Definition 5.80 For a sequence $\underline{a} := (a_1, a_2, \ldots, a_n)$ of elements in A, we define the complex $K_\bullet = K_\bullet(\underline{a})$ of A-modules as follows, and call it the *Koszul complex* of A with respect to $\underline{a}$.

(1) $K_\bullet = [0 \to K_n \to K_{n-1} \to \ldots \to K_0 \to 0]$, $K_r = \bigwedge^r E$.
(2) $\partial = \partial_r : K_r \to K_{r-1}$ is the A-homomorphism given by

$$\partial(T_I) = \sum_{j=1}^{r} (-1)^{j-1} a_{i_j} T_{I \setminus \{i_j\}}$$

for $I = \{i_1, i_2, \ldots, i_r\}$ with $i_1 < i_2 < \cdots < i_r$. (It is easy to check that $\partial_{r-1} \circ \partial_r = 0$, and hence $K_\bullet$ forms a complex with respect to ∂.)
(3) We denote the rth homology group of $K_\bullet(\underline{a})$ by $\mathrm{H}_r(\underline{a})$. Namely,

$$\mathrm{H}_r(\underline{a}) = \mathrm{H}_r(K_\bullet(\underline{a})) = \mathrm{Ker}(\partial_r)/\,\mathrm{Im}(\partial_{r+1}).$$

(4) For an A-module M, we define $K_\bullet(\underline{a}, M) = K_\bullet(\underline{a}) \otimes_A M$ and $K^\bullet(\underline{a}, M) = \mathrm{Hom}_A(K_\bullet(\underline{a}), M)$, where the grading is given by $K^r(\underline{a}, M) = \mathrm{Hom}_A(K_r(\underline{a}), M)$. We define $K^\bullet(\underline{a}) := K^\bullet(\underline{a}, A)$.

We list some basic properties of Koszul complexes, where $\mathfrak{a} = (\underline{a}) := (a_1, a_2, \cdots, a_n)$ is the ideal generated by $\underline{a}$.

Proposition 5.81

(1) $\mathrm{H}_0(\underline{a}, M) \cong M/\mathfrak{a}M$, *and* $\mathrm{H}_n((\underline{a}), M) \cong [0 :_M \mathfrak{a}]$.
(2) *For each integer r such that $0 \le r \le n$, we have* $\mathrm{H}_r(\underline{a}, M) \cong \mathrm{H}^{n-r}(\underline{a}, M)$.
(3) *For each integer r such that $0 \le r \le n$,* $\mathfrak{a}\,\mathrm{H}_r(\underline{a}, M) = 0$.

Proof As (1) and (2) follow from the definition immediately, we prove (3). It suffices to show that for $z \in K_r(\underline{a}, M)$, if $\partial_r(z) = 0$, then $a_i z \in \mathrm{Im}(\partial_{r+1})$ for each i such that $1 \le i \le n$. Indeed, we have

$$\partial_{r+1}(T_i \wedge z) = a_i z - T_i \wedge \partial_r(z) = a_i z.$$

□

The following is the main theorem of this section.

Theorem 5.82 *Let M be an R-module.*

(1) *If $\underline{a}$ is an M-sequence, then we have* $\mathrm{H}_r(\underline{a}, M) = 0$ *($\forall r > 0$).*
(2) *Conversely, if R is Noetherian, $\underline{a} \subseteq \mathrm{J}(R)$, M is nonzero and finitely generated, and* $\mathrm{H}_1(\underline{a}, M) = 0$, *then $\underline{a}$ is an M-sequence. In particular, any permutation $\sigma\underline{a} := (a_{\sigma^{-1}1}, \ldots, a_{\sigma^{-1}n})$ of $\underline{a}$ is again an M-sequence for any permutation σ.*

To prove the theorem, we consider the following decomposition. For a (homological) chain complex $C_\bullet$ of R-modules, we define the *r-shift* of $C_\bullet$, denoted by $C[r]_\bullet$ by $C[r]_n = C_{r+n}$ for $n \in \mathbb{Z}$, and $d_n^{C[r]_\bullet} = (-1)^r d_{r+n}^{C_\bullet}$. Clearly, $\mathrm{H}_i(C[r]_\bullet) \cong \mathrm{H}_{r+i}(C_\bullet)$ for $i \in \mathbb{Z}$.

Definition 5.83 (Mapping Cone) For a chain map $\varphi_\bullet : E_\bullet \to F_\bullet$ between complexes of R-modules, we define the chain complex $C_\bullet = \mathrm{Cone}(\varphi_\bullet)$ as follows. $C_n = F_n \oplus E_{n-1}$. The differential map $d_n^{C_\bullet} : C_n = F_n \oplus E_{n-1} \to C_{n-1} = F_{n-1} \oplus E_{n-2}$ is given by

$$d_n^{C_\bullet}(f_n, e_{n-1}) = d_n^{F_\bullet}(f_n) + \varphi_{n-1}(e_{n-1}) - d_{n-1}^{E_\bullet}(e_{n-1}).$$

We call $C_\bullet = \mathrm{Cone}(\varphi_\bullet)$ the *mapping cone* of $\varphi_\bullet$.

Note that $\iota_\bullet : F_\bullet \to C_\bullet$ given by $\iota_n(f_n) = (f_n, 0) \in F_n \oplus E_{n-1} = C_n$ and $\pi_\bullet : C_\bullet \to E_\bullet[-1]$ given by $\pi_n(f_n, e_{n-1}) = e_{n-1}$ are chain maps, and

$$0 \to F_\bullet \xrightarrow{\iota_\bullet} C_\bullet \xrightarrow{\pi_\bullet} E_\bullet[-1] \to 0$$

is a short exact sequence of complexes such that

$$0 \to F_n \xrightarrow{\iota_n} C_n \xrightarrow{\pi_n} E_{n-1} \to 0$$

is split exact for all $n \in \mathbb{Z}$. The associated long exact sequence is as follows:

$$\cdots \to \mathrm{H}_n(E_\bullet) \xrightarrow{\mathrm{H}_n(\varphi_\bullet)} \mathrm{H}_n(F_\bullet)$$
$$\xrightarrow{\mathrm{H}_n(\iota_\bullet)} \mathrm{H}_n(C_\bullet) \xrightarrow{\mathrm{H}_n(\pi_\bullet)} \mathrm{H}_{n-1}(E_\bullet) \xrightarrow{\mathrm{H}_{n-1}(\varphi_\bullet)} \mathrm{H}_{n-1}(F_\bullet) \to \cdots .$$

Namely, the connecting homomorphism $\delta_n : \mathrm{H}_n(E_\bullet) \to \mathrm{H}_n(F_\bullet)$ agrees with $\mathrm{H}_n(\varphi_\bullet)$. Indeed, if $z + B_n(E_\bullet) \in \mathrm{H}_n(E_\bullet)$, where $z \in Z_n(E_\bullet)$, then $(0, z) \in C_{n+1} = F_{n+1} \oplus E_n$ maps to z by π_{n+1}. As $d_{n+1}^{C_\bullet}(0, z) = (\varphi(z), 0) = \iota_n(\varphi(z)) \in Z_n(C_\bullet)$, we have $\delta_n(z + B_n(E_\bullet)) = \varphi(z) + B_n(F_\bullet) = \mathrm{H}_n(\varphi_\bullet)(z + B_n(E_\bullet))$.

Let $\underline{a} = a_1, \ldots, a_m$ and $\underline{b} = b_1, \ldots, b_n$ be two sequences in A. Assume that $b_j = \sum_i c_{ij} a_i$ for some $c_{ij} \in A$. Let $T_1', \ldots, T_n'$ be symbols, and let E be the free A-module with the basis $T_1', \ldots, T_n'$. Let $T_1, \ldots, T_m$ be symbols, and let F be the free A-module with the basis $T_1, \ldots, T_m$. Let $\rho : E \to F$ be the A-linear map given by $\rho(T_j') = \sum_i c_{ij} T_i$. Then $\bigwedge \rho : \bigwedge E = K_\bullet(\underline{b}) \to \bigwedge F = K_\bullet(\underline{a})$ is a chain map. Since $\bigwedge \mathrm{id} = \mathrm{id}$ and $\bigwedge \rho \bigwedge \psi = \bigwedge(\rho\psi)$, it follows that $\bigwedge \rho$ is a chain isomorphism if ρ is an isomorphism. As ρ and $\bigwedge \rho$ are determined by the matrix $C = (c_{ij})$, we denote $\bigwedge \rho$ by $\phi_C : K_\bullet(\underline{b}) \to K_\bullet(\underline{a})$. In particular, considering the case that $m = n$, $b_j = a_{\sigma(j)}$ $(i = 1, \ldots, n)$ for some permutation $\sigma \in \mathfrak{S}_n$, and $c_{ij} = \delta_{i\sigma(j)}$, we have that $\bigwedge \rho = \phi_{(\delta_{i\sigma(j)})} : K_\bullet(a_{\sigma 1}, \ldots, a_{\sigma n}) \to K_\bullet(a_1, \ldots, a_n)$ is an isomorphism.

Lemma 5.84 *When we set* $\underline{a'} = (a_2, a_3, \cdots, a_n)$, $K_\bullet(\underline{a})$ *is identified with*

$$\mathrm{Cone}(a_1 : K_\bullet(\underline{a}') \to K_\bullet(\underline{a}'))$$

in a natural way. In particular, there is a long exact sequence of the form

$$\cdots \to \mathrm{H}_{r+1}(\underline{a}', M) \to \mathrm{H}_{r+1}(\underline{a}, M) \to \mathrm{H}_r(\underline{a}', M)$$
$$\xrightarrow{a_1} \mathrm{H}_r(\underline{a}', M) \to \mathrm{H}_r(\underline{a}, M) \to \cdots .$$

Proof Set $L_\bullet = K_\bullet(\underline{a}')$, and set $C_\bullet = \mathrm{Cone}(a_1 : L_\bullet \to L_\bullet)$. Then we define $h : C_\bullet \to K_\bullet(\underline{a})$ by $h(T_J, 0) = T_J$ and $h(0, T_J) = T_1 \wedge T_J$ for $J \subseteq \{2, 3, \ldots, n\}$. It is straightforward to check that h is a chain isomorphism. □

Now we prove the main theorem.

Proof of Theorem 5.82 Induction on n. If $n = 1$, then

$$K_\bullet(\underline{a}, M) = [0 \to M \xrightarrow{a_1} M \to 0].$$

So both of the assertions (1) and (2) are obvious.

Next, we consider the case that $n \geq 2$, and we assume that the assertions are true for smaller n.

(1) Since a_1 is M-regular, $0 \to M \xrightarrow{a_1} M \to M/a_1M \to 0$ is exact. Taking the tensor product with the free complex $K_\bullet(\underline{a}')$, where $\underline{a}'$ is the sequence $a_2, \ldots, a_n$, we get a short exact sequence

$$0 \to K_\bullet(\underline{a}', M) \xrightarrow{a_1} K_\bullet(\underline{a}', M) \to K_\bullet(\underline{a}', M/a_1M) \to 0.$$

Since $\mathrm{H}_i(\underline{a}', M/a_1M) = 0$ for $i > 0$ by the induction assumption, $a_1 : \mathrm{H}_i(\underline{a}', M) \to \mathrm{H}_i(\underline{a}', M)$ is injective for all $i \geq 0$ and surjective for all $i > 0$. So by the long exact sequence in Lemma 5.84, we get $\mathrm{H}_i(\underline{a}, M) = 0$ for $i > 0$, as desired.

(2) Assume that $\underline{a} \subseteq \mathrm{J}(R)$, and $\mathrm{H}_1(\underline{a}, M) = 0$. Considering the sequence $\underline{c} = (a_n, a_1, a_2, \ldots, a_{n-1})$, we also have $\underline{c} \subseteq \mathrm{J}(R)$, and $\mathrm{H}_1(\underline{c}, M) = 0$. Since the sequence

$$\mathrm{H}_1(\underline{a}', M) \xrightarrow{a_n} \mathrm{H}_1(\underline{a}', M) \to \mathrm{H}_1(\underline{c}, M) \to \mathrm{H}_0(\underline{a}', M) \xrightarrow{a_n} \mathrm{H}_0(\underline{a}', M)$$

is exact, $\mathrm{H}_1(\underline{c}, M) = 0$ implies that $\mathrm{H}_1(\underline{a}', M) = 0$ by NAK, Theorem 1.86, where $\underline{a}'$ is the sequence $a_1, a_2, \ldots, a_{n-1}$. By induction, $\underline{a}' = a_1, \ldots, a_{n-1}$ is an M-sequence. We also have that $a_n : \mathrm{H}_0(\underline{a}', M) \to \mathrm{H}_0(\underline{a}', M)$ is injective. In other words, a_n is $M/\underline{a}'M$-regular. By NAK again, $M \neq 0$ implies that $M/\underline{c}M \neq 0$, and so $\underline{c}$ is an M-sequence.

Finally, as $K_\bullet(\sigma\underline{a}) \cong K_\bullet(\underline{c})$, we have $\mathrm{H}_1(\sigma\underline{a}, M) \cong \mathrm{H}_1(\underline{c}, M) = 0$ for any permutation $\sigma \in \mathfrak{S}_n$. As $\sigma\underline{a} \subseteq \mathrm{J}(R)$, we have that $\sigma\underline{a}$ is also an M-sequence, by what we have already proved. In particular, $\underline{a}$ is an M-sequence. □

Proposition 5.85 *Let $(A, \mathfrak{m})$ be a Noetherian local ring, and $\underline{a}$ and $\underline{b}$ form two minimal sets of generators of an ideal $\mathfrak{a}$ of A. Then for any A-module M, we have that $K_\bullet(\underline{a}, M) \cong K_\bullet(\underline{b}, M)$. In particular, for any $r \geq 0$, we have $\mathrm{H}_r(\underline{a}, M) \cong \mathrm{H}_r(\underline{b}, M)$.*

Proof Let $\underline{a} = (a_1, \ldots, a_m)$, and $\underline{b} = (b_1, \ldots, b_n)$. We have $n = m = \dim_{A/\mathfrak{m}} \mathfrak{a}/\mathfrak{m}\mathfrak{a}$ by Corollary 1.87. Let $\mathbb{M} = (m_{ij})$ be the matrix such that $b_i = \sum_{j=1}^n m_{ij}a_j$ (it exists, since $\underline{\mathfrak{a}}$ is a set of generators). Then the image of $\det M$ in $A/\mathfrak{m}$ is nonzero, since the image of $\underline{b}$ in $\mathfrak{a}/\mathfrak{m}\mathfrak{a}$ is linearly independent. So $\det \mathbb{M} \in A \setminus \mathfrak{m} = A^\times$. It follows that $\mathbb{M}$ is an invertible matrix, and it has an inverse $\mathbb{N} = (n_{ij})$, and we have $a_i = \sum_{j=1}^n n_{ij}b_j$. Since the two maps $\phi_\mathbb{M}$ and $\phi_\mathbb{N}$ between the Koszul complexes $K_\bullet(\underline{b}, M)$ and $K_\bullet(\underline{a}, M)$ are inverses of each other, they are isomorphisms. So we also have $\mathrm{H}_r(\underline{a}, M) \cong \mathrm{H}_r(\underline{b}, M)$ for any $r \geq 0$. □

We have the following *depth sensitivity* of Koszul complexes.

Theorem 5.86 *Let A be a Noetherian ring, and M be a finitely generated A-module, and $I = (a_1, a_2, \ldots, a_n)$ be an ideal of A. We set $r = \mathrm{depth}(I, M)$. Then*

$$n - r = \sup\{i \in \mathbb{Z} \mid \mathrm{H}_i(\underline{a}, M) \neq 0\}. \tag{5.1}$$

Proof Consider $P \in \operatorname{Spec} A$. If $P \notin \operatorname{Supp} M$, then $K_\bullet(\underline{a}, M) = 0$, and hence $\mathrm{H}_i(\underline{a}, M)_P = 0$ for all $i \in \mathbb{Z}$. If $P \notin \mathrm{V}(I)$, then we can take $a \in I \setminus P$. Then $a : \mathrm{H}_i(\underline{a}, M)_P \to \mathrm{H}_i(\underline{a}, M)_P$ is a zero map, but it is an isomorphism. Namely, $\mathrm{H}_i(\underline{a}, M)_P = 0$. So if $M = IM$, then $\mathrm{Supp}(M) \cap \mathrm{V}(I) = \emptyset$, and hence $\mathrm{H}_i(\underline{a}, M) = 0$ for all $i \in \mathbb{Z}$. So (5.1) holds in the sense that $n - \infty = -\infty = \sup \emptyset$. Now assume that $M \neq IM$. Let $s = s(M)$ be the right-hand side of (5.1). Then $r = \mathrm{depth}(I, M) < \infty$, and $0 \leq s \leq n$, since $\mathrm{H}_0(\underline{a}, M) \cong M/IM \neq 0$.

We prove $n - r = s$ by induction on r. If $r = 0$, then there exists some $P \in \operatorname{Ass} M \cap \mathrm{V}(I) = \operatorname{Ass} \operatorname{Hom}_A(A/I, M)$. Let $m \in M$ such that $P = \operatorname{Ann}_A m$. Then $0 \neq m \in 0 :_M P \subseteq 0 :_M I = \mathrm{H}_n(\underline{a}, M)$, and hence $s = n$, and we have $n - r = s$.

Next, we consider the case that $r > 0$, and we assume that the theorem is true for smaller r. Let $a \in I$ be an M-regular element. Then by the short exact sequence $0 \to M \xrightarrow{a} M \to M/aM \to 0$, we get an exact sequence

$$0 \to \mathrm{H}_i(\underline{a}, M) \to \mathrm{H}_i(\underline{a}, M/aM) \to \mathrm{H}_{i-1}(\underline{a}, M) \to 0$$

for each $i \in \mathbb{Z}$, since $a \in I$ annihilates the Koszul homology groups $\mathrm{H}_*(\underline{a}, M)$. So $s(M/aM) = s(M) + 1$, and $\mathrm{depth}(I, M/aM) = r - 1$, and $n - r = n - \mathrm{depth}(I, M/aM) - 1 = s(M/aM) - 1 = s + 1 - 1 = s$, by the induction assumption. □

Example 5.87 Let $\underline{x} = (x_1, \ldots, x_d)$ and $\underline{y} = (y_1, \ldots, y_d)$ be systems of parameters of a Noetherian local ring $(A, \mathfrak{m})$ such that $(\underline{x}) \supseteq (\underline{y})$. Then we can write $y_i = \sum_{j=1}^{d} m_{ij}x_j$. We set $\mathbb{M} = (m_{ij})$. Then the chain map $\phi = \phi_{\mathbb{M}} : K_\bullet(\underline{y}) \to K_\bullet(\underline{x})$ between the Koszul complexes is defined. If A is Cohen–Macaulay, then any system of parameters is an A-sequence. So ϕ is a map between the projective resolutions, covering the canonical surjection $A/(\underline{y}) \to A/(\underline{x})$.

Taking the dual, we get $\phi^* : K^\bullet(\underline{x}) \to K^\bullet(\underline{y})$, and

$$\begin{aligned}\mathrm{H}^d(\phi^*) : \mathrm{H}^d(K^\bullet(\underline{x})) &\cong \mathrm{Ext}^d_A(A/(\underline{x}), A)\\ &\cong A/(\underline{x}) \to \mathrm{H}^d(K^\bullet(\underline{y}))) \cong \mathrm{Ext}^d_A(A/(\underline{y}), A) \cong A/(\underline{y}).\end{aligned}$$

By definition, $\phi_d : A \cong \bigwedge^d A^d \xrightarrow{\bigwedge^d \Psi} \bigwedge^d A^d \cong A$ is the multiplication by $\det \mathbb{M}$, where $\Psi : A^d \to A^d$ is the A-linear map whose matrix is $\mathbb{M}$. This shows that $\mathrm{H}^d(\phi^*) : \mathrm{H}^d(\underline{x}) \to \mathrm{H}^d(\underline{y})$ is also the multiplication by $\det \mathbb{M}$.

If A is Cohen–Macaulay, then $\mathrm{Ext}^i_A(A/\mathfrak{m}, A) = 0$ for $i < d$. Hence by the short exact sequence $0 \to (\underline{x})/(\underline{y}) \to A/(\underline{y}) \to A/(\underline{x}) \to 0$, we get the exact sequence

$$0{=}\mathrm{Ext}^{d-1}_A((x)/(y), A) \to \mathrm{Ext}^d_A(A/(\underline{x}), A) \cong A/(\underline{x}) \to \mathrm{Ext}^d_A(A/(\underline{y}), A) \cong A/(\underline{y}).$$

So we have that $\det M : A/(\underline{x}) \to A/(\underline{y})$ is an injective map. So this map induces an isomorphism between the socles of these modules. In particular, if A is regular and $\mathfrak{m} = (\underline{x})$, then the socle of $A/(\underline{y})$ is generated by $\det \mathbb{M}$.

Next, we introduce the Čech complex. This complex is useful in calculating the local cohomology module.

Definition 5.88 (Čech Complex) Let $\underline{a} := (a_1, a_2, \cdots, a_n)$ be a sequence of elements in A. For $I \subseteq \{1, \ldots, n\}$, we set $a_I = \prod_{i \in I} a_i$. Let A_{a_I} denote the localization of A by a_I. Let $E = \bigoplus_{i=1}^{n} Ae_i$ be the A-free module with the A-basis e_i $(i = 1, \ldots, n)$. For $I = \{i_1, \ldots, i_r\} \subseteq \{1, \ldots, n\}$, where $i_1 < i_2 < \ldots < i_r$, we set $e_I = e_{i_1} \wedge e_{i_2} \wedge \cdots \wedge e_{i_r}$.

Then we define $C^\bullet = C^\bullet(\underline{a})$ to be the following complex, and call it the *Čech complex* (also called the *modified Čech complex*) of A with respect to the sequence $\underline{a}$.

(1) $C^\bullet = [0 \to C^0 \to C^1 \to \ldots \to C^n \to 0]$, where $C^r = \bigoplus_{\sharp I = r} A_{a_I} e_I$.
(2) $\delta = \delta^r : C^r \to C^{r+1}$ is the A-linear map which is given by $\delta(xe_I) = \sum_{j \notin I} \iota_{I,j}(x) e_j \wedge e_I$ for $I = \{i_1, i_2, \ldots, i_r\}$ and $x \in A_{a_I}$, where $\iota_{I,j}$ is the canonical localization $A_{a_I} \to A_{a_{I \cup \{j\}}}$. Similarly to Koszul complexes, we can prove $\delta^{r+1} \circ \delta^r = 0$ easily. So $C^\bullet$ is a complex with ∂ its coboundary maps.
(3) For an A-module M, we define $C^\bullet(\underline{a}, M) = C^\bullet(\underline{a}) \otimes_A M$.
(4) The rth cohomology group of $C^\bullet(\underline{a}, M)$ is denoted by $\mathrm{H}^r(C^\bullet(\underline{a}, M))$.

In the next section, we will see that if $\mathfrak{a}$ is the ideal generated by $\underline{a}$, then $\mathrm{H}^r(C^\bullet(\underline{a}, M)) \cong \mathrm{H}^r_{\mathfrak{a}}(M)$ is the local cohomology group, which depends only on the radical $\sqrt{\mathfrak{a}}$.

Example 5.89 Let $(A, \mathfrak{m})$ be a d-dimensional Noetherian local ring, and $\underline{a} = (a_1, \ldots, a_d)$ a system of parameters. Then $\mathrm{H}^d(C^\bullet(\underline{a}))=A_{a_1 \cdot a_2 \cdots a_d}/[\sum_{\sharp I = d-1} A_{a_I}]$. In particular, if $d = 1$, then $\mathrm{H}^1(C^\bullet(\underline{a})) = A_{a_1}/A$.

We can express a Čech complex as an inductive limit of Koszul complexes. For a sequence $\underline{a} = (a_1, \ldots, a_n)$ of elements of A, we define $\{e_I \mid \sharp I = r\}$ to be the dual basis of the basis $\{T_I \mid \sharp I = r\}$ of $K_r(\underline{a})$. So it is a free basis of $K^r(\underline{a}) = \mathrm{Hom}_A(K_r(\underline{a}), A)$. We identify $K^\bullet(\underline{a})$ by $\bigwedge E^*$, where $E = \bigoplus_{i=1}^n A \cdot T_i$, and so $E^* = \bigoplus_{i=1}^n A \cdot e_i$. Namely, for $I = \{i_1 < i_2 < \cdots < i_r\}$, $e_I \in K^\bullet(\underline{a})$ corresponds to $e_{i_1} \wedge \cdots \wedge e_{i_r}$. Thus the coboundary map $\delta^r = d_r^* : K^r(\underline{a}) \to K^{r+1}(\underline{a})$ is given by $\delta^r(\alpha) = \eta \wedge \alpha$ for $\alpha \in K^r(\underline{a}) = \bigwedge^r E^*$, where $\eta = \sum_{i=1}^n e_i$.

Definition 5.90 Let $\underline{a} = (a_1, \ldots, a_n)$ be a sequence of elements of A.

(1) We define a chain map $\phi^\bullet : K^\bullet(\underline{a}) \to C^\bullet(\underline{a})$ by $\phi^r(xe_I) = (x/a_I)e_I$. Then

$$\phi^{r+1} \circ \delta^r(xe_I) = \phi^{r+1}(\sum_{j=1}^n a_j x e_j \wedge e_I) = \sum_{j=1}^n (a_j x/a_j a_I) e_j \wedge e_I$$

$$= \sum_{j=1}^n (x/a_I) e_j \wedge e_I,$$

and hence $\phi^{r+1} \circ \delta^r = \delta^{r+1} \circ \phi^r$ has been verified.

(2) For an A-module M, we define $\phi_M : K^\bullet(\underline{a}, M) \to C^\bullet(\underline{a}, M)$ by $\phi_M = \phi \otimes_A 1_M$, where $1_M : M \to M$ denotes the identity map of M, and we also identify $K^\bullet(\underline{a}, M) = \mathrm{Hom}_A(K_\bullet(\underline{a}), M)$ by $\mathrm{Hom}_A(K_\bullet(\underline{a}), A) \otimes_A M = K^\bullet(\underline{a}) \otimes_A M$.

We write the Čech complex as the inductive limit of Koszul complexes. For a sequence $\underline{a} = (a_1, \ldots, a_n)$ of elements of A, we abbreviate the sequence $(a_1^m, \ldots, a_n^m)$ as $\underline{a^m}$.

Definition 5.91 For two Koszul complexes $K^\bullet(\underline{a^m})$ and $K^\bullet(\underline{a^l})$ with $l \le m$, we define $\phi_{m,l} : K^\bullet(\underline{a^l}) \to K^\bullet(\underline{a^m})$ as follows. For the basis element e_I, we define $\phi^r_{m,l}(e_I) = a_I^{m-l} e_I$. Then it is easy to check that $\phi_{m,l}$ is a cochain map, and for $\phi_m : K^\bullet(\underline{a^m}) \to C^\bullet(\underline{a^m}) = C^\bullet(\underline{a})$, we have $\phi_l = \phi_m \circ \phi_{m,l}$. Moreover, it is easy to see that $\phi_{s,l} = \phi_{s,m} \circ \phi_{m,l}$ for $l < m < s$. So $\{(K^\bullet(\underline{a^l}), \phi_{m,l})\}$ forms an inductive system, and ϕ_m induces a unique map $\varinjlim K^\bullet(\underline{a^m}) \to C^\bullet(\underline{a})$.

Theorem 5.92 *Let $\underline{a} = (a_1, \ldots, a_n)$ be a sequence of elements of A. Then the map $\phi : \varinjlim K^\bullet(\underline{a^m}) \to C^\bullet(\underline{a})$ induced by ϕ_m is a chain isomorphism.*

Proof As each element of $C^r(\underline{a})$ is written as the finite sum $\sum_I (b_I/a_I^{m_I}) e_I$, replacing m_I by $\max\{m_I \mid \sharp I = r\}$, we can rewrite it as $\sum_I (b'_I/a_I^m) e_I$, and this element is in $\phi_m(K^\bullet(\underline{a^m})) \subseteq \operatorname{Im}\phi$. So ϕ is surjective. We prove the injectivity. It suffices to show that for each $m \geq 0$, each r, and each $x \in \operatorname{Ker}\phi_m \cap K^r(\underline{a^m})$, there exists some $s \geq 0$ such that $\phi_{s+m,m}(x) = 0$. We can write $x = \sum_{\sharp I = r} b_I e_I$, and for each I with $\sharp I = r$, we have $a_I^{s_I} b_I = 0$ for some $s_I \geq 0$ by the definition of localization. Letting $s = \max\{s_I \mid \sharp I = r\}$, we have $\phi_{s+m,m}(x) = 0$, as desired, and hence ϕ is injective. □

As the tensor product $- \otimes_A M$ is compatible with the inductive limit, we have the following.

Corollary 5.93 *For any A-module M,*

$$\mathrm{H}^r(C^\bullet(\underline{a}, M)) \cong \varinjlim_m \mathrm{H}^r(\underline{a^m}, M).$$

5.7 Local Cohomology

In this section, we only consider Noetherian rings.

Definition 5.94 Let $\mathfrak{a}$ be an ideal of A. For an A-module M, we define

$$\Gamma_{\mathfrak{a}}(M) = \{x \in M \mid \exists n\ \mathfrak{a}^n x = 0\}.$$

Or equivalently, $\Gamma_{\mathfrak{a}}(M) = \{x \in M \mid \operatorname{Supp}_A(Ax) \subseteq \mathrm{V}(\mathfrak{a})\}$. Note that $\Gamma_{\mathfrak{a}}(M)$ is an A-submodule of M. For an A-linear map $f : M \to N$, we have $f(\Gamma_{\mathfrak{a}}(M)) \subseteq \Gamma_{\mathfrak{a}}(N)$, and so $\Gamma_{\mathfrak{a}}(-)$ is a functor from the category of A-modules to itself.

5.95 Since we have $\operatorname{Hom}_A(A/\mathfrak{a}^n, M) = [0 :_M \mathfrak{a}^n]$, we can also think that

$$\Gamma_{\mathfrak{a}}(M) = \lim_{n\to\infty} \operatorname{Hom}_A(A/\mathfrak{a}^n, M).$$

Since $\operatorname{Hom}_A(A/\mathfrak{a}^n, -)$ is left exact and the inductive limit is exact, we have that $\Gamma_{\mathfrak{a}}(-)$ is left exact. Namely, for a short exact sequence $0 \to M' \to M \to M''$ of A-modules, $0 \to \Gamma_{\mathfrak{a}}(M') \to \Gamma_{\mathfrak{a}}(M) \to \Gamma_{\mathfrak{a}}(M'')$ is exact.

On the other hand, for a short exact sequence $0 \to M' \to M \to M'' \to 0$, we have a long exact sequence

$$\begin{aligned} 0 \to \operatorname{Hom}_A(A/\mathfrak{a}^n, M') &\to \operatorname{Hom}_A(A/\mathfrak{a}^n, M) \to \operatorname{Hom}_A(A/\mathfrak{a}^n, M'') \\ &\to \operatorname{Ext}^1_A(A/\mathfrak{a}^n, M') \to \operatorname{Ext}^1_A(A/\mathfrak{a}^n, M) \to \operatorname{Ext}^1_A(A/\mathfrak{a}^n, M'') \to \cdots \end{aligned}$$

for each n. So for $m \geq n$, the canonical projection $A/\mathfrak{a}^m \to A/\mathfrak{a}^n$ induces a map of δ-functors $\operatorname{Ext}^i_A(A/\mathfrak{a}^n, -) \to \operatorname{Ext}^i_A(A/\mathfrak{a}^m, -)$.

Definition 5.96 We define $\mathrm{H}^i_{\mathfrak{a}}(M) := \lim_{m\to\infty} \mathrm{Ext}^i_A(A/\mathfrak{a}^m, M)$, which we call the ith *local cohomology* of M with respect to $\mathfrak{a}$. By definition, $\mathrm{H}^0_{\mathfrak{a}}(M) = \Gamma_{\mathfrak{a}}(M)$. Since each element of $\mathrm{Ext}^i_A(A/\mathfrak{a}^n, M)$ is annihilated by $\mathfrak{a}^n$, for each element $x \in \mathrm{H}^i_{\mathfrak{a}}(M)$, there exists some n such that $\mathfrak{a}^n x = 0$.

Proposition 5.97 *For a ring A, an ideal $\mathfrak{a}$, and a short exact sequence $0 \to M' \to M \to M'' \to 0$ of A-modules, there exists a long exact sequence of local cohomology modules*

$$0 \to \mathrm{H}^0_{\mathfrak{a}}(M') \to \mathrm{H}^0_{\mathfrak{a}}(M) \to \mathrm{H}^0_{\mathfrak{a}}(M'') \\ \to \mathrm{H}^1_{\mathfrak{a}}(M') \to \mathrm{H}^1_{\mathfrak{a}}(M) \to \mathrm{H}^1_{\mathfrak{a}}(M'') \to \mathrm{H}^2_{\mathfrak{a}}(M') \to \dots.$$

By the definition of the local cohomology using the inductive limit of Ext modules, the functor $\mathrm{H}^i_{\mathfrak{a}}(-) : \mathrm{Mod}\,A \to \mathrm{Mod}\,A$ is A-linear (cf. Definition C.5). The next long exact sequence is frequently used.

Example 5.98 If $a \in A$ is M-regular, then there is a long exact sequence

$$0 \to \mathrm{H}^0_{\mathfrak{a}}(M) \to \mathrm{H}^0_{\mathfrak{a}}(M) \to \mathrm{H}^0_{\mathfrak{a}}(M/aM) \\ \to \mathrm{H}^1_{\mathfrak{a}}(M) \to \mathrm{H}^1_{\mathfrak{a}}(M) \to \mathrm{H}^1_{\mathfrak{a}}(M/aM) \to \mathrm{H}^2_{\mathfrak{a}}(M) \to \cdots.$$

Proposition 5.99 *Let M be a finitely generated A-module, and $\mathfrak{a} \subseteq \mathfrak{m}$ an ideal of A. Then we have* $\mathrm{depth}(\mathfrak{a}, M) = \inf\{i \in \mathbb{Z} \mid \mathrm{H}^i_{\mathfrak{a}}(M) \neq 0\}$.

Proof We set $r(M) = \mathrm{depth}(\mathfrak{a}, M)$ and $s(M) = \inf\{i \in \mathbb{Z} \mid \mathrm{H}^i_{\mathfrak{a}}(M) \neq 0\}$. If $M = \mathfrak{a}M$, $\mathrm{Ext}^i_A(A/\mathfrak{a}^n, M) = 0$ for $i \in \mathbb{Z}$ and $n \geq 0$. So $s(M) = \infty = r(M)$. Now assume that $M \neq \mathfrak{a}M$ so that $r(M) < \infty$. We prove that $r(M) = s(M)$ by induction on $r(M)$. If $r(M) = 0$, then $0 \neq \mathrm{Hom}_A(A/\mathfrak{a}, M) \subseteq \mathrm{H}^0_{\mathfrak{a}}(M)$, and hence $s(M) = 0$. Assume that $r(M) > 0$, and take an M-regular element $a \in I$. Then we have a long exact sequence

$$0 \to \mathrm{H}^0_{\mathfrak{a}}(M) \to \mathrm{H}^0_{\mathfrak{a}}(M) \to \mathrm{H}^0_{\mathfrak{a}}(M/aM) \\ \to \mathrm{H}^1_{\mathfrak{a}}(M) \to \mathrm{H}^1_{\mathfrak{a}}(M) \to \mathrm{H}^1_{\mathfrak{a}}(M/aM) \to \mathrm{H}^2_{\mathfrak{a}}(M) \to \cdots.$$

Since $s(M/aM) = r(M/aM) = r(M) - 1$, we have that $a : \mathrm{H}^i_{\mathfrak{a}}(M) \to \mathrm{H}^i_{\mathfrak{a}}(M)$ is injective for $i < r(M)$. Since each element x of $\mathrm{H}^i_{\mathfrak{a}}(M)$ is annihilated by some power of $\mathfrak{a}$, there exists some $t \geq 1$ such that $a^t x = 0$. Since a^t is $\mathrm{H}^i_{\mathfrak{a}}(M)$-regular, we have that $x = 0$. So $\mathrm{H}^i_{\mathfrak{a}}(M) = 0$ for $i < r(M)$. So $0 \to \mathrm{H}^{r(M)-1}_{\mathfrak{a}}(M/aM) \to \mathrm{H}^{r(M)}_{\mathfrak{a}}(M)$ is exact. Since $\mathrm{H}^{r(M)-1}_{\mathfrak{a}}(M/aM) \neq 0$, we have that $\mathrm{H}^{r(M)}_{\mathfrak{a}}(M) \neq 0$, and hence $s(M) = r(M)$, as desired. □

In general, it is very difficult to compute $\mathrm{Ext}^i_A(N, M)$, but local cohomology modules are easy to compute in many cases. The following example is very important from the theoretical point of view.

Example 5.100 Let $\mathfrak{p}$ be a prime ideal of A, and let $E = E_A(A/\mathfrak{p})$ be the injective envelope of $A/\mathfrak{p}$. Then:

(1) If $\mathfrak{a} \subseteq \mathfrak{p}$, then $\mathrm{H}^0_\mathfrak{a}(E) = E$, and $\mathrm{H}^i_\mathfrak{a}(E) = 0$ for $i > 0$.
(2) If $\mathfrak{a} \not\subseteq \mathfrak{p}$, then $\mathrm{H}^i_\mathfrak{a}(E) = 0$ for $i \geq 0$.

Proof Since E is an injective module, $\mathrm{Ext}^i_A(A/\mathfrak{a}^n, E) = 0$, and hence $\mathrm{H}^i_\mathfrak{a}(E) = 0$ for $i > 0$. For $0 \neq x \in E$, $\mathrm{Ann}_A(x)$ is $\mathfrak{p}$-primary. So if $\mathfrak{a} \subseteq \mathfrak{p}$, then $E = \mathrm{H}^0_\mathfrak{a}(E)$. If $\mathfrak{a} \not\subseteq \mathfrak{p}$, then letting $a \in \mathfrak{a}$, $a \notin \mathfrak{p}$, by Lemma 5.43, the multiplication map $a_E : E \to E$ is bijective. On the other hand, for each $x \in \mathrm{H}^0_\mathfrak{a}(E)$, there exists some n such that $a^n x = 0$. Hence $\mathrm{H}^0_\mathfrak{a}(E) = 0$ in this case. □

Hence we get the following (it can also be proved using the uniqueness of the derived functor).

Corollary 5.101 *If $I^\bullet$ is an injective resolution of M, then we have $\mathrm{H}^i_\mathfrak{a}(M) = \mathrm{H}^i(\mathrm{H}^0_\mathfrak{a}(I^\bullet))$. Namely, $\mathrm{H}^i_\mathfrak{a}(M)$ is the ith cohomology group of the subcomplex $\mathrm{H}^0_\mathfrak{a}(I^\bullet)$ of $I^\bullet$.*

Proof Let $d^i : I^i \to I^{i+1}$ be the differential map, and set $Z^i = \mathrm{Ker}(d^i) = \mathrm{Im}(d^{i-1})$ (note that $M = Z^0$). Then we can split the exact sequence $0 \to M \to I^0 \to I^1 \to \ldots$ into short exact sequences $0 \to Z^i \to I^i \to Z^{i+1} \to 0$ $(i \geq 0)$. Taking the long exact sequence of cohomology groups, we get:

$$0 \to \mathrm{H}^0_\mathfrak{a}(Z^i) \to \mathrm{H}^0_\mathfrak{a}(I^i) \to \mathrm{H}^0_\mathfrak{a}(Z^{i+1}) \to \mathrm{H}^1_\mathfrak{a}(Z^i) \to 0,$$

$$\mathrm{H}^j_\mathfrak{a}(Z^i) \cong \mathrm{H}^{j-1}_\mathfrak{a}(Z^{i+1}) \quad (j \geq 2),$$

$$\mathrm{H}^0_\mathfrak{a}(Z^i) \cong \mathrm{Ker}[\mathrm{H}^0_\mathfrak{a}(I^i) \to \mathrm{H}^0_\mathfrak{a}(I^{i+1})].$$

So $\mathrm{H}^0_\mathfrak{a}(M) \cong \mathrm{H}^0(\mathrm{H}^0_\mathfrak{a}(I^\bullet))$, and $\mathrm{H}^j_\mathfrak{a}(M) \cong \mathrm{H}^{j-1}_\mathfrak{a}(Z^1) \cong \cdots \cong \mathrm{H}^1_\mathfrak{a}(Z^{j-1}) \cong \mathrm{H}^j(\mathrm{H}^0_\mathfrak{a}(I^\bullet))$ for $j \geq 1$. □

The following method generalizing this is a useful way of computing local cohomology.

Definition 5.102 For an ideal $\mathfrak{a}$ of A and an A-module M, we define the following conditions:

(1) If $\mathrm{H}^0_\mathfrak{a}(M) = M$, then we say that M satisfies the condition $(*)$ with respect to the ideal $\mathfrak{a}$.
(2) If we can write $M = \bigoplus_{\lambda \in \Lambda} M_\lambda$, and for each λ, we can take $a_\lambda \in \mathfrak{a}$ such that $a_\lambda : M_\lambda \to M_\lambda$ is bijective, then we say that M satisfies the condition $(**)$ with respect to $\mathfrak{a}$.

Proposition 5.103 *If M satisfies the condition $(*)$, then $\mathrm{H}^0_{\mathfrak{a}}(M) = M$ and $\mathrm{H}^i_{\mathfrak{a}}(M) = 0$ $(i > 0)$. If M satisfies the condition $(**)$, then we have $\mathrm{H}^i_{\mathfrak{a}}(M) = 0$ $(i \geq 0)$.*

Proof First, assume that M satisfies the condition $(*)$, then by definition, $\mathrm{H}^0_{\mathfrak{a}}(M) = M$. Take a minimal injective resolution $I^\bullet$ of M, and localize by a prime ideal P which does not contain $\mathfrak{a}$. Then $I^\bullet_P$ is a minimal injective resolution of $M_P = 0$, and hence $I^\bullet_P = 0$. So each I^i is a direct sum of a bunch of $E_A(A/P)$ with $P \in \mathrm{V}(\mathfrak{a})$. So $\mathrm{H}^0_{\mathfrak{a}}(I^\bullet) = I^\bullet$ by Example 5.100. Hence $\mathrm{H}^i(M) \cong \mathrm{H}^i(\mathrm{H}^0(I^\bullet)) = \mathrm{H}^i(I^\bullet) = 0$ for $i > 0$. Next, we assume that M satisfies $(**)$, and hence there is a decomposition $M = \bigoplus_\lambda M_\lambda$ such that for each λ, there exists some $a_\lambda \in \mathfrak{a}$ such that $a_\lambda : M_\lambda \to M_\lambda$ is bijective. Then $a_\lambda : \mathrm{H}^i_{\mathfrak{a}}(M_\lambda) \to \mathrm{H}^i_{\mathfrak{a}}(M_\lambda)$ is bijective. Since each element of $\mathrm{H}^i_{\mathfrak{a}}(M_\lambda)$ is annihilated by some power of a_λ, we have that $\mathrm{H}^i_{\mathfrak{a}}(M_\lambda) = 0$ $(i \geq 0)$ for each λ. Since the local cohomology is compatible with the direct sum, we have that $\mathrm{H}^i_{\mathfrak{a}}(M) = 0$ $(i \geq 0)$. □

Lemma 5.104 *Let M be an A-module, and $\mathfrak{a}$ be an ideal of A. Assume that $C^\bullet$ satisfies the following properties:*

(1) *$C^i = 0$ for $i < 0$, and $C^0 = M$.*
(2) *For $i \geq 1$, C^i satisfies the condition $(**)$ with respect to $\mathfrak{a}$.*
(3) *For each i, $\mathrm{H}^i(C^\bullet)$ satisfies the condition $(*)$ with respect to $\mathfrak{a}$.*

Then $\mathrm{H}^i_{\mathfrak{a}}(M) \cong \mathrm{H}^i(C^\bullet)$ for $i \geq 0$.

Proof We split the complex $C^\bullet$ into short exact sequences. Letting $H^i = \mathrm{H}^i(C^\bullet)$, $Z^i = \mathrm{Ker}(d^i : C^i \to C^{i+1})$, and $B^{i+1} = \mathrm{Im}(d^i : C^i \to C^{i+1})$, we get short exact sequences

$$0 \to Z^i \to C^i \to B^{i+1} \to 0 \tag{5.2}$$

and

$$0 \to B^i \to Z^i \to H^i \to 0. \tag{5.3}$$

Note that for $i \geq 1$, $\mathrm{H}^j_{\mathfrak{a}}(C^i) = 0$ for all $j \in \mathbb{Z}$.

As $B^i \subseteq Z^i \subseteq C^i$, we have that $\mathrm{H}^0_{\mathfrak{a}}(B^i) = \mathrm{H}^0_{\mathfrak{a}}(Z^i) = 0$ for $i \geq 1$. Since $M = C^0$, $B^0 = 0$, and $H^0 = Z^0$, we have $H^0 = \mathrm{H}^0_{\mathfrak{a}}(H^0) = \mathrm{H}^0_{\mathfrak{a}}(Z^0) \cong \mathrm{H}^0_{\mathfrak{a}}(C^0) = \mathrm{H}^0_{\mathfrak{a}}(M)$. Moreover, from the exact sequence (5.2), we have that $\mathrm{H}^{j+1}_{\mathfrak{a}}(Z^i) \cong \mathrm{H}^j_{\mathfrak{a}}(B^{i+1})$ for $i \geq 1$ and $j \in \mathbb{Z}$. So in particular, we have that $\mathrm{H}^1_{\mathfrak{a}}(Z^i) = 0$ for $i \geq 1$. Hence, from the short exact sequence (5.3), we get $H^i \cong \mathrm{H}^0_{\mathfrak{a}}(H^i) \cong \mathrm{H}^1_{\mathfrak{a}}(B^i)$ and $\mathrm{H}^j_{\mathfrak{a}}(B^i) \cong \mathrm{H}^j_{\mathfrak{a}}(Z^i)$ $(i \geq 2)$. Since $Z^0 = H^0$ satisfies $(*)$, $\mathrm{H}^j_{\mathfrak{a}}(M) = \mathrm{H}^j_{\mathfrak{a}}(C^0) \cong \mathrm{H}^j_{\mathfrak{a}}(B^1)$ for $j \geq 1$ by (5.2) for $i = 0$. Hence

$$H^i \cong \mathrm{H}^1_{\mathfrak{a}}(B^i) \cong \mathrm{H}^2_{\mathfrak{a}}(Z^{i-1}) \cong \mathrm{H}^2_{\mathfrak{a}}(B^{i-1}) \cong \mathrm{H}^3_{\mathfrak{a}}(Z^{i-2}) \cong \cdots \cong \mathrm{H}^i_{\mathfrak{a}}(B^1) \cong \mathrm{H}^i_{\mathfrak{a}}(M)$$

for $i \geq 1$. □

From this, we can deduce the following theorem. Indeed, local cohomology in practice is often expressed in terms of the following.

Theorem 5.105 *Let A be a Noetherian ring, and $\mathfrak{a}$ an ideal of A. If $\underline{f} = (f_1, f_2, \ldots, f_n) \subseteq A$ satisfies $\sqrt{\mathfrak{a}} = \sqrt{(f_1, f_2, \ldots, f_n)}$, then for any integer $p \in \mathbb{Z}$, we get an isomorphism of δ-functors on the category of A-modules*

$$\mathrm{H}^p_{\mathfrak{a}}(*) = \mathrm{H}^p(C^\bullet(\underline{f}, (*))).$$

Proof Verify the conditions of Lemma 5.104. □

Corollary 5.106 *Let $\mathfrak{a}$ be an ideal of A, and $\underline{f} = (f_1, f_2, \ldots, f_n) \subseteq A$ be a sequence such that $\sqrt{\mathfrak{a}} = \sqrt{(f_1, f_2, \ldots, f_n)}$. Then for any A-module M and $p > n$, we have $\mathrm{H}^p_{\mathfrak{a}}(M) = 0$. In particular, if $(A, \mathfrak{m})$ is a local ring with $\dim A = d$, then we have $\mathrm{H}^p_{\mathfrak{m}}(M) = 0$ for any $p > d$, and hence the functor $\mathrm{H}^d_{\mathfrak{m}}(*)$ is right exact.*

Proof This is because $C^p(\underline{f}, M) = 0$ if $p > n$ by definition. □

As $\mathrm{H}^d_{\mathfrak{m}}(*)$ is right exact, A-linear, and compatible with the inductive limits, the following holds (cf. Exercise 6.1).

Proposition 5.107 *For a finitely generated A-module M, we have $\mathrm{H}^d_{\mathfrak{m}}(M) \cong \mathrm{H}^d_{\mathfrak{m}}(A) \otimes_A M$.*

Proposition 5.108 *Let $f : A \to B$ be a ring homomorphism, and M be a B-module. Then for an ideal $\mathfrak{a}$ of A and $p \in \mathbb{Z}$, we have an isomorphism $\mathrm{H}^p_{\mathfrak{a}B}(M) \cong \mathrm{H}^p_{\mathfrak{a}}(M)$, where the left-hand side is the local cohomology of the B-module M, and the right-hand side is the local cohomology of M viewed as an A-module.*

Proof When we take a sequence $\underline{f} = (f_1, f_2, \ldots, f_n) \subseteq A$ such that $\sqrt{\mathfrak{a}} = \sqrt{(f_1, f_2, \ldots, f_n)}$ and consider the Čech complex of M, then this complex is the same whether we consider M as a B-module or an A-module, and the conclusion is obvious. □

Remark 5.2 Let $(A, \mathfrak{m})$ be a Noetherian local ring, and M a d-dimensional finitely generated A-module. Then the following hold:

(1) For any $p \geq 0$, we have $\mathrm{H}^p_{\mathfrak{m}}(M) = \mathrm{H}^p_{\mathfrak{m}\widehat{A}}(\widehat{M})$. Namely, the local cohomology is stable under completion.
(2) If $\dim M = r$, then for any $p > r$, we have $\mathrm{H}^p_{\mathfrak{m}}(M) = 0$.
(3) Using the local duality theorem, we will see that the local cohomology module $\mathrm{H}^p_{\mathfrak{m}}(M)$ is an Artinian A-module for every p. We will also show that $\mathrm{H}^d_{\mathfrak{m}}(M)$ is not finitely generated as an $\widehat{A}$-module if $d > 0$, see Lemma 6.24 and Proposition 6.36.

5.8 What Makes a Complex Exact?

Let A be a ring. We give a criterion for a finite complex to be exact. The main theorem of this section (Theorem 5.115) is due to D. A. Buchsbaum and D. Eisenbud[7] [BE1].

Lemma 5.109 *Let M be a finitely presented flat A-module. Then M is projective.*

Proof Take any surjective A-linear map $\psi : A^r \to M$. It suffices to show that $\psi^* : \operatorname{Hom}_A(M, A^r) \to \operatorname{Hom}_A(M, M)$ is surjective. As M is finitely presented, it suffices to show that

$$\psi_P^* : \operatorname{Hom}_{A_P}(M_P, A_P^r) \to \operatorname{Hom}_{A_P}(M_P, M_P)$$

is surjective for each $P \in \operatorname{Spec} A$. So we may assume that $(A, \mathfrak{m})$ is local. This case is already done in Proposition 5.6. □

Proposition 5.110 *Let A be semi-local, that is, A has only finitely many maximal ideals. Then any finitely generated projective module with constant rank (see* (D.8) *for the definition) is free.*

Proof Let P be a finitely generated projective A-module with constant rank n. If $\mathfrak{m} \in \operatorname{Max} A$, then since $P/\mathfrak{m}P = P_\mathfrak{m}/\mathfrak{m}P_\mathfrak{m}$, we have $\dim_{A/\mathfrak{m}} P/\mathfrak{m}P = \operatorname{rank} P = n$. So

$$P/JP = \bigoplus_{\mathfrak{m}\in\operatorname{Max} A} P/\mathfrak{m}P = \bigoplus_{\mathfrak{m}\in\operatorname{Max} A} (A/\mathfrak{m})^n = (A/J)^n.$$

So there is an A-linear map $\varphi : A^n \to P$ such that $(A/J) \otimes \varphi$ is an isomorphism. By NAK, Theorem 1.86, φ is surjective. When we set $K = \operatorname{Ker} \varphi$, then the sequence $0 \to K \to A^n \xrightarrow{\varphi} P \to 0$ splits and $K/JK = 0$. Hence we have $K = 0$. □

The following lemma, which is due to Peskine–Szpiro, is called the *acyclicity lemma* [PS].

Lemma 5.111 (Acyclicity Lemma) *Let $(A, \mathfrak{m})$ be a Noetherian local ring, and*

$$L_\bullet : 0 \to L_s \xrightarrow{\partial_s} L_{s-1} \to \cdots \to L_2 \xrightarrow{\partial_2} L_1 \xrightarrow{\partial_1} L_0$$

be a complex of finitely generated A-modules. If the following two conditions hold, then $L_\bullet$ is exact (acyclic):

(1) *For $1 \le i \le s$,* $\operatorname{depth}_A L_i \ge i$.
(2) *If $1 \le i \le s$ and* $\operatorname{H}_i(L_\bullet) \ne 0$, *then* $\operatorname{depth}_A \operatorname{H}_i(L_\bullet) = 0$,

[7] David Eisenbud, 1947–.

where $\mathrm{H}_i(L_\bullet)$ *denotes the* i*th homology group of* $L_\bullet$.

Proof Assume that $L_\bullet$ is not exact, and let $t = \max\{i \in \{1, \ldots, s\} \mid \mathrm{H}_i(L_\bullet) \neq 0\}$. We set $Z_i = \operatorname{Ker} \partial_i$ and $B_i = \operatorname{Im} \partial_{i+1}$. As Z_t is a submodule of L_t, we have that $\operatorname{depth}_A Z_t > 0$. As $\mathrm{H}_s(L_\bullet) = Z_s$, we have $1 \leq t < s$. By the depth lemma (Lemma 5.9) and the exact sequence

$$0 \to B_t \to Z_t \to \mathrm{H}_t(L_\bullet) \to 0,$$

we have $\operatorname{depth}_A B_t = 1$. As the sequence

$$0 \to L_s \to \cdots \to L_{t+1} \to B_t \to 0$$

is exact, we have $\operatorname{depth}_A L_s = s - t < s$, and this is a contradiction. □

Let F and G be finitely generated free A-modules, and $m = \operatorname{rank}_A G$ and $n = \operatorname{rank}_A F$. Let $\varphi : F \to G$ be an A-linear map. For $0 \leq t \in \mathbb{Z}$, we define $\mathrm{I}_t = \mathrm{I}_t(\varphi)$ to be the ideal of A generated by all the t-minors of the matrix of φ, where we understand that $\mathrm{I}_0 = A$ and $\mathrm{I}_i = 0$ for $i > \min(m, n)$. If M ($\neq 0$) is an A-module, then we set

$$\operatorname{rank}(\varphi, M) = \sup\{t \geq 0 \mid \mathrm{I}_t(\varphi)M \neq 0\},$$

and $\mathrm{I}(\varphi, M) = \mathrm{I}_t(\varphi)$, where $t = \operatorname{rank}(\varphi, M)$. If $M = A$, then we simply write $\operatorname{rank}(\varphi, A)$ and $\mathrm{I}(\varphi, A)$ by $\operatorname{rank} \varphi$ and $\mathrm{I}(\varphi)$, respectively.

Lemma 5.112 *The following are equivalent:*

(1) $\mathrm{I}(\varphi) = A$.
(2) $C = \operatorname{Coker} \varphi$ *is an* A*-projective module with a constant rank.*

Proof Set $r = \operatorname{rank} \varphi$.

(1) ⇒(2). Let $\mathfrak{p} \in \operatorname{Spec} A$. As $\mathrm{I}_{r+1}(\varphi) = 0$ and $\mathfrak{p} \not\supseteq \mathrm{I}_r(\varphi)$, If we choose the free bases of $F_\mathfrak{p}$ and $G_\mathfrak{p}$ appropriately, then the matrix of $\varphi_\mathfrak{p} = A_\mathfrak{p} \otimes \varphi$ is of the form

$$\left(\begin{array}{ccc|c} 1 & & & \\ & \ddots & & \mathbf{0} \\ & & 1 & \\ \hline & 0 & & 0 \end{array}\right).$$

Hence $C_\mathfrak{p} = \operatorname{Coker} \varphi_\mathfrak{p} \cong A_\mathfrak{p}^{m-r}$. Since C is finitely presented and flat, it is projective by Lemma 5.109, and it has constant rank $m - r$.

(2) ⇒(1). Assume that C is projective with constant rank s. For $\mathfrak{p} \in \operatorname{Spec} A$, since we have $C_\mathfrak{p} \cong A_\mathfrak{p}^s$, in the exact sequence

$$F_\mathfrak{p} \xrightarrow{\varphi_\mathfrak{p}} G_\mathfrak{p} \to C_\mathfrak{p} \to 0,$$

$\operatorname{Im}\varphi_{\mathfrak{p}}$ is a direct summand of $G_{\mathfrak{p}}$, and $\operatorname{Im}\varphi_{\mathfrak{p}} \cong A_{\mathfrak{p}}^{m-s}$. Hence $\varphi_{\mathfrak{p}}$ has a matrix

$$\left(\begin{array}{ccc|c} 1 & & & \\ & \ddots & & 0 \\ & & 1 & \\ \hline & 0 & & 0 \end{array}\right).$$

Hence $\mathrm{I}_{m-s}(\varphi_{\mathfrak{p}}) = A_{\mathfrak{p}}$ and $\mathrm{I}_{m-s+1}(\varphi_{\mathfrak{p}}) = 0$. As the integer s is independent of $\mathfrak{p} \in \operatorname{Spec} A$, we have that $\mathrm{I}_{m-s}(\varphi) = A$ and $\mathrm{I}_{m-s+1}(\varphi) = 0$, and $r = m - s$. □

Lemma 5.113 *Let M be a nonzero finitely generated A-module. Let $F \xrightarrow{\varphi} G \xrightarrow{\psi} H$ be a sequence of finitely generated free A-modules such that $\psi\varphi = 0$. If $\mathrm{I}(\varphi, M) = \mathrm{I}(\psi, M) = A$, then the following are equivalent:*

(1) *The sequence $F \otimes M \xrightarrow{\varphi\otimes M} G \otimes M \xrightarrow{\psi\otimes M} H \otimes M$ is exact.*
(2) $\operatorname{rank}(\varphi, M) + \operatorname{rank}(\psi, M) = \operatorname{rank} G$.

Proof Replacing A by $A/\operatorname{Ann}_A M$ if necessary, we may assume that M is faithful. So $\mathrm{I}(\varphi) = \mathrm{I}(\psi) = A$, $\operatorname{rank}(\varphi, M) = \operatorname{rank}\varphi$, and $\operatorname{rank}(\psi, M) = \operatorname{rank}\psi$. Also, it suffices to assume $(A, \mathfrak{m})$ is local.

Also, since $\mathrm{I}(\varphi) = A$ and $0 \to \operatorname{Im}(\varphi) \to G \to \operatorname{Coker}(\varphi) \to 0$ is exact, G is isomorphic to direct sum of free A-modules $\operatorname{Im}(\varphi)$ and $\operatorname{Coker}(\varphi)$. Also, since $0 \to \operatorname{Ker}(\psi) \to G \to \operatorname{Im}(\psi) \to 0$ is exact and since $\mathrm{I}(\psi) = A$, $\operatorname{Im}(\psi)$ is a free A-module of $\operatorname{rank}(\psi)$ and then $\operatorname{Ker}(\psi)$ is also a direct summand of G. Since $\psi\varphi = 0$, we have $\operatorname{Im}(\varphi) \subset \operatorname{Ker}(\psi)$. Since both are direct summands of G, $\operatorname{Im}(\varphi) = \operatorname{Ker}(\psi)$ if and only if $\operatorname{rank}(\varphi) = \operatorname{rank}(\operatorname{Ker}(\psi)) = \operatorname{rank} G - \operatorname{rank}(\psi)$. □

By Lemma 5.113, we have the following corollary immediately.

Corollary 5.114 *Let F and G be nonzero finitely generated free A-modules, and $\varphi : F \to G$ an A-linear map. Let M be a nonzero finitely generated A-module such that $\mathrm{I}(\varphi, M) = A$. Then the following are equivalent:*

(1) $0 \to F \otimes M \xrightarrow{\varphi\otimes M} G \otimes M$ *is exact.*
(2) $\operatorname{rank}\varphi = \operatorname{rank}_A F$.

The following is the main theorem of this section.

Theorem 5.115 *Let A be a Noetherian ring, and*

$$F_\bullet : 0 \to F_n \xrightarrow{\varphi_n} F_{n-1} \xrightarrow{\varphi_{n-1}} F_{n-2} \to \cdots \to F_1 \xrightarrow{\varphi_1} F_0$$

be a complex of finitely generated free A-modules. Let M be a nonzero finitely generated A-module. Then the following conditions are equivalent:

(1) *The sequence*

$$0 \to F_n \otimes M \xrightarrow{\varphi_n \otimes M} F_{n-1} \otimes M \xrightarrow{\varphi_{n-1} \otimes M} F_{n-2} \otimes M \longrightarrow$$
$$\cdots \longrightarrow F_1 \otimes M \xrightarrow{\varphi_1 \otimes M} F_0 \otimes M$$

is exact.

(2) *For* $1 \le i \le n$, *the following two conditions are satisfied:*

  (a) $\operatorname{depth}(\mathrm{I}(\varphi_i, M), M) \ge i$.
  (b) $\operatorname{rank}(\varphi_i, M) + \operatorname{rank}(\varphi_{i+1}, M) = \operatorname{rank}_A F_i$,

  where we put $\operatorname{rank}(\varphi_{n+1}, M) = 0$.

Proof Replacing A by $A/\operatorname{Ann}_A M$, we may assume that M is faithful.
(1) $\Rightarrow$ (2). We first show the following. □

Claim 5.1 For $1 \le i \le n$, $\operatorname{depth}(\mathrm{I}(\varphi_i), M) \ge 1$.

Proof In order to show this, letting $S = A \setminus \bigcup_{P \in \operatorname{Ass}_A M} P$ and replacing A by $S^{-1}A$, we may assume that $\operatorname{Max} A \subseteq \operatorname{Ass}_A M$. In particular, A is semi-local. It suffices to show that $\mathrm{I}(\varphi_i) = A$ for $1 \le i \le n$.

We assume that $\mathrm{I}(\varphi_i) \ne A$ for some $1 \le i \le n$, and take a complex $F_\bullet$ of counterexample such that n is small as possible. For this $F_\bullet$, let k be the maximum value of i such that $\mathrm{I}(\varphi_i) \ne A$. Then we have $k = n$. Indeed, if $k < n$, then $\mathrm{I}(\varphi_{k+1}) = A$, and hence $C = \operatorname{Coker} \varphi_{k+1}$ is a finitely generated projective A-module with a constant rank. As A is semi-local, C is free by Proposition 5.110. Letting $\varepsilon : F_k \to C$ be the canonical map and $f : C \to F_{k-1}$ be the map such that $f\varepsilon = \varphi_k : F_k \to F_{k-1}$, we have $\operatorname{rank} f = \operatorname{rank} \varphi_k$ and $\mathrm{I}(f) = \mathrm{I}(\varphi_k) \subsetneq A$. On the other hand,

$$0 \to C \otimes M \xrightarrow{f \otimes M} F_{k-1} \otimes M \to \cdots \to F_0 \otimes M$$

is exact. This contradicts the minimality of n. Hence $k = n$.

We take $P \in \operatorname{Max} A$ such that $\mathrm{I}(\varphi_n) \subseteq P$. Since $P \in \operatorname{Ass}_A M$, there is an injective map $A/P \xrightarrow{\iota} M$. Then we get a commutative diagram

$$\begin{array}{ccc} F_n \otimes A/P & \xrightarrow{\varphi_n \otimes A/P} & F_{n-1} \otimes A/P \\ {\scriptstyle F_n \otimes \iota} \big\downarrow & & \big\downarrow {\scriptstyle F_{n-1} \otimes \iota} \\ F_n \otimes M & \xhookrightarrow{\varphi_n \otimes M} & F_{n-1} \otimes M \end{array},$$

and we have that $\varphi_n \otimes A/P : F_n \otimes A/P \to F_{n-1} \otimes A/P$ is also injective. Since A/P is a field, when we set $r = \operatorname{rank}_A F_n$, then $\mathrm{I}_r(\varphi_n) \not\subseteq P$, and we have $r = \operatorname{rank} \varphi_n$. This is impossible, since we assume that $\mathrm{I}(\varphi_n) \subseteq P$. Hence we have $\operatorname{depth}(\mathrm{I}(\varphi_1), M) \ge 1$ for each $1 \le i \le n$. □

Claim 5.2 For any $1 \leq i \leq n$, we have $\operatorname{rank}(\varphi_i, M) + \operatorname{rank}(\varphi_{i+1}, M) = \operatorname{rank}_A F_i$, where $\operatorname{rank}(\varphi_{n+1}, M) = 0$.

Proof If $P \in \operatorname{Ass}_A M$, then by Claim 5.1, $P \not\supseteq \mathrm{I}(\varphi_i, M)$. Hence $\operatorname{rank}([\varphi_i]_P, M_P) = \operatorname{rank}(\varphi_i, M)$. As $\mathrm{I}([\varphi_i]_P, M_P) = A_P$, we have that $\operatorname{rank}(\varphi_i, M) + \operatorname{rank}(\varphi_{i+1}, M) = \operatorname{rank}([\varphi_i]_P, M_P) + \operatorname{rank}([\varphi_{i+1}]_P, M_P) = \operatorname{rank}_{A_P}[F_i]_P = \operatorname{rank}_A F_i$ by Lemma 5.112 and Lemma 5.113. □

Claim 5.3 For each $1 \leq i \leq n$, we have $\operatorname{depth}(\mathrm{I}(\varphi_i), M) \geq i$.

Proof Assuming that this claim is not true, take its counterexample such that the length n of $F_\bullet$ is small as possible. If $\operatorname{depth}(\mathrm{I}(\varphi_i), M) < i$, then we can take $\mathfrak{p} \in \mathrm{V}(\mathrm{I}(\varphi_i)) \cap \operatorname{Supp}_A M$ such that $\operatorname{depth}_{A_\mathfrak{p}} M_\mathfrak{p} < i$. So replacing A by $A_\mathfrak{p}$, we may assume that $(A, \mathfrak{m})$ is local and $\operatorname{depth}_A M < i$ for some i. Let k be the largest integer such that $1 \leq k \leq n$, $\operatorname{depth}_A M < k$, and $\mathrm{I}(\varphi_k) \neq A$. Then by Claim 5.1, we have $k > 1$. By the same reasoning as in the discussion of the proof of Claim 5.1, we have $k = n$ by the minimality of n. Indeed, if $k < n$, then we have $\mathrm{I}(\varphi_{k+1}) = A$.

Set $t = \operatorname{depth}_A M$. Then $1 \leq t < n$. Take an M-sequence $a_1, a_2, \ldots, a_t \in \mathfrak{m}$, and set $Q = (a_1, a_2, \ldots, a_t)$. Then by the exact sequence

$$0 \to F_n \otimes M \to \cdots \to F_1 \otimes M \to F_0 \otimes M,$$

we get the exact sequence

$$0 \to F_n \otimes M/QM \to \cdots \to F_t \otimes M/QM.$$

As $n > n - t$, when we set $r = \operatorname{rank}_A F_n$, then

$$\operatorname{depth}(\mathrm{I}_r(\varphi_n), M/QM) \geq 1.$$

Since we have $\operatorname{depth}_A M/QM = 0$, this shows $\mathrm{I}(\varphi_n) = \mathrm{I}_r(\varphi_n) = A$. But this contradicts the choice of k and the fact that $k = n$. □

Proof of Theorem 5.115, (2) $\Rightarrow$ (1) Assume that

$$F_\bullet \otimes M : 0 \to F_n \otimes M \to \cdots \to F_2 \otimes M \to F_1 \otimes M \to F_0 \otimes M$$

is not exact. We take $F_\bullet$ such that n is as small as possible. Then clearly

$$0 \to F_n \otimes M \to \cdots \to F_2 \otimes M \to F_1 \otimes M$$

is exact, and hence $\mathrm{H}_1(F_\bullet \otimes M) \neq 0$. Let $\mathfrak{p}$ be a minimal prime of $\mathrm{H}_1(F_\bullet \otimes M)$, and we localize by $\mathfrak{p}$. So we may further assume that $(A, \mathfrak{m})$ is local and $\mathrm{H}_1(F_\bullet \otimes_A M)$ is of finite length and nonzero. Set $t = \operatorname{depth}_A M$. By the acyclicity lemma, we have that $t < n$. As $\operatorname{depth}(\mathrm{I}(\varphi_n), M) \geq n > t$, we have that $\mathrm{I}(\varphi_n) = A$. So $C = \operatorname{Coker} \varphi_n$ is a free module. By the minimality of n,

$$0 \to C \otimes M \to F_{n-2} \otimes M \to \cdots \to F_0 \otimes M$$

is exact. But

$$0 \to F_n \otimes M \to F_{n-1} \otimes M \to C \otimes M \to 0$$

is also exact by Corollary 5.114, and this is impossible. Hence (2) $\Rightarrow$ (1) is true, and the proof of Theorem 5.115 has been completed. □

5.9 Hilbert–Burch Theorem

There is a precise theory on determinantal ideals (i.e., ideals generated by minors of matrices). We introduce a basic result here.

Proposition 5.116 *Let A be a Noetherian ring, and $M = [a_{ij}]$ be an (r, s) matrix whose coefficients are in A. Let $\mathfrak{a} = \mathrm{I}_t(M)$ $(1 \leq t \leq \min\{r, s\})$. If $\mathfrak{p} \in \mathrm{Min}_A A/\mathfrak{a}$, then $\mathrm{ht}_A \mathfrak{p} \leq (r - t + 1)(s - t + 1)$. In particular, if $\mathfrak{a} \subsetneq A$, then*

$$\mathrm{ht}_A \mathfrak{a} \leq (r - t + 1)(s - t + 1).$$

Proof Induction on t. If $t = 1$, then this is Krull's altitude theorem (Theorem 2.36). So we assume that $t > 1$, and the proposition is true for smaller t. Replacing A by $A_\mathfrak{p}$ if necessary, we may assume that $(A, \mathfrak{m})$ is local and $\sqrt{\mathfrak{a}} = \mathfrak{m}$. If there exists some (i, j) such that $a_{ij} \notin \mathfrak{m}$, then without loss of generality, we may assume that $a_{11} \notin \mathfrak{m}$, and after elementary operations on rows and columns, we may assume that M is of the form

$$M = \left[\begin{array}{c|c} 1 & 0 \\ \hline 0 & N \end{array}\right].$$

Then we have $\mathfrak{a} = \mathrm{I}_{t-1}(N)$, and hence

$$\dim A \leq [(r-1) - (t-1) - 1][(s-1) - (t-1) + 1] = (r - t + 1)(s - t + 1).$$

So we assume that $a_{ij} \in \mathfrak{m}$ for all (i, j). Let $A[X]$ be a polynomial ring, and define $M' = [b_{ij}]$ by $b_{11} = a_{11} + X$ and $b_{ij} = a_{ij}$ $((i, j) \neq (1, 1))$. We set $\mathfrak{a}' = \mathrm{I}_t(M')(\subseteq A[X])$. Then by Lemma 5.117 below, we have that $\mathfrak{m}A[X]$ is a minimal prime of $\mathfrak{a}'$. In the local ring $B = A[X]_{\mathfrak{m}A[X]}$, $\mathfrak{a}' = \mathrm{I}_t(M')$ is $\mathfrak{m}A[x]$-primary, and has height at most $(r - t + 1)(s - t + 1)$, since $b_{11} \notin \mathfrak{m}A[X]$. So we have $\dim A = \dim B \leq (r - t + 1)(s - t + 1)$, as desired. □

Lemma 5.117 *Let $(A, \mathfrak{m})$ be a Noetherian local ring, $\mathfrak{a} \subseteq A$ be an ideal of A such that $\sqrt{\mathfrak{a}} = \mathfrak{m}$. Let $A[X]$ be a polynomial ring, and $\mathfrak{a}'$ be an ideal of $A[X]$. If $\mathfrak{a}' \subseteq \mathfrak{m}A[X]$ and $\mathfrak{a}' + (X) = \mathfrak{a}A[X] + (X)$, then $\mathfrak{m}A[X]$ is a minimal prime of $\mathfrak{a}'$.*

Proof Let $\mathfrak{p} \in \operatorname{Spec} A[X]$ such that $\mathfrak{a}' \subseteq \mathfrak{p} \subseteq \mathfrak{m}A[X]$. Then $\mathfrak{a}' + (X) \subseteq \mathfrak{p} + (X) \subseteq \mathfrak{m}A[X] + (X)$. When we take $\mathfrak{p}' \in \operatorname{Spec} A[X]$ such that

$$\mathfrak{p} + (X) \subseteq \mathfrak{p}' \subseteq \mathfrak{m}A[X] + (X),$$

then we have $\mathfrak{p}' \supseteq \mathfrak{a}' + (X) = \mathfrak{a}A[X] + (X)$, and hence $\mathfrak{p}' \supseteq \mathfrak{m}A[X]$. So we have $\mathfrak{p}' = \mathfrak{m}A[X] + (X)$. Hence $\mathfrak{m}A[X] + (X)$ is a minimal prime of $\mathfrak{p} + (X)$. Hence $\operatorname{ht}_{A[X]/\mathfrak{p}}[\mathfrak{m}A[X] + (X)]/\mathfrak{p} \le 1$. Since we have $\mathfrak{p} \subseteq \mathfrak{m}A[X] \subsetneq \mathfrak{m}A[X] + (X)$, it follows that $\mathfrak{p} = \mathfrak{m}A[X]$. □

We describe some applications.

Let A be a Noetherian ring, J be an ideal of A, and the A-module A/J has a free resolution of the form

$$0 \to A^n \xrightarrow{f} A^{n+1} \xrightarrow{g} A \to A/J \to 0, \tag{5.4}$$

where f is represented by an $(n+1) \times n$ matrix: $f = [a_{\alpha\beta}]$. For each $1 \le i \le n+1$, we set $d_i = (-1)^i \det[a_{\alpha\beta}]_i$, where $[a_{\alpha\beta}]_i$ is the $n \times n$ matrix obtained by removing the ith row from $[a_{\alpha\beta}]$. We set $I = (d_1, d_2, \dots, d_{n+1})$, and let $h : A^{n+1} \to A$ be the A-linear map which is determined by $h(\mathbf{e}_i) = d_i$ $(1 \le i \le n+1)$, where $\{\mathbf{e}_i\}_{1 \le i \le n+1}$ is the standard basis of A^{n+1}.

Lemma 5.118 *The sequence* $0 \to A^n \xrightarrow{f} A^{n+1} \xrightarrow{h} A \to A/I \to 0$ *is exact.*

Proof It is easy to see that this is a complex. By the exact sequence (5.4) and Theorem 5.115, we have $n = \operatorname{rank} f$, and hence $I = \mathrm{I}(f)$. Since $\mathrm{I}_1(h) = I$ and $\operatorname{rank} h = 1$,

$$0 \to A^n \xrightarrow{f} A^{n+1} \xrightarrow{h} A \to A/I \to 0$$

is exact by Theorem 5.115 again. □

Proposition 5.119 *There is a unique non-zerodivisor* $t \in A$ *such that* $g(\mathbf{e}_i) = td_i$ *for each* i.

Proof Let $g(\mathbf{e}_i) = x_i$. Then for all $1 \le i, j \le n+1$, we have $d_i x_j = d_j x_i$. To prove this, we may assume that $i < j$. For each $1 \le k \le n$, let Δ_k denote the $(n-1)$-minor of $(a_{\alpha\beta})$ obtained by removing the ith row, the jth row, and the kth column. Then

$$f \begin{pmatrix} (-1)^{i+j}\Delta_1 \\ (-1)^{i+j+1}\Delta_2 \\ \vdots \\ (-1)^{i+j+n-1}\Delta_n \end{pmatrix} = d_i \mathbf{e}_j - d_j \mathbf{e}_i.$$

Hence $d_i x_j = d_j x_i$. Since depth$(I, A) \geq 2$, we can take a non-zerodivisor $u \in I$, and write $u = \sum_{i=1}^{n+1} t_i d_i$ ($t_i \in A$). Then we have

$$ux_j = \sum_{i=1}^{n+1} t_i(d_i x_j) = \sum_{i=1}^{n+1} t_i(d_j x_i) = \left(\sum_{i=1}^{n+1} t_i x_i\right) d_j$$

for each $1 \leq j \leq n+1$. Hence letting $x = \sum_{i=1}^{n+1} t_i x_i$, we have $ux_j = xd_j$. Hence we have $uJ = xI$. Note that x is a non-zerodivisor. Indeed, if $a \in A$ satisfies $xa = 0$, then $(ua)J = 0$. Since $J = \mathrm{I}(g)$ and depth$(\mathrm{I}(g), A) > 0$ by Theorem 5.115, we have $ua = 0$, and hence $a = 0$ follows. When we set $t = \frac{x}{u}$ ($\in \mathrm{U}(\mathrm{Q}(A))$), then we have $J = tI$. It remains to show that $t \in A$. Let $a, b \in I$ be a weak A-sequence. Then there exist some $\alpha, \beta \in J$ such that $u\alpha = xa$ and $u\beta = xb$. Hence $u\alpha\beta = xa\beta = x\alpha b$. Since x is a non-zerodivisor, we have that $a\beta = \alpha b$. As a, b is an A-sequence, we can write $\alpha = a\gamma$ ($\gamma \in A$). Hence $u\alpha = a(u\gamma) = xa$, and hence $x = u\gamma$. This shows $t = \gamma \in A$. If $t' \in A$ and $g(\boldsymbol{e}_i) = t'd_i$ for $1 \leq i \leq n+1$, then we have $t'u = x$, and hence $t' = t$. □

Under the notation of Proposition 5.119, if $\mathrm{ht}_A J \geq 2$, then $t \in \mathrm{U}(A)$. Combining all, we have the following.

Theorem 5.120 (Hilbert–Burch) *Let A be a Noetherian ring, and J be an ideal of A. Assume that the A-module A/J has a free resolution of the form*

$$0 \to A^n \xrightarrow{\partial_2} A^{n+1} \xrightarrow{\partial_1} A \to A/J \to 0$$

for some $n \geq 1$. If $\mathrm{ht}_A J \geq 2$, then $J = \mathrm{I}_n(\partial_2)$, and we can adopt $\partial_1 = (d_1, d_2, \cdots, d_{n+1})$ as

$$\partial_1 : A^{n+1} \to A,$$

where $d_k = (-1)^k \Delta_k$, and Δ_k is the determinant of the matrix obtained by removing the kth row of the representation matrix of ∂_2.

Example 5.121 Let $A = k[X, Y, Z, W]$ be the polynomial ring over a field k, and let $f = \begin{bmatrix} X & Y \\ Y & Z \\ Z & W \end{bmatrix}$, and $I = \mathrm{I}_2(f)$. Then $I = (Z^2 - YW, XW - YZ, Y^2 - XZ)$, and the A-module A/I has the free resolution

$$0 \to A^2 \xrightarrow{\partial_2 = f} A^3 \xrightarrow{\partial_1} A \to A/I \to 0,$$

where $\partial_1 = \begin{pmatrix} Z^2 - YW & XW - YZ & Y^2 - XZ \end{pmatrix}$.

Proof As $\partial_1\partial_2 = 0$ is checked easily, it suffices to show that $\mathrm{depth}(I, A) = \mathrm{ht}_A I \geq 2$, by Theorem 5.115. Since $Z^2 - YW$ is a prime element of A and $XW - YZ \notin (Z^2 - YW)$, this is trivial. □

Exercise 5.4 Let k be a Noetherian ring, $r, s \geq 1$, and let

$$A = k[X_{ij} \mid 1 \leq i \leq r,\ 1 \leq j \leq s]$$

be a polynomial ring. Prove that

$$\mathrm{ht}_A \mathrm{I}_t(M) = (r - t + 1)(s - t + 1)$$

for all $1 \leq t \leq \min\{r, s\}$, where $M = [X_{ij}]$, which is an $r \times s$ matrix whose coefficients are in A. In particular, when $r = s + 1$, the sequence

$$0 \to A^s \xrightarrow{M} A^{s+1} \xrightarrow{[d_1\ d_2\ \cdots\ d_r]} A \to A/\mathrm{I}_s(M) \to 0$$

is exact, where $d_k = (-1)^k \Delta_k$, and Δ_k is the $s \times s$ square matrix obtained by removing the kth row from $M = [X_{ij}]$.

Chapter 6
Gorenstein Rings

Now we define Gorenstein rings.

The name "Gorenstein ring" derives from Daniel Gorenstein's theorem on plane curves; connections with injective dimension found by H. Bass[1] made this concept much more important. Also the local duality found by A. Grothendieck is very important.

In the theory of Cohen–Macaulay rings and modules, the concept of the canonical module K_A is very important and Gorenstein rings are characterized by the property $K_A \cong A$.

Also the finiteness of injective dimension is essential.

6.1 Bass Numbers

Let A be a Noetherian ring. For an A-module M, we denote by

$$0 \to M \to \mathrm{E}_A^0(M) \to \mathrm{E}_A^1(M) \to \cdots \to \mathrm{E}_A^i(M) \to \cdots$$

the minimal injective resolution of M. We put $\mathrm{E}_A^i(M) = 0$ for $i < 0$.

For $\mathfrak{p} \in \operatorname{Spec} A$ and $i \in \mathbb{Z}$, we define

$$\mu^i(\mathfrak{p}, M) = \dim_{A_\mathfrak{p}/\mathfrak{p}A_\mathfrak{p}} \operatorname{Ext}_{A_\mathfrak{p}}^i(A_\mathfrak{p}/\mathfrak{p}A_\mathfrak{p}, M_\mathfrak{p})$$

and call $\mu^i(\mathfrak{p}, M)$ the ith *Bass number* of M with respect to $\mathfrak{p}$. If M is finitely generated, $\mu^i(\mathfrak{p}, M)$ is a non-negative integer. If $(A, \mathfrak{m})$ is local and if M is a Cohen–

[1] Hyman Bass, 1932–.

S. Goto, K. Watanabe, *Commutative Algebra*, Graduate Texts in Mathematics 309,
https://doi.org/10.1007/978-981-95-4227-7_6

Macaulay A-module of dimension d, then the Bass number $\mu_A^d(\mathfrak{m}, M)$ is equal to the *Cohen–Macaulay type* $r_A(M)$ of M (see Definition 6.8 below).

Theorem 6.1 *For* $\mathfrak{p} \in \operatorname{Spec} A$, *the Bass number* $\mu^i(\mathfrak{p}, M)$ $(i \in \mathbb{Z})$ *is equal to the number of* $\mathrm{E}_A(A/\mathfrak{p})$ *appearing as a direct summand of the injective A-module* $\mathrm{E}_A^i(M)$.

Proof Localizing $\mathrm{E}_A^\bullet(M)$ by $\mathfrak{p}$ and considering as $A_\mathfrak{p}$-modules, we may assume that $(A, \mathfrak{m})$ is local and $\mathfrak{p} = \mathfrak{m}$.

Decompose the minimal resolution of M as $0 \to M \to \mathrm{E}_A^0(M) \to \mathrm{E}_A^1(M) \to \cdots \to \mathrm{E}_A^{i-1}(M) \to X \to 0$ and $0 \to X \to \mathrm{E}_A^i(M) \to \mathrm{E}_A^{i+1}(M) \to \cdots$. Since X is essential in $\mathrm{E}_A^i(M)$, we have that $\operatorname{Hom}_A(A/\mathfrak{m}, X) \to \operatorname{Hom}_A(A/\mathfrak{m}, \mathrm{E}_A^i(M))$ is an isomorphism. So $\operatorname{Hom}_A(A/\mathfrak{m}, \mathrm{E}_A^i(M)) \to \operatorname{Hom}_A(A/\mathfrak{m}, \mathrm{E}_A^{i+1}(M))$ is zero for $i \geq 0$. Hence we get an isomorphism

$$\operatorname{Ext}_A^i(A/\mathfrak{m}, M) \cong \operatorname{Hom}_A(A/\mathfrak{m}, \mathrm{E}_A^i(M))$$

as an $A/\mathfrak{m}$-module and the equality $\mu_A^i(\mathfrak{m}, M) = \dim_{A/\mathfrak{m}} \operatorname{Hom}_A(A/\mathfrak{m}, \mathrm{E}_A^i(M))$ follows. □

Lemma 6.2 (Bass) *Let* $\mathfrak{p} \subseteq \mathfrak{q}$ *be prime ideals in a Noetherian ring A and assume that* $\dim A_\mathfrak{q}/\mathfrak{p}A_\mathfrak{q} = 1$. *If M is a finitely generated A-module and if* $\mu^i(\mathfrak{p}, M) > 0$ *for some* $i \in \mathbb{Z}$, *then we have* $\mu^{i+1}(\mathfrak{q}, M) > 0$.

Proof We can localize at $\mathfrak{q}$ and may assume that $(A, \mathfrak{q})$ is local. Take $x \in \mathfrak{q} \setminus \mathfrak{p}$. Then we have an exact sequence

$$0 \to A/\mathfrak{p} \xrightarrow{x} A/\mathfrak{p} \to A/(\mathfrak{p} + xA) \to 0.$$

Here note that the length $\ell_A(A/(\mathfrak{p}+xA))$ is finite. From the long exact sequence of $\operatorname{Ext}_A^*(*, M)$ we get

$$\ldots \to \operatorname{Ext}_A^i(A/\mathfrak{p}, M) \xrightarrow{x} \operatorname{Ext}_A^i(A/\mathfrak{p}, M) \to \operatorname{Ext}_A^{i+1}(A/(\mathfrak{p} + xA), M) \to \ldots.$$

Here, if we assume that $\operatorname{Ext}_A^{i+1}(A/(\mathfrak{p} + xA), M) = 0$, since $\operatorname{Ext}_A^i(A/\mathfrak{p}, M) \xrightarrow{x} \operatorname{Ext}_A^i(A/\mathfrak{p}, M)$ is surjective, we have $\operatorname{Ext}_A^i(A/\mathfrak{p}, M) = 0$ by NAK (Proposition 1.86), which contradicts our assertion $\mu^i(\mathfrak{p}, M) > 0$. Hence we must have $\operatorname{Ext}_A^{i+1}(A/(\mathfrak{p} + xA), M) \neq 0$. Since $A/(\mathfrak{p} + xA)$ has finite length, we easily get $\operatorname{Ext}_A^{i+1}(A/\mathfrak{m}, M) \neq 0$ by induction on $\ell_A(A/(\mathfrak{p} + xA))$. □

Corollary 6.3 *If M is a finitely generated A-module, then* $\dim_A M \leq \operatorname{id}_A M$. *If* $(A, \mathfrak{m})$ *is a Noetherian local ring and* $r = \operatorname{id}_A M < \infty$, *then* $\mu_A^r(\mathfrak{m}, M) > 0$.

Let $(A, \mathfrak{m})$ be a Noetherian local ring. If $\operatorname{id}_A M < \infty$, the value of $\operatorname{id}_A M$ has unique possibility.

Proposition 6.4 *Let* $(A, \mathfrak{m})$ *be a Noetherian local ring and* M *be a nonzero finitely generated* A*-module with* $\mathrm{id}_A M < \infty$. *Then we have* $\mathrm{id}_A M = \operatorname{depth} A$.

Proof Assume that $\operatorname{depth} A = r$ and $\mathrm{id}_A M = q$. Let $(x_1, \ldots, x_r)$ be a maximal regular A-regular sequence in $\mathfrak{m}$ and put $A/(x_1, \ldots, x_r) = \overline{A}$. Then by Proposition 5.81, we have

$$\operatorname{Ext}_A^r(\overline{A}, M) \cong M/(x_1, \ldots, x_r)M \neq 0 \text{ and } \operatorname{Ext}_A^i(\overline{A}, M) = 0\ (i > r).$$

Hence we have $q \geq r$. On the other hand, if we assume $q > r$, then since $\mathfrak{m} \in \operatorname{Ass}_A(\overline{A})$, we can take an exact sequence $0 \to A/\mathfrak{m} \to \overline{A} \to Y \to 0$. Then we have

$$0 = \operatorname{Ext}_A^q(\overline{A}, M) \to \operatorname{Ext}_A^q(A/\mathfrak{m}, M) \to \operatorname{Ext}_A^{q+1}(Y, M) = 0$$

and we get $\operatorname{Ext}_A^q(A/\mathfrak{m}, M) = 0$, which contradicts the fact that $\mathrm{id}_A M = q$. Hence we have $\mathrm{id}_A(M) = \operatorname{depth} A$. □

From Proposition 6.4 and Corollary 6.3, we get the following.

Corollary 6.5 *If* $(A, \mathfrak{m})$ *is a Noetherian local ring and* $\mathrm{id}_A A < \infty$, *then* A *is a Cohen–Macaulay ring and* $\dim A = \mathrm{id}_A A$ *holds.*

It was shown by Paul Roberts[2] that if there exists a finitely generated A-module M with $\mathrm{id}_A M < \infty$, then A is a Cohen–Macaulay local ring, see [Rob] and references therein.

The following proposition holds for a finitely generated A-module M with $\mathrm{id}_A M = \infty$.

Proposition 6.6 *Let* $(A, \mathfrak{m})$ *be a Noetherian local ring and* M *be a finitely generated* A*-module. If* $\mathrm{id}_A M = \infty$, *then for all* $i \geq \dim M$, *we have* $\mu^i(\mathfrak{m}, M) > 0$.

Proof We show the result by induction on $d = \dim A$. If $d = 0$, then since any injective A-module is a direct sum of $\mathrm{E}_A(A/\mathfrak{m})$, our conclusion is trivial.

Assume $d > 0$ and if $\mathrm{id}_{A_\mathfrak{q}}(M_\mathfrak{q}) = \infty$ for a prime ideal $\mathfrak{q}$ with $\operatorname{ht} \mathfrak{q} = d - 1$, then by our induction hypothesis, $\mu^i(\mathfrak{q}, M) = \mu^i(\mathfrak{q}A_\mathfrak{q}, M_\mathfrak{q}) > 0$ $(\forall i \geq \dim M_\mathfrak{q} = \dim M - 1)$. Hence our assertion follows from Lemma 6.2.

If $\mathrm{id}_{A_\mathfrak{q}}(M_\mathfrak{q}) < \infty$ for every prime ideal $\mathfrak{q}$ with $\operatorname{ht} \mathfrak{q} = d - 1$, then since $\mathrm{id}_{A_\mathfrak{q}}(M_\mathfrak{q}) = \operatorname{depth} A_\mathfrak{q} \leq d-1$ by Proposition 6.4, the only indecomposable injective A-module appearing after degree d is $\mathrm{E}_A(A/\mathfrak{m})$. Hence we have $\mu^i(\mathfrak{m}, M) > 0$ for every $i \geq d$. □

The next lemma is very important in the theory of Gorenstein rings.

Lemma 6.7 *Let* $(A, \mathfrak{m})$ *be a Noetherian local ring and* M *be a finitely generated* A*-module. If* $x \in \mathfrak{m}$ *is both* M*-regular and* A*-regular, then we have* $\mu^{i+1}(\mathfrak{m}, M) = \mu^i(\mathfrak{m}/xA, M/xM)$ *for every* $i \in \mathbb{Z}$.

---

[2] Paul Roberts, 1945–.

Proof Take a minimal injective resolution

$$\mathbb{I} : 0 \to M \xrightarrow{\varepsilon} I^0 \xrightarrow{\partial^0} I^1 \xrightarrow{\partial^1} \cdots \xrightarrow{\partial^{i-1}} I^i \xrightarrow{\partial^i} I^{i+1} \xrightarrow{\partial^{i+1}} \cdots \qquad (*)$$

of M over A, and consider the sequence

$$0 \to M/xM \xrightarrow{\varepsilon} \operatorname{Hom}_A(A/xA, I^1) \xrightarrow{\partial^1} \operatorname{Hom}_A(A/xA, I^2) \to$$
$$\cdots \xrightarrow{\partial^{i-1}} \operatorname{Hom}_A(A/xA, I^i) \xrightarrow{\partial^i} \operatorname{Hom}_A(A/xA, I^{i+1}) \xrightarrow{\partial^{i+1}} \cdots . \qquad (\clubsuit)$$

Since x is A-regular, calculating $\operatorname{Ext}^i_A(A/xA, M)$ using the projective resolution $0 \to A \xrightarrow{x} A \to A/xA \to 0$ of A/xA, we have

$$\mathrm{H}^i(\operatorname{Hom}_A(A/xA, \mathbb{I})) \cong \operatorname{Ext}^i_A(A/xA, M) = \begin{cases} 0 & (i \geq 2) \\ M/xM & (i = 1). \end{cases}$$

Since x is also M-regular, $x \notin P$ for any $P \in \operatorname{Ass} I^0 = \operatorname{Ass} M$. So $x : I^0 \to I^0$ is an isomorphism, and hence $\operatorname{Hom}_A(A/xA, I^0) = 0$. So we have that the one shift $\operatorname{Hom}_A(A/xA, \mathbb{I})[1]$, where $\operatorname{Hom}_A(A/xA, \mathbb{I})[1]^i = \operatorname{Hom}_A(A/xA, I^{i+1})$, is a resolution of M/xM, and hence ($\clubsuit$) is exact.

If I is an injective A-module and $\mathcal{S}$ is an exact sequence of A/xA-modules, then $\operatorname{Hom}_{A/xA}(\mathcal{S}, \operatorname{Hom}_A(A/xA, I)) \cong \operatorname{Hom}_A(\mathcal{S}, I)$ is exact, and hence $\operatorname{Hom}_A(A/xA, I)$ is an injective A/xA-module. So ($\clubsuit$) is an injective resolution of M/xM. So

$$\mu^i(\mathfrak{m}/xA, M/xM) = \dim_{A/\mathfrak{m}} \mathrm{H}^i(\operatorname{Hom}_{A/xA}(A/\mathfrak{m}, \operatorname{Hom}_A(A/xA, \mathbb{I})[1]))$$
$$= \dim_{A/\mathfrak{m}} \mathrm{H}^{i+1}(\operatorname{Hom}_A(A/\mathfrak{m}, \mathbb{I})) = \mu^{i+1}(\mathfrak{m}, M),$$

as desired. □

6.2 Gorenstein Local Rings

We introduce the notion of "Cohen–Macaulay type" for a Cohen–Macaulay ring.

Definition 6.8 Let $(A, \mathfrak{m})$ be a Noetherian local ring, and M a Cohen–Macaulay A-module of dimension d. We call $\mu^d(\mathfrak{m}, A)$ the *Cohen–Macaulay type* of M and denote it by

$$\mathrm{r}(M).$$

By Lemma 6.7, if $x \in \mathfrak{m}$ is both A-regular and M-regular, then we have $\mathrm{r}(M) = \mathrm{r}(M/xM)$.

If $d = 0$, then $\mathrm{r}(M)$ agrees with the type of the Artinian A-module M defined in Definition 5.52.

Let $(A, \mathfrak{m})$ be a d-dimensional Noetherian local ring. We say that A is a *Gorenstein local ring* (or, simply A is *Gorenstein*) if $\mathrm{id}_A A < \infty$. If A is Gorenstein, then A is Cohen–Macaulay (Corollary 6.5) and $\mathrm{id}_A A = d$ holds. Also, for any $\mathfrak{p} \in \mathrm{Spec}\, A$, $A_\mathfrak{p}$ is also Gorenstein.

Theorem 6.9 *The following conditions on a Noetherian local ring $(A, \mathfrak{m})$ are equivalent: For the definition of irreducible ideals, see Definition 2.30.*

(1) *A is Gorenstein. Namely,* $\mathrm{id}_A A < \infty$.
(2) $\mu^i(\mathfrak{m}, A) = 0$ *for every $i > d$.*
(3) $\mu^i(\mathfrak{m}, A) = 0$ *for some $i > d$.*
(4) $\mu^i(\mathfrak{m}, A) = 1$ *if $i = d$, and* 0 *otherwise.*
(5) *$\widehat{A}$ is a Gorenstein ring.*
(6) *For every $i \in \mathbb{Z}$ and $\mathfrak{p} \in \mathrm{Spec}\, A$, $\mu^i(\mathfrak{p}, A) = 1$ if $i = \mathrm{ht}_A \mathfrak{p}$ and* 0 *otherwise.*
(7) *A is Cohen–Macaulay and every system of parameters $(x_1, \ldots, x_d)$ of A generates an irreducible ideal.*
(8) *A is Cohen–Macaulay and for a system of parameters $(x_1, \ldots, x_d)$ of A, $(x_1, \ldots, x_d)$ is an irreducible ideal.*
(9) *A is Cohen–Macaulay and* $\mathrm{r}(A) = 1$.

The Cohen–Macaulay type $\mathrm{r}(A)$ of a Cohen–Macaulay local ring A is the measure of the "distance" from Gorenstein local rings. Before proving this theorem, let us look at the case $d = 0$.

Proposition 6.10 *The following conditions on an Artinian local ring $(A, \mathfrak{m})$ are equivalent:*

(1) *A is Gorenstein. Namely, A is an injective A-module.*
(2) *The ideal* (0) *is irreducible in A.*
(3) $\mathrm{Hom}_A(A/\mathfrak{m}, A) \cong A/\mathfrak{m}$.
(4) *For every ideal I of A, the equality $\ell_A(A/I) = \ell_A([0 :_A I])$ holds.*
(5) *For any ideal I of A, the equality $I = [0 :_A [0 :_A I]]$ holds.*
(6) *For every finitely generated A-module M, the equality $\ell_A(M) = \ell_A(\mathrm{Hom}_A(M, A))$ holds.*

Proof We put $E = \mathrm{E}_A(A/\mathfrak{m})$. From Theorem 5.53, (5), $\ell_A(A) = \ell_A(E)$. Hence (1) is equivalent to saying that $A \cong E$. By Lemma 5.43, (3), (2) implies $\mathrm{E}_A(A) = E$ and conditions (1), (2) are equivalent. Since $\mathrm{Hom}_A(A/\mathfrak{m}, A) \cong [0 :_A \mathfrak{m}]$, (3) is equivalent to (2).

Let I be an ideal of A. Since $\mathrm{Hom}_A(A/I, A) \cong [0 :_A I]$ by Proposition 1.103, (2), (4) follow from (1) via Matlis duality.

Next, we show (4) $\Rightarrow$(5) $\Rightarrow$(3). It is obvious that $I \subseteq [0 :_A [0 :_A I]]$ and by (4), we have (5).

If we assume (3) does not hold, then there exists an ideal J satisfying $0 \subsetneq J \subsetneq [0 :_A \mathfrak{m}]$. But then we have $[0 : J] = \mathfrak{m}$ and $J \subsetneq [0 :_A \mathfrak{m}] = [0 :_A [0 :_A J]]$, which contradicts (5). Hence (1) to (5) are all equivalent.

If $A \cong E$, then (6) follows from Matlis duality. Conversely, if we put $M = A/\mathfrak{m}$ in (6), we get (2). Hence (6) is also equivalent to (1)–(5). □

Proof of Theorem 6.9 (3) ⇒(1) follows from Proposition 6.6. (1) ⇒(2) follows from Corollary 6.5. (2) ⇒(3) is trivial. So (1), (2), and (3) are equivalent.

(1) ⇔(5). This is immediate, since $\mu^i(\mathfrak{m}\widehat{A}, \widehat{A}) = \mu^i(\mathfrak{m}, A)$ for all i.

(6) ⇒(4) is clear, since (4) is the special case of (6) where $\mathfrak{p} = \mathfrak{m}$. (4) ⇒(2) is trivial.

So from any of (1)–(9), the Cohen–Macaulay property of A follows.

For a non-zerodivisor $x \in \mathfrak{m}$, we have $\mu^i(\mathfrak{m}/xA, A/xA) = \mu^{i+1}(\mathfrak{m}, A)$ for all $i \in \mathbb{Z}$ by Lemma 6.7. So A is Gorenstein if and only if A/xA is so.

(8) ⇒(4). The zero ideal of $\overline{A} = A/(x_1, \ldots, x_d)$ is irreducible. So by Proposition 6.10, we have $\mu^i_{\overline{A}}(\mathfrak{m}\overline{A}, \overline{A}) = \delta_{i0}$ for $i \in \mathbb{Z}$, where δ_{i0} is Kronecker's delta. Since A is Cohen–Macaulay, $x_1, \ldots, x_d$ is a regular sequence, and hence $\mu^i_A(\mathfrak{m}, A) = \mu^{i-d}_{\overline{A}}(\mathfrak{m}\overline{A}, \overline{A}) = \delta_{id}$ for $i \in \mathbb{Z}$, and this proves (4).

(1) ⇒(7). Since $\overline{A} = A/(x_1, \ldots, x_d)$ is Gorenstein, the zero ideal (0) of $\overline{A}$ is irreducible by Proposition 6.10. So $(x_1, \ldots, x_d)$ is an irreducible ideal. (7) ⇒(8) is obvious.

(1) ⇒(6). By Corollary 5.28, if $\mathrm{id}_{A_\mathfrak{p}} A_\mathfrak{p} \leq \mathrm{id}_A A < \infty$, and hence $A_\mathfrak{p}$ is Gorenstein. By (1) ⇒(4), which has already been proved, we have the assertion.

(4) ⇒(9) is obvious.

(9) ⇒(1). Letting $\overline{A} = A/(x_1, \ldots, x_n)$, we have $\mathrm{r}(\overline{A}) = \mathrm{r}(A) = 1$. By Proposition 6.10, we have that $\overline{A}$ is Gorenstein, and hence A is also Gorenstein. □

From the proof above, we derive the following statements.

Corollary 6.11 *Let $(A, \mathfrak{m})$ be a Noetherian local ring,*

(1) *If $x \in \mathfrak{m}$ is an A-regular element, A is Gorenstein if and only if A/xA is Gorenstein.*
(2) *If A is Gorenstein, then $A_\mathfrak{p}$ is Gorenstein for any prime ideal $\mathfrak{p}$ of A.*

If A is not a local ring, the Gorenstein property is defined by localization.

Definition 6.12 Let A be a Noetherian ring. Then we say that A is a *Gorenstein ring* if for every prime ideal $\mathfrak{p}$ of A, $A_\mathfrak{p}$ is Gorenstein as defined above. By Theorem 6.9, it is equivalent to say that for every maximal ideal $\mathfrak{m}$ of A, $A_\mathfrak{m}$ is Gorenstein.

Definition 6.13 Let $(A, \mathfrak{m})$ be a Noetherian local ring. If $\widehat{A} \cong S/(x_1, \ldots, x_r)S$ for some *regular local ring* S and an S-regular sequence $x_1, \ldots, x_r$ of S, we say that A is a *local complete intersection*.

Regular local rings are examples of Gorenstein rings since $\mathrm{id}_A A < \infty$.

Example 6.14 Regular local rings are local complete intersections, and local complete intersections are Gorenstein.

Proposition 6.15 *If $(A, \mathfrak{m})$ is a Cohen–Macaulay local ring of dimension d, then for every system of parameters $J = (x_1, \ldots, x_d)$ of A, $\mathrm{r}(A)$ equals the length of $[J : \mathfrak{m}]/J$. Also, if $d = 0$, $\mathrm{r}(A)$ equals the number of minimal generators of $E = \mathrm{E}_A(A/\mathfrak{m})$.*

Proof Since $\mathrm{r}(A) = \mathrm{r}(A/J)$, we may assume $d = 0$. We put $E = \mathrm{E}_A(A/\mathfrak{m})$. Then by Theorem 5.46, $\mathrm{r}(A) = \mu^0(\mathfrak{m}, A) = \ell_A([0 :_A \mathfrak{m}])$. Also, by Matlis Duality, $\mathrm{Hom}_A([0 :_A \mathfrak{m}], E) \cong E/\mathfrak{m}E$. □

We discuss the Cohen–Macaulay type of tensor products.

Theorem 6.16 (Tensor Product of Gorenstein Rings) *Let $f : (A, \mathfrak{m}) \to (B, \mathfrak{n})$ be a flat local ring homomorphism between Cohen–Macaulay rings, and M be a Cohen–Macaulay A-module. Then we have*

$$\mathrm{r}(B \otimes_A M) = \mathrm{r}(M) \cdot \mathrm{r}(B/\mathfrak{m}B).$$

In particular, B is Gorenstein if and only if both A an $B/\mathfrak{m}B$ are Gorenstein.

Proof Since Cohen–Macaulay type is invariant by factoring by regular elements, we may assume $\dim M = 0$ and $\dim B/\mathfrak{m}B = 0$, see Corollary 5.66. Then $B \otimes_A M$ is also zero-dimensional. Put $\mathrm{r}(M) = r$ and $\mathrm{r}(B/\mathfrak{m}B) = s$. Then

$$\begin{aligned}\mathrm{Hom}_B(B/\mathfrak{n}, B \otimes_A M) &\cong \mathrm{Hom}_B(B/\mathfrak{n}, \mathrm{Hom}_B(B/\mathfrak{m}B, B \otimes_A M))\\ &\cong \mathrm{Hom}_B(B/\mathfrak{n}, \mathrm{Hom}_A(A/\mathfrak{m}A, M) \otimes_A B)\\ &\cong \mathrm{Hom}_B(B/\mathfrak{n}, (A/\mathfrak{m})^r \otimes_A B) \cong \mathrm{Hom}_B(B/\mathfrak{n}, B/\mathfrak{m}B)^r \cong (B/\mathfrak{n})^{rs}\end{aligned}$$

and we have the first assertion. The second assertion follows from the first one, letting $M = A$. We have $\mathrm{r}(B) = 1$ if and only if $\mathrm{r}(A) = \mathrm{r}(B/\mathfrak{m}B) = 1$. □

6.3 Canonical Module and the Local Duality Theorem

Let $(A, \mathfrak{m})$ be a d-dimensional Noetherian local ring, and set $E = E_A(A/\mathfrak{m})$.

Lemma 6.17 *If A is Gorenstein, then $\mathrm{H}^d_{\mathfrak{m}}(A) \cong E$. In particular, we have $\mathrm{H}^d_{\mathfrak{m}}(A)^\vee \cong \widehat{A}$, where $\widehat{A}$ is the completion of A.*

Proof Let $I^\bullet$ be the minimal injective resolution of A. Since $\mu^i_A(\mathfrak{m}, A) = \delta_{id}$ by Theorem 6.9, $\Gamma_{\mathfrak{m}}(I^\bullet)$ is concentrated in degree d, and $\Gamma_{\mathfrak{m}}(I^d) = E$. The result follows from Corollary 5.101. The last assertion follows immediately from the Matlis duality (Theorem 5.53, (1)) $E^\vee \cong \widehat{A}$. □

6.18 Let $(A, \mathfrak{m})$ be a d-dimensional Cohen–Macaulay local ring, and K be a Cohen–Macaulay A-module of dimension d. We say that $K = K_A$ is a *canonical module* of A if $\mathrm{H}^d_{\mathfrak{m}}(K)^\vee \cong \widehat{A}$, where $\widehat{A}$ is the completion. By Lemma 6.17, the

A-module A is a canonical module if A is Gorenstein (see also 6.26). Note that the canonical module $K = K_A$ is unique up to isomorphisms by the next lemma.

Lemma 6.19 *Let $(A, \mathfrak{m})$ be a Noetherian local ring and M, N be finitely generated A-modules. If $\widehat{M} \cong \widehat{N}$ as $\widehat{A}$-modules, then $M \cong N$ as A-modules.*

Proof Let $\widehat{\psi} \in \mathrm{Hom}_{\widehat{A}}(\widehat{M}, \widehat{N})$ be an isomorphism. Since we have

$$\widehat{A} \otimes_A \mathrm{Hom}_A(M, N) \cong \mathrm{Hom}_{\widehat{A}}(\widehat{M}, \widehat{N})$$

by Proposition 1.120, we can write $\widehat{\psi} = \psi + \sum_{i=1}^{n} \widehat{\alpha}_i \psi_i$ for some $\psi, \psi_i \in \mathrm{Hom}_A(M, N)$ and $\widehat{\alpha}_i \in \widehat{\mathfrak{m}}$. So $\widehat{N} = \widehat{\psi}(\widehat{M}) = \psi(\widehat{M}) + \widehat{\mathfrak{m}}\widehat{N}$. By NAK, we have that $\psi : \widehat{M} \to \widehat{N}$ is surjective. Applying Lemma 3.4 to $\widehat{\psi}^{-1}\psi : \widehat{M} \to \widehat{M}$, we have that $\psi : \widehat{M} \to \widehat{N}$ is an isomorphism. Since $\widehat{A}$ is faithfully flat over A, we have that $\psi : M \to N$ is also an isomorphism, as desired. □

The local duality theorem (Theorem 6.22 and Corollary 6.23) is a very important property of the canonical module K_A. To prove this, we need some basics on δ-functors and the universality of derived functors.

6.20 Let $\mathcal{A}$ be an abelian category with enough projectives, and $\mathcal{B}$ be an abelian category. A contravariant cohomological *δ-functor* from $\mathcal{A}$ to $\mathcal{B}$ is a pair $T = ((T^n), (\delta^n))$ such that $(T^n)_{n\in\mathbb{Z}}$ is a collection of contravariant additive functors $T^n : \mathcal{A} \to \mathcal{B}$ for $n \in \mathbb{Z}$ such that $T^n = 0$ for $n < 0$, and $\delta^n = \delta^n(\mathcal{S}) : T^n(L) \to T^{n+1}(N)$ is a morphism defined for each short exact sequence

$$\mathcal{S} : 0 \to L \xrightarrow{f} M \xrightarrow{g} N \to 0,$$

such that the following two conditions are satisfied:

(1) (Long exact sequence) The sequence

$$\cdots \to T^{n-1}(L) \xrightarrow{\delta^{n-1}} T^n(N) \xrightarrow{T(g)} T^n(M) \xrightarrow{T(f)} T^n(L) \xrightarrow{\delta^n} T^{n+1}(N) \to \cdots$$

is exact.

(2) Given a chain map from $\mathcal{S}$ to $\mathcal{S}' : 0 \to L' \xrightarrow{f'} M' \xrightarrow{g'} N' \to 0$, the diagram

$$\begin{array}{ccc} T^n(L') & \xrightarrow{\delta^n} & T^{n+1}(N') \\ \downarrow & & \downarrow \\ T^n(L) & \xrightarrow{\delta^n} & T^{n+1}(N) \end{array}$$

is commutative.

A *morphism* between two (contravariant cohomological) δ-functors $F : S \to T$ is a collection of natural transformations $F^n : S^n \to T^n$ which induces chain maps between long exact sequences. We say that S is a *universal δ-functor* if for every δ-functor T and any natural transformation $F^0 : S^0 \to T^0$, there exists a unique morphism $F : S \to T$ that extends F^0. If $\mathcal{A}$ has enough projectives and $G : \mathcal{A} \to \mathcal{B}$ is a contravariant left exact functor, then the right derived functor RG is universal [Wei, Theorem 2.4.6]. Using this, it is easy to show the following lemma, whose proof we leave for interested readers.

Lemma 6.21 (The Uniqueness of the Right Derived Functor) *Let $\mathcal{A}$ have enough projectives and $G : \mathcal{A} \to \mathcal{B}$ be a contravariant left exact functor. Let T be a contravariant cohomological δ-functor, and $f : G \to T^0$ be a natural isomorphism, and $F : RG \to T$ be the morphism such that $F^0 = f$, which comes from the universality of RG. Let $r \geq 1$. If $T^i(P) = 0$ for every $1 \leq i \leq r$ and every projective object P of $\mathcal{A}$, then $F^i : R^iG \to T^i$ is a natural isomorphism for $i = 0, 1, \dots, r$. In particular, if $T^i(P) = 0$ for every $i > 0$ and every projective object P of $\mathcal{A}$, then $F : RG \to T$ is an isomorphism.*

Theorem 6.22 *Let $(A, \mathfrak{m})$ be a Noetherian local ring of dimension d with the canonical module K_A. We fix an isomorphism $h : K_A \otimes_A \widehat{A} \to \mathrm{H}^d_{\mathfrak{m}}(A)^\vee$. Let $r \geq 0$, and assume that $\mathrm{H}^{d-i}_{\mathfrak{m}}(A) = 0$ for $1 \leq i \leq r$. Let M be a finitely generated A-module. Then there exists a unique morphism of covariant homological δ-functors on M*

$$\Psi_i : \mathrm{H}^{d-i}_{\mathfrak{m}}(M) \to \mathrm{Ext}^i_A(M, K_A)^\vee \tag{6.1}$$

such that $\Psi_0(A)$ is given by

$$\Psi_0(A) : \mathrm{H}^d_{\mathfrak{m}}(A) \xrightarrow{\phi} \mathrm{H}^d_{\mathfrak{m}}(A)^{\dagger\dagger} \xrightarrow{h^\dagger} (K_A \otimes_A \widehat{A})^\dagger = K_A^\vee \cong \mathrm{Ext}^0_A(A, K_A)^\vee,$$

where $(-)^\dagger = \mathrm{Hom}_{\widehat{A}}(-, E_A(A/\mathfrak{m}))$, and ϕ is the canonical map to the double Matlis dual. Moreover, Ψ_i is an isomorphism for $0 \leq i \leq r$.

We also have a unique morphism of contravariant cohomological δ-functors

$$\Phi^i : \mathrm{Ext}^i_A(M, K_A) \otimes_A \widehat{A} \to \mathrm{H}^{d-i}_{\mathfrak{m}}(M)^\vee \tag{6.2}$$

such that $\Phi^0 : K_A \otimes_A \widehat{A} = \mathrm{Ext}^0_A(A, K_A) \otimes_A \widehat{A} \to \mathrm{H}^d_{\mathfrak{m}}(M)^\vee$ is h. Φ^i is an isomorphism for $0 \leq i \leq r$.

Proof If a δ-functor Φ^i uniquely exists and is an isomorphism for $0 \leq i \leq r$, then Ψ_i is defined to be the composition

$$\mathrm{H}^{d-i}_{\mathfrak{m}}(M) \xrightarrow{\phi} \mathrm{H}^{d-i}_{\mathfrak{m}}(M)^{\dagger\dagger} \xrightarrow{(\Phi^i)^\dagger} (\mathrm{Ext}^i_A(M, K_A) \otimes_A \widehat{A})^\dagger \cong \mathrm{Ext}^i_A(M, K_A)^\vee,$$

and it is also a δ-functor such that $\Psi_0(A)$ is the required map. Moreover, since Φ^i is an isomorphism, $\mathrm{H}_{\mathfrak{m}}^{d-i}(M)$ is Artinian for $0 \le i \le r$, by Lemma 5.55. Then the canonical map ϕ is an isomorphism by the Matlis duality, and hence Ψ_i is an isomorphism for $0 \le i \le r$. Since we get Φ^i from Ψ_i applying the faithful functor $(-)^\vee$, the uniqueness of Ψ_i (for the required choice of $\Psi_0(A)$) follows from the uniqueness of Φ^i. So we only need to consider the problems on Φ^i in (6.2). For this, we may assume that A is complete.

Since $\mathrm{H}_{\mathfrak{m}}^i(M) = 0$ for $i > d$, $\mathrm{H}_{\mathfrak{m}}^d(M)$ is a right exact functor on M. Taking the Matlis dual, $\mathrm{H}_{\mathfrak{m}}^d(M)^\vee$ is a contravariant left exact functor.

Applying the result of Exercise 6.1 to the functor $T^0 = \mathrm{H}_{\mathfrak{m}}^d(*)^\vee$, we have $T^0(M) = \mathrm{H}_{\mathfrak{m}}^d(M)^\vee \cong \mathrm{Hom}_A(M, \mathrm{H}_{\mathfrak{m}}^d(A)^\vee) = \mathrm{Hom}_A(M, K_A)$. This gives an isomorphism of functors $\Phi^0 : \mathrm{Hom}_A(-, K_A) \to T^0$ on $(\mathrm{mod}\, A)$ to $(\mathrm{Mod}\, A)$ with the required $\Phi^0(A)$. So a morphism Φ^i between δ-functors extending Φ^0 uniquely exists, by the universality of the right derived functor $\mathrm{Ext}_A^*(-, K_A)$. Since a projective object of $(\mathrm{mod}\, A)$ is of the form A^n and we assume that $\mathrm{H}_{\mathfrak{m}}^{d-i}(A) = 0$ for $1 \le i \le r$, we have $T^i(P) = 0$ for every projective object P of $(\mathrm{mod}\, A)$ and $1 \le i \le r$. By Lemma 6.21, Φ^i is an isomorphism for $0 \le i \le r$, and we are done. □

Exercise 6.1 Let A be a Noetherian ring, and let T be an additive functor from $(\mathrm{mod}\, A)$ to $(\mathrm{Mod}\, A)$, where $(\mathrm{mod}\, A)$ and $(\mathrm{Mod}\, A)$ denote the category of finitely generated A-modules and the category of A-modules, respectively. For $a \in A$ and an A-module M, denote by $\mu_M(a)$ the homothety map $(m \mapsto am) \in \mathrm{Hom}_A(M, M)$. We assume that $T(\mu_A(a)) = \mu_{T(A)}(a)$ for $a \in A$. Prove the following isomorphisms of functors:

(1) If T is a covariant right exact functor, then there is a unique natural isomorphism $\eta : T \cong T(A) \otimes_A -$ such that $\eta_A : T(A) \cong T(A) \otimes_A A$ is the map given by $x \mapsto x \otimes 1$.
(2) If T is a contravariant left exact functor, then there is a unique natural isomorphism $\zeta : \mathrm{Hom}_A(-, T(A)) \cong T$ such that $\zeta_A : \mathrm{Hom}_A(A, T(A)) \cong T(A)$ is given by $\psi \mapsto \psi(1)$.

In particular, in (1) (resp. (2)), T is an *A-linear functor* in the sense that for every $M, N \in (\mathrm{mod}\, A)$, the map $T : \mathrm{Hom}_A(M, N) \to \mathrm{Hom}_A(T(M), T(N))$ (resp. $F : \mathrm{Hom}_A(M, N) \to \mathrm{Hom}_A(T(N), T(M))$) is an A-linear map.

Corollary 6.23 (Local Duality Theorem for Cohen–Macaulay Rings) *Let $(A, \mathfrak{m})$ be a Cohen–Macaulay local ring of dimension d with the canonical module K_A. Let M be a finitely generated A-module. Then we have the following isomorphism of covariant homological δ-functors on M:*

$$\Psi_i : \mathrm{H}_{\mathfrak{m}}^{d-i}(M) \to \mathrm{Ext}_A^i(M, K_A)^\vee.$$

We also have the following isomorphism of contravariant cohomological δ-functors:

$$\Phi^i : \mathrm{Ext}^i_A(M, K_A) \otimes_A \widehat{A} \to \mathrm{H}^{d-i}_{\mathfrak{m}}(M)^\vee.$$

Proof This is because $\mathrm{H}^{d-i}_{\mathfrak{m}}(A) = 0$ for all $i \geq 1$ by the Cohen–Macaulay property. □

If A is Gorenstein, then A is Cohen–Macaulay, K_A exists, and $K_A = A$. Utilizing this, we get the following corollary. It is also a powerful application of Cohen's structure theorem.

Corollary 6.24 *Let $(A, \mathfrak{m})$ be a Noetherian local ring, and M be a finitely generated A-module. Then $\mathrm{H}^i_{\mathfrak{m}}(M)$ is an Artinian A-module for $i \geq 0$.*

Proof Since $\mathrm{H}^i_{\mathfrak{m}}(M) \cong \mathrm{H}^i_{\mathfrak{m}\widehat{A}}(\widehat{M})$, and an Artinian A-module and an Artinian $\widehat{A}$-module are the same thing, we may assume that A is complete. By Cohen's structure theorem (Theorem 4.40) and Proposition 5.108, we may further assume that A is a regular local ring. Then by Corollary 6.23 (where $K_A = A$), $H^i_{\mathfrak{m}}(M) \cong \mathrm{Ext}^{d-i}_A(M, A)^\vee$ is an Artinian A-module by the Matlis duality (Theorem 5.53), since $\mathrm{Ext}^{d-i}_A(M, A)$ is a Noetherian A-module, where $d = \dim A$ and $(-)^\vee = \mathrm{Hom}_A(-, E_A(A/\mathfrak{m}))$ is the Matlis dual. □

Lemma 6.25 *Let A be a commutative ring, B a faithfully flat A-algebra, and M an A-module. If $M \otimes_A B$ is a finitely generated B-module, M is a finitely generated A-module.*

Proof Let $n_1, \ldots, n_r$ be the generators of $M \otimes_A B$. We can write $n_j = \sum_i m_{ij} \otimes b_{ij}$ with $m_{ij} \in M$ and $b_{ij} \in B$. Let N be the finitely generated A-submodule of M generated by m_{ij}. Then the inclusion $N \otimes_A B \to M \otimes_A B$ is surjective. Since B is faithfully flat over A, we have that $N = M$. □

Definition 6.26 Let $(A, \mathfrak{m})$ be a d-dimensional Noetherian local ring, and K be an A-module. We say that $K = K_A$ is a *canonical module of A if*

$$K \otimes_A \widehat{A} \cong [\mathrm{H}^d_{\mathfrak{m}}(A)]^\vee,$$

where $\widehat{A}$ is the completion. By Corollary 6.23, this definition agrees with that of 6.18 when A is Cohen-Macaulay.

By Lemma 6.25, K_A is a finitely generated A-module. Note that the canonical module $K = K_A$ is unique up to isomorphisms by the next lemma.

6.27 Let C be a finite A-algebra. The functor $(-)^\vee = \mathrm{Hom}_A(-, E)$ is a contravariant exact functor from the category of C-modules to itself, and it is $\mathrm{Hom}_C(-, E')$, where $E' = \mathrm{Hom}_A(C, E)$. Since E' is C-injective, $\mathrm{Hom}_C(\kappa(P), E') \cong \kappa(P)^\vee$ is 0 if $P \in \mathrm{Spec}\, C$ is not maximal, while it is $\mathrm{Hom}_A(\kappa(P), \kappa(\mathfrak{m})) \cong \kappa(P)$ if P is maximal. So we have that $E' = \bigoplus_{\mathfrak{l} \in \mathrm{Max}\, C} E_C(C/\mathfrak{l})$. In particular, if $(C, \mathfrak{l})$ is local, then the matlis duality $\mathrm{Hom}_C(-, E_C(C/\mathfrak{l}))$ of C is the same as $(-)^\vee$.

Lemma 6.28 *Let C be a finite A-algebra, and M be a finitely generated C-module. Then the natural transformations Ψ_i in* (6.1) *and Φ^i in* (6.2) *of Theorem 6.22 are C-linear, and give morphisms between δ-functors* $(\operatorname{mod} C) \to (\operatorname{Mod} C)$.

Proof In view of (6.27), it suffices to show that $\Phi^i(M) : T_1^i(M) \to T^i(M)$ is C-linear, where $T_1^i = \operatorname{Ext}_A^i(-, K_A) \otimes_A \widehat{A}$ and $T^i = \mathrm{H}_{\mathfrak{m}}^{d-i}(-)^\vee$. For each M and $c \in C$,

$$\Phi^i(M)\mu_{T_1^i(M)}(c) = \Phi^i(M)T_1^i(\mu_M(c)) = T^i(\mu_M(c))\Phi^i(M) = \mu_{T^i(M)}(c)\Phi^i(M)$$

by the naturality of Φ^i, and hence $\Phi^i(M)$ is a C-linear map, where $\mu_M(c) : M \to M$ is the homothety map $m \mapsto cm$. □

The following formula is called the *adjunction theorem* and is very important. It shows that every homomorphic image of a Gorenstein local ring has a canonical module.

Theorem 6.29 (Adjunction Theorem) *Let $(A, \mathfrak{m})$ be a d-dimensional Noetherian local ring with the canonical module K_A. Let C be a finite A-algebra, $\mathfrak{l}$ a maximal ideal of C, $D = C_{\mathfrak{l}}$, and $\mathfrak{n} = \mathfrak{l}D$. Assume that $\mathrm{H}_{\mathfrak{m}}^{d-i}(A) = 0$ for $1 \le i \le t$, where $t = \dim A - \dim D$. Then $\operatorname{Ext}_A^t(C, K_A)_{\mathfrak{l}}$ is the canonical module of D. In particular, if A is Gorenstein, then $\operatorname{Ext}_A^t(C, A)_{\mathfrak{l}}$ is the canonical module of D.*

Proof We set $L = \operatorname{Ext}_A^t(C, K_A)$. Let $\widehat{(-)}$ denote the $\mathfrak{m}$-adic completion. Note that $\widehat{D}$ is also the $\mathfrak{n}$-adic completion of D. Note also that $\widehat{C}$ is the finite direct product of complete local rings, and $\widehat{D}$ is one of them by Corollary 4.37.

$$L_{\mathfrak{l}} \otimes_D \widehat{D} = L \otimes_C D \otimes_D \widehat{D} = L \otimes_C \widehat{C} \otimes_{\widehat{C}} \widehat{D} = \widehat{L} \otimes_{\widehat{C}} \widehat{D} = (\widehat{L})_{\mathfrak{l}\widehat{C}}.$$

So we may assume that A is complete and D is finite over A. Let $r = \dim D$. Then by Theorem 6.22, we have $\mathrm{H}_{\mathfrak{n}}^r(D)^\vee = \mathrm{H}_{\mathfrak{m}D}^r(D)^\vee = \mathrm{H}_{\mathfrak{m}}^r(D)^\vee = \operatorname{Ext}_A^{d-r}(D, K_A)$, which are isomorphisms of D-modules by Lemma 6.28. □

Let $(A, \mathfrak{m})$ be a d-dimensional Noetherian local ring, and K an A-module. We say that K is a *dualizing module* of A if K is finitely generated, and $\mu^i(\mathfrak{m}, K) = \delta_{id}$, Kronecker's delta. For example, if A is Gorenstein, the A-module A is dualizing.

Proposition 6.30 *Let $(A, \mathfrak{m})$ be a Noetherian local ring. Then the following hold:*

(1) *If K is a dualizing A-module, then K is a Cohen–Macaulay A-module, and* $\operatorname{depth}_A K = \dim_A K = \operatorname{id}_A K = \dim A$.
(2) *A finitely generated A-module K is dualizing if and only if A is Cohen–Macaulay and K is the canonical module of A.*

Proof Induction on d. We may assume that A is complete. Note that A has the canonical module $K_A = \mathrm{H}_{\mathfrak{m}}^d(A)^\vee$.

(1) Since $\mu^d(\mathfrak{m}, K) = 1$ and $\mu^i(\mathfrak{m}, K) = 0$ for $i < d$, we have that $\operatorname{depth}_A K = d$. Hence K is a Cohen–Macaulay module of dimension d. By Proposition 6.6, we have $\operatorname{id}_A K < \infty$. Then by Corollary 6.3, $\operatorname{id}_A K = d$.
(2) If A is Cohen–Macaulay and $K = K_A$, then $\operatorname{Ext}^i_A(A/\mathfrak{m}, K) \cong \mathrm{H}^{d-i}_{\mathfrak{m}}(A/\mathfrak{m})^\vee$ by the local duality. Hence K is dualizing.

Conversely, assume that K is dualizing. Then by (1) and Proposition 6.4, $\operatorname{depth} A = \operatorname{id}_A K = d$, and hence A is Cohen–Macaulay. If $d = 0$, then $K \cong E_A(A/\mathfrak{m}) \cong K_A$. Assume that $d > 0$. By the local duality, we have $\operatorname{Ext}^1_A(K, K_A) \cong \mathrm{H}^{d-1}_{\mathfrak{m}}(K)^\vee = 0$. We can take an $(A \oplus K \oplus K_A)$-regular element x. Then

$$0 \to \operatorname{Hom}_A(K, K_A) \xrightarrow{x} \operatorname{Hom}_A(K, K_A) \to \operatorname{Hom}_A(K, K_A/xK_A) \to 0 \qquad (6.3)$$

is exact. By Lemma 6.7, we have that both K/xK and K_A/xK_A are dualizing, and hence isomorphic to $K_{A/xA}$ by induction. Let $\overline{\varphi} : K/xK \to K_A/xK_A$ be an isomorphism. By the exact sequence (6.3), $\overline{\varphi}$ is liftable to $\varphi : K \to K_A$. By NAK, φ is surjective. Let $L = \operatorname{Ker}\varphi$. Then it is easy to see that $\mu^i(\mathfrak{m}, L) = 0$ for $i \neq d, d+1$. If $L \neq 0$, then $\operatorname{id}_A L = \operatorname{depth} A \leq d$, and hence $\mu^{d+1}(\mathfrak{m}, L) = 0$. So $\mu^d(\mathfrak{m}, L) = \mu^d(\mathfrak{m}, K) - \mu^d(\mathfrak{m}, K_A) = 1 - 1 = 0$. Hence $\operatorname{depth}_A L = \infty$, and $L = 0$. It follows that $\varphi : K \to K_A$ is an isomorphism. □

Corollary 6.31 *Let $(A, \mathfrak{m})$ be a Noetherian local ring. Then A is Gorenstein if and only if A is Cohen–Macaulay and the A-module A is the canonical module of A.*

Corollary 6.32 *Let $f : (A, \mathfrak{m}) \to (B, \mathfrak{n})$ be a flat local homomorphism between Cohen–Macaulay local rings. For a finitely generated A-module K, the following are equivalent:*

(1) *K is the canonical module of A, and $B/\mathfrak{m}B$ is Gorenstein.*
(2) *$K \otimes_A B$ is the canonical module of B.*

Proof Since A and B are Cohen–Macaulay, the canonical modules and the dualizing modules are the same thing. So factoring out by regular sequences, we may assume that $\dim B = 0$.

(1) ⇒(2). By assumption and Theorem 6.16, $\mathrm{r}_B(K \otimes_A B) = \mathrm{r}_A(K) \cdot \mathrm{r}_B(B/\mathfrak{m}B) = 1$. So the injective envelope of $K \otimes_A B$ is $E_B(B/\mathfrak{n})$. Since

$$\ell_B(K \otimes_A B) = \ell_A(K) \cdot \ell_B(B/\mathfrak{m}B) = \ell_A(A) \cdot \ell_B(B/\mathfrak{m}B) = \ell_B(B) = \ell_B(E_B(B/\mathfrak{n})),$$

we have that $K \otimes_A B = E_B(B/\mathfrak{n}) = K_B$.

(2) ⇒(1). $\mathrm{r}_A(K) \cdot \mathrm{r}_B(B/\mathfrak{m}B) = \mathrm{r}_B(K \otimes_A B) = 1$, and hence $\mathrm{r}_A(K) = \mathrm{r}_B(B/\mathfrak{m}B) = 1$. Hence $B/\mathfrak{m}B$ is Gorenstein, and the injective envelope of K is $E_A(A/\mathfrak{m})$. Since

$$\ell_A(E_A(A/\mathfrak{m})) \cdot \ell_B(B/\mathfrak{m}B) = \ell_A(A)\ell_B(B/\mathfrak{m}B)$$
$$= \ell_B(B) = \ell_B(K_B) = \ell_B(K \otimes_A B) = \ell_A(K)\ell_B(B/\mathfrak{m}B),$$

we have that $\ell_A(K) = \ell_A(E_A(A/\mathfrak{m}))$, and hence $K = E_A(A/\mathfrak{m})$, as desired. □

Theorem 6.33 *Let $(A, \mathfrak{m})$ be a d-dimensional Cohen–Macaulay local ring with the canonical module K_A. Then the following hold:*

(1) $\mathrm{Ext}^i_A(K_A, K_A) = 0$ *for $i > 0$, and the canonical map $\phi : A \to \mathrm{Hom}_A(K_A, K_A)$ is an isomorphism. In particular, K_A is a faithful module, and* $\mathrm{H}^d_{\mathfrak{m}}(K_A) \cong E_A(A/\mathfrak{m})$.
(2) (I. Reiten[3]) *A is a homomorphic image of a Gorenstein local ring.*
(3) *For each $\mathfrak{p} \in \mathrm{Spec}\, A$, $(K_A)_\mathfrak{p}$ is the canonical module $K_{A_\mathfrak{p}}$ of the local ring $A_\mathfrak{p}$.*
(4) *For every $i \in \mathbb{Z}$ and $\mathfrak{p} \in \mathrm{Spec}\, A$, we have $\mu^i_A(\mathfrak{p}, K_A) = \delta_{i,\mathrm{ht}\,\mathfrak{p}}$, where δ is Kronecker's delta.*
(5) *If $x \in \mathfrak{m}$ is a non-zerodivisor of A, then $K_A/xK_A \cong K_{A/xA}$.*
(6) *The number of minimal generators $\mu_A(K_A)$ of K_A agrees with $\mathrm{r}(A)$.*
(7) *For every finitely generated A-module $M \neq 0$, the equality*

$$\mathrm{depth}_A M = d - \max\{i \mid \mathrm{Ext}^i_A(M, K_A) \neq 0\}$$

holds. In particular, if $\mathrm{depth}_A(M) = d$, then we have $\mathrm{Ext}^i_A(M, K_A) = 0$ $(i > 0)$.

Proof We prove (1). We may assume that A is complete. Let $I^\bullet$ be the minimal injective resolution of K_A. Since K_A is dualizing, $\Gamma_\mathfrak{m}(I^i) = 0$ if $i \neq d$ and is $E = E_A(A/\mathfrak{m})$ if $i = d$. So $\mathrm{H}^d_\mathfrak{m}(K_A) = E$, and $\mathrm{H}^i_\mathfrak{m}(K_A) = 0$ for $i \neq 0$. By the local duality, $\mathrm{Hom}_A(K_A, K_A) \cong A$, and $\mathrm{Ext}^i_A(K_A, K_A) = 0$ for $i > 0$. Since $\phi(1)$ is the identity map of K_A, we have $\phi(1) \notin \mathfrak{m}\,\mathrm{Hom}_A(K_A, K_A)$, and $\phi : A \to \mathrm{Hom}_A(K_A, K_A) \cong A$ is surjective. By Corollary 3.4, it is an isomorphism. So K_A is a faithful module.

(2) Set $B = A \ltimes K_A$, the idealization of K_A, see 1.89 for definition. Since B is a finite extension of A, $\dim B = \dim A$. Since K_A is a square-zero ideal of B and $B/K_A \cong A$, B is local. Note that a system of parameters $x_1, \ldots, x_d$ of A is also that of B, which is a B-sequence. So B is Cohen–Macaulay. We have $K_B = \mathrm{Hom}_A(B, K_A) = \mathrm{Hom}_A(K_A, K_A) \oplus \mathrm{Hom}_A(A, K_A) = A \oplus K_A$. It is not so difficult to show that $K_B = B$ as a B-module. By Corollary 6.31, B is Gorenstein. Since $A = B/K_A$, A is the homomorphic image of a Gorenstein local ring.

(3) Write $A = B/I$, where B is a Gorenstein local ring, and I an ideal of B. Then $K_A = \mathrm{Ext}^t_B(A, B)$, where $t = \dim B - \dim A$. For every minimal prime Q of I, we have $\mathrm{ht}\, Q = \dim B - \dim B/Q = t$. So for every $P \in V(I)$, we have $\dim B_P - \dim A_P = \mathrm{ht}\, IB_P = t$. So $(K_A)_P = \mathrm{Ext}^t_{B_P}(A_P, B_P) = K_{A_P}$, as desired.

(4) By (3), we may assume that $\mathfrak{p} = \mathfrak{n}$, and this case is trivial.

[3] Idun Reiten, 1942–2025.

(5) Since $0 \to A \xrightarrow{x} A \to A/xA \to 0$ is a minimal free resolution of A/xA, we have $K_{A/xA} = \mathrm{Ext}^1_A(A/xA, K_A) \cong K_A/xK_A$ by Theorem 6.29.

(6) By (5), we may assume that A is Artinian. Then $K_A = E_A(A/\mathfrak{m})$, and the assertion is clear from the Matlis duality.

(7) is immediate from the local duality and Proposition 5.99. □

The duality by the canonical module plays important roles in the theory of maximal Cohen–Macaulay modules.

Definition 6.34 Let $(A, \mathfrak{m})$ be Cohen–Macaulay local ring of dimension d.

(1) A finitely generated A-module M is said to be *maximal Cohen–Macaulay* (sometimes called *MCM* for short) if $M = 0$ or $\mathrm{depth}_A M = d$.
(2) If M is an MCM A-module, then we define

$$D(M) = \mathrm{Hom}_A(M, K_A).$$

We will show that $D(M)$ is also an MCM A-module. We call $D(M)$ the *canonical dual* of M.

Proposition 6.35 *Let $(A, \mathfrak{m})$ be a Cohen–Macaulay local ring of dimension d. Then the following statements hold for MCM A-modules:*

(1) *If M is an MCM A-module, then so is $D(M)$.*
(2) *If $0 \to M' \to M \to M'' \to 0$ is an exact sequence of MCM A-modules, then the dual sequence $0 \to D(M'') \to D(M) \to D(M') \to 0$ is also an exact sequence.*
(3) *The natural map $h_M : M \to D(D(M))$ is an isomorphism.*
(4) $\mathrm{r}(M) = \mu_A(D(M))$.

Proof (1), (2) If M is an MCM A-module, we have $\mathrm{Ext}^i_A(M, K_A) = 0$ $(i > 0)$ by Theorem 6.22. Hence D is an exact functor on the category of MCM A-modules. If $F_\bullet \to M \to 0$ is a minimal free resolution of an MCM A-module M, then $0 \to D(M) \to D(F_\bullet)$ is also an exact sequence. Since $D(A^n) \cong K_A^n$ is an MCM A-module, $D(M)$ is also an MCM A-module.

(3) Consider a short exact sequence of MCM A-modules $\mathcal{S} : 0 \to L \to M \to N \to 0$. Consider the chain map $h_{\mathcal{S}} : \mathcal{S} \to DD\mathcal{S}$, where $h_M : M \to D(D(M))$ is the map given by $h_M(x)(\psi) = \psi(x)$ $(x \in M, \psi \in D(M))$. Thus we obtain the following commutative diagram:

$$\begin{array}{ccccccccc}
\mathcal{S}: & 0 & \longrightarrow & L & \longrightarrow & M & \longrightarrow & N & \longrightarrow & 0 \\
\downarrow h_{\mathcal{S}} & & & \downarrow h_L & & \downarrow h_M & & \downarrow h_N & & \\
DD\mathcal{S}: & 0 & \longrightarrow & D(D(L)) & \longrightarrow & D(D(M)) & \longrightarrow & D(D(N)) & \longrightarrow & 0
\end{array}$$

In particular, for an MCM A-module M, take a free A-module F so that $0 \to N \to F \to M \to 0$ is a short exact sequence. Then N is also an MCM A-module and since $h_A : A \to D(D(A))$ is an isomorphism, we deduce that h_M is a surjection for every MCM A-module M. On the other hand, since h_N is an injection, h_N is a bijection. Hence the natural map $h_M : M \to D(D(M))$ is an isomorphism for every MCM A-module M.

(4) We may assume that A is complete. By the local duality theorem, we have $D(M) \cong [\mathrm{H}^d_{\mathfrak{m}}(M)]^\vee$. By Theorem 5.53, we have

$$\mu_A(D(M)) = \mu_A([\mathrm{H}^d_{\mathfrak{m}}(M)]^\vee) = \mathrm{r}_A(\mathrm{H}^d_{\mathfrak{m}}(M)) = \mathrm{r}_A(M).$$

□

A very interesting theory has been developed concerning MCM modules. The references [Yo] and [LW] are very good introductions.

Let A be a Noetherian ring, M a finitely generated A-module, and N an A-submodule of M. Let $N = Q_1 \cap \cdots \cap Q_n$ be a minimal primary decomposition of N in M such that $\sqrt{\mathrm{Ann}_A(M/Q_i)} \in \mathrm{Assh}_A(M/N)$ if and only if $i \le r$. Then we define $U_M(N) = Q_1 \cap \cdots \cap Q_r \subseteq M$. Note that $\mathrm{Ass}_A(M/U_M(N)) = \mathrm{Assh}_A(M/N)$ and $\mathrm{Ass}_A(U_M(N)/N) = \mathrm{Ass}_A(M/N) \setminus \mathrm{Assh}_A(M/N)$. So $U_M(N)/N$ is the largest submodule of M/N such that $\dim U_M(N)/N < \dim M/N$.

Proposition 6.36 *Let $(A, \mathfrak{m})$ be a Noetherian local ring, and M a finitely generated A-module of dimension d. We set $N = \mathrm{H}^d_{\mathfrak{m}}(M)^\vee$, where $(-)^\vee = \mathrm{Hom}_A(-, E_A(A/\mathfrak{m}))$ is the Matlis dual. Then the following hold:*

(1) *N is a finitely generated $\widehat{A}$-module, and* $\mathrm{Supp}_{\widehat{A}} N = \bigcup_{P \in \mathrm{Assh}_{\widehat{A}} \widehat{M}} V(P)$.
(2) $\dim_{\widehat{A}} N = d$.
(3) *$\mathrm{H}^d_{\mathfrak{m}}(M)$ is finitely generated as an $\widehat{A}$-module if and only if $d = 0$.*

Proof It is easy to see that we may assume that A is a complete regular local ring again. We set $s = \dim A$. Then by Theorem 6.22, $N \cong \mathrm{Ext}^{s-d}_A(M, A)$, which is a finitely generated A-module.

(1) Let $P \in \mathrm{Assh}_A M$. Then we have $\mathrm{ht}\, P = s - d$ by Lemma 5.24. Note that M_P is a module of finite length over the $(s-d)$-dimensional regular local ring A_P. So we have

$$N_P \cong \mathrm{Ext}^{s-d}_{A_P}(M_P, A_P) \neq 0,$$

and $P \in \mathrm{Supp}_A N$. Hence $\bigcup_{P \in \mathrm{Assh}_A M} V(P) \subseteq \mathrm{Supp}_A N$. Let $L = U_M(0)$. Then $\dim_A L < d$, and $\mathrm{Ass}_A(M/L) = \mathrm{Assh}_A M$. Since

$$0 = \mathrm{H}^d_{\mathfrak{m}}(L) \to \mathrm{H}^d_{\mathfrak{m}}(M) \to \mathrm{H}^d_{\mathfrak{m}}(M/L) \to 0$$

is exact, $N \cong \mathrm{H}^d_{\mathfrak{m}}(M/L)^\vee$. So if $P \in \mathrm{Supp}_A N$, then $P \supseteq \mathrm{Ann}_A N \supseteq \mathrm{Ann}_A(M/L)$, and hence $P \in \mathrm{Supp}_A(M/L) = \bigcup_{P \in \mathrm{Assh}_A M} V(P)$, as desired.

(2) follows immediately from (1).

(3) $\mathrm{H}^d_{\mathfrak{m}}(M)$ is finitely generated if and only if N is Artinian. This is equivalent to $d = 0$ by (2).

□

If A is not Cohen–Macaulay, then the local duality theorem (for arbitrary i, as in Corollary 6.23) does not hold in full generality. If the conclusion of Corollary 6.23 holds, then $\mathrm{H}^{d-i}_{\mathfrak{m}}(A)^\vee \cong \mathrm{Ext}^i_A(A, K_A) \otimes_A \widehat{A} = 0$ for $i > 0$, and A must be Cohen–Macaulay. But as in Theorems 6.22 and 6.29, some part of the theorem still holds and the canonical module K_A plays an important role frequently. We will see to what extent the results on K_A for Cohen–Macaulay cases above hold true.

Until the end of this section, $(A, \mathfrak{m})$ is a d-dimensional Noetherian local ring with the canonical module K_A.

First, note the following:

Proposition 6.37 *Let $I = U_A(0)$, and set $A' = A/I$. Then we have $K_A \cong K_{A'}$.*

Proof Look at the exact sequence $0 \to I \to A \to A' \to 0$. Then since $\dim I < d$, $\mathrm{H}^d_{\mathfrak{m}}(I) = \mathrm{H}^{d+1}_{\mathfrak{m}}(I) = 0$. So $\mathrm{H}^d_{\mathfrak{m}}(A) \to \mathrm{H}^d_{\mathfrak{m}}(A')$ is an isomorphism. Taking the Matlis dual, we get $\widehat{K}_{A'} \cong \widehat{K}_A$. So $K_{A'} \cong K_A$. □

Lemma 6.38 *Let $\mathfrak{p} \in \operatorname{Spec} A$. If either (i) $A = B/\mathfrak{a}$ is a homomorphic image of a Gorenstein local ring $(B, \mathfrak{n})$; or (ii) $A_\mathfrak{p}$ is Cohen–Macaulay, then we have $(K_A)_\mathfrak{p} \cong K_{A_\mathfrak{p}}$ if $\mathfrak{p} \in \operatorname{Supp} K_A = \bigcup_{\mathfrak{q}\in \operatorname{Assh} A} V(\mathfrak{q})$, while $(K_A)_\mathfrak{p} = 0$ if $\mathfrak{p} \notin \operatorname{Supp} K_A$.*

Proof Set $t = \operatorname{ht} \mathfrak{a}$.

The assertion $\operatorname{Supp} K_A = \bigcup_{\mathfrak{q}\in \operatorname{Assh} A} V(\mathfrak{q})$ follows from Proposition 6.36, and the last assertion is trivial. We want to show that for every $\mathfrak{p} \in \operatorname{Supp} K_A$, we have $(K_A)_\mathfrak{p} \cong K_{A_\mathfrak{p}}$.

First, assume (i). We can take $\mathfrak{q} \in \operatorname{Assh} A$ such that $\mathfrak{p} \supseteq \mathfrak{q}$. Let $P = \pi^{-1}(\mathfrak{p})$ and $Q = \pi^{-1}(\mathfrak{q})$, where $\pi : B \to B/\mathfrak{a} = A$ is the canonical map. Then

$$\dim B_P - \dim A_\mathfrak{p} = \operatorname{ht} \mathfrak{a} B_P = \operatorname{ht} Q = t = \dim B - \dim A.$$

So $(K_A)_\mathfrak{p} = \mathrm{Ext}^t_B(A, B)_P \cong \mathrm{Ext}^t_{B_P}(A_\mathfrak{p}, B_P) = K_{A_\mathfrak{p}}$.

Next, assume (ii). Let $P \in \operatorname{Min}_{\widehat{A}}(\widehat{A}/\mathfrak{p}\widehat{A})$. Then $\widehat{A}_P$ is Cohen–Macaulay, since $A_\mathfrak{p}$ and $\widehat{A}_P/\mathfrak{p}\widehat{A}_P$ are. Moreover, $(K_A)_\mathfrak{p} \otimes_{A_\mathfrak{p}} \widehat{A}_P = (K_{\widehat{A}})_P = K_{\widehat{A}_P}$ by case (i) and Cohen's structure theorem. By Corollary 6.32, we are done. □

From Proposition 6.37, to investigate the properties of K_A, we can assume that $\operatorname{Ass}_A A = \operatorname{Assh}_A A$, or equivalently, $U_A(0) = 0$, replacing A by $A/U_A(0)$.

Let $(R, \mathfrak{m})$ be a Noetherian local ring, and M a finite R-module. We say that M is *unmixed* if $\operatorname{Ass}_{\widehat{R}} \widehat{M} = \operatorname{Assh}_{\widehat{R}} \widehat{M}$, where $\widehat{R}$ and $\widehat{M}$ denote the $\mathfrak{m}$-adic completion of R and M, respectively. We say that the local ring R is unmixed if it is so as an R-module.

Theorem 6.39 *The following properties hold true:*

(1) *If $f \in \mathfrak{m}$ satisfies $\dim A/(f) = d - 1$, then f is a K_A-regular element.*

(2) *If* $(f, g) \subseteq \mathfrak{m}$ *satisfies* $\dim A/(f, g) = d - 2$, *then* (f, g) *is a* K_A*-regular sequence. Moreover, if for* $r \geq 3$ *and* $\mathrm{H}^{d-i}_{\mathfrak{m}}(A) = 0$ $(i = 1, \ldots, r-2)$ *holds, then every subsystem of parameters* $f_1, \ldots, f_r$ *of* A *is a* K_A*-regular sequence.*

(3) *The following conditions are equivalent:*

   (a) *A satisfies* $\operatorname{Ass} A = \operatorname{Assh} A$, *or equivalently,* $U_A(0) = 0$.
   (b) K_A *is a faithful module, or equivalently, the canonical map* $\phi_A : A \to \operatorname{Hom}_A(K_A, K_A)$ *is injective.*
   (c) *A is unmixed, that is,* $\operatorname{Ass} \widehat{A} = \operatorname{Assh} \widehat{A}$.

(4) *The following conditions are equivalent:*

   (a) *A satisfies Serre's condition* (S_2).
   (b) *The natural map* $\phi : A \to \operatorname{Hom}_A(K_A, K_A)$, $\phi(a)(x) = ax$ *is an isomorphism.*
   (c) $\operatorname{Hom}_A(K_A, K_A) \cong A$ *as an A-module.*
   (d) $\mathrm{H}^d_{\mathfrak{m}}(K_A) \cong \mathrm{E}_A(A/\mathfrak{m})$.

Proof

(1) Replacing A by A' (Proposition 6.37), we may assume that $\operatorname{Ass} A = \operatorname{Assh} A$. Then f is A-regular. Consider the long exact sequence

$$\mathrm{H}^{d-1}_{\mathfrak{m}}(A/fA) \to \mathrm{H}^d_{\mathfrak{m}}(A) \xrightarrow{f} \mathrm{H}^d_{\mathfrak{m}}(A) \to 0 = \mathrm{H}^d_{\mathfrak{m}}(A/fA)$$

induced by the short exact sequence $0 \to A \xrightarrow{f} A \to A/fA \to 0$. Taking the Matlis dual, we get an exact sequence

$$0 \to K_{\widehat{A}} \xrightarrow{f} K_{\widehat{A}} \to K_{\widehat{A}/f\widehat{A}}. \qquad (**)$$

Hence f is a K_A-regular element, and $(K_A/fK_A) \otimes_A \widehat{A}$ is isomorphic to an $\widehat{A}$-submodule of $K_{\widehat{A}/f\widehat{A}}$.

(2) Since g is $K_{\widehat{A}/f\widehat{A}}$-regular by (1), it is also $(K_A/fK_A) \otimes_A \widehat{A}$-regular. Hence it is also K_A/fK_A-regular, and (f, g) is a K_A-sequence. Now assume that $r \geq 3$ and $\mathrm{H}^{d-i}_{\mathfrak{m}}(A) = 0$ for $i = 1, \ldots, r-2$. Set $I_j = (f_1, \ldots, f_j)$, and $A_j = A/I_j$. We have that $\operatorname{Ext}^i_A(A_{r-2}, K_A) \otimes_A \widehat{A} \cong \mathrm{H}^{d-i}_{\mathfrak{m}}(A_{r-2})^\vee = 0$ for $d - i > \dim A_{r-2} = d - r + 2$, that is, for $i < r - 2$. So $\operatorname{depth}(I_{r-2}, K_A) \geq r - 2$, and the Koszul complex $K_\bullet(f_1, \ldots, f_{r-2}; K_A)$ is acyclic by the depth sensitivity (Theorem 5.86), and hence $f_1, \ldots, f_{r-2}$ is K_A-regular by Theorem 5.82. Since

$$\begin{aligned}\operatorname{Ext}^{r-2}_A(A_{r-2}, K_A) &\cong \mathrm{H}^{r-2}(f_1, \ldots, f_{r-2}, K_A) \\ &\cong \mathrm{H}_0(f_1, \ldots, f_{r-2}, K_A) \cong K_A/I_{r-2}K_A\end{aligned}$$

is the canonical module of A_{r-2} by the adjunction theorem (Theorem 6.29), f_{r-1}, f_r is $K_A/I_{r-2}K_A$-regular by what we have proved, and hence $f_1, \dots, f_r$ is a K_A-sequence.

(3) (a) $\Rightarrow$(b). By Lemma 6.38, we have that $\operatorname{Supp} K_A = \operatorname{Spec} A$. To prove that K_A is faithful, we prove that $\phi : A \to \operatorname{Hom}_A(K_A, K_A)$ is injective. Since A satisfies (S_1), it suffices to show that $\phi_{\mathfrak{p}} : A_{\mathfrak{p}} \to \operatorname{Hom}_{A_{\mathfrak{p}}}((K_A)_{\mathfrak{p}}, (K_A)_{\mathfrak{p}})$ is injective for each minimal prime $\mathfrak{p}$ of A by Lemma 3.48. Since $A_{\mathfrak{p}}$ is Cohen–Macaulay, we can replace $(K_A)_{\mathfrak{p}}$ by $K_{A_{\mathfrak{p}}}$ by Lemma 6.38. Then the assertion is nothing but the Matlis duality, since $K_{A_{\mathfrak{p}}} = E_{A_{\mathfrak{p}}}(\kappa(\mathfrak{p}))$.

(b) $\Rightarrow$(c). Since $\phi_A : A \to \operatorname{Hom}_A(K_A, K_A)$ is injective, $\phi_{\widehat{A}} \to \operatorname{Hom}_{\widehat{A}}(K_{\widehat{A}}, K_{\widehat{A}})$ is injective. So $U_{\widehat{A}}(0) \subseteq \operatorname{Ker} \phi_{\widehat{A}} = 0$, and hence A is unmixed.

(c) $\Rightarrow$(a). By (a) $\Rightarrow$(b) above, $\operatorname{Ann}_{\widehat{A}} K_{\widehat{A}} = 0$, and hence $U_A(0) \subseteq \operatorname{Ann}_A K_A = 0$. So $\operatorname{Ass} A = \operatorname{Assh} A$.

(4) Statement (2) shows that K_A satisfies the condition (S_2) as an A-module. Hence so does $\operatorname{Hom}_A(K_A, K_A)$. This proves (b) $\Rightarrow$(a).

Next, assume (a). Since A satisfies (S_1), we have $\operatorname{Ass} A = \operatorname{Assh} A$. To show that ϕ is an isomorphism, since both A and $\operatorname{Hom}_A(K_A, K_A)$ satisfy (S_2), it suffices to show that $\phi_{\mathfrak{p}} : A_{\mathfrak{p}} \to \operatorname{Hom}_{A_{\mathfrak{p}}}((K_A)_{\mathfrak{p}}, (K_A)_{\mathfrak{p}})$ is an isomorphism for every $\mathfrak{p} \in \operatorname{Spec} A$ with $\operatorname{ht} \mathfrak{p} \leq 1$. By the (S_2) assumption, $A_{\mathfrak{p}}$ is Cohen–Macaulay. Hence $(K_A)_{\mathfrak{p}} \cong K_{A_{\mathfrak{p}}}$, and the result follows from Proposition 6.35. Thus we have shown that (a) and (b) are equivalent.

(b) $\Rightarrow$(d). We have $\mathrm{H}^d_{\mathfrak{m}}(K_A) \cong A^{\vee} = \mathrm{E}_A(A/\mathfrak{m})$ by Theorem 6.22.

(d) $\Rightarrow$(c). By (3), $\operatorname{Hom}_A(K_A, K_A) \otimes_A \widehat{A} \cong \mathrm{H}^d_{\mathfrak{m}}(K_A)^{\vee} \cong \mathrm{E}_A(A/\mathfrak{m})^{\vee} \cong A^{\vee\vee} \cong \widehat{A}$. So $\operatorname{Hom}_A(K_A, K_A) \cong A$ by Lemma 6.19.

(c) $\Rightarrow$(b). Since $\phi(1) = 1_{K_A} \notin \mathfrak{m} \operatorname{Hom}_A(K_A, K_A)$ and $\operatorname{Hom}_A(K_A, K_A)$ is cyclic, ϕ is surjective. By Corollary 3.4, ϕ is an isomorphism.

□

Definition 6.40

(1) We call the natural map $\phi : A \to \operatorname{Hom}_A(K_A, K_A)$ the *(S_2)-ification* of A. We have a natural A-algebra structure of $\operatorname{Hom}_A(K_A, K_A)$ defined by the composition of maps. Being a subring of

$$\prod_{\mathfrak{p} \in \operatorname{Assh} A} \operatorname{Hom}_{A_{\mathfrak{p}}}((K_A)_{\mathfrak{p}}, (K_A)_{\mathfrak{p}}) \cong \prod_{\mathfrak{p} \in \operatorname{Assh} A} A_{\mathfrak{p}}$$

by Lemma 6.38, this is a commutative ring, and is a finite A-algebra.

(2) If A is a normal domain, by Theorem 6.39, (1) and Proposition 3.54, K_A is isomorphic to a divisorial ideal. Hence we can define the isomorphism class $\operatorname{cl}(K_A) \in \operatorname{Cl}(A)$, which we call the *canonical class* of A.

(3) If A is the canonical module of A, we say that A is *quasi-Gorenstein*. A Noetherian ring R is said to be quasi-Gorenstein if $R_{\mathfrak{m}}$ is quasi-Gorenstein for every maximal ideal $\mathfrak{m}$ of R. Note that R is Gorenstein if and only if R is quasi-Gorenstein and Cohen–Macaulay.

(4) We say that A is $\mathbb{Q}$-*Gorenstein* if A is a normal domain and $\mathrm{cl}(K_A) \in \mathrm{Cl}(A)$ has finite order.

The following result was found by P. Murthy.

Corollary 6.41 (Murthy's[4] Theorem) *A Noetherian local UFD with a canonical module is quasi-Gorenstein. Let R be a Cohen–Macaulay ring which is a homomorphic image of a Gorenstein ring. If R is a UFD, then R is Gorenstein.*

Proof A UFD is a Krull domain whose class group is trivial. So if the local UFD has a canonical module K_A, its canonical class is trivial. Hence $A \cong K_A$, and A is quasi-Gorenstein.

For the second assertion, we may localize at maximal ideals, and the problem is reduced to the first assertion. □

Not all Noetherian local rings have the canonical modules. There is an example of a Cohen–Macaulay local UFD which is not Gorenstein [Og]. Such a ring cannot have a canonical module, see Theorem 6.33 and Corollary 6.41 above. Even if a Noetherian local ring A has a canonical module, it may not be a homomorphic image of a Gorenstein local ring, see [Og2].

We set $D = \mathrm{Hom}_A(-, K_A)$, $A' = A/U_A(0)$, and

$$A'' = \mathrm{End}_A K_A = \mathrm{Hom}_A(K_A, K_A).$$

Note that D is a functor from (Mod A) to (Mod A'') and (mod A) to (mod A'').

Lemma 6.42 *The map $h_{K_A} : K_A \to DDK_A$ is an isomorphism.*

Proof We may assume that A is complete. To prove that h_{K_A} is an isomorphism, we may assume that $A = A'$. Then since A satisfies (S_1), and for $\mathfrak{p} \in \mathrm{Spec}\, A$ with $\mathrm{ht}\,\mathfrak{p} \leq 1$, $A_\mathfrak{p}$ is Cohen–Macaulay. Hence $(K_A)_\mathfrak{p} = K_{A_\mathfrak{p}}$. So $(h_{K_A})_\mathfrak{p} : (K_A)_\mathfrak{p} \to D_\mathfrak{p} D_\mathfrak{p} (K_A)_\mathfrak{p}$ is an isomorphism, where $D_\mathfrak{p} = \mathrm{Hom}_{A_\mathfrak{p}}(-, (K_A)_\mathfrak{p})$. Hence h_{K_A} is an isomorphism by Proposition 3.49. □

Lemma 6.43 *For a finitely generated A-module M, the following are equivalent:*

(1) *There is an injective A-linear map $M \to K_A^m$ for some m.*
(2) $\mathrm{Ass}_A M \subseteq \mathrm{Assh}_A A$.
(3) *The canonical map $h_M : M \to DDM$ is injective.*
(4) $\mathrm{Ass}_{\widehat{A}} \widehat{M} \subseteq \mathrm{Assh}_{\widehat{A}} \widehat{A}$.

If these conditions are satisfied, M is an A'-module.

Proof (1) ⇒(2). $\mathrm{Ass}_A M \subseteq \mathrm{Ass}_A K_A^m = \mathrm{Assh}_A A$.

(2) ⇒(3). Since $\dim U_A(0)M < d$,

$$\mathrm{Ass}_A \mathrm{Hom}(U_A(0)M, M) \subseteq \mathrm{Supp}_A(U_A(0)M) \cap \mathrm{Assh}_A A = \emptyset.$$

[4] Pavaman Murthy, 1935–.

So M is an A'-module, where $A' = A/U_A(0)$. So we may assume that $A = A'$. By Lemma 3.48, it suffices to show that $(h_M)_\mathfrak{p}$ is injective for each $\mathfrak{p} \in \operatorname{Min} A$. This follows from Proposition 6.35, since $A_\mathfrak{p}$ is Cohen–Macaulay, and $(K_A)_\mathfrak{p} = K_{A_\mathfrak{p}}$.

(3) $\Rightarrow$(1). Take a surjective map $\pi : A^m \to DM$. Then we get an injective map $\pi^* : DDM \hookrightarrow DA^m = K_A^m$. So $\pi^* h_M : M \to K_A^m$ is an injective A-linear map.

(2) $\Leftrightarrow$(4). Condition (3) holds if and only if $\widehat{h_M} = h_{\widehat{M}} : \widehat{M} \to \widehat{D}\widehat{D}\widehat{M}$ is injective, where $\widehat{D} = \operatorname{Hom}_{\widehat{A}}(-, K_{\widehat{A}})$. So by (2) $\Leftrightarrow$(3), we have the equivalence.

The last assertion is shown in the proof of (2) $\Rightarrow$(3). □

Theorem 6.44 (Aoyama[5]) *For a finitely generated A-module M, the following are equivalent:*

(1) *M satisfies Serre's (S_2) condition.*[6]
(2) *The canonical map $h_M : M \to DDM$ is an isomorphism.*
(3) *There is an exact sequence of A-modules of the form*

$$0 \to M \to K_A^m \to K_A^n.$$

(4) *$\widehat{M}$ satisfies Serre's (S_2) condition.*

If the conditions are satisfied, then M is an A''-module. The category $(S_2)(A)$ of A-modules that satisfy the (S_2) condition is equivalent to $(S_2)(A'')$.

Proof (1) $\Rightarrow$(2). By Lemma 6.43, M and DDM are A'-modules, and hence we may assume that $U_A(0) = 0$. Then localizing at a prime $\mathfrak{p}$ with $\operatorname{ht} \mathfrak{p} \leq 1$, we have that $(h_M)_\mathfrak{p}$ is an isomorphism by Proposition 6.35 and Lemma 6.38. So h_M is an isomorphism by Proposition 3.49.

(2) $\Rightarrow$(3). Take a presentation

$$A^n \to A^m \to DM \to 0$$

and apply the contravariant left exact functor D.

(3) $\Rightarrow$(1) follows from the fact that K_A satisfies (S_2).

(1) $\Rightarrow$(4). Condition (2) is equivalent to the condition $\widehat{h_M} = h_{\widehat{M}} : \widehat{M} \to \widehat{D}\widehat{D}\widehat{M}$ is an isomorphism. So by (1) $\Leftrightarrow$(2), we have (1) $\Leftrightarrow$(4).

Since $M \cong DDM$ and DDM has an A''-module structure which extends the A-module structure, we have that M is also an A''-module. The restriction of scalars $(S_2)(A'') \to (S_2)(A)$ is an equivalence, whose quasi-inverse is $DD : (S_2)(A) \to (S_2)(A'')$. □

Corollary 6.45 *Let $(B, \mathfrak{n})$ be a Noetherian local ring, and K a finitely generated B-module. Then the following are equivalent:*

---

[5] Yoichi Aoyama, 1948–.

[6] The statement of this theorem is seemingly slightly different from that in the original paper [Aoy]. This is because the difference of the definition of Serre's (S_k) condition for a module.

(1) *The trivial extension $C = B \ltimes K$ is quasi-Gorenstein.*
(2) *B satisfies Serre's (S_2) condition, and K is the canonical module of B.*

Proof Since K is a square-zero ideal of C, ${}^a\pi : \operatorname{Spec} B \to \operatorname{Spec} C$ given by $P \mapsto \pi^{-1}(P)$ is a bijection, and this preserves the height of a prime ideal. It follows that C satisfies (S_2) as a C-module if and only if it satisfies (S_2) as a B-module.

(1) $\Rightarrow$(2). Since $B \oplus K$ satisfies (S_2), both B and K satisfy (S_2) as B-modules. Note that K annihilates K in C, and K is a faithful B-module. It follows that $K_B = \operatorname{Hom}_C(B, C) = [0 :_C K] = K$.

(2) $\Rightarrow$(1). This is proved similarly to Theorem 6.33, (2). □

Lemma 6.46 *For a finitely generated A-module K, the following are equivalent:*

(1) $K \cong K_A$.
(2) *For each maximal ideal $\mathfrak{n}$ of A'', $K_{\mathfrak{n}} \cong K_{A''_{\mathfrak{n}}}$.*

Proof We may assume that A is complete. Then we have

$$K_A \cong DDK_A \cong DDDA \cong DA'' \cong \mathrm{H}^d_{\mathfrak{m}}(A'')^\vee$$
$$\cong \bigoplus_{\mathfrak{n}\in\operatorname{Max} A''} \operatorname{Hom}_{A''_{\mathfrak{n}}}(\mathrm{H}^d_{\mathfrak{n}}(A''_{\mathfrak{n}}), E_{A''_{\mathfrak{n}}}(A''/\mathfrak{n})) \cong \bigoplus_{\mathfrak{n}\in\operatorname{Max} A''} K_{A''_{\mathfrak{n}}}.$$

Hence (1) $\Rightarrow$(2) is trivial. If we assume (2), then $K \cong \bigoplus_{\mathfrak{n}} K_{\mathfrak{n}} \cong \bigoplus_{\mathfrak{n}} K_{A'_{\mathfrak{n}}} \cong K_A$. □

Theorem 6.47 (Aoyama) *Let $f : (B, \mathfrak{n}) \to (C, \mathfrak{l})$ be a flat local homomorphism between Noetherian local rings. Let K be a B-module, and assume that $C \otimes_B K$ is the canonical module of C. Then K is the canonical module of B, and $C_P/\mathfrak{n}C_P$ is Gorenstein for every $P \in \operatorname{Min}(C/\mathfrak{n}C)$.*

Proof Replacing B and C by the completions $\widehat{B}$ and $\widehat{C}$, respectively, we may assume that both B and C are complete. Then taking a minimal prime P of $\mathfrak{n}C$ and replacing C by the completion of C_P, we may further assume that $\mathfrak{n}C$ is $\mathfrak{l}$-primary, and $P = \mathfrak{l}$. We set $d = \dim B = \dim C$. Since C is faithfully flat over B, K is finitely generated by Lemma 6.25. Since $C \otimes_B K \cong K_C$ satisfies (S_2), it is easy to see that K satisfies (S_2). So K is a B''-module, where $B'' = \operatorname{End}_B K_B$, and $C \otimes_B K = K_C$ is a $C'' = \operatorname{End}_C K_C$-module. So by Lemma 6.46, replacing B by $B''_{\mathfrak{N}}$ for a maximal ideal $\mathfrak{N}$ of B'' and replacing C by $(C \otimes_B B'')_{\mathfrak{L}}$ for a maximal ideal $\mathfrak{L}$ of $C \otimes_B B''$ lying over $\mathfrak{N}$, we may further assume that B and C satisfy (S_2). By Corollary 6.45, replacing B and C by $B \ltimes K$ and $C \otimes_B (B \ltimes K) = C \ltimes K_C$, respectively, we may further assume that C is quasi-Gorenstein, and $B = K$.

Now we have

$$(C/\mathfrak{l})^{\mathrm{r}_C(C/\mathfrak{n}C)\cdot \mathrm{r}_B(\mathrm{H}^d_{\mathfrak{n}}(B))} = \operatorname{Hom}_C(C/\mathfrak{l}, (C/\mathfrak{n}C))^{\mathrm{r}_B(\mathrm{H}^d_{\mathfrak{n}}(B))}$$
$$\cong \operatorname{Hom}_C(C/\mathfrak{l}, C/\mathfrak{n}C \otimes_{B/\mathfrak{n}} \operatorname{Hom}_B(B/\mathfrak{n}, \mathrm{H}^d_{\mathfrak{n}}(B)))$$

$$= \mathrm{Hom}_C(C/\mathfrak{l}, \mathrm{Hom}_C(C/\mathfrak{n}C, \mathrm{H}^d_{\mathfrak{l}}(C))$$

$$= \mathrm{Hom}_C(C/\mathfrak{l}, \mathrm{H}^d_{\mathfrak{l}}(C)) \cong \mathrm{Hom}_C(C/\mathfrak{l}, E_C(C/\mathfrak{l})) \cong C/\mathfrak{l}.$$

Hence $\mathrm{r}_C(C/\mathfrak{n}C) = 1$ and $\mathrm{r}_B(\mathrm{H}^d_{\mathfrak{n}}(B)) = 1$. Since $C/\mathfrak{n}C$ is Artinian, $C/\mathfrak{n}C$ is Gorenstein. Since $E_B(B/\mathfrak{n})$ is the injective envelope of $\mathrm{H}^d_{\mathfrak{n}}(B)$, we have that $K_B = \mathrm{H}^d_{\mathfrak{n}}(B)^\vee$ is a homomorphic image of $E_B(B/\mathfrak{n})^\vee = B$. Namely, K_B is a cyclic module. Since K_B is faithful, $K_B \cong B = K$, as desired. □

Corollary 6.48 (Aoyama) *Let $(A, \mathfrak{m})$ be a Noetherian local ring with the canonical module K_A, and $\mathfrak{p} \in \mathrm{Supp}\, K_A$. Then $(K_A)_{\mathfrak{p}}$ is the canonical module of $A_{\mathfrak{p}}$.*

Proof Let $P \in \mathrm{Min}\,(\widehat{A}/\mathfrak{p}\widehat{A})$. Then since $\widehat{A}$ is a homomorphic image of a Gorenstein local ring, we have $(K_A \otimes_A \widehat{A}) \otimes_{\widehat{A}} \widehat{A}_P \cong (K_A)_{\mathfrak{p}} \otimes_{A_{\mathfrak{p}}} \widehat{A}_P$ is the canonical module of $\widehat{A}_P$ by Lemma 6.38. By Theorem 6.47, the result follows. □

Corollary 6.49 *Let A be a quasi-Gorenstein Noetherian local ring, and $\mathfrak{p} \in \mathrm{Spec}\, A$. Then $A_{\mathfrak{p}}$ is quasi-Gorenstein.*

Remark 6.1 If $(A, \mathfrak{m})$ is Cohen–Macaulay, then we have the isomorphism of functors

$$\mathrm{H}^{d-j}_{\mathfrak{m}}(*) \cong [\mathrm{Ext}^j_A(*, \mathrm{K}_A)]^\vee$$

on the category of finitely generated A-modules.

Conversely, if the isomorphism above holds true, then we can assert that A is Cohen–Macaulay. In fact, if $\mathrm{H}^{d-j}_{\mathfrak{m}}(A) \cong [\mathrm{Ext}^j_A(A, \mathrm{K}_A)]^\vee$ holds for every j, since the right-hand side is 0 for $j > 0$, A must be Cohen–Macaulay.

6.50 We say that a Noetherian ring A satisfies the (T_n) condition if A_P is Gorenstein for any $P \in \mathrm{Spec}\, A$ with $\mathrm{ht}\, P \le n$. As a regular local ring is Gorenstein, the (R_n) condition implies (T_n) condition. We say that a Noetherian ring A is *generically Gorenstein* if the total quotient ring $\mathrm{Q}(A)$ is Gorenstein.

Lemma 6.51 *For a Noetherian ring A, the following are equivalent:*

(1) *A is generically Gorenstein.*
(2) *A satisfies (T_0) and (S_1).*

Proof Note that $\mathrm{Q}(A)$ is the localization of A by $A \setminus \bigcup_{P \in \mathrm{Ass}\, A} P$. In particular, if $P \in \mathrm{Ass}\, A$, then $P\,\mathrm{Q}(A) \in \mathrm{Ass}\, \mathrm{Q}(A)$.

(1) ⇒(2). If $P \in \mathrm{Ass}\, A$, then $P\,\mathrm{Q}(A) \in \mathrm{Ass}\, \mathrm{Q}(A)$ is a minimal prime. Hence P is also a minimal prime, and hence A satisfies (S_1). For a minimal prime P of A, $A_P \cong \mathrm{Q}(A)_{P\,\mathrm{Q}(A)}$ is Gorenstein, and hence A satisfies (T_0).

(2) ⇒(1). Since $\mathrm{Ass}\, A = \mathrm{Min}\, A$, $\mathrm{Q}(A) = \prod_{P \in \mathrm{Min}\, A} A_P$. Since each A_P is Gorenstein by the (T_0) assumption, we have that $\mathrm{Q}(A)$ is Gorenstein. □

A similar proof shows that A is reduced if and only if $\mathrm{Q}(A)$ is reduced if and only if $\mathrm{Q}(A)$ is regular if and only if A satisfies (R_0) and (S_1).

Lemma 6.52 *Let A be a generically Gorenstein Noetherian commutative ring, and M a finite module. Then the following are equivalent:*

(1) *M is torsionless.*
(2) *M is a first syzygy in the sense that there exists some injective map $M \to F_0$ such that F_0 is a finite free A-module.*
(3) *M satisfies Serre's (S_1) condition.*

Proof (1) $\Rightarrow$(2). There is a surjective A-linear map $\varphi : A^n \to M^*$, which induces an injective A-linear map $\varphi^* : M^{**} \to A^n$. So the composite $M \hookrightarrow M^{**} \xrightarrow{\varphi^*} A^n$ is also injective, and hence M is a first syzygy.

(2) $\Rightarrow$(3). Since A satisfies (S_1), we have that F_0 satisfies (S_1). The (S_1) condition, which says that any associated prime is of height zero, is inherited by a submodule, and hence M satisfies (S_1).

(3) $\Rightarrow$(1). By Lemma 3.48, we may assume that A is an Artinian local ring. Then $A = \mathrm{Q}(A)$, and hence A is Gorenstein. So A is the canonical module of A and M is a maximal Cohen–Macaulay module, and by Proposition 6.35, (3), we have that $M \to M^{**}$ is injective. □

Proposition 6.53 *Let A be a Noetherian commutative ring which satisfies (T_1) and (S_2), and M its finite module (e.g., A is normal). Then the following are equivalent:*

(1) *M is reflexive.*
(2) *M is a second syzygy in the sense that there exists some exact sequence $0 \to M \to F_0 \to F_1$ such that F_0 and F_1 are finite free A-modules.*
(3) *M satisfies Serre's (S_2) condition.*

Proof This is proved similarly to Proposition 3.51. Use the fact that a maximal Cohen–Macaulay module over a Gorenstein local ring is reflexive in proving (3) $\Rightarrow$(1). □

6.4 Structure Theorem for Gorenstein Rings of Codimension 3

Let A be a Noetherian ring, and M be a finitely generated A-module. We define the *grade* of M to be $\operatorname{grade}_A M = \inf\{n \in \mathbb{Z} \mid \operatorname{Ext}^n_A(M, A) \neq 0\}$, and the *codimension* of M to be $\operatorname{codim}_A M = \operatorname{ht}_A(\operatorname{Ann}_A M)$. Note that for an ideal I of A, we have $\operatorname{grade}_A A/I = \operatorname{depth}_A(I, A)$. (This number is often denoted by $\operatorname{grade}_A I$.)

Lemma 6.54 *Let A be a Noetherian ring, and M be a nonzero finitely generated A-module. Then*

$$\operatorname{grade}_A M \le \operatorname{codim}_A M \le \operatorname{hd}_A M.$$

Proof If the first inequality does not hold, then there exists some $P \in \operatorname{Min}_A M$ such that $\operatorname{ht} P < \operatorname{grade}_A M$. Then it is easy to see that $\operatorname{Ext}^r_{A_P}(M_P, A_P) \neq 0$ for

$r = \operatorname{depth} A_P$. So $\operatorname{grade}_A M \leq \operatorname{depth} A_P \leq \operatorname{ht} P < \operatorname{grade}_A M$, and this is a contradiction.

We prove the second inequality. We may assume that $\operatorname{hd}_A M < \infty$. Take a minimal prime P of M such that $\operatorname{ht} P = \operatorname{codim}_A M$. Now M_P is a nonzero module of finite length of finite projective dimension. If such a module exists, then A_P is Cohen–Macaulay (an easy consequence of Bass's conjecture or the (new) intersection conjecture, solved by Paul Roberts [Rob]). So $\operatorname{ht} P = \operatorname{depth} A_P = \operatorname{hd}_{A_P} M_P \leq \operatorname{hd}_A M$, as desired. □

The inequality $\operatorname{grade}_A M \leq \operatorname{hd}_A M$ (for $M \neq 0$) is proved without difficulty. We say that M is a *perfect module* if $\operatorname{hd}_A M = \operatorname{grade}_A M$ holds. If so, $\operatorname{codim}_A M = \operatorname{hd}_A M$. Usually, we call these values the *codimension* of M. This is finite, since $\operatorname{hd}_A 0 = -\infty$ and $\operatorname{grade}_A 0 = \infty$, and if $M \neq 0$, then $\operatorname{grade}_A M$ is always finite (see the proof of Lemma 6.54 above). If S is a multiplicatively closed subset of A, then $\operatorname{hd}_{S^{-1}A} S^{-1}M \leq \operatorname{hd}_A M$ and $\operatorname{grade}_A M \leq \operatorname{grade}_{S^{-1}A} S^{-1}M$. So a nonzero localization of a perfect module is perfect of the same codimension. If I is an ideal of A such that A/I is a perfect module, we say that I is a *perfect ideal* (of codimension $\operatorname{codim}_A A/I = \operatorname{ht} I$). If, moreover, $\operatorname{Ext}_A^h(A/I, A)$ is a projective A/I-module of rank one, then we say that I is a *Gorenstein ideal.*

Proposition 6.55 *Let A be a Cohen–Macaulay ring, and M a nonzero finitely generated A-module of finite projective dimension. If M is perfect, then M is Cohen–Macaulay. The converse is true if* $\operatorname{Supp}_A M$ *is connected.*

Proof First, we assume that $(A, \mathfrak{m})$ is local. Since A is Cohen–Macaulay, we have that $\operatorname{grade}_A M = \operatorname{codim}_A M$. We also have $\operatorname{codim}_A M + \dim_A M = \dim A$ and $\operatorname{hd}_A M + \operatorname{depth}_A M = \operatorname{depth} A$. Since $\dim A = \operatorname{depth} A$, we have that $\operatorname{grade}_A M = \operatorname{hd}_A M$ if and only if $\dim_A M = \operatorname{depth}_A M$. If M is perfect, then M_P is perfect and hence is Cohen–Macaulay by the local case for any $P \in \operatorname{Spec} A$. Conversely, if M is Cohen–Macaulay, then $h(P) = \operatorname{hd}_{A_P} M_P = \operatorname{grade}_{A_P} M_P < \infty$ for every $P \in \operatorname{Supp}_A M$. Then $h(P) = \operatorname{hd}_{A_P} M_P$ is upper-semicontinuous, and $h(P) = \operatorname{grade}_{A_P} M_P$ is lower-semicontinuous on $\operatorname{Spec} A$. Since $\operatorname{Supp}_A M$ is connected, then $h(P)$ has the constant value h on $\operatorname{Supp}_A M$, and M is perfect of codimension h. □

Proposition 6.56 *Let $(B, \mathfrak{n})$ be a Noetherian local ring, and I an ideal of B. Let $A = B/I$, and assume that $h = \operatorname{hd}_B A < \infty$. Let*

$$\mathbb{F} : 0 \to F_h \to \cdots \to F_1 \to F_0 \to A \to 0 \qquad (\#)$$

be the minimal free resolution as a B-module. Then the following are equivalent:

(1) *I is a Gorenstein ideal.*
(2) *I is perfect, and $\mathbb{F}^* = \operatorname{Hom}_B(\mathbb{F}, B) \cong \mathbb{F}[-h]$ as complexes of B-modules, where $\mathbb{C}[-h]^i = \mathbb{C}^{i-h}$ for a (cohomological) complex $\mathbb{C}$.*
(3) *I is perfect, and* $\operatorname{rank}_B F_h = 1$.

If, moreover, B is Gorenstein, then this is also equivalent to

(4) *A is Gorenstein.*

Proof (1) ⇒(2). I is perfect by the definition of Gorenstein ideals. Since $\mathrm{H}^i(\mathbb{F}^*) = \mathrm{Ext}^i_B(A, B)$ is 0 for $i \neq h$ and A for $i = h$ by the definition of Gorenstein ideals, we have that

$$\mathbb{F}^*[h] : 0 \to F_0^* \to \cdots \to F_h^* \to A \to 0$$

is a minimal free resolution of A. So (2) follows, by the uniqueness of the minimal free resolution (C.14).

(2) ⇒(3) is easy. (3) ⇒(1). Since I is perfect, $\mathbb{F}^*[h]$ is a minimal free resolution of $\mathrm{Ext}^h_B(A, B)$. Since $\mathrm{rank}_B F_h^* = 1$, $\mathrm{Ext}^h_B(A, B)$ is a cyclic module. Since $Ext^h_B(\mathrm{Ext}^h_B(A, B), B) = \mathrm{H}^h((\mathbb{F}^*[h])^*) = \mathrm{H}^h(\mathbb{F}^{**}[-h]) = \mathrm{H}^0(\mathbb{F}) = A$, we have that $\mathrm{Ext}^h_B(A, B)$ is a faithful module, and hence $\mathrm{Ext}^h_B(A, B) \cong A$, as desired.

Now assume that B is Gorenstein. Then I is Gorenstein if and only if A is Cohen–Macaulay and $K_A = \mathrm{Ext}^h_B(A, B)$ is isomorphic to A, by Proposition 6.55. This is equivalent to saying that A is Gorenstein. □

Let B be a Noetherian local ring. We want to know Gorenstein ideals I of low codimensions and their minimal free resolutions. If $\mathrm{ht}(I) \leq 3$, then we can determine the structure of I.

To begin with, if $\mathrm{ht}(I) = 1$, then $\mathrm{hd}_A I = \mathrm{hd}_A A/I - 1 = 1 - 1 = 0$, and I is a projective (or equivalently, free) B-module, and is a principal ideal.

Next, let us consider the case $\mathrm{ht}(I) = 2$.

Theorem 6.57 (Serre) *Let B be a Noetherian local ring, and I be a Gorenstein ideal of B of codimension two. Then I is generated by a B-regular sequence of length* 2.

Proof Consider the minimal free resolution $F_\bullet$ of $A = B/I$ over B. Since $\mathrm{hd}_B A = 2$, $F_\bullet$ looks like

$$0 \to F_2 \to F_1 \to B \to A \to 0.$$

Since I is Gorenstein, $\mathrm{rank}_B F_2 = \mu_B(\mathrm{Ext}^2_B(A, B)) = \mu_B(A) = 1$. Hence $F_1 \cong B^2$ and I is generated by two elements, say a, b. Then the Koszul complex $K_\bullet(a, b)$ is also a minimal free resolution of A by the depth sensitivity (Theorem 5.86). Since B is local, a, b is a regular sequence of length 2 by Theorem 5.82. □

Next, let us assume $\mathrm{ht}\, I = 3$. In this case, a skew-symmetric matrix plays an essential role. Let $M = [a_{ij}]$ be a skew-symmetric $n \times n$ matrix over a commutative ring A. That is, $a_{ji} = -a_{ij}$ for $1 \leq i < j \leq n$, and $a_{ii} = 0$ for $i = 1, \ldots, n$. If n is odd, then $\det M = 0$. This is easy for the ring $\tilde{A} = \mathbb{Z}[x_{ij}]$ and $\tilde{M} = (x_{ij})$, where x_{ij} is the indeterminates for $1 \leq i < j \leq n$, and $x_{ji} = -x_{ij}$ and $x_{ii} = 0$. Indeed, we have

$$\det \tilde{M} = \det {}^t\tilde{M} = \det(-\tilde{M}) = (-1)^n \det \tilde{M} = -\det \tilde{M}.$$

Since $2\det\tilde{M} = 0$ and 2 is a non-zerodivisor in $\tilde{A}$, we have $\det\tilde{M} = 0$. For the general case, we take a unique ring homomorphism $\psi : \tilde{A} \to A$ such that $\psi(x_{ij}) = a_{ij}$, and then $\det M = \det[\psi(x_{ij})] = \psi(\det\tilde{M}) = 0$. If $n = 2m$ is even, we define the *Pfaffian* of M, denoted by $\operatorname{pf} M$, by

$$\operatorname{pf} M = \sum_{\sigma\in\Gamma_{2m}} (-1)^{\sigma} a_{\sigma 1\,\sigma 2} a_{\sigma 3\,\sigma 4} \cdots a_{\sigma(2m-1)\,\sigma(2m)},$$

where $\Gamma_{2m} = \{\sigma \in \mathfrak{S}_{2m} \mid \sigma(2i-1) < \sigma(2i)\ (i = 1, \dots, m),\quad \sigma 1 < \sigma 3 < \sigma 5 < \cdots < \sigma(2m-1)\}$.

Lemma 6.58 *Let $n = 2m$ be an even positive integer, $M = [a_{ij}]$ be an $n \times n$ skew symmetric matrix over A, and $Q \in M_n(A)$. Then the following hold:*

(1) $\operatorname{pf}(QM^t Q) = (\det Q)(\operatorname{pf} M)$.
(2) $\det M = (\operatorname{pf} M)^2$.
(3) *(Expansion formula) For each $1 \le i \le n$, we have*

$$\sum_{j=1}^{n} (-1)^j a_{ij} \operatorname{pf} M_{ij} = (-1)^{i-1} \operatorname{pf} M,$$

where M_{ij} is the $(2m-2) \times (2m-2)$ skew symmetric matrix obtained by deleting the ith and jth rows and columns from M.
(4) *If $i \ne j$, and the ith and the jth rows of M agree (hence the ith and the jth columns also agree), then* $\operatorname{pf} M = 0$.

The proof is left to the reader.

For an *odd integer* $n \ge 3$, let $M = [a_{ij}]$ be a skew-symmetric $n \times n$ matrix with coefficients in a Noetherian ring A. For every $1 \le j \le n$, we denote by M_j the skew-symmetric $(n-1) \times (n-1)$ matrix obtained by deleting ith row and ith column of M, and set $p_j = (-1)^j \operatorname{pf}(M_j)$.

Let $\operatorname{Pf}_{n-1}(M)$ be the ideal of A generated by $p_1, p_2, \dots, p_n$. Then by Lemma 6.58, we have

$$\sum_{j=1}^{n} a_{ij} p_j = 0.$$

(Consider the $(n+1) \times (n+1)$ alternating matrix $N = [b_{lj}]$ defined by $b_{lj} = a_{lj}$ for $1 \le l, j \le n$, $b_{n+1,n+1} = 0$, $b_{l,n+1} = a_{l,i} = -b_{n+1,l}$ for $1 \le l \le n$, and expand $\operatorname{pf}(N) = 0$.) Hence if we put $P = {}^t(p_1, p_2, \dots, p_n)$, we have a complex of free A-modules

$$0 \to A \xrightarrow{P} A^n \xrightarrow{M} A^n \xrightarrow{{}^tP} A.$$

Theorem 6.59 *Let B be a Noetherian ring, $n \geq 3$ be an odd integer, and $M = [a_{ij}]$ be an $n \times n$ skew symmetric matrix. Let $\mathfrak{a} = \mathrm{Pf}_{n-1}(M)$, and assume that* depth $(\mathfrak{a}, B) \geq 3$. *Then the following statements hold true:*

(1)

$$0 \to B \xrightarrow{P} B^n \xrightarrow{M} B^n \xrightarrow{{}^tP} B \to B/\mathfrak{a} \to 0 \qquad (\heartsuit)$$

is a free resolution of $B/\mathfrak{a}$.

(2) *If $B \neq \mathfrak{a}$, then $\mathfrak{a}$ is a Gorenstein ideal of codimension three.*

Proof Since $\sqrt{\mathrm{I}_{n-1}(M)} \supseteq \mathfrak{a} = \mathrm{Pf}_{n-1}(M)$, (1) follows from Theorem 5.115.

We prove (2). By (1), $3 \leq \operatorname{depth}(\mathfrak{a}, B) = \operatorname{grade}_B(B/\mathfrak{a}) \leq \operatorname{hd}_B(B/\mathfrak{a}) \leq 3$. So $\mathfrak{a}$ is perfect of codimension 3. Since the resolution $\mathbb{F}$ in (1) is self-dual, so $\operatorname{Ext}_B^3(B/\mathfrak{a}, B) \cong \mathrm{H}^3(\mathbb{F}^*) = \mathrm{H}^0(\mathbb{F}) \cong B/\mathfrak{a}$, which is rank one projective as a $B/\mathfrak{a}$-module. Hence $\mathfrak{a}$ is Gorenstein. □

Example 6.60 Let k be a field and $n \geq 3$ be an *odd* integer. Let

$$B = k[X_{ij} \mid 1 \leq i < j \leq n]$$

the polynomial ring over k and let $M = [a_{ij}]$ be a skew-symmetric matrix defined by $a_{ij} = X_{ij}$ $(i < j)$, then we have depth $(\mathrm{Pf}_{n-1}(M), B) = 3$ and the ring $B/\mathrm{Pf}_{n-1}(M)$ is a UFD and is Gorenstein.

The structure theorem for height 3 Gorenstein ideals, saying that such ideals are only the ones given in Theorem 6.59, was proved by D. A. Buchsbaum and D. Eisenbud. Part (1) of Theorem 6.61 below was independently proved by J. Watanabe in [WJ].

Theorem 6.61 (Buchsbaum–Eisenbud [BE2]) *Let $(A, \mathfrak{m})$ be a Noetherian local ring, and $\mathfrak{a} \subsetneq A$ be a Gorenstein ideal of A of codimension three. Then the following statements hold true:*

(1) *The minimal number of generators of $\mathfrak{a}$ is an* odd *integer $n \geq 3$.*
(2) *There exists an odd integer $n \geq 3$ and $n \times n$ skew-symmetric matrix M with coefficients in $\mathfrak{m}$ and $\mathfrak{a} = \mathrm{Pf}_{n-1}(M)$.*

For the proof, we refer the reader to [BE2] or [BH]. We give an example.

Example 6.62 Let k be a field and put $A = k[t^5, t^6, t^7, t^8]$. Then we can assert that $A \cong k[X, Y, Z, W]/\mathfrak{p}$, $\mathfrak{p} = (Y^2 - XZ, YZ - XW, Z^2 - YW, ZW - X^3, W^2 - X^2Y)$.[7]

[7] This example is one of the examples of numerical semigroup rings which we treat in Chap. 7, Sect. 7.5. Since the semigroup $S = \langle 5, 6, 7, 8 \rangle$ is symmetric, $A = k[S]$ is a Gorenstein ring. If S is generated by 4 elements and is symmetric, then it was proved by H. Bresinsky that the defining ideal of $k[S]$ is minimally generated by 3 or 5 elements in [Bre]. A simpler proof using the structure Theorem 6.61 can be found in [Wa7].

We align the relations between these 5 generators of $\mathfrak{p}$ in a suitable order and get a skew-symmetric matrix M:

$$M = \begin{pmatrix} 0 & X^2 & 0 & -W & Z \\ -X^2 & 0 & -W & Z & -Y \\ 0 & W & 0 & -Y & X \\ W & -Z & Y & 0 & 0 \\ -Z & Y & -X & 0 & 0 \end{pmatrix}.$$

Let M_1 be the matrix obtained by deleting the 1st row and the 1st column of M, the determinant of M_1 is $(Y^2 - XZ)^2$, corresponding the 1st generator of $\mathfrak{p}$. The other generators are obtained in the same manner.

Chapter 7
Graded Rings

Sometimes it is very useful to put "gradings" on rings and modules. Also, like the concept of "Hilbert series" or "Poincaré series," grading gives us a new powerful tool.

Also, in many cases, if we want to investigate graded rings or modules, it is only necessary to treat "graded" ideals and submodules, so we can restrict our considerations to very limited objects (see Appendix F).

There are several theorems which are proved only for graded rings and modules. In this chapter, we introduce the concept of grading and its power.

7.1 Graded Rings and Graded Modules

Definition 7.1

(1) We say that $R = \bigoplus_{n \in \mathbb{Z}} R_n$ is a *graded ring* if every element x is written

$$x = \sum x_n, \ x_n \in R_n$$

in a unique manner, where each R_n is an additive subgroup of R satisfying $R_n R_m \subseteq R_{m+n}$. We write $R = \bigoplus_{n \geq 0} R_n$ if $R_n = 0$ for $n < 0$.

By the definition, R_0 is a subring of R and every R_n is an R_0 submodule of R.

We say that $a \in R$ is a *homogeneous element* of degree n if $a \in R_n$.

(2) If $M = \bigoplus_{n \in \mathbb{Z}} M_n$ is an R-module satisfying $R_n M_m \subseteq M_{m+n}$ for every $m, n \in \mathbb{Z}$, we say that M is a *graded R-module*. We say an element $x \in M_n$ is a *homogeneous element* of degree n. In this case we say

$$\deg x = n.$$

S. Goto, K. Watanabe, *Commutative Algebra*, Graduate Texts in Mathematics 309,
https://doi.org/10.1007/978-981-95-4227-7_7

Also, every M_n is an R_0-submodule of M.

(3) If an ideal I of R is generated by homogeneous elements, we say that I is a *graded ideal* or a *homogeneous ideal* of R. Likewise, if an R-submodule N of a graded R-module M is generated by homogeneous elements, we say N is a *graded R-submodule of M*.

A homogeneous ideal of R is a graded R-submodule of the graded R-module R.

Remark 7.1 An R-submodule N of a graded R-module M is a graded R-submodule of M if and only if it satisfies the condition

$$\text{If} \quad x = \sum x_n \in N \ (x_n \in M_n) \text{ then } \forall n, x_n \in N. \tag{7.1}$$

In fact, if N is generated by homogeneous elements $\{x_i\}$, then every homogeneous component of $\sum a_i x_i$ $(a_i \in R)$ is in N.

Example 7.2 Let $S := k[X, Y, Z]$ be a polynomial ring of 3 variables over a field k. The natural way to consider S as a graded ring is to define $X, Y, Z \in S_1$. But by this grading, $f = X^2 + Y^3 + Z^5$ is not a homogeneous element. But if we define $X \in S_{15}, Y \in S_{10}, Z \in S_6$, then $f \in S_{30}$ is homogeneous and we can consider $k[X, Y, Z]/(f)$ a graded ring. In this manner, we can use the theory of graded rings to define a suitable grading for the objects we want to study.

Example 7.3 Let $R = \bigoplus_{n \geq 0} R_n$ be a graded ring and $M = \bigoplus_{n \in \mathbb{Z}} M_n$ be a graded R-module. Then

$$M_{\geq n} := \bigoplus_{i \geq n} M_i, \quad M_{>n} = \bigoplus_{i>n} M_i$$

are graded R submodules of M. In particular, we write

$$R_+ = R_{>0} = \bigoplus_{i>0} R_i.$$

If R_0 is a field, then R_+ is the unique maximal homogeneous ideal of R and we write $R_+ = \mathfrak{m}$.

We will show, for many important properties **P**, that R satisfies the property **P** if and only if so does the local ring $R_{\mathfrak{m}}$, see Appendix F.

If we define $M_{\leq n} = \bigoplus_{i \leq n} M_i$, $M_{\leq n}$ is not an R-submodule of M but we can think that

$$M_{\leq n} = M/M_{>n}$$

and a factor module of M.

If $R = \bigoplus_{n \in \mathbb{Z}} R_n$, $R_{\geq 0}$, $R_{\leq 0}$ are subrings of R.

Let us discuss the condition for a graded ring R to be Noetherian.

Proposition 7.4

(1) *For a graded ring $R = \bigoplus_{n \geq 0} R_n$, the following conditions are equivalent:*

(a) *R is Noetherian.*
(b) *R_0 is Noetherian and R is a finitely generated R_0-algebra.*
(c) *R_0 is Noetherian and R_+ is a finitely generated ideal of R.*

(2) *$R = \bigoplus_{n \in \mathbb{Z}} R_n$ is Noetherian if and only if both $R_{\geq 0}$ and $R_{\leq 0}$ are Noetherian.*

Before proving this proposition, let us introduce the notion of *pure subrings*.

Definition 7.5 Assume that R is a subring of S. Then we say R is a *pure subring* of S if for any R-module M, the map $M \cong M \otimes_R R \to M \otimes_R S, x \to x \otimes 1$ is injective.

In particular, if we put $M = R/I$, the condition "$R/I \to R/I \otimes_R S \cong S/IS$ is injective" is equivalent to $IS \cap R = I$. If this condition holds for every ideal I of R, we say that the ring extension $R \subseteq S$ is *cyclically pure*. The notions pure and cyclically pure are equivalent conditions under hypotheses discussed in [Ho2].

If R is a pure subring of S, then for every multiplicatively closed set $T \subseteq R$, the ring extension $T^{-1}R \subseteq T^{-1}S$ is also pure.

Graded rings have several natural pure subrings.

Example 7.6 Let R be a pure subring of S and let $i : R \to S$ be the natural injection.

(1) If S has a decomposition $S = R \oplus M$ as R-modules, or, in other words, there is a homomorphism of R-modules $\phi : S \to R$ such that $\phi \circ i = 1_R$, then R is a pure subring of S. In particular, if $R = \bigoplus_{n \geq 0} R_n$ is a graded ring, then R_0 is a pure subring of R.
(2) For a $\mathbb{Z}$-graded ring $R = \bigoplus_{n \in \mathbb{Z}} R_n$, the subrings $R_{\geq 0}$, $R_{\leq 0}$ are isomorphic to a pure subring of $R[T]$ with new variable T. In fact, if we define $T \in (R[T])_{-1}$ in the graded ring $R[T]$, then $R[T]_0 = \bigoplus_{n \geq 0} R_n T^n \cong R_{\geq 0}$ and if $T \in (R[T])_1$, $R[T]_0 = \bigoplus_{n \leq 0} R_n T^n \cong R_{\leq 0}$. Hence by (1) we have our results.

The following properties of pure subrings are very useful.

Lemma 7.7 *Let R be a pure subring of S. Then the following properties hold:*

(1) *If S is Noetherian, then so is R.*
(2) *If S is an integrally closed domain, so is R.*
(3) *If S is Noetherian and Cohen–Macaulay, and $S_{\mathfrak{m}}/\mathfrak{m}S_{\mathfrak{m}}$ is Artinian for any maximal ideal $\mathfrak{m}$ of R, then R is Cohen–Macaulay.*

Proof

(1) Take the ascending chain $I_1 \subseteq I_2 \subseteq \dots$ of ideals of R. Since S is Noetherian, there is n such that $I_n S = I_{n+1} S = \dots$. Since $I_n = I_n S \cap R$, we get $I_n = I_{n+1} = \dots$.

(2) Let K (resp. L) be the field of fractions of R (resp. S). Assume that $a/b \in K$ $(a, b \in R)$ is integral over R. Then a/b is integral over S and since S is integrally closed, $a/b \in S$, or $a \in bS$. Hence $a \in bS \cap R = bR$ and $a/b \in R$.
(3) See Appendix G, Theorem G.5.

□

Remark 7.2 Statement (3) of Lemma 7.7 does not hold when S is not a finitely generated R-module. Such examples are given in Sect. 7.3 in the case of "Segre products." Also, with the same assumption, "S is Gorenstein" does not imply "R is Gorenstein."

Let us come back to the proof of Proposition 7.4.

Proof of Proposition 7.4 (a) ⇒(c) is clear from Lemma 7.7.

To prove (c) ⇒(a), let $(f_1, \ldots, f_s)$ be a homogeneous generating set of R_+. Then it is easy to show that every homogeneous element $g \in R_+$ is written as $g = \sum a_i f_i$. Hence we can show $R_0[f_1, \ldots, f_s] \supseteq R_n$ by induction on n.

(b) ⇒(a) is the "Basis Theorem" (Chap. 1, Theorem 1.38).

(2) If R is Noetherian, then so are $R_{\geq 0}$, $R_{\leq 0}$ by Lemma 7.7. Conversely, if $R_{\geq 0}, R_{\leq 0}$ are Noetherian, then so is R_0 by (1) and we have shown in (1) that $R_{\geq 0} = R_0[f_1, \ldots, f_s]$, $R_{\leq 0} = R_0[g_1, \ldots g_t]$. Hence we can write

$$R = R_0[f_1, \ldots, f_s, g_1, \ldots g_t].$$

□

The following theorem is very important in the theory of graded rings.

Theorem 7.8 *Let $R = \bigoplus_{n\in\mathbb{Z}} R_n$ be a graded ring and $M = \bigoplus_{n\in\mathbb{Z}} M_n$ be a graded R-module. Then every $\mathfrak{p} \in \mathrm{Ass}_R(M)$ is a homogeneous ideal.*

Proof If $\mathfrak{p} \in \mathrm{Ass}_R(M)$, then by Definition 2.14, $\mathfrak{p} = \mathrm{Ann}_R(x)$ for some $x \in M, x \neq 0$. If x is a homogeneous element, then $\mathfrak{p} = \mathrm{Ann}_R(x)$ is a homogeneous ideal. Otherwise, take any element $a = \sum a_n \in \mathfrak{p}$.

Since our conclusion is clear if x or a is homogeneous, we show by induction on the number of nonzero homogeneous components.

Put a' (resp. x') be the homogeneous component of a (resp. x) of lowest degree and write $a = a' + a'', x = x' + x''$. Then since $a'x'$ is the unique homogeneous element of lowest degree of $ax = 0$, we must have $a'x' = 0$. If $a'x = 0$, then $a' \in \mathfrak{p}$ by definition and also $a'' \in \mathfrak{p}$. Hence by our induction hypothesis, every homogeneous component of a'' is in $\mathfrak{p}$, too.

If $a'x \neq 0$, on the contrary, we have $a' \notin \mathfrak{p}$. Then assume $ba'x = 0$ for some $b \in R$. Then $ba' \in \mathfrak{p}$ but since $a' \notin \mathfrak{p}$, we must have $b \in \mathfrak{p}$. In particular, we must have $a(a'x) = 0$. But the number of nonzero homogeneous components of $a'x$ is strictly less than that of x and by induction hypothesis, we have $a \in \mathfrak{p}$. □

Corollary 7.9 *Let $R = \bigoplus_{n\in\mathbb{Z}} R_n$ be a graded ring. Then R is an integral domain if and only if any product of nonzero homogeneous elements is nonzero.*

Proof If $\mathrm{Ass}_R(R)$ contains some $\mathfrak{p} \neq 0$, then there is a nonzero homogeneous zerodivisor by Theorem 7.8. □

Let us see homogeneous localizations of graded rings and graded modules.

Definition 7.10 Let R be a graded ring, $S \subseteq R$ be a multiplicatively closed set and M be a graded R-module.

(1) If every element of S is homogeneous, we can think of $S^{-1}M$ as a graded $S^{-1}R$-module putting $a/s \in (S^{-1}M)_{n-t}$ for $a \in M_n, s \in S \cap R_t$.
(2) In particular, if $\mathfrak{p}$ is a homogeneous prime ideal of R, then let $M_{(\mathfrak{p})}$ be the localization of M by the homogeneous elements of $R \setminus \mathfrak{p}$. We call $M_{(\mathfrak{p})}$ the *homogeneous localization* of M at $\mathfrak{p}$.
(3) We call R an *H-local ring* if R has unique homogeneous maximal ideal.[1] Also we say "Let $(R, \mathfrak{m})$ be an H-local ring" if $\mathfrak{m}$ is the unique graded maximal ideal of R. In particular, $(R_{(\mathfrak{p})}, \mathfrak{p}R_{(\mathfrak{p})})$ is an H-local ring.

We will introduce the "category of graded R-modules" $\mathfrak{M}_H(R)$.

Definition 7.11 Let R be a graded ring.

(1) We will write $M \in \mathfrak{M}_H(R)$ if M is a graded R-module.
(2) For $M, N \in \mathfrak{M}_H(R)$, $f : M \to N$ is a morphism in $\mathfrak{M}_H(R)$ if for every $n \in \mathbb{Z}$, $f(M_n) \subseteq f(N_n)$. We also write $f \in \mathrm{Hom}_R(M, N)$.
(3) For $M \in \mathfrak{M}_H(R)$, we define $M(n) \in \mathfrak{M}_H(R)$ by $M(n)_m = M_{n+m}$ (M and $M(n)$ are the same as R-modules). In particular, $R(n)$ is a free R-module whose generator is of degree $-n$.

Theorem 7.12 *The following conditions on a graded ring* $R = \bigoplus_{n\in\mathbb{Z}} R_n \neq R_0$ *are equivalent and we say that* R *is an H-simple ring if* R *satisfies these conditions.*

(1) (0) *and* R *are the only graded ideals of* R.
(2) *Any nonzero homogeneous element of* R *is a unit in* R.
(3) *Any graded* R*-module is free.*
(4) R_0 *is a field and either* $R = R_0$ *or for some* $T \in R_n$ *with* $n \geq 1$, $R = R_0[T, T^{-1}]$. *In particular, if* R *is an H-simple graded ring which is not a field, then* R *is a PID and* $\dim R = 1$.

Proof We prove the equivalence of the conditions. If we assume (1), then (2) is clear. Assume (2) and assume R is not a field. Let n be minimal positive such that $R_n \neq 0$. If $0 \neq T \in R_n$, then by our assumption, $R_n = TR_0$ and if $R_m \neq 0$, then $m \in n\mathbb{Z}$. Hence we have $R = R_0[T, T^{-1}]$ and the converse will be clear.

Assume R_0 is a field and $R = R_0[T, T^{-1}]$ for some $T \neq 0 \in R_n$. If $M \in \mathfrak{M}_H(R)$, then any free basis of $\bigoplus_{i=0}^{n-1} M_i$ over R_0 is a free basis of M over R. □

Example 7.13 If $(R, \mathfrak{m})$ is an H-local graded ring, then $R/\mathfrak{m}$ is H-simple. In particular, if $\mathfrak{p}$ is a homogeneous prime ideal of R, then $R_{(\mathfrak{p})}/\mathfrak{p}R_{(\mathfrak{p})}$ is H-simple

[1] H stands for "homogeneous."

We will show that the properties of a graded ring R are determined by looking at homogeneous prime ideals.

Theorem 7.14 *Let $R = \bigoplus_{n\in\mathbb{Z}} R_n$ be a graded ring and assume $\mathfrak{p}$ is a prime ideal of R which is not homogeneous.*

In this case, let $\mathfrak{p}^*$ be the homogeneous ideal of R generated by all the homogeneous elements of $\mathfrak{p}$. Then the following statements hold:

(1) *$\mathfrak{p}^*$ is a prime ideal and there are no prime ideals between $\mathfrak{p}^*$ and $\mathfrak{p}$.*
(2) *If $M \in \mathfrak{M}_H(R)$ and $M_\mathfrak{p} \neq 0$, then $M_{\mathfrak{p}^*} \neq 0$ and we have*

$$\dim M_\mathfrak{p} = M_{\mathfrak{p}^*} + 1.$$

(3) *Under the assumption above, for every $n \in \mathbb{Z}$, we have*

$$\mu_n(\mathfrak{p}, M) = \mu_{n-1}(\mathfrak{p}^*, M).$$

Proof We define $\iota : R \to R[t, t^{-1}]$ to be the inclusion map, where t is a new variable. We also define $\omega, \omega^- : R \to R[t, t^{-1}]$ by $\omega(a) = \sum_i a_i \otimes t^i$ and $\omega^-(a) = \sum_i a_i \otimes t^{-i}$ for $a = \sum_i a_i$, where $a_i \in R_i$ for $i \in \mathbb{Z}$. We also define $\Box$ and $\Box^-$ from $R[t, t^{-1}]$ to itself by $\Box(\sum_i a_i t^i) = \sum_i \omega(a_i) t^i$ and $\Box^-(\sum_i a_i t^i) = \sum_i \omega^-(a_i) t^i$, where $a_i \in R$. It is easy to see that these are all ring homomorphisms, $\Box$ and $\Box^-$ are inverses of each other, and $\omega = \Box\iota$. Since $\Box$ is an isomorphism and ι is faithfully flat, ω is also faithfully flat.

We prove (1). As $\omega^{-1}(\mathfrak{p}R[t, t^{-1}]) = \mathfrak{p}^*$ and $\mathfrak{p}R[t, t^{-1}]$ is a prime ideal, $\mathfrak{p}^*$ is a prime ideal. To prove that there is no prime ideal between $\mathfrak{p}^*$ and $\mathfrak{p}$, replacing R by $R_{(\mathfrak{p}^*)}/\mathfrak{p}^* R_{(\mathfrak{p}^*)}$, we may assume that R is H-simple. So $\dim R \leq 1$, and $\mathfrak{p} \neq \mathfrak{p}^*$. The assertion is clear.

We prove (2). If $\mathfrak{p}^* = 0$, then the assertion is true by (1). We consider the general case. Replacing R by $R/\mathrm{Ann}_R M$, we may assume that M is faithful. By Theorem 5.70, applied to the map of local rings $A = R_\mathfrak{p} \to A' = R_\mathfrak{p}(t) = R[t, t^{-1}]_{\mathfrak{p}R[t,t^{-1}]}$ induced by ι, we have that $\dim M_\mathfrak{p} = \dim A = \dim A' = \dim(M_\mathfrak{p} \otimes_A A')$, since $A'/\mathfrak{m}A' = \dim(A/\mathfrak{m})(t)$ is a field, and hence is zero-dimensional, where $\mathfrak{m} = \mathfrak{p}R_\mathfrak{p} = \mathfrak{p}A$. Moreover, by the same theorem applied to the map of local rings $A = R_{\mathfrak{p}^*} \to A' = R_\mathfrak{p}(t)$ induced by ω, $\dim M_{\mathfrak{p}^*} = \dim R_{\mathfrak{p}^*} = \dim R_\mathfrak{p}(t) - \dim(R_\mathfrak{p}/\mathfrak{p}^* R_\mathfrak{p})(t) = \dim R_\mathfrak{p} - 1 = \dim M_\mathfrak{p} - 1$, by the case that $\mathfrak{p}^* = 0$.

(3) Recall that $\mu_n(\mathfrak{p}, M) = \dim_{\kappa(\mathfrak{p})} \mathrm{Ext}_{R_\mathfrak{p}}(\kappa(\mathfrak{p}), M_\mathfrak{p})$. To compute this number, we can do homogeneous localization by $\mathfrak{p}^*$. Hence it suffices to consider the case that $(R, \mathfrak{p}^*)$ is an H-local ring. Then since $R/\mathfrak{p}^*$ is a PID, we have $\mathfrak{p} = (\mathfrak{p}^*, f)$ for some element $f \in \mathfrak{p}$. Then taking $\mathrm{Ext}^*_R(*, M)$ to the short exact sequence

$$0 \to R/\mathfrak{p}^* \xrightarrow{f} R/\mathfrak{p}^* \to R/\mathfrak{p} \to 0,$$

we get a long exact sequence

$$\cdots \to \operatorname{Ext}_R^{n-1}(R/\mathfrak{p}^*, M) \xrightarrow{f} \operatorname{Ext}_R^{n-1}(R/\mathfrak{p}^*, M)$$
$$\to \operatorname{Ext}_R^n(R/\mathfrak{p}, M) \to \operatorname{Ext}_R^n(R/\mathfrak{p}^*, M) \xrightarrow{f} \operatorname{Ext}_R^n(R/\mathfrak{p}^*, M) \to \cdots .$$

Since $\operatorname{Ext}_R^i(R/\mathfrak{p}^*, M)$ is a graded $R/\mathfrak{p}^*$-module, is a free $R/\mathfrak{p}^*$-module. Hence we get $\operatorname{Ext}_R^n(R/\mathfrak{p}, M) \cong \operatorname{Ext}_R^{n-1}(R/\mathfrak{p}^*, M)/f\operatorname{Ext}_R^{n-1}(R/\mathfrak{p}^*, M)$, which gives our desired result. □

Corollary 7.15 *Let $R = \bigoplus_{n\in\mathbb{Z}} R_n$ be a graded ring and $\mathfrak{p}$ be a prime ideal, which is not homogeneous. Then putting $\mathfrak{p}^*$ as the ideal generated by all the homogeneous elements of $\mathfrak{p}$, the following statements hold:*

(1) *$R_{\mathfrak{p}}$ is a regular local ring if and only if so is $R_{\mathfrak{p}^*}$.*
(2) *$R_{\mathfrak{p}}$ is Cohen–Macaulay if and only if so is $R_{\mathfrak{p}^*}$.*
(3) *$R_{\mathfrak{p}}$ is Gorenstein if and only if so is $R_{\mathfrak{p}^*}$.*

7.2 H-Graded Rings

Let $S = k[X_1, \ldots, X_n]$ be a polynomial ring over k. Then the subrings of S generated by monomials of $X_1, \ldots, X_n$ are called *semigroup rings*. Also, the factor ring S/I, where the ideal I is generated by square-free monomials is called a *Stanley–Reisner ring*. To study such rings, a grading by $\mathbb{Z}^n$ is very useful. In this section, we study the general theory of such graded rings.

The proofs of the theorems in this section are essentially the same as the ones in the previous section, so readers only need to pay attention to the difference between $\mathbb{Z}$ and H.

Let H be an Abelian group. In this chapter, we always assume that H is finitely generated and torsion free. Namely, $H \cong \mathbb{Z}^n$ for some n.

Definition 7.16

(1) We call a ring R an *H-graded ring* if $R = \bigoplus_{h\in H} R_h$ is a direct sum of $\{R_h \mid h \in H\}$, where R_h is an R_0-module for every $h \in H$ and $R_h \cdot R_{h'} \subseteq R_{h+h'}$ as for the product of the ring. An element $x \in R_h$ is called a *homogeneous element* of degree h.
(2) Let M be an R-module. If $M = \bigoplus_{h\in H} M_h$ is a direct sum of R_0-modules and if $R_h M_{h'} \subseteq M_{h+h'}$ for every $h, h' \in H$, then we say that M is an H-graded R-module. Likewise, an R-submodule M' of M is called an H-graded submodule of M if M' is generated by homogeneous elements of M. This condition is equivalent to saying that if $x = \sum_{h\in H}$ with $x_h \in M_h$, $x \in M'$ if and only if $x_h \in M'$ for every $x_h \in M'$.

An ideal I of R is called an *H-graded ideal*, *H-ideal*, or a *homogeneous ideal* of R if I is an H-graded submodule of R.

(3) If $I \subseteq R$ is an H-ideal, then the residue class ring R/I is also an H-graded ring by defining

$$(R/I)_h = R_h/I_h.$$

In the following, let R be a Noetherian H-graded ring.
We denote by

$$\mathfrak{M}_H(R)$$

the category of H-graded R-modules. A morphism

$$f : M \to N$$

in $\mathfrak{M}_H(R)$ is a homomorphism $f : M \to N$ of R-modules satisfying $f(M_h) \subseteq N_h$ for every $h \in H$.

Theorem 7.17 *If R is an H-graded ring and if M is an H-graded R-module, then every $\mathfrak{p} \in \mathrm{Ass}_R(M)$ is an H-ideal.*

Proof Let $x = \sum x_h \in M$ and assume that $\mathfrak{p} \in \mathrm{Ass}_R(M)$ and $\mathfrak{p} = \mathrm{Ann}_R(x)$. We will show that if $a = \sum a_h \in \mathfrak{p}$, then every $a_h \in \mathfrak{p}$. Since our assumption is clear if a or x is homogeneous, we prove by induction on pair (m, n), where m (resp. n) is the number of nonzero homogeneous elements of a (resp. x).

In the homogeneous components of $0 = ax = \sum\sum a_g x_h$, we can choose g, h such that $g + h$ is different from any other $g' + h'$ appearing in this sum (in usual $\mathbb{Z}$ grading, if g and h are minimal, then $g + h$ is uniquely minimal among $\{g' + h'\}$).

Then since $ax = 0$, we must have $a_g x_h = 0$. Now, we have $a(a_g x) = 0$. If $a_g x = 0$, then $a_g \in \mathfrak{p}$ and $a - a_g \in \mathfrak{p}$. Since $a - a_g$ has a strictly smaller number of nonzero homogeneous elements, we have $a_{g'} \in \mathfrak{p}$ for every $a_{g'}$ by our induction hypothesis. □

Corollary 7.18 *If R is an H-graded ring, R is an integral domain if and only if every product of nonzero homogeneous elements is nonzero.*

Remark 7.3 We assumed that our H is torsion free. This assumption is necessary for Theorem 7.17 and Corollary 7.18. In fact, assume that $H = \mathbb{Z}/(2) = \{0, 1\}$, $R = \mathbb{Z}[\sqrt{2}]/(7) = \mathbb{Z}/(7) \bigoplus (\mathbb{Z}/(7))\sqrt{2}$ and regard R as an H-graded ring, putting $R_0 = \mathbb{Z}/(7)$, $R_1 = (\mathbb{Z}/(7))\sqrt{2}$. Then R does not have a homogeneous zerodivisor. But we have $(5 + 3\sqrt{2})(5 - 3\sqrt{2}) = 0$ in R and R is *not* an integral domain. We do not go into this here but we can get some results defining an analogous concept of prime ideals for that grading [Ka].

What are the "simplest" H-graded rings? We can give a complete classification, which is useful in our further arguments.

Theorem 7.19 *Let H be as in Definition 7.16. Then the following conditions on an H-graded ring R are equivalent. We call R an H-simple ring if R satisfies these conditions.*

(1) *Every nonzero homogeneous element of R is a unit.*
(2) *R and (0) are the only H-graded ideals of R.*
(3) *Any H-graded R-module is free.*

Proof (1) $\Leftrightarrow$(2). Let $I \neq 0$ be an H-graded ideal of R. Then I contains a nonzero homogeneous element, which is a unit by (1). Hence $I = R$. Conversely, let x be a nonzero homogeneous element. Then xR is an H ideal which is not (0). Hence $xR = R$ and x is a unit.

Note that R_0 is a field if R satisfies these conditions.

(2) $\Rightarrow$(3). Assume (2) and let M be an H-graded R-module. Define $L := \{h \in H | R_h \neq 0\}$. Also, put $L_M := \{h \in H | M_h \neq 0\}$ and let $\{h_s | s \in L_M/L\}$ be the set of representatives of the set L_M/L.

Then for every nonzero homogeneous element $x \in M$, there exists a unit $u \in R$ such that $ux \in M_{h_s}$ for some h_s.

For any h_s, let B_{h_s} be a basis of M_{h_s} over $k = R_0$ and if we put $B = \bigcup_{s \in L_M/L} B_s$, then M is a free R-module with basis B.

Conversely, if $I \neq R$, (0) is an H-ideal, then R/I is a non-free H-graded R-module. Thus conditions (2) and (3) are equivalent. □

The proof of Theorem 7.19 shows that if R is an H-simple ring, then $L = \{h \in H | R_h \neq 0\}$ is an abelian group isomorphic to $\mathbb{Z}^r$ for some r and putting $k = R_0$,

$$R \cong k[L] \cong k[X_1, X_1^{-1}, \ldots, X_r, X_r^{-1}].$$

In particular, R is regular with $\dim R = r$.

Definition 7.20 An ideal $\mathfrak{m}$ of R is called an *H-maximal* if $R/\mathfrak{m}$ is H-simple. Since an H-simple ring is an integral domain, an H-maximal ideal is a prime ideal. R is called an *H-local H-graded ring* if R has a unique H-maximal ideal. We also say that $(R, \mathfrak{m})$ is an H-local R-graded ring if $\mathfrak{m}$ is the unique H-maximal ideal of R. R is H-simple if and only if $(R, 0)$ is H-local.

Localization is very important also in the category of H-graded rings. If a multiplicatively closed set S consists of H-homogeneous elements, then we can consider $S^{-1}R$ as an H-graded ring defining $x/s \in S^{-1}R_{h'-h}$ if $s \in R_h, x \in R_{h'}$. In the same manner, $S^{-1}M$ is an H-graded module if M is an H-graded R-module.

For an H-prime ideal $\mathfrak{p}$ of R, let $R_{(\mathfrak{p})}$ be the localization of R by homogeneous elements of $R \setminus \mathfrak{p}$. Then since $\mathfrak{p}$ is the unique maximal ideal not intersecting $R \setminus \mathfrak{p}$, $R_{(\mathfrak{p})}$ is an H-local ring with the H-maximal ideal $\mathfrak{p}R_{(\mathfrak{p})}$.

Let us discuss dimension theory of H-graded rings. Our goal is to know the height of ideals from the information of H-prime ideals.

Let $\mathfrak{p}$ be a prime ideal of R which is not homogeneous; we define $\mathfrak{p}^*$ to be the homogeneous ideal generated by the homogeneous elements of $\mathfrak{p}$. Then $\mathfrak{p}^*$ is also a prime ideal.[2] To see the height of $\mathfrak{p}/\mathfrak{p}^*$ in $R/\mathfrak{p}^*$, it suffices to see in $R_{(\mathfrak{p}*)}/\mathfrak{p}^* R_{(\mathfrak{p}*)}$. Since this latter ring is H-simple, the dimension is at most the rank of the group generated by $\{h \in H | (R/\mathfrak{p}^*)_h \neq 0\}$. In particular, if $H = \mathbb{Z}$, there is no prime ideal strictly between $\mathfrak{p}$ and $\mathfrak{p}^*$.

Theorem 7.21 *Let $M \in \mathfrak{M}_H(R)$ and $\mathfrak{p}$ be a prime ideal of R with $M_\mathfrak{p} \neq 0$. Then $M_{\mathfrak{p}^*} \neq 0$ and the following equality holds:*

$$\dim M_\mathfrak{p} = \dim M_{\mathfrak{p}^*} + \operatorname{ht} \mathfrak{p}/\mathfrak{p}^*. \tag{7.2}$$

Proof A minimal element of $\operatorname{Supp}_R(M)$ is in $\operatorname{Ass}_R(M)$ and by Theorem 7.17, an element of $\operatorname{Ass}_R(M)$ is an H-prime ideal. Since $\mathfrak{p}^*$ is the maximal H-prime contained in $\mathfrak{p}$, for any $\mathfrak{q} \in \operatorname{Ass}_R(M)$, $\mathfrak{q} \subseteq \mathfrak{p}^*$ and we see $\mathfrak{p}^* \in \operatorname{Supp}_R(M)$.

Take $\mathfrak{q} \in \operatorname{Ass}_R(M)$ so that $\mathfrak{q} \subseteq \mathfrak{p}$ and the length of the prime chain from $\mathfrak{p}$ to $\mathfrak{q}$ is maximal. Since we are only interested in the length of prime chains, we may assume $M = R/\mathfrak{q}$, that is, $M = R$ is a domain.

We will prove by induction on $s := \dim M_{\mathfrak{p}^*}$. If $s = 0$, then our assertion is trivial. If $s > 0$, take a homogeneous element $0 \neq f \in \mathfrak{p}^*$ then we have $\dim(R/fR)_\mathfrak{p} = \dim R_\mathfrak{p} - 1$, $\dim(R/fR)_{\mathfrak{p}^*} = \dim R_{\mathfrak{p}^*} - 1$, we have the desired result by the induction hypothesis. □

We defined the category $\mathfrak{M}_H(R)$. We want to investigate projective resolutions in this category.

Definition 7.22 Let $M \in \mathfrak{M}_H(R)$ and $h \in H$.

(1) We define $M(h) \in \mathfrak{M}_H(R)$ by

$$M(h)_{h'} = M_{h+h'}.$$

In particular, $R(h)$ is a principally generated R-module generated by $1 \in R_0 = R(h)_{-h}$.

(2) For $M, N \in \mathfrak{M}_H(R)$, we denote by $\operatorname{Hom}_R(M, N)_h$ the set of R-homomorphisms $f : M \to N$ such that $f(M_{h'}) \subseteq N_{h+h'}$ (or morphism $f : M \to N(h)$ in $\mathfrak{M}_H(R)$).

If M is finitely generated, since $\phi \in \operatorname{Hom}_R(M, N)$ is determined by the images of the homogeneous generators $\{x_i\}_{i=1}^n$ of M and since each $\phi(x_j)$ is a finite sum of homogeneous elements of N, we have

$$\operatorname{Hom}_R(M, N) = \bigoplus_{h \in H} \operatorname{Hom}_R(M, N)_h.$$

In general, we define $\underline{\operatorname{Hom}}_R(M, N) = \bigoplus_{h \in H} \operatorname{Hom}_R(M, N)_h$, which is an H-graded module by $\underline{\operatorname{Hom}}_R(M, N)_h = \operatorname{Hom}_R(M, N)_h$. In particular, we have

[2] Since the product of homogeneous elements not in $\mathfrak{p}^*$ are also not in $\mathfrak{p}^*$.

$$\underline{\mathrm{Hom}}_R(R(h), N) \cong N(-h).$$

Remark 7.4 If $M \in \mathfrak{M}_H(R)$ is generated by homogeneous elements $x_i \in M_{h_i}$ $(i = 1, \dots, n)$, we have a surjective R homomorphism

$$f : R(-h_1) \oplus R(-h_2) \oplus \dots \oplus R(-h_n) \to M,$$

mapping the generator of $R(-h_i)$ to x_i. Since we can do the same process to $\mathrm{Ker}\,(f) \in \mathfrak{M}_H(R)$, we can construct a projective decomposition

$$\cdots \to F_n \to \cdots \to \cdots F_0 \to M \to 0$$

of M in $\mathfrak{M}_H(R)$. Here each F_i is a direct sum of modules of the form $R(h)$. Hence we can state the following on Ext modules:

Corollary 7.23 *For $M, N \in \mathfrak{M}_H(R)$, we have* $\mathrm{Ext}^i_R(M, N) \in \mathfrak{M}_H(R)$.

Definition 7.24 (Graded Ext Module) If we consider $\mathrm{Ext}^i_R(M, N)$ as an H-graded module as we have seen in Corollary 7.23, we write $\underline{\mathrm{Ext}}^i_R(M, N)$.

The properties of R like regular, Gorenstein, Cohen–Macaulay are determined by the information concerning only H-graded prime ideals. For that purpose, we show the following theorem concerning $\mu_n(\mathfrak{p}, M)$ defined in Chap. 6.

Theorem 7.25 *Let $M \in \mathfrak{M}_H(R)$, $\mathfrak{p}$ be a prime ideal of R with* $\mathrm{ht}\,\mathfrak{p}/\mathfrak{p}^* = r$. *Then for every $n \geq 0$, the following equalities hold:*

$$\mu_n(\mathfrak{p}, M) = \mu_{n-r}(\mathfrak{p}^*, M). \tag{7.3}$$

Proof Since $\mu_n(\mathfrak{p}, M) = \dim_{\kappa(\mathfrak{p})} \mathrm{Ext}^n(\kappa(\mathfrak{p}), M_\mathfrak{p})$, we may assume $R = R_{(\mathfrak{p}^*)}$, $M = M_{(\mathfrak{p}^*)}$. Hence we assume that $(R, \mathfrak{p}^*)$ is an H local ring.

Since $R/\mathfrak{p}^*$ is H-simple, hence is regular, $(R/\mathfrak{p}^*)_\mathfrak{p}$ is a regular local ring. Hence $(\mathfrak{p}/\mathfrak{p}^*)_\mathfrak{p}$ is generated by a regular sequence $(f_1, \dots, f_r)$ $(f_1, \dots, f_r \in R)$. If we put $\mathfrak{q}_i = ((f_1, \dots, f_i) + \mathfrak{p})R$, we have short exact sequences

$$0 \to (R/\mathfrak{q}_i)_\mathfrak{p} \xrightarrow{f_{i+1}} (R/\mathfrak{q}_i)_\mathfrak{p} \to (R/\mathfrak{q}_{i+1})_\mathfrak{p} \to 0. \tag{*}$$

On the other hand, since each $\mathrm{Ext}^i_R(R/\mathfrak{p}^*, M)$ is a graded $R/\mathfrak{p}^*$-module, it is a free $R/\mathfrak{p}^*$-module. Looking at the long exact sequences of $\mathrm{Hom}_R(\ *\ , M)$ obtained by $(*)$, we obtain if $i = 0$

$$\begin{aligned}\cdots \to\ & \mathrm{Ext}^j_R(R/\mathfrak{q}_1, M) \to \mathrm{Ext}^j_R(R/\mathfrak{p}^*, M) \xrightarrow{f_1} \mathrm{Ext}^j_R(R/\mathfrak{p}^*, M)\\ &\to \mathrm{Ext}^{j+1}_R(R/\mathfrak{q}_1, M) \to \dots\end{aligned}$$

Since $\mathrm{Ext}_R^j(R/\mathfrak{p}^*, M)$ is a free $R/\mathfrak{p}^*$-module for every j, we obtain

$$\mathrm{Hom}_R(R/\mathfrak{q}_1, M) = 0,\ \mathrm{Ext}_R^{j+1}(R/\mathfrak{q}_1, M)_\mathfrak{p} \cong (\mathrm{Ext}_R^j(R/\mathfrak{p}^*, M)\otimes_R R/\mathfrak{q}_1)_\mathfrak{p}\ (j \geq 0).$$

Since the latter is a free $(R/\mathfrak{q}_1)_\mathfrak{p}$-module, repeating the same operation for $i = 1, 2, \ldots, r-1$ we obtain

$$\mathrm{Ext}_R^i(R/\mathfrak{p}, M)_\mathfrak{p} = 0 \quad (i = 0, 1, \ldots r-1),$$
$$\mathrm{Ext}_R^{j+r}(R/\mathfrak{p}, M)_\mathfrak{p} \cong (\mathrm{Ext}_R^j(R/\mathfrak{p}^*, M) \otimes_R R/\mathfrak{p})_\mathfrak{p} \quad (j \geq 0).$$

We get the desired equation from the definition of μ. □

Corollary 7.26 *Let R be an H-graded ring. Then R is a Cohen–Macaulay (resp. Gorenstein) ring if and only if for each H-prime ideal $\mathfrak{p}$, $R_\mathfrak{p}$ is a Cohen–Macaulay (resp. Gorenstein) ring.*

7.3 The Canonical Module and the Invariant $a(R)$

If R is a graded ring, then the canonical module K_R of R has a natural structure of a graded R-module. When R is a Gorenstein ring, K_R is a free R-module, but as a graded module, we potentially have a difference of gradings. The invariant $a(R)$ defined in [GW1] is defined by the isomorphism $K_R \cong R(a(R))$ as a graded R-module, and called the "*a-invariant*" of R. This invariant $a(R)$ is closely related to the Gorenstein property and rational singularities.

In Sect. 7.2 we introduced the category $\mathfrak{M}_H(R)$ and projective resolutions in it. Also, we can define injective object and injective resolutions in $\mathfrak{M}_H(R)$. But we must note that injective objects in $\mathfrak{M}_H(R)$ are *not* injective as R-modules in the usual sense in general, while a projective object in $\mathfrak{M}_H(R)$ is also projective as a usual R-module.

Definition 7.27 Let R be an H-graded ring and M, N be H-graded R-modules.

(1) We put

$$\underline{\mathrm{Hom}}_R(M, N) = \bigoplus_{h\in H} \mathrm{Hom}_R(M, N)_h.$$

For example, if $M = R(h)$, then $\underline{\mathrm{Hom}}_R(R(h), N) \cong N(-h)$. If R is Noetherian and M is finitely generated, then we can consider $\underline{\mathrm{Hom}}_R(M, N)$ a graded module giving the natural structure of a graded module on $\mathrm{Hom}_R(M, N)$.

(2) We say that M is *H-injective* if M is injective in the category $\mathfrak{M}_H(R)$. This is equivalent to saying that the functor $\underline{\mathrm{Hom}}_R(*, M)$ is an exact functor on $\mathfrak{M}_H(R)$.

Example 7.28

(1) If R is H-simple, then every H-graded R-module is a free module as well as an H-injective module. But if $R = k[H]$ (k is a field), then R is not an injective R-module in the usual sense.
(2) If $(R, \mathfrak{m})$ is an H-local ring and if $R/\mathfrak{m}$ is a field, then the injective envelope E of $R/\mathfrak{m}$ in $\mathfrak{M}_H(R)$ is also an injective R-module, since $\mathrm{Ext}^i_R(R/\mathfrak{m}, E) = 0$ $(\forall i > 0)$.

The existence of injective envelopes in $\mathfrak{M}_H(R)$ can be proved from the existence of injective envelopes as usual R-modules (Theorem 5.34).

Proposition 7.29 *If R is an H-graded ring and M is an H-graded R-module, then M is an H-graded submodule of an H-injective module.*

Proof For simplicity, first we give the proof of the case $H = \mathbb{Z}$.[3]

Let $R = \bigoplus_{n\in\mathbb{Z}} R_n$ be a graded ring. We denote it by R', if we forget the graded structure of R. By "R'-module," we mean an R-module without graded structure.

We denote by (Mod R') the category of R'-modules and by $\mathfrak{M}_\mathbb{Z}(R)$ the category of graded R-modules. We introduce 2 functors F and G,

$$F : \mathfrak{M}_\mathbb{Z}(R) \to (\mathrm{Mod}\, R'), \quad G : (\mathrm{Mod}\, R') \to \mathfrak{M}_\mathbb{Z}(R),$$

where F is the "forgetful functor" (forget the graded module structure) and we define for an R'-module N,

$$G(N) = \bigoplus_{n\in\mathbb{Z}} NX^n,$$

where X^n denotes a symbol to represent graded; namely, we define the grading on $G(N)$ by $G(N)_n = NX^n$. Then we get an isomorphism for a graded R-module $M = \bigoplus_{n\in\mathbb{Z}} M_n$,

$$\mathrm{Hom}_{R'}(F(M), N) \cong \underline{\mathrm{Hom}}_R(M, G(N)).$$

Indeed, to $\phi \in \mathrm{Hom}_{R'}(F(M), N)$ we assign $\psi \in \underline{\mathrm{Hom}}_R(M, G(N))$ given by $\psi(x) = \phi(x)X^n$ for $x \in M_n$ and conversely, given $\psi \in \underline{\mathrm{Hom}}_R(M, G(N))$, we assign $\phi \in \mathrm{Hom}_{R'}(F(M), N)$ given by $\phi(x) = \sum_{n\in\mathbb{Z}} y_n$ if $\psi(x) = \sum_{\in\mathbb{Z}} y_n X^n$ for $x \in F(M)$. It will be clear that this correspondence gives the desired isomorphism. By this isomorphism, we can assert that if I is an injective R'-module, then $G(I) \in \mathfrak{M}_\mathbb{Z}(R)$ is an injective object in $\mathfrak{M}_\mathbb{Z}(R)$ since $\mathrm{Hom}_{R'}(\bullet, I)$ is an exact functor.

Now, given $M \in \mathfrak{M}_\mathbb{Z}(R)$ we can embed M into an injective R'-module I. Then we can embed M in $G(I)$.

[3] The proof of this proposition was missing in the original Japanese book. This proof is due to Mitsuyasu Hashimoto.

In the above, we assumed that $H = \mathbb{Z}$. For general H, it suffices to define $G(N) = \bigoplus_{h \in H} NX^h$. □

Now we can define the injective envelope $\underline{E}_R(M)$ of M as the unique injective object $\underline{E}_R(M)$, $M \subseteq \underline{E}_R(M)$ and this inclusion is essential as in Theorem 5.34.

Definition 7.30 Let R be an H-graded ring and M be an H-graded R-module.

(1) We write $\underline{E}_R(M)$ for the injective envelope of M in $\mathfrak{M}_H(R)$. Namely, M is a graded submodule of $\underline{E}_R(M)$, which is an injective object of $\mathfrak{M}_H(R)$ and the injection $M \to \underline{E}_R(M)$ is essential in $\mathfrak{M}_H(R)$.
(2) The *minimal injective resolution*

$$0 \to M \to I^0 \to I^1 \to \cdots I^i \to I^{i+1} \to \cdots$$

of M is defined by $I^0 = \underline{E}_R(M)$, $I^1 = \underline{E}_R(\underline{E}_R(M)/M)$, $d^0 : I^0 \to I^1$ is defined as the composition of surjection and embedding $I^0 = \underline{E}_R(M) \to \underline{E}_R(M)/M \hookrightarrow I^1$ and if $I^\bullet$ is defined until $d^{i-1} : I^{i-1} \to I^i$, then we define $I^{i+1} = \underline{E}_R(\mathrm{Coker}\,(d^{i-1}))$.

If we treat M as an object of $\mathfrak{M}_H(R)$, the term "injective resolution" means the injective resolution in $\mathfrak{M}_H(R)$.

Example 7.31 (1) Let k be a field, $H = \mathbb{Z}$ and consider $R = k[X]$ as an H-graded ring putting $\deg(X) = 1$. Then the injective resolution of R in $\mathfrak{M}_H(R)$ is

$$0 \to I^0 = k[X, X^{-1}] \to I^1 = I^0/R = X^{-1}k[X^{-1}] \to 0.$$

Remark 7.5 We can assert if R is a Noetherian H-graded ring, then each indecomposable injective object in $\mathfrak{M}_H(R)$ is of the form $\underline{E}_R(R/\mathfrak{p})$ for an H-prime ideal $\mathfrak{p}$.

Remark 7.6 If R is a Noetherian H-graded ring, an indecomposable H-injective R-module is isomorphic to $\underline{E}_R(R/\mathfrak{p})$ for an H-prime ideal $\mathfrak{p}$ of R. This fact can be shown in the same manner as in the case of decompositions of injective R-modules over a Noetherian ring.

The theory of graded rings finitely generated over a field is very important in the sense that they give very abundant examples of rings from algebraic geometry.

In the following, we assume that our graded ring is an H-graded H-local ring $(R, \mathfrak{m})$ such that $R_0 = k$ is a field and the composition $k \hookrightarrow R \to R/\mathfrak{m}$ is an isomorphism.

Matlis duality for graded rings is very important:

Definition 7.32 (Matlis Duality) Let R be an H-graded ring and M be a graded R-module which is either Noetherian or Artinian as a graded R-module. Then we define a graded R-module M^* by

$$M^* = \bigoplus_{h \in H} \mathrm{Hom}_k(M_{-h}, k)$$

and call it the *dual* of M, where we define $M^*_h = \mathrm{Hom}_k(M_{-h}, k)$, and the multiplication of $x \in R_g$, $x_{M^*} : {M^*}_{-h-g} \to {M^*}_{-h}$ is defined as the dual of $x_M : R_g \to R_{g+h}$.

Since we assumed that M is either Noetherian or Artinian, each M_h is a finite-dimensional k-vector space and hence we have a natural isomorphism

$$M^{**} = (M^*)^* \cong M.$$

The functor $M \to M^*$ is a contravariant exact functor and the following statements hold.

Theorem 7.33

(1) *If M is projective (resp. injective), then M^* is injective (resp. projective).*
(2) *M is a Noetherian (resp. an Artinian) R-module, then M^* is an Artinian (resp. a Noetherian) R-module.*

Proof

(1) Since

$$\begin{aligned}\mathrm{Hom}_R(M, R(h)^*) &= \mathrm{Hom}_R(M, \mathrm{Hom}_k(R(h), k)) \\ &\cong \mathrm{Hom}_k(M \otimes_R R(h), k) = M(h)^*,\end{aligned}$$

for a finitely generated graded R-module M, the functor $\mathrm{Hom}_R(*, R(h)^*)$ is an exact functor. Hence $R(h)^*$ is an injective R-module.
(2) Let $N \in \mathfrak{M}(R)$ be an Artinian module and put $M = N^*$. If N' is a graded submodule of N, then $M' = \mathrm{Ker}\,(N^* \to (N')^*)$ is a graded R submodule of M. Thus from a descending chain of graded R submodules N, we get an ascending chain of graded submodules of M and vice versa. Hence if N is Artinian, then M is Noetherian and hence is finitely generated as a graded R-module.

In particular, the dual of the natural surjection $R \to R/\mathfrak{m} = k$, $k^* = k \to R^*$ is essential by Lemma 7.34, R^* is an injective envelope of $R/\mathfrak{m}$. □

Lemma 7.34 *Let $(R, \mathfrak{m})$ be an H-local ring and assume $k = R/\mathfrak{m}$ is a field. Then for finitely generated graded R-modules M, N and a surjection $f : M \to N$, f is minimal (namely,* $\mathrm{Ker}\,(f) \subseteq \mathfrak{m}M$*) if and only if $f^* : N^* \to M^*$ is essential.*

Proof Recall that the following three conditions on f are equivalent:

(1) f is minimal.
(2) For every proper submodule M' of M, $f(M') \subsetneq N$.
(3) $f^* : N^* \to M^*$ is essential.

□

Now, we define the canonical module K_R and the invariant $a(R)$ of a graded ring R. Recall that (Theorem 6.39) the canonical module K_A of a local ring $(A, \mathfrak{m})$ of dimension d is defined as

$$\widehat{K_A} \cong \mathrm{Hom}_A(\mathrm{H}^d_{\mathfrak{m}}(A), E_A(A/\mathfrak{m})).$$

Note that if $(R, \mathfrak{m})$ is an H-local H-graded ring of dimension d, then $\mathrm{H}^d_{\mathfrak{m}}(R)$ is a graded Artinian R-module by the computation of local cohomology groups.

Definition 7.35 (Graded Canonical Module) Let $(R, \mathfrak{m})$ be an H-local H-graded ring of $\dim R = d$.

(1) We call $K_R = [\mathrm{H}^d_{\mathfrak{m}}(R)]^*$ the (graded) *canonical module* of R.
(2) If $K_R \cong R(a)$, namely K_R is generated by an element of degree $-a$, then we define $a = a(R) \in H$ and $a(R)$ is called the *the a-invariant* of R.
(3) If $H = \mathbb{Z}$, we define

$$a(R) = \max\{n \mid [\mathrm{H}^d_{\mathfrak{m}}(R)]_n \neq 0\} = -\min\{n \mid [K_R]_n \neq 0\}.$$

Example 7.36 Let k be a field.

(1) If $R = k[X_1, \ldots, X_d]$ is a polynomial ring over k and assume that $X_i \in R_{h_i}$ $(i = 1, 2, \ldots, d)$. We define

$$C_i : 0 \to C_i^0 = k[X_i, X_i^{-1}] \to C_i^1 = k[X_i, X_i^{-1}]/k[X_i] = X_i^{-1}k[X_i^{-1}] \to 0.$$

Then we have $\mathrm{H}^0(C_i) = k[X_i]$, $\mathrm{H}^j(C_i) = 0$ $(j \neq 0)$. Next, if we define

$$C = C_1 \otimes_k C_2 \otimes_k \ldots \otimes_k C_d,$$

the complex C is a resolution of R. Then since the multiplication of X_j on $k[X_j, X_j^{-1}]$ is bijection, C^i $(0 \leq i < d)$ satisfies the condition $(**)$ of Definition 5.28 with respect to $\mathfrak{m}$. Also, $C^d = \bigotimes_{1 \leq j \leq d} X_j^{-1}k[X_j^{-1}] \cong (X_1X_2\cdots X_d)^{-1}k[X_1^{-1}, X_2^{-1}, \ldots, X_d^{-1}]$ satisfies the condition $(*)$ of Definition 5.102. Hence we conclude that $\mathrm{H}^i_{\mathfrak{m}}(R) = 0$ for $i \neq d$ and

$$\begin{aligned}\mathrm{H}^d_{\mathfrak{m}}(R) \cong C^d &= (X_1X_2\cdots X_d)^{-1}k[X_1^{-1}, X_2^{-1}, \ldots, X_d^{-1}] \\ &\cong R^*(h_1 + h_2 + \ldots + h_d).\end{aligned}$$

Hence we conclude

$$K_R \cong R(-h_1 - h_2 - \cdots - h_d) \quad \text{and} \quad a(R) = -\sum_{i=1}^{d} h_i.$$

If $H = \mathbb{Z}$ and we put standard grading on R, namely $X_i \in R_1$ ($i = 1, 2, \ldots, d$), then we have $a(R) = -d$.

(2) (Semigroup ring) Let $H = \mathbb{Z}$ and let $S \subseteq \mathbb{N}$ be a submonoid of $\mathbb{N}$ such that $\mathbb{N} \setminus S$ is a finite set (note that we define $\mathbb{N} = \{0, 1, 2, \ldots\}$). Then the semigroup ring

$$R = k[S] = k[T^s | s \in S]$$

is a graded ring, defining $T^s \in R_s$. Then from the short exact sequence

$$0 \to R \to k[T, T^{-1}] \to k[T, T^{-1}]/R, \to 0$$

we have $\mathrm{H}^1_{\mathfrak{m}}(R) \cong k[T, T^{-1}]/R$ and we get

$$a(R) = \max\{n \in \mathbb{Z} \mid n \notin S\}.$$

We will study such rings in detail in Sect. 7.5.

Assume that $K_R \cong R(a)$, and $f \in R_h$ is a non-zerodivisor. The short exact sequence

$$0 \to R(-h) \xrightarrow{f} R \to R/fR \to 0$$

induces the cohomology long exact sequence

$$\cdots \to \mathrm{H}^{d-1}_{\mathfrak{m}}(R) \to \mathrm{H}^{d-1}_{\mathfrak{m}}(R/fR) \to \mathrm{H}^d(R(-h)) \xrightarrow{f} \mathrm{H}^d(R) \to 0.$$

Taking the dual of this sequence, we get the following proposition.

Proposition 7.37 *If $K_R \cong R(a)$, $f \in R_h$ is a non-zerodivisor and if $\mathrm{H}^{d-1}_{\mathfrak{m}}(R) = 0$, then we have $K_{R/fR} \cong (R/fR)(a + h)$.*

The following theorem for graded rings holds as well as in the local case.

Theorem 7.38 *Let $(R, \mathfrak{m})$ be an H-local H-graded ring with $R_0 = k$, a field. Then the following hold true:*

(1) *R is Cohen–Macaulay if and only if $\mathrm{H}^i_{\mathfrak{m}}(R) = 0$ ($\forall i < d$).*
(2) *R is Gorenstein if and only if R is Cohen–Macaulay and $K_R \cong R(a(R))$.*

The proof is the same as in Theorem 6.33 and the other properties stated there also hold. In particular, the local duality theorem holds also in the graded version. The proof is the same as in Theorem 6.22.

Theorem 7.39 (Local Duality, Graded Version) *Let $(R, \mathfrak{m})$ be a Cohen–Macaulay H-local H-graded ring with $\dim R = d$. Then the following statement holds for every i and for every finitely generated R-module M:*

$$\mathrm{H}_{\mathfrak{m}}^{d-i}(M) \cong [\underline{\mathrm{Ext}}_R^i(M, K_R)]^*. \tag{7.1}$$

Taking the Matlis duality, we have the following:

$$\underline{\mathrm{Ext}}_R^i(M, K_R) \cong \mathrm{H}_{\mathfrak{m}}^{d-i}(M)^*. \tag{7.2}$$

Remark 7.7 If we do not assume $(R, \mathfrak{m})$ is Cohen–Macaulay, the following holds true:

$$\underline{\mathrm{Hom}}_R(M, K_R) \cong \mathrm{H}_{\mathfrak{m}}^d(M)^*.$$

Let us observe two examples, where the invariant $a(R)$ plays a crucial role.

Definition 7.40 Let R be an H-graded ring.

(1) If $H = \mathbb{Z}$, for a positive integer r, we put

$$R^{(r)} = \bigoplus_{n \in \mathbb{Z}} R_{rn}$$

and call it the r*th Veronese subring* of R. Also, for a graded R-module M, we put $M^{(r)} = \bigoplus_{n\in\mathbb{Z}} M_{rn}$.

(2) In general, for a subgroup L of H, we put

$$R^{(L)} = \bigoplus_{h \in L} R_h \quad \text{and} \quad M^{(L)} = \bigoplus_{h \in L} M_h.$$

Remark 7.8 The Veronese subring $R^{(L)}$ of R is clearly a direct summand of R and hence a pure subring of R. Hence the Noetherian property or normality property is inherited from R to $R^{(L)}$.

Proposition 7.41 *Let R be an H-graded ring and M be a graded R-module.*

(1) *The functor $M \to M^{(d)}$ (or $M \to M^{(L)}$) is a covariant exact functor.*

(2) *If $H = \mathbb{Z}$, for every finitely generated R-module M, we have an isomorphism*

$$\mathrm{H}_{\mathfrak{m}^{(d)}}^i(M^{(d)}) \cong [\mathrm{H}_{\mathfrak{m}}^i(M)]^{(d)}.$$

Hence, if M is a Cohen–Macaulay R-module, so is $M^{(d)}$.

If L is a subgroup of H with $[H : L] < \infty$, then for any finitely generated H-graded R-module M, we have

$$\mathrm{H}_{\mathfrak{m}^{(L)}}^i(M^{(L)}) \cong [\mathrm{H}_{\mathfrak{m}}^i(M)]^{(L)}.$$

Hence if M is an equi-dimensional Cohen–Macaulay module, so is $M^{(L)}$.

(3) $K_{R^{(d)}} \cong [K_R]^{(d)}$ *holds and if* L *is a subgroup of* H *with finite index, then we have* $K_{R^{(L)}} \cong [K_R]^{(L)}$.

Proof

(1) is clear from the definition.
(2) Let

$$0 \to M \to I^0 \to I^1 \to \cdots \to I^n \to \cdots$$

be an injective resolution of M in the category $\mathfrak{M}(R)$. Then we have $\mathrm{H}^i_{\mathfrak{m}}(M) \cong \mathrm{H}^i(\mathrm{H}^0_{\mathfrak{m}}(I^\bullet))$, where $I^\bullet = [I^0 \to I^1 \to \cdots]$. Then the sequence

$$0 \to M^{(L)} \to [I^0]^{(L)} \to [I^1]^{(L)} \to \cdots$$

is also exact and each I^i is a direct sum of indecomposable modules satisfying either condition $(*)$ or $(**)$ of Definition 5.102. Hence from Lemma 7.42 below, we have

$$\mathrm{H}^i_{\mathfrak{m}^{(L)}}(R) \cong \mathrm{H}^i(\mathrm{H}^0_{\mathfrak{m}^{(L)}}(I^\bullet)^{(L)}) \cong [\mathrm{H}^0_{\mathfrak{m}}(I^\bullet)]^{(L)} \cong [\mathrm{H}^i_{\mathfrak{m}}(R)]^{(L)}.$$

□

Lemma 7.42 *Let* $L \subseteq H$ *be a subgroup of finite index. If a graded* R*-module* M *satisfies condition* $(*)$ *or* $(**)$ *with respect to a* H*-graded ideal* $\mathfrak{a}$*, so does* $M^{(L)}$ *with respect to* $\mathfrak{a}^{(L)}$.

Proof Assume M satisfies the condition $(*)$ with respect to $\mathfrak{a}$. Then for any homogeneous element $M^{(L)}$, if $\mathfrak{a}^n x = 0$, then clearly $((\mathfrak{a})^{(L)})^n x = 0$. Also, if some homogeneous element $a \in \mathfrak{a}_h$ acts on M bijectively, then we can take a positive integer n so that $a^n \in \mathfrak{a}^{(L)}$ and a^n acts bijectively on $M^{(L)}$. □

Corollary 7.43 *Let* R *be an* H*-graded Gorenstein ring and let* L *be a subgroup of* H *of finite index. If* $a(R) \in L$*, then* $R^{(L)}$ *is Gorenstein.*

Proof If $w \in [K_R]_{-a(R)}$ is a generator of K_R as an R-module, then since $a(R) \in L$, we can take w as the generator of $(K_R)^{(L)} = K_{R^{(L)}}$. □

We can show the "converse" of Corollary 7.43 in the case where R is "standard graded."

Theorem 7.44 *Let* $R = \bigoplus_{n \geq 0} R_n$ *be a graded Cohen–Macaulay integral domain generated by* R_1 *over* $k = R_0$ *with* $d = \dim R \geq 2$*. If* $R^{(r)}$ *is a Gorenstein ring for some* $r \geq 2$*, then* R *is Gorenstein and* $r \mid a(R)$.

Proof We consider $K_{R^{(r)}} = [K_R]^{(r)}$ as a submodule of K_R. Assume that $w \in (K_R)_{nr}$ generates $[K_R]^{(d)}$. Then, in particular, we have $\dim_k [K_R]_{nr} = 1$. It suffices to show that the same w generates K_R as an R-module.

We know that K_R is a torsion free R-module. First, let us show that $a(R) = -rn$. It is clear that $a(R) \geq -rn$ by definition. If $0 \neq x \in [K_R]_{nr-s}$ for some $s > 0$, then since $sR \subseteq K_R$, we have $1 = \dim_k(K_R)_{nr} \geq \dim_k R_s$, which is impossible since $\dim R \geq 2$ and $R = k[R_1]$.

We show that K_R is generated by w. Since w is a generator of $[K_R]^{(r)}$, we have $(K_R/wR)^{(r)} = 0$. Let m be the minimal degree such that $(K_R/wR)_m \neq 0$ and assume $0 \neq u \in (K_R/wR)_m$. If $m \equiv -s \pmod r$, then we must have $(R_s)u = 0$. Thus we must have depth $(K_R/wR) = 0$. But on the other hand, we must have depth $(K_R/wR) = d - 1 \geq 1$. Contradiction! This shows that R is Gorenstein. □

Example 7.45 Theorem 7.44 does not hold if R is not generated by R_1. For example, let $R = k[X, Y, Z]/(X^2 + Y^3 + Z^5)$ with $X \in R_{15}, Y \in R_{10}, Z \in R_6$. Then we have $a(R) = -15 - 10 - 6 + \deg(X^2 + Y^3 + Z^5) = -1$. But we can show that $R^{(r)}$ is Gorenstein if $r = 1, 2, 3, 4, 5, 6, 8, 9, 12$ and moreover, $R^{(r)}$ is a polynomial ring if $d = 2, 3, 5, 6$.

Next, we introduce the concept of the *Segre product* of two graded rings over a field k. Geometrically, this concept corresponds to the product $X \times Y$ of two projective varieties X, Y over the field k. This concept is very convenient for making various examples and researchers of commutative ring theory think "To make a counterexample, consider Segre products!"

Definition 7.46 Let $R = \bigoplus_{n \geq 0} R_n$, $S = \bigoplus_{n \geq 0} S_n$ be graded rings with $R_0 = S_0 = k$, a field. Then we define

$$R \sharp S = \bigoplus_{n \geq 0} R_n \otimes_k S_n,$$

which we consider as a subring of $R \otimes_k S$, the *Segre product* of R and S. Likewise, for an R-module M (resp. an S-module N), we define

$$M \sharp N = \bigoplus_{n \in \mathbb{Z}} M_n \otimes_k N_n$$

and call it the Segre product of M and N. For $a \in R_n$ and $b \in S_n$, we write $a \sharp b$ as the image of $a \otimes b$ in $[R \sharp S]_n$. In the same manner we define $x \sharp y \in [M \sharp N]_m$ for $x \in M_m$ and $y \in N_m$. We consider $M \sharp N$ as a graded $R \sharp S$ module by

$$(a \sharp b)(x \sharp y) = (ax) \sharp (by).$$

In the following, we denote the graded maximal ideal of R (resp. S, resp. $T = R \sharp S$) by $\mathfrak{m}$ (resp. $\mathfrak{n}$, resp. $\mathfrak{M}$).

Let us note that if $\dim_k M_n, \dim N_n < \infty$ for every n, then we have a natural isomorphism $[M \sharp N]^* \cong M^* \sharp N^*$ as graded $M \sharp N$ modules.

Remark 7.9 In the situation in Definition 7.46, we can think of $R \otimes_k S$ as a $\mathbb{Z}^2$ graded ring putting $[R \otimes_k S]_{(m,n)} = R_m \otimes_k S_n$. If we put $L = \{(n, n) \mid n \in \mathbb{Z}\}$

as the diagonal subgroup of $\mathbb{Z}^2$, we can regard the Segre product $R\sharp S$ as a kind of Veronese subring, since we have

$$R\sharp S = [R \otimes_k S]^{(L)}.$$

Hence the Noetherian property and the normality are inherited from $R \otimes_k S$ to $R\sharp S$ (Lemma 7.7). But since L is not of finite index in $\mathbb{Z}^2$, the Cohen–Macaulay property need not be inherited by $R\sharp S$ even if so are R and S.

The next lemma will be useful to compute local cohomology groups of Segre products.

Lemma 7.47 *Let M and N be a graded R-module and a graded S-module, respectively.*

(1) *If either M or N satisfies the condition $(*)$ of Definition 5.102, then so does $M\sharp N$, too.*
(2) *If both M and N satisfy the condition $(**)$, then so does $M\sharp N$.*

Proof

(1) If M satisfies the condition $(*)$, then for every $x \in M$, we can take n so that $\mathfrak{m}^n x = 0$. Namely, for sufficiently large n, we may assume $R_n x = 0$. Then for every homogeneous element $y \in N$ with $\deg y = \deg x$, we have $[R\sharp S]_n(x\sharp y) = 0$ and $M\sharp N$ satisfies the condition $(*)$.
(2) If the action of $a \in R_n$ (resp. $b \in S_m$) on M (resp. on N) is bijective, then the action of $a^m\sharp b^n$ on $M\sharp N$ is bijective. Hence $M\sharp N$ satisfies the condition (**).

□

Various properties of Segre products are obtained from the following formulas of local cohomology.

Theorem 7.48 *If both R and S have depth at least 2, then so does $T = R\sharp S$ and the following isomorphisms hold true:*

$$\mathrm{H}^m_{\mathfrak{M}}(R\sharp S) \cong \bigoplus_{i+j=m+1} [\mathrm{H}^i_{\mathfrak{m}}(R)\sharp \mathrm{H}^j_{\mathfrak{n}}(S)] \oplus [R\sharp \mathrm{H}^m_{\mathfrak{n}}(S)] \oplus [\mathrm{H}^m_{\mathfrak{m}}(R)\sharp S].$$

Before proving this theorem, let us state the applications.

Theorem 7.49 *Let $R = \bigoplus_{n\geq 0} R_n$, $S = \bigoplus_{n\geq 0} S_n$ be graded rings with $R_0 = S_0 = k$, a field.*

(1) *If $\dim R = r$ and $\dim S = s$, then $\dim R\sharp S = r + s - 1$.*
(2) *If $\dim R, \dim S \geq 2$, then $K_T \cong K_R\sharp K_S$.*
(3) *If R and S are Cohen–Macaulay and $a(R), a(S) < 0$, then $R\sharp S$ is also Cohen–Macaulay.*
(4) *Conversely, let R, S be Cohen–Macaulay rings of dimension ≥ 2 generated by R_1 (resp. S_1). If $R\sharp S$ is Cohen–Macaulay, then $a(R), a(S)$ are both negative.*

(5) *If* $\dim R, \dim S \geq 2$, depth R, depth $S \geq 2$ *and* $K_R \cong R(a)$, $K_S \cong S(a)$, *then* $K_T \cong T(a)$. *In particular, if both R and S are Gorenstein with* $a(R) = a(S) < 0$, *then* $R\sharp S$ *is also a Gorenstein ring.*

(6) *Conversely, assume R and S are Cohen–Macaulay rings of* $\dim R, \dim S \geq 2$ *and R (resp. S) is generated by* R_1 *(resp.* S_1*). If* $R\sharp S$ *is a Gorenstein ring, then so are R and S with* $a(R) = a(S) < 0$.

Proof

(1) r, s are the maximal i, j such that $H^i_{\mathfrak{m}}(R)$ (resp. $H^j_{\mathfrak{n}}(S)$) is nonzero. By Theorem 7.48, $H^{r+s-1}_{\mathfrak{M}}(T) \cong H^r_{\mathfrak{m}}(R)\sharp H^s_{\mathfrak{n}}(S) \neq 0$ and $H^n_{\mathfrak{M}}(T) = 0$ for $n > r+s-1$. Hence we have $\dim T = r+s-1$. We assumed, for simplicity, depth $R, S \geq 2$ in 7.48. But this condition is not necessary when we compute the local cohomology of highest degree.

(2) By Definition 7.35, we have

$$K_T = [H^{r+s-1}_{\mathfrak{M}}(T)]^* \cong [H^r_{\mathfrak{m}}(R)\sharp H^s_{\mathfrak{n}}(S)]^* \cong [H^r_{\mathfrak{m}}(R)]^* \sharp [H^s_{\mathfrak{n}}(S)]^* \cong K_R \sharp K_S.$$

It is clear that (3) and (5) follow from (1) and (2).

(4) If we assume that $a(R) \geq 0$, for some $n \geq 0$, we have $H^r_{\mathfrak{m}}(R)_n \neq 0$. Since $S_n \neq 0$ by our assumption, $H^r_{\mathfrak{m}}(R)_n \otimes S_n \neq 0$ is included in $H^r_{\mathfrak{M}}(T)$ and T is not Cohen–Macaulay since $r < \dim T = r+s-1$.

(6) We will show that if $K_T \cong K_R\sharp K_S$ is principally generated, then we must have $a(R) = a(S)$. On the contrary, assume that $a(R) < a(S)$. Let $w \in [K_S]_{-a(S)}$, $w \neq 0$. Then since $wS \cong S(s(S)) \subseteq K_S$ and $\dim S \geq 2$, we have $\dim_k [wS]_{-a(R)} = \dim_k S_{-a(R)+a(S)} \geq 2$. Since the minimal nonzero degree of $K_T \cong K_R\sharp K_S$ is $-a(R)$ and since $\dim_k [K_T]_{-a(R)} \geq 2$, K_T cannot be principally generated.

□

Now let us prove Theorem 7.48.

Proof of Theorem 7.48 Let $\mathcal{I}$ (resp. $\mathcal{J}$) be minimal injective resolutions of R (resp. S) in $\mathfrak{M}_H(R)$ (resp. $\mathfrak{M}_H(S)$). Put $'\mathcal{I} = H^0_{\mathfrak{m}}(\mathcal{I}), '\mathcal{J} = H^0_{\mathfrak{n}}(\mathcal{J})$ and $''\mathcal{I} = \mathcal{I}/'\mathcal{I}$, $''\mathcal{J} = \mathcal{J}/'\mathcal{J}$ be quotient complexes. Then we have the exact sequences of complexes:

$$0 \to' \mathcal{I} \to \mathcal{I} \to'' \mathcal{I} \to 0, \quad 0 \to' \mathcal{J} \to \mathcal{J} \to'' \mathcal{J} \to 0.$$

Since $H^0(\mathcal{I}) = R$, $H^i(\mathcal{I}) = 0\,(i > 0)$, from the long exact sequence of cohomology, we get

$$H^i_{\mathfrak{m}}(R) \cong H^i('\mathcal{I}) \quad \text{and} \quad H^j_{\mathfrak{n}}(S) \cong H^j('\mathcal{J}).$$

Moreover, since $'\mathcal{I}^i = 0\ (i = 0, 1), '\mathcal{J}^j = 0\ (j = 0, 1)$ by our assumption, we get the following isomorphisms:

$$\mathrm{H}^0(''\mathcal{I}) \cong R, \quad \mathrm{H}^i(''\mathcal{I}) \cong \mathrm{H}^{i+1}_{\mathfrak{m}}(R)\ (i > 0),$$
$$\mathrm{H}^0(''\mathcal{J}) \cong S, \quad \mathrm{H}^j(''\mathcal{J}) \cong \mathrm{H}^{i+1}_{\mathfrak{n}}(S)\ (j > 0).$$

Now, since the functor $\sharp$ is an exact functor, $\mathcal{I}\sharp\mathcal{J}$ gives a resolution of $R\sharp S$. We put $\mathcal{K} = \mathcal{I}\sharp\mathcal{J}$. For each degree i, $\mathcal{K}^i$ decomposes into direct sums of 4 components coming from $'\mathcal{I}\sharp'\mathcal{J},''\mathcal{I}\sharp'\mathcal{J},'\mathcal{I}\sharp''\mathcal{J},''\mathcal{I}\sharp''\mathcal{J}$, in which the direct sum of 3 components $'\mathcal{I}\sharp'\mathcal{J},''\mathcal{I}\sharp'\mathcal{J},'\mathcal{I}\sharp''\mathcal{J}$ forms a subcomplex of $\mathcal{I}\sharp\mathcal{J}$. We denote by $'\mathcal{K}$ this direct sum and the quotient complex by $''\mathcal{K}$, we have the following short exact sequence of complexes:

$$0 \to' \mathcal{K} \to \mathcal{K} \to'' \mathcal{K} \to 0, \quad ''\mathcal{K} \cong'' \mathcal{I}\sharp''\mathcal{J}.$$

By Lemma 7.47, each component of $'\mathcal{K}$ satisfies $(*)$ and each component of $''\mathcal{K}$ satisfies $(**)$. Since $\mathrm{H}^i(\mathcal{K}) = 0$ $(i > 0)$, we have

$$\mathrm{H}^i_{\mathfrak{M}}(T) \cong \mathrm{H}^i('\mathcal{K}) \cong \mathrm{H}^{i-1}(''\mathcal{K}).$$

Since each component of $'\mathcal{K}^i = 0$ $(i = 0, 1)$ C is a direct sum of each direct summands, we have

$$\mathrm{H}^{m-1}(''\mathcal{K}) \cong \bigoplus_{i+j=m-1} \mathrm{H}^i(''\mathcal{I})\sharp\mathrm{H}^j(''\mathcal{J})$$

and we get our conclusion. □

Example 7.50 The statement of Theorem 7.49(4) is not correct if R, S are *not* generated by R_1, S_1, respectively.

Let $R = k[X, Y, Z]/(X^2 + Y^3 + Z^7)$, $S = k[U, V, W]/(U^3 + V^4 + W^4)$. Then R, S are graded rings, giving X, Y, Z, U, V, W the degrees 21, 14, 6, 4, 3, 3, respectively. Then we have $a(R) = 1$ and $a(S) = 2$ but since both $\mathrm{H}^2_{\mathfrak{m}}(R)\sharp S$ and $R\sharp\mathrm{H}^2_{\mathfrak{n}}(S)$ are 0, $R\sharp S$ is a Cohen–Macaulay ring.

7.4 Poincaré Series

Let us introduce the Poincaré series of a graded ring R. By introducing this concept, it frequently happens that a ring-theoretic problem is translated to a numerical problem. Also, commutative ring theory plays a very important role in combinatorial problems, and in that case, the Poincaré series is very important.

Definition 7.51 Let R be a H-graded ring finitely generated over a field k. We fix an isomorphism $H \cong \mathbb{Z}^n$ and introduce n variables $T_1, \ldots, T_n$ so that for $h = (a_1, \ldots, a_n)$, we write $T^h = T_1^{a_1} T_2^{a_2} \cdots T_n^{a_n}$.

(1) If $\dim_k R_h$ is finite for every $h \in H$, we define

$$P(R, T) = \sum_{h \in H} \dim_k R_h T^h$$

and call this series the *Poincaré series* of R.

(2) If M is a finitely generated graded R-module, then since $\dim_k M_h$ is finite for every $h \in H$, we define

$$P(M, T) = \sum_{h \in H} \dim_k M_h T^h$$

and call this series the Poincaré series of M.

The graded rings we treat in this section satisfy the condition "$\dim_k R_h$ **is finite for every** $h \in H$."

Remark 7.10 Let the homogeneous generators of R be $h_1, \ldots, h_m$. If the inner product $a.h_i$ is positive for all i and for some $a \in \mathbb{Z}^n$, then $\dim_k R_h < \infty$ for every $h \in H$.

On the other hand, if we put the standard grading on $S = k[X_1, \ldots, X_n]$, then each homogeneous component of the localization S_f for $f \in S_d, d > 0$ has infinite dimension over k.

Let us investigate the properties of the Poincaré series. First, we will show the additivity. **Until the end of this section, all modules are graded modules over an** H**-graded ring** R.

Lemma 7.52

(1) *Let* $0 \to M_n \to M_{n-1} \to \cdots \to M_1 \to M_0 \to 0$ *be an exact sequence of graded* R*-modules. Then we have*

$$\sum_{i=0}^{n} (-1)^i P(M_i, T) = 0.$$

In particular, for a short exact sequence $0 \to M' \to M \to M'' \to 0$, *we have*

$$P(M, T) = P(M', T) + P(M'', T).$$

(2) *For* $h \in H$, *we have* $P(M(h), T) = T^{-h} P(M, T)$.

(3) *If* $x \in R_h$ *is an* M*-regular element, then we have*

$$P(M/xM, T) = (1 - T^h) P(M, T). \tag{7.3}$$

In particular, if $(x_1, \ldots, x_r)$ *is an* M*-regular sequence with* $x_i \in R_{h_i}$, *then we have*

$$P(R,t) = \frac{P(R/(x_1,\dots,x_r)R, T)}{(1-T^{h_1})\cdots(1-T^{h_r})}. \tag{7.4}$$

Proof

(1) For a short exact sequence $0 \to M' \to M \to M'' \to 0$, since we have $\dim_k M_h = \dim_k M'_h + \dim_k M''_h$ for every $h \in H$, our result follows easily and for a long exact sequence, the result will be clear if we decompose it into short exact sequences.

(2) For every $h' \in H$, since we have $\dim_k M(h)_{h'} = \dim_k M_{h+h'}$, we have

$$P(M(h),T) = \sum_{h' \in H} \dim_k M_{h+h'} T^{h'}$$
$$= T^{-h} \sum_{h'+h' \in H} \dim_k M_{h+h'} T^{h+h'} = T^{-h} P(M,T).$$

(3) follows from (2) and the short exact sequence $0 \to M(-h) \xrightarrow{x} M \to M/xM \to 0$.

□

Let us examine some examples.

Example 7.53

(1) If $R = k[X_1,\dots,X_n]$ is a polynomial ring over k with $\deg(X_i) = h_i$, then we have

$$P(R,T) = \frac{1}{(1-T^{h_1})\cdots(1-T^{h_n})}.$$

(2) If $R = k[X_1,\dots,X_{n+r}]/(f_1,\dots,f_r)$ is a complete intersection with $\deg(X_i) = g_i$, $(f_1,\dots,f_r)$ a regular sequence with $\deg f_i = h_i$, then we have

$$P(R,T) = \frac{(1-T^{h_1})\cdots(1-T^{h_r})}{(1-T^{g_1})\cdots\cdots(1-T^{g_{n+r}})}.$$

(3) Next, we will look at a more complicated example. Let s, t, X, Y, Z, W be variables and $R = k[s^3, s^2t, st^2, t^3] \cong k[X,Y,Z,W]/(Y^2 - XZ, Z^2 - YW, XW - YZ)$.

If we put $H = \mathbb{Z}$ and if the degree of X, Y, Z, W is 1, then we have

$$P(R,T) = \frac{1+2T}{(1-T)^2}.$$

If we put $H = \mathbb{Z}^2$, $\deg s = (1, 0)$, $\deg t = (0, 1)$, then we have

$$P(R, T) = \frac{1 + T_1^2 T_2 + T_1 T_2^2}{(1 - T_1^3)(1 - T_2^3)}.$$

Proof (1) If $R = k[X_1, \ldots, X_n]$, since $(X_1, \ldots, X_n)$ form a regular sequence and since $R/(X_1, \ldots, X_n) = k$, $P(k, T) = 1$. Hence from Lemma 7.52, (1) follows. We can prove (2) in the same manner.

(3) Note that R is normal with $\dim R = 2$, hence is Cohen–Macaulay. Then (X, W) forms an R-regular sequence.

Since $R/(X, W)R \cong k[Y, Z]/(Y^2, YZ, Z^2)$, we have $P(R/(X, W)R, t) = 1 + 2t$ and from Lemma 7.52 we get our desired result. □

For a while we assume that R is a Noetherian graded ring and we assume $H = \mathbb{Z}$. Then the Poincaré series $P(R, t)$ is a rational function.

Proposition 7.54

(1) *If* $\dim R = n$, *then for some positive integers* $d_1, \cdots, d_n$ *and a polynomial* $f(t)$ *whose coefficients are in* $\mathbb{Z}$,

$$P(R, t) = \frac{f(t)}{(1 - t^{d_1}) \cdots (1 - t^{d_n})}. \tag{7.5}$$

In particular, n is the order of the pole of the rational function $P(R, t)$ *at* $t = 1$.

(2) *If R is a Cohen–Macaulay ring, then the coefficients of* $f(t)$ *are non-negative integers.*

(3) *If R is Cohen–Macaulay, then we have* $a(R) = \deg(f) - (d_1 + d_2 + \cdots + d_n)$. *Namely,* $a(R)$ *is the order of the pole of* $f(t)$ *at* ∞.

Proof

(1) We show our conclusion by the induction on $n = \dim R$. If $n = 0$, then since $R_d = 0$ for large enough d, $P(R, t)$ is a polynomial whose coefficients are non-negative integers.

If depth $R > 0$, then we can take a regular element $x \in R_d$ $(d > 0)$. Then by Theorem 7.17, we have $P(R, t) = P(R/xR, t)/(1 - t^d)$. Since our assertion holds for $P(R/xR, t)$ by the induction hypothesis, so does $P(R, t)$.

If $n > 0$ and depth $R = 0$, then putting $\mathfrak{a} = \mathrm{H}^0_{\mathfrak{m}}(R)$, we have depth $R/\mathfrak{a} > 0$ and $P(R, t) = P(R/\mathfrak{a}, t) + P(\mathfrak{a}, t)$. Since our assertion holds true for $P(R/\mathfrak{a}, t)$ and $P(\mathfrak{a}, t)$ is a polynomial with non-negative integer coefficient, we are done.

(2) Since R is Cohen–Macaulay, we can take a regular sequence $(x_1, \ldots, x_n)$ with $x_i \in R_{d_i}$ $(d_i > 0)$. Then we have

$$P(R, t) = \frac{P(R/(x_1, \ldots, x_n), t)}{(1 - t^{d_1}) \cdots (1 - t^{d_n})}.$$

by Lemma 7.52. Since $P(R/(x_1, \ldots, x_n), t) = f(t)$ is a polynomial whose coefficients are non-negative integers, we are done.

(3) If n (= dim R) = 0, then we have $a(R) = \max\{n | R_n \neq 0\} = \deg(f)$. Assume our assertion is true for dim $R = n - 1$ and take a non-zerodivisor $x \in R_d$. Then since $P(R/xR, t) = (1 - t^d)P(R, t)$ and since our assertion is true for $P(R/xR, t)$, we can assert that $a(R) = a(R/xR) - d$ and the order of the pole of $P(R/xR, t) = P(R, t)/(1 - t^d)$ is $d+$ order of the pole of $P(R, t)$ at ∞, we are done.

□

The following "duality theorem" for Poincaré series is called *Stanley's theorem* and is very important for determining the Gorenstein property.

Theorem 7.55 (Stanley) [4] *Let $(R, \mathfrak{m})$ be an H-local H-graded ring finitely generated over a field $k \cong R/\mathfrak{m}$ and let M be a finitely generated R-module. Then the following equality holds true:*

$$P(M, t^{-1}) = \sum_{i=0}^{d} (-1)^i P((\mathrm{H}^i_{\mathfrak{m}}(M))^*, t). \tag{7.4}$$

Proof Since R is a homomorphic image of a polynomial ring $S = k[X_1, \ldots, X_N]$ over k, and since we can consider both sides as graded S-modules, we can assume $R = S$. Then since S is regular, M has a graded finite free resolution

$$0 \to F_r \to F_{r-1} \to \ldots \to F_1 \to F_0 \to M \to 0$$

over S. Since we have additivity of Poincaré series for exact sequences, we have

$$P(M, t) = \sum_{i=0}^{r} (-1)^i P(F_i, t).$$

First, we show the theorem when M is a graded free module. Since a graded free S module is a direct sum of $S(n)$, we can assume that $M = S(n)$, since the Poincaré series of a direct sum is a sum of each component.

Then we have $P(S(n), t) = \dfrac{t^{-n}}{(1 - t^{d_1})(1 - t^{d_2}) \cdots (1 - t^{d_N})}$ and the left-hand side of the theorem is

[4] Richard Peter Stanley, 1944–.

$$P(S(n), t^{-1}) = \frac{t^n}{(1-t^{-d_1})(1-t^{-d_2})\cdots(1-t^{-d_N})}$$
$$= (-1)^N \frac{t^{n+d_1+\ldots+d_N}}{(1-t^{d_1})(1-t^{d_2})\cdots(1-t^{d_N})}.$$

On the other hand, we have seen in Example 7.36 that $H^i_{\mathfrak{m}}(S(n)) = 0$ $(i \neq N)$, $H^N_{\mathfrak{m}}(S(n)) \cong S^*(d_1 + d_2 + \ldots + d_N + n)$ and $P(H^N_{\mathfrak{m}}(S(n)), t) = \dfrac{t^{d_1+d_2+\cdots+d_N+n}}{(1-t^{d_1})(1-t^{d_2})\cdots(1-t^{d_N})}$. Hence our theorem holds if $M = S(n)$.

Now, for any graded S-module M, since $\operatorname{hd}_S M$ is finite, we show the theorem by induction on $\operatorname{hd}_S M$. If M is free, then M is a direct sum of $S(n)$ and the theorem is true. If $\operatorname{hd}_S M > 0$, then take a graded free S module F so that we have an exact sequence $0 \to N \to F \to M \to 0$. Since $\operatorname{hd}_S N = \operatorname{hd}_S M - 1$, the theorem holds for N. We have $P(M, t) = P(F, t) - P(N, t)$ and from the long exact sequence of local cohomology and Matlis duality, we have

$$\sum_{i=0}^{d}(-1)^i P((H^i_{\mathfrak{m}}(M))^*, t) = \sum_{i=0}^{d}(-1)^i P((H^i_{\mathfrak{m}}(F))^*, t) - \sum_{i=0}^{d}(-1)^i P((H^i_{\mathfrak{m}}(N))^*, t).$$

This proves our theorem. □

Remark 7.11 Proposition 7.54 does not hold in the same manner for $H = \mathbb{Z}^n$, $n \geq 3$. On the other hand, we can see from the proof that Theorem 7.55 holds true for any H-grading, although we gave the proof for $H = \mathbb{Z}$ for simplicity.

As a corollary of Theorem 7.55, we have the following important result.

Theorem 7.56 (Criterion for Gorenstein Property) *Let R be a Cohen–Macaulay $\mathbb{Z}$-graded ring of dimension d. Then we have the following:*

(1) *We have $P(K_R, t) = (-1)^d P(R, t^{-1})$.*
(2) *If R is Gorenstein, then we have*

$$P(R, t^{-1}) = (-1)^d t^{-a} P(R, t) \qquad (*)$$

for $a = a(R)$. Conversely, if $P(R, t^{-1}) = (-1)^d t^{-a} P(R, t)$ holds and if R is an integral domain, then R is Gorenstein.

Proof Assertion (1) follows directly from equation (7.4).

(2) If $K_R \cong R(a)$, since $P(K_R, t) = t^{-a} P(R, t)$, we have the equation $(*)$. Conversely, if R is a Cohen–Macaulay domain and if we have $(*)$, $(K_R)_a \neq 0$. Take $0 \neq w \in (K_R)_a$. Since K_R is a torsion-free R-module, $P(wR, t) = t^{-a} P(R, t) = P(K_R, t)$, we have $K_R = wR$, showing that $K_R \cong R(a)$. Hence R is Gorenstein.

□

Example 7.57 Put $R = A\sharp B$, $A \cong k[X, Y, Z]/(X^3 + Y^3 + Z^3) \cong B$. Then since $P(A, t) = \dfrac{1 - t^3}{(1 - t)^3}$, putting $H(A, n) = \dim_k A_n$, we get easily, $H(A, n) = 3n$ $(n \geq 1)$, $H(A, 0) = 1$. Since we have $H(R, n) = H(A, n)H(B, n)$, we get $H(R, n) = 9n^2$ $(n \geq 1)$, $H(R, 0) = 1$. Hence we get the Poincaré series

$$P(R, t) = \frac{1 + 6t + 12t^2 - t^3}{(1 - t)^3}.$$

The coefficient -1 of t^3 shows that R is not Cohen–Macaulay. Also, by Theorem 7.49 (5), we have $K_R \cong R$. Comparing

$$P(R, t^{-1}) = -\frac{-1 + 12t + 6t^2 + t^3}{(1 - t)^3}$$

with equation (7.4), $P(R, t^{-1}) + P(K_R, t) = \dfrac{2 - 6t + 6t^2 - 2t^3}{(1 - t)^3} = 2$ shows that $\mathrm{H}^2_{\mathfrak{m}}(R) \cong k^2$.

Remark 7.12 (Hilbert–Samuel Polynomial) If R is generated by R_1 over $R_0 = k$, we can take a parameter system from R_1,[5] hence by Proposition 7.54, we can put $P(R, t) = \dfrac{f(t)}{(1 - t)^n}$. Then we can expand

$$P(R, t) = \frac{a_n}{(1 - t)^n} + \frac{a_{n-1}}{(1 - t)^{n-1}} + \ldots + \frac{a_1}{1 - t} + g(t) \text{ ($g(t)$ is a polynomial).}$$

Since $\dfrac{a_k}{(1 - t)^k} = a_k \sum_{m \geq 0} \begin{pmatrix} m + k - 1 \\ k - 1 \end{pmatrix} t^m$, the coefficient of t^m of $P(R, t)$ is a polynomial of m. If we put

$$\mathrm{H}(R.m) = \dim_k R_m,$$

when $m > \deg g(t)$, $\mathrm{H}(R, m) = \sum_{k \geq 1} a_k \begin{pmatrix} m + k - 1 \\ k - 1 \end{pmatrix}$ is a polynomial of m.

In particular, if R is a Cohen–Macaulay ring, then the following conditions are equivalent:

(1) For every $m \geq 0$, $\mathrm{H}(R, m)$ is a polynomial of m.
(2) $a(R) < 0$.

In other words, we can say, "$\mathrm{H}(R, m)$ is a polynomial of m for $m \geq a(R)$".

[5] If necessary, take a field extension—$P(R, t)$ does not change under a field extension.

7.5 Semigroup Rings

It is very natural to consider the semigroup ring of a semigroup. This procedure reflects properties of each side and is very useful to investigate. Also, it is a very good way to make nice examples of rings.

7.58 (Notation) In this section, a *semigroup* means a subset S of $\mathbb{Z}^n$, closed under addition and containing $\mathbf{0}$. Also, we assume always that all semigroups are finitely generated. We write $\langle a_1, a_2, \dots, a_m \rangle$ for the semigroup generated by $a_1, a_2, \dots, a_m \in \mathbb{Z}^n$. Namely,

$$\langle a_1, a_2, \dots, a_n \rangle = \{\sum n_i a_i \mid n_i \in \mathbb{N}\}.$$

Recall that we always assume $0 \in \mathbb{N}$.

For a semigroup $S \subseteq \mathbb{Z}^n$ and a ring A, we put

$$A[S] = A[T^s \mid s \in S] \subseteq A[T_1, T_1^{-1}, \dots, T_n, T_n^{-1}],$$

where we put $T^s = T_1^{s_1} \cdots T_n^{s_n}$ for $s = (s_1, \dots, s_n) \in \mathbb{Z}^n$, and call it the *semigroup ring* of S over A. In this section we treat the case where $A = k$ is a field[6] for simplicity.

If k is a field, for property $\mathbf{P}$ of a ring treated in this section, "$k[S]$ satisfies the property $\mathbf{P}$" does not depend on k.

If $S = \langle a_1, a_2, \dots, a_n \rangle$, then we have

$$k[S] = k[T^{a_1}, \dots, T^{a_n}].$$

If we put

$$H = S - S := \{s - s' \mid s, s' \in S\},$$

the subgroup of $\mathbb{Z}^n$ generated by S, then the semigroup ring $R = k[S]$ can be considered as an H-graded ring, putting $T^s \in R_s$. We consider $R = k[S]$ always as an H-graded ring in this manner.

Let us look at some examples.

Example 7.59

(1) $\mathbb{N}^n, \mathbb{Z}^n$ are finitely generated semigroups and we see that $k[\mathbb{N}^n] = k[T_1, \dots, T_n]$ (resp. $k[\mathbb{Z}^n] = k[T_1, T_1^{-1}, \dots, T_n, T_n^{-1}]$) is a polynomial ring (resp. Laurent polynomial ring) of n variables over k.

[6] If A contains a field k, then since $k[S] \subseteq A[S]$ is a faithfully flat extension of rings, for most instances of property $\mathbf{P}$ treated in this book, $A[S]$ satisfies the property $\mathbf{P}$ if and only if both $k[S]$ and A satisfy the property $\mathbf{P}$.

(2) For a positive integer r, if we put

$$S = \{(m_1, \ldots, m_n) \in \mathbb{N}^n \mid \sum_{i=1}^{n} m_i \equiv 0 \pmod{r}\},$$

then $k[S] = k[T_1, \ldots, T_n]^{(r)}$ is the rth Veronese subring of $k[T_1, \ldots, T_n]$.

(3) If $S = \{0, 2, 3, \ldots\} \subseteq \mathbb{N}$, then we can write $S = \langle 2, 3 \rangle$. Then we have $k[S] = k[T^2, T^3] \cong k[X, Y]/(X^3 - Y^2)$.

(4) If $S = \{0, 3, 4, 5, \ldots\} \subseteq \mathbb{N}$, then we can write $S = \langle 3, 4, 5 \rangle$ and we have $k[S] = k[T^3, T^4, T^5] \cong k[X, Y, Z]/(XZ - Y^2, X^3 - YZ, X^2Y - Z^2)$. (See Chap. 1, Example 1.46.)

First, let us look at subsemigroups of $\mathbb{N}$.

Definition 7.60 (Numerical Semigroups) Let S be a subsemigroup of $\mathbb{N}$. If S has a common divisor $a > 1$, then putting $S' = \{s/a \mid s \in S\}$, the map $\phi : S' \to S$, $\phi(x) = ax$ is a bijection preserving addition. Hence up to isomorphism, we may reduce to the case when GCD of all elements of S is 1. This condition is equivalent to saying that for some $N > 0$, $S \supseteq \{n \in \mathbb{N} \mid n \geq N\}$. We call such S a *numerical semigroup*.

Let us review some notations and properties of numerical semigroups.

Proposition–Definition 7.61 Let $S \subseteq \mathbb{N}$ be a numerical semigroup.

(1) We put $F(S) = \max\{n \in \mathbb{Z} \mid n \notin S\}$ and call it the *Frobenius number* of S. Also, we put $c(S) = F(S) + 1 = \min\{n \in \mathbb{N} \mid n + \mathbb{N} \subseteq S\}$ and call it the *conductor* of S.

(2) We put $g(S) = \sharp\{n \in \mathbb{N} \mid n \notin S\}$ and call it the *genus of* S.[7]

(3) The inequality $2g(S) \geq c(S)$ always holds and the equality holds if and only if the following property (∇) holds:

$$\forall n \in \mathbb{Z}, n \notin S, F(S) - n \in S. \tag{∇}$$

If this property holds, we say that S is *symmetric*.

Proof We see that $c(S) - g(S) = \sharp\{s \in S \mid 0 < s < F(S)\}$. Hence it suffices to show that $\sharp\{s \in S \mid s < F(S)\} \leq g(H)$. For a given $s \in S, s < F(S)$, if $F(S) - s \in S$ holds, then we have a contradiction $F(S) \in S$, which shows that $F(S) - s \notin S$. Hence we have an injection

$$\{s \in S \mid s < F(S)\} \to \{n \in \mathbb{N} \mid n \notin S\}$$

[7] There is a strong connection between $g(S)$ and the genus of a smooth curve over $\mathbb{C}$ via the concept of "Weierstrass semigroup" of a point P on a curve C.

and we get $c(S) - g(S) \leq g(S)$. If the equality holds, then the injection is a bijection and the condition (∇) holds. □

Theorem 7.62 *Let $S \subseteq \mathbb{N}$ be a numerical semigroup and $R = k[S]$ be the semigroup ring of S. We consider that R is a $\mathbb{Z}$-graded ring by $T^s \in R_s$ and we put $\mathfrak{m} = R_+$. The following statements hold:*

(1) *R is a Cohen–Macaulay graded domain with $a(R) = F(S)$.*
(2) *R is a Gorenstein ring if and only if S is a symmetric semigroup.*

Proof (1) Let a be the minimal positive integer in S. Then since $k[T^a] \subseteq R \subseteq k[T]$, we have $\dim R = \dim k[T^a] = 1$. Being a one-dimensional domain, R is Cohen–Macaulay.

Next, look at the following exact sequence of graded R-modules:

$$0 \to R \to k[T, T^{-1}] \to k[T, T^{-1}]/R \to 0.$$

For any $s \in S$, $s \neq 0$, the multiplication by T^s acts bijectively on $k[T, T^{-1}]$ and the action is nilpotent on $k[T, T^{-1}]/R$. Hence $k[T, T^{-1}]$ (resp. $k[T, T^{-1}]/R$) satisfies the condition $(**)$ (resp. $(*)$) of Definition 5.28. Hence from the long exact sequence of local cohomology groups, we get $\mathrm{H}^i_{\mathfrak{m}}(R) = 0$ $(i \neq 1)$ and

$$\mathrm{H}^1_{\mathfrak{m}}(R) \cong k[T, T^{-1}]/R.$$

Since the basis of $k[T, T^{-1}]/R$ over k is $\{T^n \mid n \in \mathbb{Z}, n \notin S\}$, we get

$$a(R) = \max\{n \mid \mathrm{H}^1_{\mathfrak{m}}(R)_n \neq 0\} = F(S).$$

Also, R is Gorenstein if and only if $K_R \cong R(a(R))$, which is equivalent to $\{-n \mid n \in \mathbb{Z} \setminus S\} = \{-F(S) + s \mid s \in S\}$, namely, to the condition (∇). □

Exercise 7.1 Let S be a numerical semigroup and $R = k[S]$.

(1) If $S = \langle p, q \rangle$ $((p, q) = 1)$, show that $R \cong k[X, Y]/(Y^p - X^q)$ and $a(R) = F(S) = (p-1)(q-1) - 1$.
(2) If $S = \langle a, b, c \rangle$ is symmetric then reordering (a, b, c) if necessary, we can assume that for some positive integer d, $a = da', b = db', (d, c) = 1, c \in \langle a', b' \rangle$. In particular, if any 2 of a, b, c are relatively prime, then S is *not* symmetric. (Theorem 6.57 and [Wa1], Proposition 3).
(3) Using (2) show that $\langle 5, 6, 7 \rangle$ and $\langle 5, 6, 8 \rangle$ are not symmetric and $\langle 7, 6, 8 \rangle$ and $\langle 9, 6, 8 \rangle$ are symmetric.[8]

[8] By this condition, if a, b, c are small, then we can "instantly" know if $\langle a, b, c \rangle$ is symmetric or not.

(4) (Herzog[9]; [He]) If $S = \langle a, b, c\rangle$ is not symmetric, then we have

$$k[S] \cong k[X, Y, Z]/\mathfrak{p}(a, b, c),$$

where $\mathfrak{p}(a, b, c)$ is generated by 3 maximal minors of the following 2×3 matrix:

$$\begin{pmatrix} X^{\alpha} & Y^{\beta} & Z^{\gamma} \\ Y^{\beta'} & Z^{\gamma'} & X^{\alpha'} \end{pmatrix},$$

where $\alpha + \alpha' = \min\{n > 0 \mid na \in \langle b, c\rangle\}$ and $(\alpha + \alpha')a = \beta' b + \gamma c$ holds. Also we have the following equalities:

$$\begin{aligned} a &= \beta\gamma + \beta'\gamma + \beta'\gamma, \\ b &= \gamma\alpha + \gamma'\alpha + \gamma'\alpha', \\ c &= \alpha\beta + \alpha'\beta + \alpha'\beta'. \end{aligned}$$

(5) (Delorme, [Del])* Let $S = \langle a_1, a_2, \ldots, a_n\rangle$ and assume that R is a complete intersection. Then we can take subsets M, N of $\mathbb{N}$ so that

$$\{a_1, a_2, \ldots, a_n\} = cM \cup dN,$$

where $(c, d) = 1$ and the semigroup ring $k[M]$ (resp. $k[N]$) is a complete intersection. For example, if $S = \langle 15, 20, 14, 21\rangle$, then $\{15, 20, 14, 21\} = 5\{3, 4\} \cup 7\{2, 3\}$.

(6) (Bresinsky [Bre])* If $S = \langle a, b, c, d\rangle$ is a *symmetric* numerical semigroup generated by 4 elements and if $k[S]$ is not a complete intersection, then putting $k[S] = k[X_1, X_2, X_3, X_4]/\mathfrak{p}(a, b, c, d)$, $\mu(\mathfrak{p}(a, b, c, d)) = 5$ and $\mathfrak{p}(a, b, c, d)$ is generated by the 4×4 Pfaffians of an anti-symmetric 5×5 matrix. See also Chap. 6, Sect. 6.4 and also [BFS, Wa7].

7.63 Next, let us observe a semigroup $S \subseteq \mathbb{Z}^n$. If we let $H = S - S := \{s - s' \mid s, s' \in S\}$ be the group generated by S. Then, since H is a subgroup of $\mathbb{Z}^n$, $H \cong \mathbb{Z}^r$ for some r. Since $k[S]$ is an affine ring over k and $k[S]$ and $k[H] \cong k[T_1, T_1^{-1}, \ldots, T_r, T_r^{-1}]$ has the same field of fractions, we have $\dim k[S] = r$ (see Theorem 3.25).

Before considering general semigroups, let us explain the procedure to obtain the essential part necessary for our study.

Proposition 7.64 *Let $S \subseteq \mathbb{Z}^n$ be a semigroup and assume that there exists $s \in S$, $s \neq 0$, $-s \in S$. Then if G is a maximal group contained in S, then S decomposes as $S \cong S' \times G$ and S' does not contain a non-trivial additive group. In this case, $k[S]$ is Cohen–Macaulay (resp. Gorenstein) if and only if so is $k[S']$.*

---

[9] Jürgen Herzog, 1941–2024.

Proof Let $G \subseteq S$ be a subgroup and let $L \supseteq G$ be the inverse image of torsion of $\mathbb{Z}^n/G$ by the canonical surjection $\mathbb{Z}^n \to \mathbb{Z}^n/G$. Since G is a maximal subgroup of S, L/G is a finite group and $\mathbb{Z}^n/L$ is torsion free and we have $S \cap L = G$. Then $\mathbb{Z}^n$ decomposes as $\mathbb{Z}^n = L \oplus L'$. Let $\pi : \mathbb{Z}^n \to \mathbb{Z}^n/L \cong L'$ be the projection and put $S' = \pi(S)$. Then S' does not have a non-trivial subgroup and we have $S \cong G \times S'$. Hence we have $R = k[S] \cong k[S'] \otimes_k k[G]$ gives us the desired result. □

In the following, we assume that our semigroup S does not have a non-trivial subgroup.

Next, we introduce a criterion for $k[S]$ to be Cohen–Macaulay when $\dim k[S] \geq 2$ and S is *simplicial*.

Definition 7.65 Let $S \subseteq \mathbb{Z}^n$ be a semigroup and assume that $G = S - S \cong \mathbb{Z}^r$. If there exist $e_1, \ldots, e_r \in S$, which are linearly independent over $\mathbb{Q}$ (in $\mathbb{Q}^n = \mathbb{Z}^n \otimes_{\mathbb{Z}} \mathbb{Q}$) and if any element $s \in S$ can be written as $s = \sum_{i=1}^r a_i e_i$, $a_i \in \mathbb{Q}$, $a_i \geq 0$, then we call S a *simplicial semigroup*. In this case, if we put $x_i = T^{e_i}$ $(i = 1, \ldots, r)$, then it is easy to see that $x_1, \ldots, x_r$ make a system of parameters of $k[S]$.

Example 7.66 Let $S \subseteq \mathbb{Z}^n$ be a semigroup.

(1) If $\dim k[S] \leq 2$, then S is always simplicial.
(2) If $S = \langle (0, 0, 1), (1, 0, 1), (0, 1, 1), (1, 1, 1) \rangle \subseteq \mathbb{Z}^3$, then $\dim k[S] = 3$ and S is not simplicial. In fact, $k[S] \cong k[T_1, T_2] \sharp k[U_1, U_2]$ (Segre product of $k[T_1, T_2]$ and $k[U_1, U_2]$; see Definition 7.46).

Let us introduce a simple criterion to determine the Cohen–Macaulay property of a simplicial semigroup.

Theorem 7.67 *Let S be a simplicial semigroup and put $R = k[S]$. Let $e_1, \ldots, e_r \in S$ be a sequence of elements of S satisfying the condition in Definition 7.65. Then the following conditions are equivalent:*

(1) *R is Cohen–Macaulay.*
(2) *R satisfies Serre's condition (S_2).*
(3) *If $s, s' \in S$ and e_i, e_j $(i \neq j)$ satisfies $s + e_i = s' + e_j$, then $s - e_j = s' - e_i \in S$.*

Proof First, let us note that condition (3) is equivalent to saying that putting $x_i = T^{e_i}$ $(i = 1, \ldots, r)$, for every (i, j), (x_i, x_j) is a regular sequence.

In fact, for an element $f \in R$, assume that $x_j f \in x_i R$. Since R is an H-graded ring, we may assume that $f = T^s, s \in S$. Then there exists $s' \in S$ such that $s + e_j = s' + e_i$ and by condition (3), $s - e_i \in S$ and we show that $f \in x_i R$.

Now, since (1) $\Rightarrow$(2) $\Rightarrow$(3) is trivial, we show (3) $\Rightarrow$(1). For that purpose, it suffices to show that $(x_1, \ldots, x_r)$ is a regular sequence. Since each $(x_1, \ldots, x_i) : x_{i+1}$ is an H-graded ideal, we have only to show that if $f \in R$ is a homogeneous element and $f \in (x_1, \ldots, x_i) : x_{i+1}$, then $f \in (x_1, \ldots, x_i)$.

Let $f = T^s$ and assume that $f x_{i+1} \in (x_1, \ldots, x_i)$. Then there exists $s' \in S$ and $j \leq i$ such that $s + e_{i+1} = s' + e_j$. Then using condition (3) for e_{i+1} and e_j, we get $s - e_j \in S$ and we show that $f \in x_j R \subseteq (x_1, \ldots, x_i)$. □

Semigroup rings which are normal are extensively investigated as a local theory of "toric geometry" in algebraic geometry. We leave such theory to other literature, for example, [BH] but we only introduce Hochster's theorem for normal semigroup rings.

The original proof of Theorem 7.68 in [Ho1] was rather difficult. But nowadays, it will be easier to understand that "a pure subring of a regular ring is F-regular and hence Cohen–Macaulay" (see Chap. 10, Sect. 10.2).

The proof in this paper was rather difficult at that time. But nowadays, it will be easier to understand that "a pure subring of a regular ring is F-regular and hence Cohen–Macaulay" (see Chap. 10, Sect. 10.2).

Theorem 7.68 (Hochster[10]) *A normal semigroup ring $k[S]$ is a Cohen–Macaulay ring.*

A normal semigroup is defined by a *rational polyhedral cone*, which is defined by a finite number of linear inequalities

$$\sigma = \{(x_1, \dots, x_n) \mid l_i(x_1, \dots, x_n) \geq 0 \ (i = 1, \dots, s)\},$$

where $l_1, \dots, l_s$ are $\mathbb{Q}$ coefficient homogeneous linear forms.

Then it is easy to see that all normal semigroup rings over a field k are of the form

$$k[\sigma \cap \mathbb{Z}^n] = k[X_1^{a_1} \cdots X_n^{a_n} \mid (a_1, \dots a_n) \in \sigma \cap \mathbb{Z}^n].$$

$\tau \subseteq \sigma$ is a *face* of σ if τ is of the form $\tau = \sigma \cap H_{l'}$, where l' is a $\mathbb{Q}$ coefficient linear form defined by $H_{l'} = \{(x_1, \dots, x_n) \mid l'(x_1, \dots, x_n) \geq 0\}$. If τ is a face of σ, we write $\tau \prec \sigma$.

If σ is defined as above, we define

$$\text{relint}(\sigma) = \{(x_1, \dots, x_n) \in \sigma \mid l_i(x_1, \dots, x_n) > 0 \ (i = 1, \dots, s)\}$$

and in the same way for $\tau \prec \sigma$. If $\sigma = \{0\}$, then $\text{relint}(\sigma) = \{0\}$.

The most typical example is $k[X_1, \dots, X_n] = k[\sigma \cap \mathbb{Z}^n]$, where σ is defined by $x_i \geq 0 \ (i = 1, 2, \dots, n)$.

Now we state the canonical module formula for $R = k[\sigma \cap \mathbb{Z}^n]$.

Theorem 7.69 (Stanley's Theorem) *If $R = k[\sigma \cap \mathbb{Z}^n]$ be a normal semigroup ring. Then $K_R \cong k[\text{relint}(\sigma) \cap \mathbb{Z}^n]$.*

The detailed proof of this theorem can be found in [BH] but the following argument shows the essential part of the proof.

---

[10] Melvin Hochster, 1943–.

Proof It is easy to see that

$$P(R, (t_1, \ldots, t_n)) = \Sigma_{\tau \prec \sigma} t^{\text{relint}(\tau)\cap\mathbb{Z}^n},$$

where $t^{\text{relint}(\tau)\cap\mathbb{Z}^n} = \sum_{(a_1,\ldots,a_n)\in\text{relint}(\tau)\cap\mathbb{Z}^n} t_1^{a_1} t_2^{a_2} \cdots t_n^{a_n}$.

Since $P(K_R, t) = (-1)^d P(R, t^{-1})$ by Theorem 7.55, we get the desired result. □

Example 7.70 If $R = k[X_1, X_2] = k[\sigma \cap \mathbb{Z}^2]$, then faces of σ are $\{\sigma, \tau_1 = \sigma \cap (x_1 = 0), \tau_2 = \sigma \cap (x_2 = 0), \tau_3 = \{0\}$. Then

$$R(R, (t_1, t_2)) = 1 + \sum_{n\geq 1} t_1^n + \sum_{n\geq 1} t_2^n + \sum_{n_1,n_2\geq 1} t_1^{n_1} t_2^{n_2}$$
$$= 1 + \frac{t_1}{1-t_1} + \frac{t_2}{1-t_2} + \frac{t_1 t_2}{(1-t_1)(1-t_2)},$$

corresponding to relint τ_3, relint τ_2, relint τ_1, relint σ. Then we have

$$(-1)^2 P(R, t^{-1}) = 1 - \frac{1}{1-t_1} - \frac{1}{1-t_2} + \frac{1}{(1-t_2)(1-t_2)} = \frac{t_1 t_2}{(1-t_1)(1-t_2)}.$$

Exercise 7.2 Let $S \subseteq \mathbb{Z}^n$ be a semigroup and $H = S - S \subseteq \mathbb{Z}^n$ be the subgroup generated by S. We define that S is *normal* if for every $x \in H$, if $mx \in S$ for some positive integer, then $x \in S$. Then show that the following conditions are equivalent:

(1) S is normal.
(2) $k[S]$ is a normal domain.

Exercise 7.3 Let $S \subseteq \mathbb{Z}^n$ be a semigroup and let $H = S - S$ be the group generated by S. Let $T \subseteq \mathbb{Z}^n$ be another semigroup containing S and satisfying the condition $S = T \cap H$. Show the following statements:

(1) If T is normal, then so is S.
(2) Natural inclusion of rings $k[S] \to k[T]$ is a split monomorphism as $k[S]$-modules.
(3) If $k[T]$ is regular, then can we say that $k[S]$ is regular, too?

7.6 Simplicial Complexes and Stanley–Reisner Rings

The concept of simplicial complexes is very important in topology. For a simplicial complex Δ, we can define the *Stanley–Reisner ring* $k[\Delta]$ of Δ. Using $k[\Delta]$, we can solve many geometric or combinatorial problems.

Actually, Richard Stanley[11] [St1] solved the "Upper Bound Conjecture" concerning triangular decomposition of spheres. Also, G. A. Reisner [Rei] showed that the Cohen–Macaulay property of $k[\Delta]$ is "topological"; namely, if Δ and Δ' are triangulations of same (homeomorphic) topological spaces, then $k[\Delta]$ is Cohen–Macaulay if and only if so is $k[\Delta']$.[12]

Definition 7.71 We define $[n] = \{1, 2, \ldots, n\}$.

(1) If $\Delta \subseteq 2^{[n]}$ satisfies the condition

$$\text{If } \sigma \in \Delta \text{ and } \tau \subseteq \sigma, \text{ then } \tau \in \Delta,$$

then we say Δ is a *simplicial complex* on $[n]$.

(2) We call $\sigma \in \Delta$ a *face* of Δ and a *facet* if σ is *maximal* among the faces of Δ with respect to the inclusion relation.

(3) We define $\dim \sigma = |\sigma| - 1$ for $\sigma \in \Delta$ and define

$$\dim \Delta = \max\{\dim \sigma \mid \sigma \in \Delta\}.$$

We say $\sigma \in \Delta$ is a *vertex* of Δ if $\dim \sigma = 0$.

(4) Let k be a ring (usually we consider the case where k is a field). We put $S = k[X_1, \ldots, X_n]$, a polynomial ring over k and define $X_\tau = \prod_{i \in \tau} X_i$. Then we define I_Δ to be the ideal generated by $\{X_\tau \mid \tau \notin \Delta\}$ and

$$k[\Delta] = S/I_\Delta.$$

We call $k[\Delta]$ the *face ring* (or *Stanley–Reisner ring*) of Δ over k.

7.72 In this section, we always assume that $\Delta \subseteq [n]$, k is a field and let

$$S = k[X_1, \ldots, X_n], k[\Delta] = S/I_\Delta.$$

Also, we consider S is an H-graded ring over $H = \mathbb{Z}^n$ as in Sect. 7.2. Since I_Δ is an H-graded ideal, we can consider $k[\Delta]$ as an H-graded ring. Note that a homogeneous element of S by this grading is a monomial and hence H-graded ideals are generated by monomials. In particular H-graded *prime* ideals are of the form $(X_{i_1}, \ldots, X_{i_r})$.

We always consider $k[\Delta]$ as a $\mathbb{Z}^n$-graded ring.

Remark 7.13 For a ring $R = S/I$, the following conditions are equivalent:

(1) $R = k[\Delta]$ for some simplicial complex Δ.
(2) I is generated by monomials and R is reduced.
(3) I is generated by monomials without square factor.

---

[11] Richard Stanley, 1944–.

[12] The Gorenstein property of $k[\Delta]$ depends on the triangulation except for spheres.

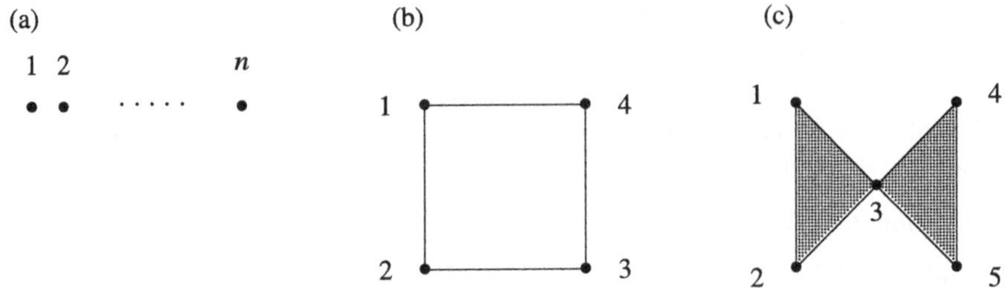

Fig. 7.1 Visualizations of simplicial complexes

Example 7.73 When we write a simplicial complex Δ, we only need to indicate the *facets* of Δ by Definition 7.71. When the set of facets of Δ is $\{\sigma_1, \sigma_2, \dots, \sigma_r\}$, then we write

$$\Delta = \langle \sigma_1, \sigma_2, \dots, \sigma_r \rangle.$$

Also a geometric description of Δ is used sometimes. For example, If Δ is as Fig. 7.1a, b, c, then $k[\Delta]$ is as follows:

(1) $\Delta = \langle \{1\}, \{2\}, \dots, \{n\} \rangle$, $I_\Delta = (X_i X_j \mid i \neq j)$,
(2) $\Delta = \langle \{1, 2\}, \{2, 3\}, \{3, 4\}, \{4, 1\} \rangle$, $I_\Delta = (X_1 X_3, X_2 X_4)$,
(3) $\Delta = \langle \{1, 2, 3\}, \{3, 4, 5\} \rangle$, $I_\Delta = (X_1 X_4, X_1 X_5, X_2 X_4, X_2 X_5)$.

Let us determine the dimension of $k[\Delta]$.

Proposition–Definition 7.74

(1) For $\sigma \in \Delta$, we define

$$\mathfrak{p}_\sigma = (X_i \mid i \notin \sigma).$$

Since $S/\mathfrak{p}_\sigma \cong k[X_i \mid i \in \sigma]$, $\mathfrak{p}_\sigma$ is a prime ideal and contains I_Δ.
(2) The set of H-graded prime ideals of $k[\Delta]$ coincides with

$$\{\mathfrak{p}_\sigma \mid \sigma \in \Delta\}$$

and if $\tau \subseteq \sigma$, then $\mathfrak{p}_\tau \supseteq \mathfrak{p}_\sigma$.
(3) $I_\Delta = \bigcap_{\sigma \in \Delta} \mathfrak{p}_\sigma$.
(4) A minimal prime ideal of I_Δ is $\mathfrak{p}_\sigma$ for some facet σ of Δ. Namely, there is a one-to-one correspondence between the minimal primes of $k[\Delta]$ and the facets of Δ.
(5) $\dim k[\Delta] = \dim \Delta + 1$.

Proof Take a generator X_τ of $I_\Delta, \tau \notin \Delta$. Then for any $\sigma \in \Delta$, there exists $i \in \tau, i \notin \sigma$. Hence $X_\tau \in \mathfrak{p}_\sigma$. Namely, we get $I_\Delta \subseteq \mathfrak{p}_\sigma$. Conversely, for $\sigma \subseteq [n]$, if $I_\Delta \subseteq \mathfrak{p}_\sigma$, then we have $\sigma \in \Delta$. Since H-homogeneous elements of S are monomials, we have also (2). The rest will follow easily. □

The aim of this section is to express the local cohomology modules of $k[\Delta]$ using the cohomology groups of Δ. For that purpose, we consider Čech complex $C^\bullet$ with respect to the sequence $(X_1, \dots, X_n)$ (see Chap. 5, Definition 5.88). Since the maximal ideal $\mathfrak{m} = (X_1, \dots, X_n)$, we have

$$\mathrm{H}^i_{\mathfrak{m}}(k[\Delta]) \cong \mathrm{H}^i(C^\bullet)$$

by Theorem 5.105. Since each X_i is an H-homogeneous element, the complex $C^\bullet$ is a complex of H-graded modules. Also, to express $\mathrm{H}^i_{\mathfrak{m}}(k[\Delta])_{\boldsymbol{a}}$ for $\boldsymbol{a} = (a_1, \dots, a_n)$, the signature of $(a_1, \dots, a_n)$ is very important.

Definition 7.75 For $\boldsymbol{a} = (a_1, \dots, a_n) \in \mathbb{Z}^n$, we define:

(1) $\mathrm{sgn}(\boldsymbol{a}) = (\mathrm{sgn}\,(a_1), \dots, \mathrm{sgn}\,(a_n)) \in \mathbb{Z}^n$, where for $a \in \mathbb{Z}$ we put $\mathrm{sgn}\,(a) = 1$ (resp. 0, resp. -1) if $a > 0$ (resp. $a = 0$, resp. $a < 0$).
(2) $\mathrm{neg}(\boldsymbol{a}) = \{i \in [n] \mid a_i < 0\}$, $\mathrm{pos}(\boldsymbol{a}) = \{i \in [n] \mid a_i > 0\} \subseteq [n]$. We write $\boldsymbol{a} \geq \boldsymbol{0}$ if $\mathrm{neg}(\boldsymbol{a}) = \emptyset$.

Let us express the Poincaré series of $k[\Delta]$. We assume that $\dim k[\Delta] = d$.

Proposition 7.76 *The Poincaré series of $k[\Delta]$ as $\mathbb{Z}^n$-graded ring is given by*

$$P(k[\Delta], T_1, \dots, T_n) = \sum_{\boldsymbol{a} \in \mathbb{Z}^n, \boldsymbol{a} \geq \boldsymbol{0}, \mathrm{pos}(\boldsymbol{a}) \in \Delta} T^{\boldsymbol{a}} = \sum_{\sigma \in \Delta} \frac{\prod_{i \in \sigma} T_i}{\prod_{i \in \sigma}(1 - T_i)}.$$

If we consider $k[\Delta]$ as a $\mathbb{Z}$-graded ring, putting $\deg X_i = 1$ ($\forall X_i$), then we have

$$P(k[\Delta], T) = \sum_{i=0}^{\dim \Delta + 1} f_{i-1} \frac{T^i}{(1 - T)^i} = (1 - T)^{-d} \sum_{i=1}^{d} f_{i-1} T^i (1 - T)^{d-i},$$

where we put $f_i = \sharp\{\sigma \in \Delta \mid \dim \sigma = i\}$ ($i \geq 0$) and $f_{-1} = 1$.

Proof For $\boldsymbol{a} \in \mathbb{Z}^n$, $\boldsymbol{a} \geq \boldsymbol{0}$, we can see that $T^{\boldsymbol{a}} \notin I_\Delta$ if and only if $\mathrm{pos}(\boldsymbol{a}) \in \Delta$. Also, for $\sigma \in \Delta$, we have $\sum_{\boldsymbol{a} \geq 0, \mathrm{pos}(\boldsymbol{a}) = \sigma} T^{\boldsymbol{a}} = \dfrac{\prod_{i \in \sigma} T_i}{\prod_{i \in \sigma}(1 - T_i)}$. Hence we have our result. □

To compute the local cohomology groups $\mathrm{H}^i_{\mathfrak{m}}(k[\Delta])$ of $k[\Delta]$, let us note the following facts.

Proposition 7.77 *The following statements hold for $\mathbb{Z}^n$-graded module $\mathrm{H}^i_{\mathfrak{m}}(k[\Delta])$:*

(1) *For $\boldsymbol{a}, \boldsymbol{b} \in \mathbb{Z}^n$, if $\mathrm{sgn}\,(\boldsymbol{a}) = \mathrm{sgn}\,(\boldsymbol{b})$, then $\mathrm{H}^i_{\mathfrak{m}}(k[\Delta])_{\boldsymbol{a}} \cong \mathrm{H}^i_{\mathfrak{m}}(k[\Delta])_{\boldsymbol{b}}$ as k-vector spaces.*
(2) *Let $\boldsymbol{a} = (a_1, \dots, a_n) \in \mathbb{Z}$. If $a_i > 0$ for some i, then for any $j \geq 0$, we have $\mathrm{H}^j_{\mathfrak{m}}(k[\Delta])_{\boldsymbol{a}} = 0$.*

Proof

(1) follows from the construction of $C^{\bullet}$. If sgn $(\boldsymbol{a})$ = sgn $(\boldsymbol{b})$, then for each i, $C^i_{\boldsymbol{a}}$ and $C^i_{\boldsymbol{b}}$ are clearly isomorphic.

(2) By Remark 5.2, $\mathrm{H}^j_{\mathfrak{m}}(k[\Delta])$ is an Artinian module for every j. If $\mathrm{H}^j_{\mathfrak{m}}(k[\Delta])_{\boldsymbol{a}} \neq 0$ for some $\boldsymbol{a} = (a_1, \ldots, a_n) \in \mathbb{Z}$ with $a_i > 0$. Then putting

$$M_n = \bigoplus_{(b_1,\ldots,b_n)\in\mathbb{Z}, b_i \geq n} \mathrm{H}^j_{\mathfrak{m}}(k[\Delta])_{(b_1,\ldots,b_n)},$$

M_n is a $k[\Delta]$ submodule of $\mathrm{H}^j_{\mathfrak{m}}(k[\Delta])$ and $\{M_n\}_{n\geq 1}$ form an infinite decreasing sequence of $\mathrm{H}^j_{\mathfrak{m}}(k[\Delta])$ and contradicts the fact that $\mathrm{H}^j_{\mathfrak{m}}(k[\Delta])$ is an Artinian module.

□

The subcomplex $\mathrm{lk}_{\Delta}(\sigma)$ (the *link* of σ) for $\sigma \in \Delta$ plays a very important role in our theory.

Definition 7.78 For $\sigma \in \Delta$, we define

$$\mathrm{lk}_{\Delta}(\sigma) = \{\tau \in \Delta \mid \sigma \cap \tau = \emptyset \text{ and } \sigma \cup \tau \in \Delta\}.$$

Note that $\mathrm{lk}_{\Delta}(\emptyset) = \Delta$.

Example 7.79 Let us see the complexes in Example 7.73. If

$$\Delta = \langle\{1, 2\}, \{2, 3\}, \{3, 4\}, \{4, 1\}\rangle$$

and $\sigma = \{1\}$, then $\mathrm{lk}_{\Delta}(\sigma) = \{\{2\}, \{4\}\}$. If $\Delta = \langle\{1, 2, 3\}, \{3, 4, 5\}\rangle$ and $\sigma = \{3\}$, then $\mathrm{lk}_{\Delta}(\sigma) = \langle\{1, 2\}, \{4, 5\}\rangle$ (Fig. 7.2).

The following Theorem gives the condition for $k[\Delta]$ to be Cohen–Macaulay.

Theorem 7.80 (Reisner's Theorem) *For* $\boldsymbol{a} \in \mathbb{Z}^n$, *we put* $\mathrm{neg}(\boldsymbol{a}) = \sigma \subseteq [n]$. *Then, for* $j \geq 0$, *putting* $\sharp(\mathrm{lk}(\sigma)) = r$, *we have the following isomorphism:*

$$\mathrm{H}^j_{\mathfrak{m}}(k[\Delta])_{\boldsymbol{a}} \cong \tilde{\mathrm{H}}^{j-r-1}(\mathrm{lk}_{\Delta}(\sigma)).$$

(If $\sigma \notin \Delta$, *we have* $\mathrm{H}^j_{\mathfrak{m}}(k[\Delta])_{\boldsymbol{a}} = 0$.*) In particular,* $k[\Delta]$ *is Cohen–Macaulay if and only if* $\forall\sigma \in \Delta, \forall j \neq \dim \mathrm{lk}_{\Delta}(\sigma)$, $\tilde{\mathrm{H}}^j(\mathrm{lk}_{\Delta}(\sigma)) = 0$.

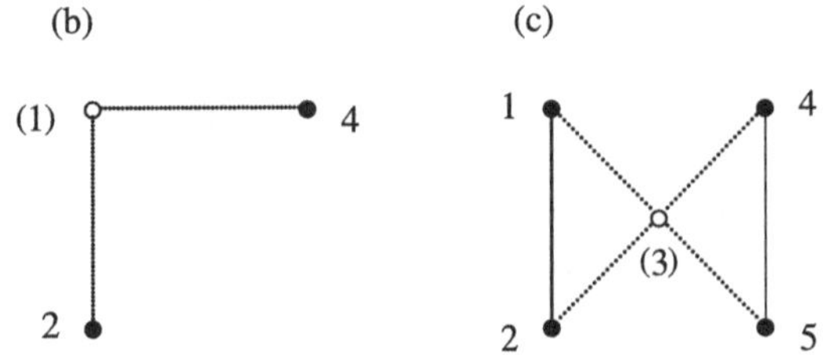

Fig. 7.2 Links of (b), (c) in Fig. 7.1

Proof Let us compute $\mathrm{H}^j_{\mathfrak{m}}(k[\Delta])_{\boldsymbol{a}} \cong \mathrm{H}^j(C^\bullet)_{\boldsymbol{a}}$ for $j < \dim k[\Delta]$. By Proposition 7.77, we may assume that if $\boldsymbol{a} = (a_1, \ldots, a_n)$, we may assume that $a_i \leq 0$ for every i.

Recall the definition of the Čech complex $C^\bullet$:

$$C^r = \oplus_{\tau \subseteq [n], \tau = r} k[\Delta]_{X_\tau}.$$

Note that for every τ and for every $\boldsymbol{a} \in \mathbb{Z}^n$, $(k[\Delta]_{X_\tau})_{\boldsymbol{a}}$ is either a 1-dimensional k-vector space or 0. When is $(k[\Delta]_{X_\tau})_{\boldsymbol{a}} \neq 0$? $k[\Delta]_{X_\tau}$ has negative degree only if $\mathrm{neg}(\boldsymbol{a}) \subseteq \tau$ and if $\tau \notin \Delta$, then $[k[\Delta]_{X_\tau}]_{\boldsymbol{a}} = 0$. Hence $[k[\Delta]_{X_\tau}]_{\boldsymbol{a}} \neq 0$ if and only if $\tau = \mathrm{neg}(\boldsymbol{a}) \cup \tau'$, $\tau' \in \mathrm{lk}_\Delta(\mathrm{neg}(\boldsymbol{a}))$. If we put $\sigma = \mathrm{neg}(\boldsymbol{a})$, $C^\bullet_{\boldsymbol{a}}$ is isomorphic to the complex computing $\tilde{\mathrm{H}}^\bullet(\mathrm{lk}_\Delta(\sigma))$, starting from degree r. Thus we have an isomorphism $\mathrm{H}^j_{\mathfrak{m}}(k[\Delta])_{\boldsymbol{a}} \cong \tilde{\mathrm{H}}^{j-r-1}(\mathrm{lk}_\Delta(\sigma))$ and our proof is complete. □

Let us see the condition $k[\Delta]$ to be Cohen–Macaulay when $\dim \Delta$ is small.

Example 7.81 Let Δ be a simplicial complex and we assume that Δ is equi-dimensional.

(1) If $\dim \Delta = 0$, then $k[\Delta]$ is a reduced ring of dimension 1, hence is Cohen–Macaulay.
(2) If $\dim \Delta = 1$, the proper links are dimension 0. Hence the only non-trivial condition in Theorem 7.80 is $\tilde{\mathrm{H}}^0(\Delta, k) = 0$, which is equivalent to saying that Δ is connected. Namely, $k[\Delta]$ is a Cohen–Macaulay ring if and only if Δ is connected.
(3) In the case $\dim \Delta = 2$, in the same manner, $k[\Delta]$ is Cohen–Macaulay if and only if the following conditions are satisfied:

    (a) Δ is connected and $\tilde{\mathrm{H}}^1(\Delta, k) = 0$.
    (b) For every vertex $\sigma = \{i\}$ of Δ, $\mathrm{lk}_\Delta(\sigma)$ is connected.

(4) Let us give two examples of $\dim \Delta = 2$.

    (a) (See Fig. 7.1c) Let Δ be a union of 2 triangles intersecting at one point x. Then since $\mathrm{lk}_\Delta(\{x\})$ is not connected, $k[\Delta]$ is not Cohen–Macaulay.
    (b) Let Δ be a triangulation of a real projective sphere $\mathbb{P}^2_{\mathbb{R}}$. Then since Δ is a manifold, the link of Δ at every point is the circle S^1. Also we know that $\tilde{\mathrm{H}}^1(\Delta, \mathbb{Z}) \cong \mathbb{Z}/(2)$. Hence $k[\Delta]$ is *not* Cohen–Macaulay if char $(k) = 2$.

An important class of Cohen-Macaulay simplicial complexes are *shellable* complexes. We only explain the definition and Cohen–Macaulay property here and refer to [BH], Chapter 5, for more detailed arguments on shellable complexes.

Example 7.82 Let $\Delta = \langle \sigma_1, \sigma_2, \ldots, \sigma_m \rangle$ be an equi-dimensional simplicial complex of dimension n, whose facets are $\{\sigma_1, \sigma_2, \ldots, \sigma_m\}$. We put $\Delta_i = \langle \sigma_1, \sigma_2, \ldots, \sigma_m \rangle$. We say that Δ is *shellable* if we can order $\sigma_1, \sigma_2, \ldots, \sigma_m$ so that

$$\dim(\Delta_i \cap \sigma_{i+1}) = n - 1$$

for every $i = 1, \dots, m-1$. If Δ is shellable then $k[\Delta]$ is Cohen–Macaulay. Actually, we have a short exact sequence

$$0 \to k[\Delta_{i+1}] \to k[\Delta_i] \oplus k[\sigma_{i+1}] \overset{(p,-p')}{\to} k[\Delta_i \cap \sigma_{i+1}] \to 0, \qquad (**)$$

where $p : k[\Delta_i] \to k[\Delta_i \cap \sigma_{i+1}]$, $p' : k[\sigma_{i+1}] \to k[\Delta_i \cap \sigma_{i+1}]$ are natural surjections. Then taking the local cohomology long exact sequence of $(**)$, since $k[\sigma_{i+1}], k[\Delta_i \cap \sigma_{i+1}]$ are Cohen–Macaulay of dimension $d, d-1$, respectively, we can deduce that $k[\Delta_{i+1}]$ is Cohen–Macaulay, if so is $k[\Delta_i]$.

Now, we see from Proposition 7.77, that if $\mathrm{H}^j_{\mathfrak{m}}(k[\Delta])_{\boldsymbol{a}} \neq 0$ for some $\boldsymbol{a}$ with $\mathrm{neg}(\boldsymbol{a}) \neq \emptyset$, then the length of $\mathrm{H}^j_{\mathfrak{m}}(k[\Delta])$ is infinite. Hence we have the following corollary.

Corollary 7.83 *For every $j < \dim k[\Delta]$, $\mathrm{H}^j_{\mathfrak{m}}(k[\Delta])$ has finite length if and only if*

$$\forall \sigma \in \Delta, \sigma \neq \emptyset, \forall j < \dim \mathrm{lk}_\Delta(\sigma), \quad \tilde{\mathrm{H}}^j(\mathrm{lk}_\Delta(\sigma)) = 0 \qquad (7.5)$$

and if $\mathrm{H}^j_{\mathfrak{m}}(k[\Delta])$ has finite length for some j, then $\mathrm{H}^j_{\mathfrak{m}}(k[\Delta])_{\boldsymbol{a}} \neq 0$ only for $\boldsymbol{a} = \boldsymbol{0}$. If $\mathrm{H}^j_{\mathfrak{m}}(k[\Delta])$ satisfies this condition for all $j < \dim k[\Delta]$, then $k[\Delta]$ is a Buchsbaum ring (see, Chap. 9).

In particular, if the geometric realization of Δ is a manifold, then every proper link of Δ is a sphere and hence $k[\Delta]$ is a Buchsbaum ring.

Next, let us introduce the notion of *order complex*, which is an important class of examples of simplicial complexes and will also play an important role in the next section.

Definition 7.84 (Order Complex) Let P be a partially ordered set. We call a finite partially ordered set a *poset* for short. We denote the order relation in P by $\geq$.

(1) We call a totally ordered subset a *chain* of P.
(2) Let $\Delta(P)$ be the set of chains (including the empty set) of P. Then $\Delta(P)$ satisfies the condition to be a simplicial complex and we call $\Delta(P)$ the *order complex* of P.
(3) We denote $k[\Delta(P)]$, the face ring of $\Delta(P)$, simply by $k[P]$.
(4) We define rank P to be the maximal of length of chains in P, where if C is a chain in P, the length of the chain C is $|C| - 1$. We call rank P the *rank* of P. Note that we have

$$\mathrm{rank}\, P = \dim k[P] - 1.$$

(5) Let $\alpha, \beta \in P$. If $\alpha \leq \beta$ or $\beta \leq \alpha$, we say α and β are *comparable*, and otherwise we write $\alpha \not\sim \beta$ and say α and β are *incomparable*.

(6) Let $\alpha, \beta \in P$. If $\alpha > \beta$ and if there is no element γ such that $\alpha > \gamma > \beta$, then we say α *covers* β.
(7) We write $x(\alpha)$ for the image of $\alpha \in P$ in $k[P]$.
(8) We express a poset P by its *Hasse diagram* $\Gamma(P)$. The vertex set of $\Gamma(P)$ is just P and for $A, B \in P$, we connect A and B so that A is above B if and only if A covers B.
    For example, if $\Gamma(P)$ is given by Fig. 7.3 (2), A covers D, E and $A \not\succ F$.

Let X be a subset of P. If X is *not* a chain of P, then for some $\alpha, \beta \in X$, we have $\alpha \not\succ \beta$. Hence we can assert:

Proposition 7.85 *Let P be a poset. Then we have*

$$k[P] \cong k[X_\alpha | \alpha \in P]/(X_\alpha X_\beta | \alpha \not\succ \beta).$$

Namely, $x(\alpha_1)x(\alpha_2)\cdots x(\alpha_n) = 0$ in $k[P]$ if and only if there exists i, j such that $\alpha_i \not\succ \alpha_j$. In particular, the defining ideal of $k[P]$ is always defined by quadrics.

Remark 7.14 For a simplicial complex Δ, to have $k[\Delta] \cong k[P]$ for a poset P, the condition "the defining ideal I_Δ is generated by quadrics" is not a sufficient condition. For example, if Δ is a pentagon, then I_Δ is generated by quadrics, but not of the form $k[P]$. Actually, for an n-gon Δ, $\Delta = \Delta(P)$ for some poset P if and only if n is even.

Remark 7.15 Let P be a poset and Q be a subset of P. We consider Q as a poset by the order induced from P. Then we have an inclusion $k[Q] \subseteq k[P]$ and a surjection $k[P] \to k[Q] \cong k[P]/(x_\alpha | \alpha \notin Q)$.

The following facts are obvious from the definition.

Proposition 7.86 *Let P be a poset and k be a field. Then for $\alpha \in P$, $x(\alpha) \in k[P]$ is a non-zerodivisor in $k[P]$ if and only if α is comparable to any element of P.*

Example 7.87 Let P, Q be posets illustrated in Fig. 7.3.

(1) $k[P] \cong k[A, B, C, D, E, F]/(AB, CD, EF)$, and $|\Delta(P)|$ is a 2-dimensional sphere.

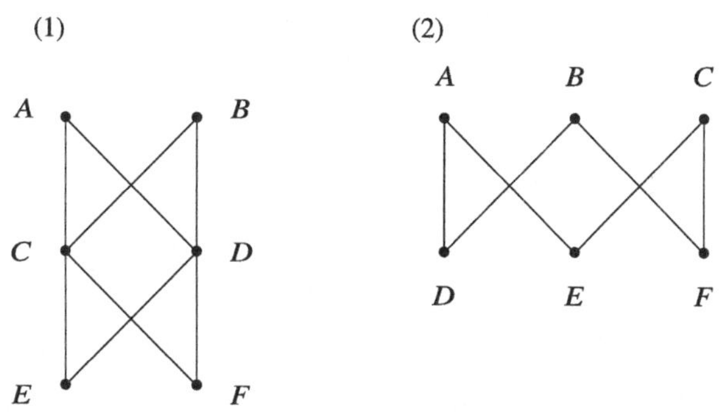

Fig. 7.3 Hasse diagrams of the posets

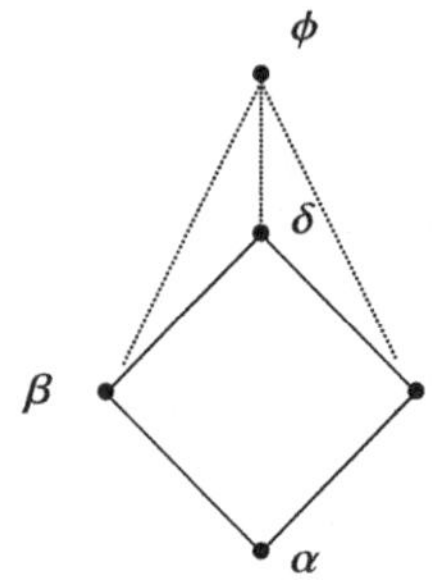

Fig. 7.4 Explanation of definition of wonderful posets

(2) $k[Q] \cong k[A, B, C, D, E, F]/(AB, AC, BC, AF, BE, CD, DE, DF, EF)$, and it is easy to see that $|\Delta(Q)|$ is a hexagon.

We want to find a class of P so that $k[P]$ is a Cohen–Macaulay ring. We have the notion of *wonderful posets.*

Definition 7.88 (Wonderful Posets) Let P be a poset. We call P a *wonderful*[13] poset if it satisfies the following condition ($\sharp$) (see Fig. 7.4).

($\sharp$) Let $\widehat{P} = P \cup \{\widehat{1}, \widehat{0}\}$, where $\widehat{1}$ (resp. $\widehat{0}$) is the largest (resp. smallest) element of $\widehat{P}$. If $\beta, \gamma \in P$ cover $\alpha \in \widehat{P}$ (including $\alpha = \widehat{0}$) and if there exists $\phi \in \widehat{P}, \phi > \beta, \gamma$, then there exists $\delta \in P$, which covers β, γ and $\delta < \phi$.

Theorem 7.89 *Let P be a poset and k be a field. If P is wonderful, then $k[P]$ is a Cohen–Macaulay ring.*[14]

Proof We will show the theorem by induction on $\operatorname{rank} P$ and the number of minimal elements of P. The conclusion is clear if $\operatorname{rank} P = 0$, since $k[P]$ is reduced and $\dim P = 1$. If $\operatorname{rank} P = 1$ and for minimal elements β, γ, from condition ($\sharp$) for $\alpha = \widehat{0}$, $\phi = \widehat{1}$, there is δ, which covers β, γ. Hence, in particular, $\Delta(P)$ is connected and $k[P]$ is Cohen–Macaulay by Example 7.81.

Assume that the conclusion is true for $\operatorname{rank} P = d - 1$ and assume that $\operatorname{rank} P = d$, and P has n minimal elements. If $n = 1$, let α be the unique minimal element of P. Then since α is compatible with every element of P, $k[P] \cong k[Q][x_\alpha]$ is a polynomial ring over $k[Q]$, where $Q = P \setminus \alpha$. Since Q satisfies the condition ($\sharp$) and $\operatorname{rank} Q < \operatorname{rank} P$, $k[Q]$, hence $k[P]$ are Cohen–Macaulay by the induction hypothesis.

If $n \geq 2$, let $\alpha, \beta_1, \ldots, \beta_{n-1}$ be the minimal elements of P and put

$$P_1 = \{\delta \in P | \delta \geq \alpha\}, \quad P_2 = \{\delta \in P | \delta \geq \beta_i (\exists \beta_i)\}.$$

[13] The name "wonderful" is due to D. Eisenbud (cf. [Ei]). This concept is also called "locally semimodular."

[14] Actually our conclusion is true for any Cohen–Macaulay ring k.

We have seen before that for each β_i, there is $\delta_i \in P$ covering α and β_i. Hence we have rank P_1 = rank $P_2 = d$, rank $(P_1 \cap P_2) = d - 1$. Also, we see easily that P_1, P_2 and $P_1 \cap P_2$ satisfy the condition (♯). Now, we have the short exact sequence

$$0 \to k[P] \to k[P_1] \oplus k[P_2] \to k[P_1 \cap P_2] \to 0,$$

where the 1st map is defined by $x \to (\pi_1(x), \pi_2(x))$ and the (resp. the 2nd) map is defined by $(x, y) \to \rho_1(x) - \rho_2(y)$ (resp. ρ_i $k[P_i] \to k[P_1 \cap P_2]$ $(i = 1, 2)$). Here $\pi_i : k[P] \to k[P_i]$, ρ_i $k[P_i] \to k[P_1 \cap P_2]$ $(i = 1, 2)$ are canonical surjections. Then $k[P_2]$ (resp. $k[P_1 \cap P_2]$) is Cohen–Macaulay by the induction hypothesis on n (resp. on the rank). Hence we get that $k[P]$ is Cohen–Macaulay, too, by a long exact sequence of local cohomology groups. □

As an important example of wonderful posets, we introduce the notion of *distributive lattices*.

Definition 7.90 (Lattices and Distributive Lattices) Let P be a poset.

(1) We say that P is a *lattice* if for every pair (α, β) of elements of P has supremum $\alpha \vee \beta$ and infimum $\alpha \wedge \beta$.[15]
(2) We call P a *distributive lattice* if P is a lattice and satisfies the *distributive law*

$$(\alpha \vee \beta) \wedge \gamma = (\alpha \wedge \gamma) \vee (\beta \wedge \gamma),$$
$$(\alpha \wedge \beta) \vee \gamma = (\alpha \vee \gamma) \wedge (\beta \vee \gamma).$$

Proposition 7.91 *If P is a distributive lattice, then P is a wonderful poset.*

Proof Let $\beta, \gamma, \alpha \in P$ satisfy the assumption of the condition (♯). Then clearly we have $\beta \wedge \gamma = \alpha$. By the definition of supremum, $\delta = \beta \vee \gamma$ satisfies $\delta \leq \phi$. To show the condition (♯), it suffices to show that δ covers β, γ. Take an element $\varepsilon \in P$ such that $\delta > \varepsilon \geq \beta$.

By the distributive law, we have $\varepsilon = \varepsilon \wedge \delta = \varepsilon \wedge (\beta \vee \gamma) = (\varepsilon \wedge \beta) \vee (\varepsilon \wedge \gamma) = \beta \vee (\varepsilon \wedge \gamma)$.

By the definition $\delta = \beta \vee \gamma$, $\varepsilon > \gamma$ is impossible. Hence $\varepsilon \wedge \gamma < \gamma$. But since $\alpha < \varepsilon, \gamma$, $\varepsilon \wedge \gamma \geq \alpha$. Since γ covers α, we must have $\varepsilon \wedge \gamma = \alpha$. Hence we have $\varepsilon = \beta \vee (\varepsilon \wedge \gamma) = \beta \vee \alpha = \beta$. Hence we have shown that δ covers β and the condition (♯) is proved. □

7.7 ASL and Hodge Algebras

To define a commutative ring, the most standard way is to give generators and relations.

$$A = k[X_1, \ldots, X_n]/(f_1, \ldots, f_r).$$

[15] $\gamma = \sup\{\alpha, \beta\}$ if and only if $\gamma \geq \alpha, \beta$ and if $\delta \in P$ satisfies $\delta \geq \alpha, \beta$, then $\delta \geq \gamma$.

But when we have 100 generators and 2000 relations, it will be very difficult to understand the properties of the ring.

The notion of *ASL* (Algebras with Straightening Laws) attaches rings to a finite poset. By that method, we unify the relations of the generators and also give geometric meanings to the ring.

Definition 7.92 Let B be a ring and P be a poset. Let $A = \bigoplus_{n\geq 0} A_n$ be a graded ring with $A_0 = B$ and we assume that P is a subset of A and every element of P is a homogeneous element of positive degree. Let $M = \alpha_1\alpha_2\cdots\alpha_n$ be a product of elements of P and assume that $\{\alpha_1, \alpha_2 \ldots, \alpha_n\}$ is a chain of P. Then we say that M is a *standard monomial*. We regard 1 as a standard monomial, too.

We say that A is an *ASL* (*algebra with straightening laws*) over B on P if the following two conditions are satisfied:

ASL1 A is a free B module and the set of standard monomials is a basis of A over B.

ASL2 If $\beta, \gamma \in P$, $\beta \not\sim \gamma$, then if we express $\beta\gamma = \sum c_M M$ as a B-linear combination of standard monomials, every M with $c_M \neq 0$ is divisible by some $\alpha \in P$ with $\alpha < \beta, \gamma$.

In what follows, we assume $B = k$ is a field.

Example 7.93 Let us list some examples of ASL.

(1) Let P be any poset. Then the ring $k[P]$ defined in Definition 7.84 is an ASL over k on P. In this case, $\beta\gamma = 0$ for all $\beta \not\sim \gamma$. $k[P]$ is called the *discrete ASL* on P over k.
(2) If $P = \{\alpha_1 < \alpha_2 < \ldots < \alpha_n\}$ is a totally ordered poset, then $k[P]$ is the only ASL on P over k and is a polynomial ring of n variables over k.
(3) Let P be the poset described in Fig. 7.5, (1). Then $A = k[\alpha, \beta, \gamma]/(\alpha\beta - \gamma^n)$ is an ASL for every $n > 0$.
(4) Let P be the poset as in Fig. 7.5, (2) and let $a_1, \ldots, a_n \in k$ be different n elements in k. Let x, T be variables over k and put $\alpha_i = (x - a_i)^{-1}T \in k(x)[T]$, $\gamma = T$. Then for $i \neq j$, we have

$$\alpha_i\alpha_j = (a_j - a_i)^{-1}(\alpha_i - \alpha_j)\gamma.$$

Hence the subring $A = k[\alpha_1, \ldots, \alpha_n, \gamma]$ of $k(x)[T]$ is an ASL on P over k.

Remark 7.16 Let A be an ASL on P over k. Then the following statements hold true:

(1) The defining ideal of A is generated by the relations in (ASL2). Namely, let $\{X_\alpha \mid \alpha \in P\}$ be the set of variables and $S = k[X_\alpha \mid \alpha \in P\}$ be the polynomial ring over k and let $\phi : S \to A$ be defined by $\phi(X_\alpha) = \alpha$. Then ϕ is surjective

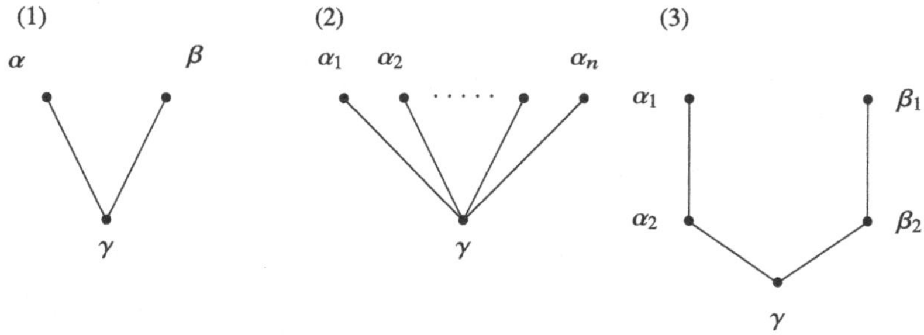

Fig. 7.5 Examples of ASL

and Ker (ϕ) is defined by the relation in (ASL2). In particular, the number of minimal generators of Ker ϕ is equal to

$$\sharp\{\{\alpha, \beta\} \mid \alpha, \beta \in P, \alpha \not\sim \beta\}.$$

(2) If $\alpha, \beta \in P, \alpha \not\sim \beta$ and if there does not exist $\gamma \in P$ with $\gamma < \alpha, \beta$, then $\alpha\beta = 0$. Hence if A is a domain, P must have *unique* minimal element.
(3) If A is an ASL on poset P, the set of standard monomials of a given degree is the same in A and $k[P]$ (we assume that $\deg \alpha$ is the same in A and $k[P]$). Hence A and $k[P]$ have the same Poincaré series. In particular, we have

$$\dim A = \operatorname{rank} P + 1.$$

Also, if $\deg(\alpha) = 1$ for every $\alpha \in P$, then the multiplicity of A is equal to the number of chains of P with length rank P.
(4) We call $I \subseteq P$ a *poset ideal* if I satisfies the condition

$$\text{If } \alpha \in I \text{ and } \beta < \alpha, \text{ then } \beta \in I.$$

Note that $P \setminus I$ is a poset by induced order from P. If A is an ASL on P over k, then A/IA is also an ASL on $P \setminus I$ (in (ASL2), if $\beta \in I$ or $\gamma \in I$, then the right-hand side of (ASL2) is also in IA).

In this manner, we get from an ASL A, many ASL's automatically.
(5) If α is the unique maximal element of P, then α is a non-zerodivisor of A and $A/(\alpha - 1)$ is an ASL on $P \setminus \{\alpha\}$.

Exercise 7.4 Prove the statements in Remark 7.16.

Exercise 7.5 Let P be the poset indicated in Fig. 7.5 (3). (We assume, for simplicity, degrees of $\alpha_1, \alpha_2, \beta_1, \beta_2, \gamma$ are all 1). Prove that if A is an ASL on P, then A is not an integral domain.

Exercise 7.6 Let P be a finite poset.

Fig. 7.6 The poset $\Gamma(5.2)$

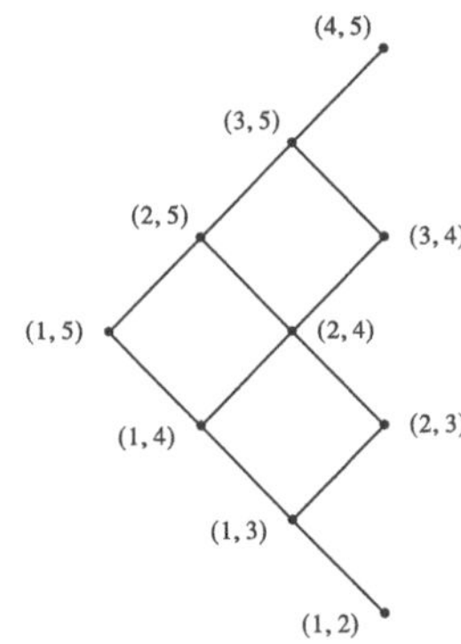

(1) Let $\mathrm{J}(P)$ be the set of poset ideals of P ordered by inclusion relation. Show that $\mathrm{J}(P)$ is a distributive lattice with respect to $\cup$ and $\cap$.

(2) Conversely, let L be a finite distributive lattice. Then show that L is naturally isomorphic to $\mathrm{J}(P)$, where P is the set of *join-irreducible elements* ($\alpha \in L$ is called join-irreducible, if $\alpha \neq \beta \vee \gamma$ for any $\beta, \gamma < \alpha$) of L, ordered by the induced order from L.

The following example is perhaps the most famous example of ASL. The proof can be found in [BV], for example. We list the important results obtained from the main theorem as problems. Interested readers are recommended to refer to [BV], [Ei], [DEP].

Example 7.94 (Plücker Embeddings of Grassmann Varieties) For positive integers $r < n$, let $X = \{X_{ij} \mid 1 \leq i \leq n, 1 \leq j \leq r\}$ be an $n \times r$ matrix whose coefficients are variables over k. We define the poset $\Gamma(n, r)$ in the following manner.

(1) We put

$$\Gamma(n.r) = \{[i_1, i_2. \dots .i_r] \mid 1 \leq i_1 < i_2 < \dots < i_r \leq n\}$$

and order them by

$$[i_1, i_2. \dots .i_r] \leq [j_1, j_2. \dots .j_r] \iff \forall k, i_k \leq j_k.$$

We illustrate the Hasse diagram of the poset $\Gamma(5.2)$ in Fig. 7.6.

(2) We denote by $[i_1, i_2, \dots, i_r]$ the determinant of the $[i_1, i_2, \dots, i_r]$ minor of $\boldsymbol{X}$.

(3) Let $G(n, r)$ be the subring of $k[\boldsymbol{X}]$ generated by all $r \times r$ minors of $\boldsymbol{X}$. Actually, $G(n, r)$ is the coordinate ring of the *Grassmann variety* $\mathrm{Grass}(n, r)$ embedded in the $\binom{n}{r} - 1$-dimensional projective space by "Plücker coordinates."

Theorem 7.95 *$G(n, r)$ is an ASL on the poset $\Gamma(n, r)$.*

7.96 The unique maximal element of $\Gamma(n,r)$ is $\alpha = [n-r+1, \ldots, n]$ and the ring homomorphism $G(n,r) \to G(n,r)/(\alpha - (-1)^{r(r-1)/2})$ corresponds to substituting the last r by r matrix for the r by r matrix

$$\begin{pmatrix} & & & 1 \\ & & 1 & \\ & \cdot^{\cdot^{\cdot}} & & \\ 1 & & & \end{pmatrix}.$$

Hence if we put $S = G(n,r)/(\alpha - (-1)^{r(r-1)/2})$, then $S = k[X_{ij}]_{1 \le i \le r, 1 \le j \le n-r}$, the polynomial ring of $(n-r) \times r$ variables over k. On the other hand, by Remark 7.16 (5), $S = k[X_{ij}]_{1 \le i \le r, 1 \le j \le n-r}$ is an ASL on the poset $\Gamma(n,r) \setminus \{\alpha\}$. Hence this correspondence gives an ASL structure on S.

Now, for any $\sigma \in \Gamma(n,r) \setminus \{\alpha\}$, define $I_\sigma := \{\gamma \in \Gamma(n,r) \mid \gamma \not\geq \sigma\}$. Then by Remark 7.16 (4), $S/I_\sigma S$ is an ASL on a poset whose unique minimal element is σ. Actually, $G(n,r)/I_\sigma G(n,r)$ is the homogeneous coordinate ring of a Schubet variety. In particular, let $S = k[\boldsymbol{X}]$, where $\boldsymbol{X} = (X_{ij})_{1 \le i \le s, 1 \le j \le t}$ with $s \le t$, and put $P = \Gamma(s+t, s)$. Then $S \cong G(s+t,s)/(\alpha - (-1)^{s(s-1)/2})$, and if we put $\sigma = (1, 2, \ldots, s-1, t+1)$, then $S/I_\sigma S \cong S/\mathrm{I}_s(\boldsymbol{X})$.

For example, if $n = 5$, $r = 2$ and $\sigma = (1,4)$ (resp. $\sigma = (2,4)$), then $I_\sigma S = \mathrm{I}_2((X_{ij}))$ (the ideal generated by the 3 maximal minors of (X_{ij})) (resp. $S/I_\sigma S \cong k[X_{12}, X_{13}, X_{22}, X_{23}]/(X_{12}X_{23} - X_{13}X_{22})$).

Exercise 7.7 Show that the graded ring $G(n,r)$ has the following properties; we assume that every element of $\Gamma(n,r)$ has degree 1 in $G(n,r)$.

(1) Let $P = [1,r] \times [1, n-r]$ be the poset defined by the order $(i,j) \le (i',j')$ if and only if $i \le i'$ and $j \le j'$. Then show that $\Gamma(n,r) \cong \mathrm{J}(P)$. Hence $\Gamma(n,r)$ is a distributive lattice by Proposition 7.6 (it is also simple to show that $\Gamma(n,r)$ is a distributive lattice).
(2) Let α be the unique maximal element of $\Gamma(n,r)$. Since α is a prime element in $G(n,r)$ and $G(n,r)[\alpha^{-1}] \cong [G(n,r)/(\alpha - 1)][\alpha, \alpha^{-1}]$, show that $G(n,r)$ is a UFD.
(3) Show that $a(G(n,r)) = -n$.

Exercise 7.8 Let $\boldsymbol{Y} = (Y_{ij} \mid 1 \le i \le s, 1 \le j \le t)$ be the set of st variables of the form $s \times t$ matrix. Then

(1) Put

$$\Gamma'(s,t) = \{[i_1, i_2, \ldots, i_r | j_1, j_2, \ldots, j_r] \\ \mid 1 \le i_1 < i_2 < \ldots < i_r \le s, 1 \le j_1 < j_2 < \ldots < j_r \le t\}$$

and define the order on $\Gamma'(s,t)$ by

$$[i_1, i_2, \ldots, i_r | j_1, j_2, \ldots, j_r] \leq [i'_1, i'_2, \ldots, i'_{r'} | j'_1, j'_2, \ldots, j'_{r'}]$$

if and only if $r' \leq r$ and $i_k \leq i'_k$, $j_k \leq j'_k (\forall k, 1 \leq k \leq r')$.

(2) Let Y' be the $(s, s+t)$ matrix, adding the (s, s) matrix $E' = (e'_{ij})$ on the right-hand side of Y, where the $e'_{ij} = 1$ if $i + j = s + 1$ and $e'_{ij} = 0$ otherwise.
Now, letting $(i_1, i_2, \ldots, i_s) \in \Gamma(s+t, s) \setminus \{(t+1, \ldots, s+t)\}$ correspond to the element in $\Gamma'(s, t)$ that agrees with $(i_1, i_2, \ldots, i_s)$ minor of Y' up to sign, show that $\Gamma(s+t, s) \setminus \{(t+1, \ldots, s+t)\} \cong \Gamma'(s, t)$.

(3) Let us correspond to $[i_1, i_2 . \ldots . i_r | j_1, j_2 . \ldots . j_r]$ the $r \times r$ minor of $(i_1, i_2 . \ldots . i_r)$ rows and $(j_1, j_2 . \ldots . j_r)$ columns of Y. Show that the polynomial ring $k[Y]$ is an ASL on the poset $\Gamma'(s, t)$, using 7.96.

(4) In particular, if we put $\sigma_k = [1, 2, \ldots, k | 1, 2, \ldots, k]$, then $I_{\sigma_k} k[Y]$ is generated by the all $(k+1)$-minors of Y, where $I_{\sigma_k} = \{\tau \in \Gamma'(s, t) \mid \tau \not\geq \sigma_k\}$. Show that $k[Y]/ \mathrm{I}_{k+1}(Y)$ is an ASL on $\Gamma'(t, s) \setminus I_{\sigma_k}$. Show also that the height of $\mathrm{I}_{k+1}(Y)$ is $(s-k)(t-k)$.

The following theorem is the most essential one in ASL theory.

Theorem 7.97 (ASL; Main Theorem) *Let A be an ASL on a poset P over a ring B. If the discrete ASL $B[P]$ is Cohen–Macaulay (resp. Gorenstein), then so is A.*

Remark 7.17 There are many questions left in the theory of ASL. For example, as far as the authors know, there is no example of a Cohen–Macaulay ASL A on a *non-Cohen–Macaulay* poset P ($k[P]$ is non-Cohen–Macaulay). Also, we do not have an example of an ASL *domain* which is not Cohen–Macaulay.

Proof of Theorem 7.97 Let Ind(A) be the set of $\alpha \in P$, which appears on the right-hand side of (ASL2). We have $\mathrm{Ind(A)} = \emptyset \Leftrightarrow A = B[P]$. We prove the theorem by induction on $\sharp(\mathrm{Ind(A)})$.

Let α be a minimal element of Ind(A). We define the ideal $I_k \subseteq A$ by $I_k = \alpha^k A$. We define the ring $\mathcal{R} \subseteq A[t, t^{-1}]$ by

$$\mathcal{R} = \ldots \oplus At^k \oplus \ldots \oplus At \oplus A \oplus I_1 t^{-1} \oplus \ldots \oplus t^{-k} I_k \oplus \ldots .$$

Then we can show the following statements:

(1) We define $i : P \to \mathcal{R}$ by $i(\beta) = \beta (\beta \in P, \beta \neq \alpha)$, $i(\alpha) = \alpha t^{-1}$. Then by this embedding, $\mathcal{R}$ is an ASL on P over $B[t]$.

(2) We put $G_I(A) := \oplus_{n \geq 0} I_n / I_{n+1} t^{-n} \cong \mathcal{R}/t\mathcal{R}$. Then $G_I(A)$ is an ASL on P over B and $\mathrm{Ind}(\mathrm{G_I(A)}) \subseteq \mathrm{Ind(A)} \setminus \{\alpha\}$.

Hence $\mathcal{R}$ is a free $B[t]$-module by the axiom of ASL, whose general fiber is A and the special fiber is $G_I(A)$. Since $\mathrm{Ind}(G_I(A)) \subsetneq \mathrm{Ind}(A)$, $G_I(A)$ is Cohen–Macaulay by the induction hypothesis and then we can conclude so is A, since A is a general fiber. □

Let us introduce another important example of ASL. This example was first introduced in the paper [Hi] of Takayuki Hibi and called the *Hibi rings*.

Exercise 7.9 (Hibi Rings[16]) Let P be a poset, $\mathrm{J}(P)$ be the set of poset ideals of P (including $\emptyset$ and P). Let $\{X_\alpha \mid \alpha \in P\} \cup \{T\}$ be the set of variables. For $I \in \mathrm{J}(P)$, put $X_I = \prod_{\alpha \in I} X_\alpha T$. Then we define

$$R(P) = k[X_I \mid I \in \mathrm{J}(P)] \subseteq k[T, X_\alpha \mid \alpha \in P].$$

(1) Show that $R(P)$ is an ASL on the distributive lattice $\mathrm{J}(P)$.
(2) Show that $R(P)$ is a normal semigroup ring.
(3) Show that $R(P)$ is a Gorenstein ring if and only if every maximal chain of P has the same length (or, equivalently, $k[P]$ is equi-dimensional). This property follows from the structure theorem of K_R for a normal semigroup ring R (Theorem 7.69).

The concept of Hodge Algebras[17] generalizes the ASL axiom to the most extent so that we can prove Theorem 7.97.

Definition 7.98 Let H be a finite set. We call an element of $\mathbb{N}^H$ "a monomial on H." We define that M divides N for monomials M, N on H in the usual way. We define

$$\mathrm{Supp}M = \{x \in H \mid M(x) > 0\}.$$

Let $\Sigma \subseteq \mathbb{N}^H$ be an ideal generated by monomials on H. We call the minimal elements of Σ the generators of Σ. We call M a *standard monomial* if $M \notin \Sigma$.

For a commutative ring A, given an injection $\phi : H \to A$, for $M \in \mathbb{N}^H$, we define $\phi(M) \in A$ as the product of elements of $\phi(H)$.

We say that A *is a Hodge algebra on H over B generated by $\phi(H)$ and governed by* Σ if A satisfies the following conditions:

Hodge1 A is a free B module whose basis over B is given by the set of standard monomials with respect to Σ.

Hodge2 If $N \in \Sigma$ is a generator of Σ and assume that

$$N = \Sigma_i b_{N,i} M_{N,i} \quad (b_{N.i} \neq 0)$$

is the representation of N using standard monomials. Then for every $x \in H$ dividing N and for every i, there exists $y \in H$ with $y < x$ dividing $M_{N,i}$.

In the case of an ASL, then we take $\Sigma = (\alpha\beta \mid \alpha \not< \beta)$.

The main theorem of the theory of Hodge algebras is the following and the proof is essentially the same as in the case of ASL.

---

[16] Takayuki Hibi, 1956–.

[17] This is named after William Vallance Douglas Hodge (1903–1975), who studied the ASL structure of $G(n, r)$ and gave the basic idea of ASL.

Theorem 7.99 (Hodge Algebra; Main Theorem) *Let A be a Hodge algebra generated by $\phi(H)$ and governed by Σ. Then if the discrete Hodge algebra is Cohen–Macaulay (resp. Gorenstein), then so is A.*

7.8 Invariant Subrings of Finite Groups

It occurs very frequently that the action of a group make the mathematics very rich and fruitful. In this section we restrict to the action of a finite group G on a ring R and consider the *invariant subring* R^G.

The theory of invariant subrings differs very much in the case where the order $|G|$ is a unit in the base field k or not. In this section, we always assume that *the order $|G|$ is a unit in the ring R*. In this case, the theory is very simple. On the other hand, if $|G|$ is *not* invertible in R, the theory becomes very complicated. In this section we discuss when the invariant subring is Cohen–Macaulay (resp. Gorenstein, regular).

Definition 7.100 Let G be a finite group and A be a commutative ring. We denote the unit of G by e.

(1) We say that "G acts on A" if every element $\sigma \in G$ defines a ring automorphism of A and we have $(\sigma\tau)(a) = \sigma(\tau(a))(\forall\sigma, \tau \in G, \forall a \in A)$. The action of e on A is the identity map.
(2) We denote the *invariant subring* of the action of G by A^G:

$$A^G = \{a \in A \mid \sigma(a) = a \ (\forall\sigma \in G)\}.$$

(3) Let G act on A and M be an A-module. If for every $\sigma \in G$, given the automorphism $\sigma_M : M \to M$ as an A-module such that $(\sigma_M\tau_M)(x) = \sigma_M(\tau_M(x))$ for every $\sigma, \tau \in G$, satisfying $\sigma_M(ax) = \sigma(a)\sigma_M(x)$ ($\forall a \in A, x \in M$), we say that M is an *(A, G)-module*.
(4) We define the invariant submodule $M^G = \{a \in M \mid \sigma(x) = x \ (\forall\sigma \in G)\}$. It will be obvious to show that M^G is an A^G-module.
(5) If M, N are (A, G)-modules and if an A-module map $f : M \to N$ satisfy the condition $f(\sigma(x)) = \sigma(f(x))$ ($\forall\sigma \in G, \forall x \in M$), then we say that f is a *map of (A, G)-modules*.
(6) If A is a k-algebra for some ring k, we say that *G acts on A over k*, if the action of $\sigma \in G$ on A is a k-algebra isomorphism.

Let us begin with some fundamental properties.

Proposition 7.101 *Let a finite group G act on a ring A.*

(1) *A is integral over A^G. Moreover, if A is a finitely generated k algebra and if G acts over k, then A is a finite A^G-module.*
(2) *Let A be a domain with the field of fractions $\mathrm{Q}(A)$. Then $A^G = A \cap \mathrm{Q}(A^G)$ and $\mathrm{Q}(A^G) = \mathrm{Q}(A)^G$, where $\mathrm{Q}(A^G)$ is the field of fractions of $\mathrm{Q}(A^G)$.*

Hence if A is integrally closed, then so is A^G.

Proof

(1) Let x be any element of A. Put $f(T) = \prod_{\sigma \in G}(T - \sigma(x))$. Then we can show that $f(T) \in A^G(T)$ and $f(x) = 0$. Hence x is integral over A^G. The second statement will be clear since A is finitely generated over k and hence over A^G.
(2) The first statement is clear. If y/x $(x, y \in A, x \neq 0)$ is G-invariant, multiplying the denominator and numerator by $\prod_{\sigma \in G, \sigma \neq e} \sigma(x)$ makes both the denominator and numerator in A^G.

□

Remark 7.18 Statement (2) of Proposition 7.101 does not hold if G is infinite. For example, assume that k is an infinite field, and let $G = k^*$ be the multiplicative group of k, acting on $A = k[X, Y]$ by $a(X) = aX, a(Y) = aY$ $(a \in k^*)$, then we have $A^G = k = \mathrm{Q}(A^G)$ but $\mathrm{Q}(A)^G = k(X/Y) \neq \mathrm{Q}(A^G)$.

Let a finite group G act on A. If the order $|G|$ of G is a unit in A, the *Reynolds operator* $\rho : A \to A^G$ plays a very important role.

Definition 7.102 If a finite group G acts on a ring A and if $|G| \in U(A)$, we define $\rho : A \to A^G$ as follows:

$$\rho(a) = \frac{1}{|G|} \sum_{\sigma \in G} \sigma(a).$$

Note that $\rho(a) = a$ if $a \in A^G$. In the same manner, for an (A, G)-module M, we define $\rho : M \to M^G$ by $\rho(x) = \frac{1}{|G|} \sum_{\sigma \in G} \sigma(x)$.

Remark 7.19 Recall that in this case, A^G is a *pure subring* of A (Definition 7.102). If $|G| \notin U(A)$, A^G is *not* a pure subring of A, in general.

Proposition 7.103 *Under the assumption of Definition 7.102, the following statements hold true:*

(1) *$\rho : M \to M^G$ is an A^G-module homomorphism.*
(2) *The functor $M \to M^G$ is an exact functor from the category of (A, G)-modules to the category of A^G-modules.*

Proof

(1) follows easily from the definition of ρ.
(2) The functor $M \to M^G$ is always left exact (we do not need the map ρ and also it holds if G is an infinite group).

Now let $f : M \to N$ be a surjective (A, G)-module homomorphism. We want to show that $f^G : M^G \to N^G$ is a surjection. If $x \in N^G$, take $y \in M, f(y) = x$. Then we have $f(\rho(y)) = \rho(f(y)) = \rho(x) = x$, which shows that $M^G \to N^G$ is surjective. □

Theorem 7.104 *Assume that a finite group G acts on a Noetherian local ring $(A, \mathfrak{m})$ and assume that $|G| \in U(A)$. Let M be an (A, G)-module. Then, the action of G on M induces an action on the local cohomology group $\mathrm{H}^i_{\mathfrak{m}}(M)$ and we have an isomorphism*

$$\mathrm{H}^i_{\mathfrak{m}^G}(M^G) \cong [\mathrm{H}^i_{\mathfrak{m}}(M)]^G \quad (\forall i),$$

where the left-hand side denotes the local cohomology groups of an A^G-module M^G with respect to $\mathfrak{m}^G$.

Before stating the proof of the theorem, we introduce some applications.

Corollary 7.105 *Under the assumptions of Theorem 7.104, put $\dim A = d$. Then the following statements hold true:*

(1) *If A is a Cohen–Macaulay ring, so is A^G.*
(2) *If A is Gorenstein and the action of G on the socle of $\mathrm{H}^d_{\mathfrak{m}}(A)$ is trivial, then A^G is Gorenstein.*

Proof

(1) By our assumption, we have $\mathrm{H}^i_{\mathfrak{m}}(A) = 0$ for $i < d$. Hence by our theorem, $\mathrm{H}^i_{\mathfrak{m}^G}(A^G) = 0$, for $i < d$, too. This also follows from Lemma 7.7 (3) (see Remark 7.19).
(2) Since A^G is Cohen–Macaulay by (1), it suffices to show that $\mathrm{H}^d_{\mathfrak{m}^G}(A^G) \cong [\mathrm{H}^d_{\mathfrak{m}}(A)]^G$ is an injective A^G-module. Let z be a generator of the socle of $\mathrm{H}^d_{\mathfrak{m}}(A)$. Since z is G-invariant by our assumption, we have $\rho(z) = z$. Now, the Gorenstein property of A implies that $\mathrm{H}^d_{\mathfrak{m}}(A) \cong E_A(A/\mathfrak{m})$ and for every $x \in [\mathrm{H}^d_{\mathfrak{m}}(A)]^G, x \neq 0$, we can take $a \in A$ so that $ax = z$. Then we have $z = \rho(z) = \rho(ax) = 1/|G| \sum_{\sigma \in G} \sigma(a)\sigma(x) = 1/|G| \sum_{\sigma \in G} \sigma(a)x = \rho(a)x$ and we show that z is the socle of $\mathrm{H}^d_{\mathfrak{m}^G}(A^G)$. Hence A^G is Gorenstein, too.

□

Proof of Theorem 7.104 Note that A^G is a local ring with maximal ideal $\mathfrak{m}^G$. Let $\boldsymbol{x} = (x_1, \ldots, x_d)$ be a system of parameters of A^G. Since $A^G \to A$ is a finite map, $\boldsymbol{x}A$ is an $\mathfrak{m}$-primary ideal. Hence we have

$$\mathrm{H}^i_{\mathfrak{m}}(A) \cong \mathrm{H}^i(C^\bullet),$$

where $C^\bullet$ is the Čech complex of A with respect to $\boldsymbol{x}$ (Definition 5.88). For every r, C^r is a direct sum of localizations of A by the products of r elements in $\boldsymbol{x}$. For $f \in A^G, x \in A$, the action of $\sigma \in G$ on C^r is defined by $\sigma(\dfrac{x}{f}) = \dfrac{\sigma(x)}{f}$, compatible with the boundary map of $C^\bullet$. Hence $C^\bullet$ is a complex of (A, G)-modules. Since the functor $M \to M^G$ is exact (Proposition 7.103), we have $\mathrm{H}^i((C^G)^\bullet) \cong \mathrm{H}^i(C^\bullet)^G$ for every i.

Now, the complex $[C^G]^\bullet$ has the following properties:

(1) $[C^G]^0 = A^G$, $\mathrm{H}^i_{\mathfrak{m}^G}([C^G]^i) = 0$ $(\forall i \in \mathbb{Z})$;
(2) $\mathrm{Supp}(\mathrm{H}^i([C^G]^\bullet) \subseteq \{\mathfrak{m}^G\}$ $(\forall i \in \mathbb{Z})$.

Hence it follows that $\mathrm{H}^i_{\mathfrak{m}^G}(M^G) \cong \mathrm{H}^i([C^G]^\bullet) \cong [\mathrm{H}^i_{\mathfrak{m}}(M)]^G$ by Lemma 5.104.

□

Remark 7.20 If A is a Gorenstein ring, the action of G preserves Soc $(\mathrm{H}^d_{\mathfrak{m}}) \cong A/\mathfrak{m}$ and if we let z be a generator of Soc $(\mathrm{H}^d_{\mathfrak{m}})$, then $\forall \sigma \in G, \sigma(z) = \xi(\sigma)z$ for some $\xi(\sigma) \in (A/\mathfrak{m})^*$. By this action, we get the character

$$\xi = \xi_A : G \to (A/\mathfrak{m})^*.$$

If $\xi = \mathbf{1}$ (unit character), then A^G is Gorenstein, too. (The converse does not hold in general.)

Let us consider the case where a linear group G acts on a polynomial ring S. We have very beautiful group and ring theoretic properties in this case. In the following, let k be a field.

Definition 7.106 Let $S := k[X_1, \cdots, X_n]$ be a polynomial ring of n variables and let G be a finite subgroup of $\mathrm{GL}(n, k)$.

For $\sigma = (a_{ij}) \in G$, we define

$$\sigma(X_i) := \textstyle\sum_{p=1}^n a_{pi} X_p$$
$$\sigma(f(X_1, \cdots, X_n)) := f(\sigma(X_1), \cdots, \sigma(X_n))$$

and then this defines a k-isomorphism and the action of G on S. **From now, until Corollary 7.117, we always assume that $|G| \neq 0$ in k.**

Since G is a finite group, $S^G \subseteq S$ is a finite extension of rings and we have $\dim S^G = \dim S = n$. From Corollary 7.105, S^G is Cohen–Macaulay.

Also, we can consider S as a graded ring with the usual grading, putting $\deg X_i = 1$ for every i. Since the action of G preserves the grading, we can consider S^G as a graded ring, also. Then we define the Poincaré series defined in Sect. 7.3, which has a very interesting property by the theorem of Molien:

Theorem 7.107 (Molien's Theorem) *The Poincaré series of the invariant subring S^G is given by the following formula:*

$$P(S^G, t) = \frac{1}{|G|} \sum_{\sigma \in G} \frac{1}{\det(\mathbf{1} - \sigma t)}.$$

Here $\mathbf{1}$ *denotes the $n \times n$ unit matrix and if the eigenvalues of σ are $(e_1, \ldots, e_n)$ (counting the multiplicity), then* $\det(\mathbf{1} - \sigma t) = (1 - e_1 t) \cdots (1 - e_n t)$.

Before proving the theorem, we state some applications.

Theorem 7.108 *If* $P(S^G, t) = \dfrac{f(t)}{(1-t^{d_1})(1-t^{d_2})\cdots(1-t^{d_n})}$, *the following statements hold true:*

(1) $|G| = (d_1 \dots d_n)/(f(1))$.
(2) *In particular, if* $S^G = k[f_1, \dots, f_n]$ *is generated by* n *algebraically independent elements* $f_1, \dots, f_n$, *then* $|G| = d_1 \dots d_n$.
(3) *In particular, if* $S^G \cong k[Y_1, \dots, Y_{n+r}]/(h_1, \dots, h_r)$ *is a complete intersection with* $\deg(Y_i) = d_i$, $\deg(h_j) = e_j$, *then* $|G| = (d_1 \dots d_{n+r})/(e_1 \dots e_r)$.
(4) *We say that* $\sigma \in G$ *is a pseudo-reflection, if* $\operatorname{rank}(\sigma - I) = 1$. *Then the number of the pseudo-reflections of* G *is given by* $\sum_{i=1}^{n}(d_i - 1) - \dfrac{2f'(1)}{f(1)}$.
(5) $a(S^G) = \deg(f(t)) - \sum_{i=1}^{n} d_i \le -n$. *The equality holds if and only if* $G \subseteq \mathrm{SL}(n, k)$.

We prove Theorem 7.108 first assuming Molien's theorem.

Proof of Theorem 7.108 Multiplying both sides of

$$P(S^G, t) = \frac{f(t)}{(1-t^{d_1})\dots(1-t^{d_n})} = \frac{1}{|G|}\sum_{\sigma\in G}\frac{1}{\det(1-\sigma t)}$$

by $(1-t)^n$ and taking the limit $t \to 1$,[18] we get (1) since $\lim_{t\to 1}(1-t)^n/\det(1-\sigma t)$ is 1 if $\sigma = \mathbf{1}$ and 0 otherwise. Hence we get

$$\frac{f(1)}{d_1 \cdots d_n} = \frac{1}{|G|}.$$

Statements (2) and (3) easily follow from (1).

We obtain (4) by comparing the coefficients of Laurent expansions of $\dfrac{1}{(1-t)^{n-1}}$ of both sides of

$$\frac{1}{|G|}\sum_{\sigma\in G}\frac{1}{\det(1-\sigma t)} = \frac{f(t)}{(1-t^{d_1})\dots(1-t^{d_n})}.$$

If σ is not a pseudo-reflection, then the coefficient of $\dfrac{1}{\det(I-\sigma t)}$ is 0 and if σ is a pseudo-reflection with $\det(\sigma) = e$, then the coefficient of $\dfrac{1}{(1-t)^{n-1}}$ in the Laurent expansion of $\dfrac{1}{\det(I-\sigma t)}$ is $\dfrac{1}{1-e}$. If $e = -1$, this value is $1/2$, if $e \neq -1$, then

[18] If denominator is $(1-t)^n g(t)$, then the "limit" is $1/g(1)$ and 0 if denominator is not divisible by $(1-t)^n$.

$\sigma \neq \sigma^{-1}$ and since $\dfrac{1}{1-e} + \dfrac{1}{1-e^{-1}} = 1$, then the coefficient of $\dfrac{1}{(1-t)^{n-1}}$ in the Laurent series of $\dfrac{1}{|G|} \sum_{\sigma \in G} \frac{1}{\det(1-\sigma t)}$ is half the number of pseudo-reflections of G.

(5) By Proposition 7.54(3), $a(S^G)$ is the order of the pole ($-$ order of zero if $a(R) < 0$) of $P(S^G, t)$ at $t = \infty$. Every term of the right-hand side of the equality in Molien's theorem is of the form $t^{-n}(\det(\sigma - t^{-1}))$ and has 0 of order n at ∞. Hence the inequality is obvious. In the expansion of right-hand side with respect to t^{-1}, the coefficient of t^{-n} is $\dfrac{1}{|G|} \sum_{\sigma \in G} \det \sigma^{-1} = \dfrac{1}{|G|} \sum_{\sigma \in G} \det \sigma$. This sum is 1 if $G \subseteq \mathrm{SL}(n, k)$ and 0 otherwise. So our conclusion follows. □

Now we will prove Molien's theorem. The next lemma is essential.

Lemma 7.109 *Assume that a finite group G acts on a k-vector space. Then we have the following equality:*

$$\dim V^G = \frac{1}{|G|} \sum_{\sigma \in G} \mathrm{Tr}\ \sigma,$$

where Tr σ *denotes the trace of the action of σ on V.*

Proof The Reynolds operator $\rho = \frac{1}{|G|} \sum_{\sigma \in G} \sigma : V \to V^G$ is the identity on V^G. Hence $\dim V^G = \mathrm{Tr}\,\rho$. On the other hand, since Tr is compatible with addition, we have $\mathrm{Tr}\,\rho = \frac{1}{|G|} \sum_{\sigma \in G} \mathrm{Tr}\ \sigma$. □

Proof of Theorem 7.107 Let $\mathrm{Tr}(\sigma, S_n)$ be the trace of the action of σ on S_n. Then by 7.109 we have

$$P(S^G, t) = \frac{1}{|G|} \sum_{n \geq 0} \sum_{\sigma \in G} (\mathrm{Tr}(\sigma, S_n)) t^n.$$

Hence it suffices to show

$$\frac{1}{\det(1 - \sigma t)} = \sum_{n \geq 0} (\mathrm{Tr}(\sigma, S_n)) t^n.$$

Each element σ can be diagonalized taking a suitable basis of S_1. If $\sigma(X_i) = e_i X_i$, then we have

$$\frac{1}{\det(1 - \sigma t)} = \frac{1}{1 - e_1 t} \cdots \frac{1}{1 - e_n t} = (\sum_{m \geq 0} e_1^m t^m) \cdots (\sum_{m \geq 0} e_n^m t^m)$$

and the coefficient of t^m of this expansion is

$$\sum_{a_1+\cdots+a_n=m} e_1^{a_1}\cdots e_n^{a_n}.$$

On the other hand, since the basis of S_m is $\{X_1^{a_1}\cdots X_n^{a_n} \mid a_1+\cdots+a_n=m\}$ and since

$$\sigma(X_1^{a_1}\cdots X_n^{a_n}) = e_1^{a_1}\cdots e_n^{a_n}X_1^{a_1}\cdots X_n^{a_n},$$

$\mathrm{Tr}(\sigma, S_m) = \sum_{a_1+\cdots+a_n=m} e_1^{a_1}\cdots e_n^{a_n}$. This proves the theorem. □

We can give the criterion for the Gorenstein property of S^G using Molien's theorem.

Theorem 7.110 (Watanabe[19], [Wa2, Wa3])

(1) *If* $G \subseteq \mathrm{SL}(n,k)$, *then* S^G *is a Gorenstein ring.*
(2) *Conversely, if* G *does not contain any pseudo-reflections and if* S^G *is a Gorenstein ring, then we have* $G \subseteq \mathrm{SL}(n,k)$.

Proof Let us give two proofs of (1). The first proof uses Corollary 7.105. Let $\mathfrak{m} = (X_1,\ldots,X_n)$ be the graded maximal ideal of S. We see the action of G on the socle of $\mathrm{H}^n_{\mathfrak{m}}(S)$. Let $(f_1,\ldots,f_n)$ be a system of parameters of S consisting of homogeneous elements of S^G. By Theorem 5.105, we have $\mathrm{H}^n_{\mathfrak{m}}(S) \cong \lim_{s\to\infty} S/(f_1^s,\ldots,f_n^s)S$ and the socle of $\mathrm{H}^n_{\mathfrak{m}}(S)$ coincides with the socle of $S/(f_1,\ldots,f_n)S$. The rest of the proof is given in the next lemma. □

Lemma 7.111 *The action of* G *on* $\mathrm{Soc}\,(\mathrm{H}^n_{\mathfrak{m}}(S))$ *is given by* $\xi = \det^{-1}$. *Hence if* $G \subseteq \mathrm{SL}(n,k)$, $\mathrm{Soc}\,(\mathrm{H}^n_{\mathfrak{m}}(S))$ *is* G*-invariant* ($\mathrm{Soc}\,(\mathrm{H}^n_{\mathfrak{m}}(S))$ *denotes the socle of* $\mathrm{H}^n_{\mathfrak{m}}(S)$).

Proof Let $\xi(\sigma)$ be the action of $\sigma \in G$ on $\mathrm{Soc}\,(\mathrm{H}^n_{\mathfrak{m}}(S))$. Then $\xi(\sigma)$ does not depend on the choice of $(f_1,\ldots,f_n)$ and G. Hence we may assume that G is a cyclic group generated by σ. Also, the action is the same if we consider in an extension field of k. Since we assume that $|G| \neq 0$ in k, we can diagonalize σ taking a suitable basis (if σ cannot be diagonalized in any extension field of k, then considering the Jordan canonical form, the order of σ is infinity if $\mathrm{char}\,(k)=0$ and divisible by p if $\mathrm{char}\,(k)=p>0$).

If $\mathrm{ord}\,(\sigma)=m$, then we can take $(X_1^m,\ldots,X_n^m)$ as a G-invariant system of parameters. Then $z=(X_1X_2\cdots X_n)^{m-1}$ is a generator of $\mathrm{Soc}\,(S/(X_1^m,\ldots,X_n^m)S)$, and we have that $\sigma(z) = (\det(\sigma))^{m-1}z = \det(\sigma)^{-1}z$. Hence we have shown $\xi = \det^{-1}$. □

As the second proof of Theorem 7.110 (1), we introduce Stanley's proof using Molien's theorem and Theorem 7.56. (In characteristic $p>0$, since in Theorem 7.107 we compute $P(S^G,t)$ modulo p, this does not give a direct proof.)

[19] Kei-ichi Watanabe, 1944–.

Proof of Theorem 7.110 (Continued) To show that S^G is Gorenstein, it suffices to show that $P(S^G, t^{-1}) = (-1)^n t^{-a} P(S^G, t)$ for $a = a(S^G)$ by Theorem 7.56(2).

By Theorem 7.107, we have $P(S^G, t^{-1}) = \frac{1}{|G|} \sum_{\sigma \in G} \dfrac{1}{\det(\mathbf{1} - \sigma t^{-1})}$.

Multiplying the denominator and numerator of $\det(\mathbf{1} - \sigma t^{-1})$ by $\det(\sigma^{-1}) t^n$, we get $\dfrac{1}{\det(\mathbf{1} - \sigma t^{-1})} = \dfrac{\det \sigma^{-1} t^n}{\det(\sigma t - \mathbf{1})}$. Hence if $G \subseteq \mathrm{SL}(n, k)$, then the right-hand side becomes $= (-1)^n \dfrac{1}{|G|} \sum_{\sigma^{-1} \in G} \frac{t^n}{\det(\mathbf{1} - \sigma t^{-1})} = (-1)^n t^n P(S^G, t)$ and we can assert that S^G is Gorenstein by Theorem 7.56 (2).

The proof of (2) reduces to the next proposition, which uses Molien's theorem. This proof works only if char $(k) = 0$ and we will explain the case char $k = p > 0$ later. (The first proof of this [Wa3] uses divisor class groups.)

By the next proposition, if G does not contain any pseudo-reflections and if S^G is Gorenstein, then $a(S^G) = -n$. Then by Theorem 7.108 (5), $G \subseteq \mathrm{SL}(n, k)$. □

Proposition 7.112 *Assume that S^G is Gorenstein. If the number of pseudo-reflections in G is $r(G)$, then we have*

$$a(S^G) = -n - r(G).$$

Proof Put $P(S^G, t) = \dfrac{f(t)}{(1 - t^{d_1}) \cdots (1 - t^{d_n})}$ with $f(t) = 1 + a_1 t + \ldots + a_i t^i + \ldots + a_d t^m$. Since S^G is Gorenstein, we have $a_i = a_{m-i}$ for every i, $0 \leq i \leq m$ by Theorem 7.56 (2). We have seen that $r(G) = \sum_{i=1}^n (d_i - 1) - \dfrac{2f'(1)}{f(1)}$ by Theorem 7.108 (4). But from the symmetry of the coefficients of $f(t)$, we have $\dfrac{2f'(1)}{f(1)} = m$. Hence we obtain $a(S^G) = m - \sum_{i=1}^n d_i = -n - r(G)$. □

The condition that S^G is regular is very famous.

Theorem 7.113 (Chevalley, Shephard–Todd; [Che], [ST]) *The following conditions are equivalent:*

(1) *S^G is isomorphic to a polynomial ring over k.*
(2) *G is generated by pseudo-reflections.*

Remark 7.21 Here we assume that $|G| \neq 0$ in k. But (1) ⇒(2) holds true even if $|G| = 0$ in k (we will not give a proof here). But if $|G| = 0$ in k, there are counter-examples to (2) ⇒(1).

To prove (2) ⇒(1), we note the following statement.

Remark 7.22 Assume that either $S = k[X_1, X_2, \ldots, X_n]$ is a polynomial ring over k and R is a graded k-subalgebra of S such that S is a finite R-module or $(S, \mathfrak{n})$ is a regular local ring finite over a subalgebra $(R, \mathfrak{m})$.

Then R is regular if and only if S is faithfully flat over R.

Proof If R is regular, then a regular parameter system of R is a system of parameters in S and is a regular sequence in S, since S is regular. Hence S is faithfully flat over R by Theorem 5.65.

Conversely, if S is flat over R, we have

$$\mathrm{Tor}_R^m(R/\mathfrak{m}, R/\mathfrak{m}) \otimes_R S \cong \mathrm{Tor}_S^m(S/\mathfrak{m}S, S/\mathfrak{m}S).$$

Since S is regular, the left-hand side is 0 if $m > \dim S$. Hence $\mathrm{Tor}_R^m(R/\mathfrak{m}, R/\mathfrak{m}) = 0$ if $m > \dim S$, which implies that R is regular (Theorem 5.58). □

We state the proof of (2) ⇒(1) in a generalized form.

Theorem 7.114 *We consider the following two cases:*

(1) *S is a graded ring over a field k with $S_0 = k$, G is a finite group acting faithfully on S as graded k isomorphisms.*
(2) *$(S, \mathfrak{n})$ is a local ring and G is a finite group acting faithfully on S and the action of G on $S/\mathfrak{n}$ is the identity.*

For $\sigma \in G$, we say that σ is a generalized reflection if there is a non-zerodivisor $l \in S$ such that $\sigma(x) - x \in (l)$ for every $x \in S$.

If G is generated by generalized reflections, then S is flat over $R = S^G$.

Remark 7.23 If $\sigma \in G \subseteq \mathrm{GL}(n, k)$ and rank $(\sigma - \mathbf{1}) = 1$, then by a linear change of generators of S, we can take $\sigma(X_i) = X_i$, $(i = 1, 2, \ldots, X_{n-1})$. Hence we have $\forall f \in S, \sigma(f) - f \in (X_n)$. Hence if $G \subseteq \mathrm{GL}(n, k)$ acts on $S = k[X_1, \ldots, X_n]$, "generalized reflection" defined in Theorem 7.114 is equivalent to the pseudo-reflections.

Remark 7.24 If G is generated by generalized reflections, since S is flat over R by Theorem 7.114, if S is Gorenstein (resp. a complete intersection), then so is S^G. Generally speaking, S^G becomes a "better" ring than S.

To prove Theorem 7.114, the following lemma is essential.

Lemma 7.115 *In the assumptions of Theorem 7.114, take $f_1, f_2, \ldots, f_r \in R$, $f_1 \notin (f_2, \ldots, f_r)R$. If for $g_1, g_2, \ldots, g_r \in S$, we have*

$$f_1 g_1 + f_2 g_2 + \ldots + f_r g_r = 0 \qquad (\dagger)$$

then we have $g_1 \in \mathfrak{m}S$, where in case (1), we put $\mathfrak{m} = R_+$ and $\mathfrak{m} = \mathfrak{n}^G$ in case (2). We also assume that $f_1, f_2, \ldots, f_r, g_1, g_2, \ldots, g_r$ are homogeneous elements in case (1).

Proof We prove statement (1). Statement (2) can be proved in the same way, considering ord (g) (the maximal r such that $g \in \mathfrak{n}^r$) instead of $\deg(g)$.

We show by induction on $\deg(g_1)$. If $g_1 \in S_0 = k$, if $g_1 \neq 0$, then the equality becomes $f_1 = -g_1^{-1}(f_2 g_2 + \ldots + f_r g_r)$. Acting ρ on both sides, we have $f_1 =$

$-g_1^{-1}(f_2\rho(g_2)+\ldots+f_r\rho(g_r)) \in (f_2, \ldots, f_r)R$, which contradicts our assumption. Hence $g_1 = 0 \in \mathfrak{m}S$.

If $\deg(g_1) > 0$, consider the action of a generalized reflection $\sigma \in G$ to (†). If $\sigma(g_1) = g_1$ for every such σ, since G is generated by generalized reflections, we have $g_1 \in R_+ = \mathfrak{m}$ and we have the lemma.

If $g_1 \notin R$, take a generalized reflection $\sigma(g_1)$ such that $\sigma(g_1) \neq g_1$ and put $\sigma(g_i) - g_i = lh_i$ $(i = 1, 2, \ldots, r)$. Taking the difference of $(\sigma(\dagger))$ and (†) and divide by l, we get $f_1h_1 + f_2h_2 + \ldots + f_rh_r = 0$ with $\deg(h_1) < \deg(g_1)$. Hence we have $h_1 \in \mathfrak{m}S$ by induction hypothesis, it follows that $g_1 = lh_1 \in \mathfrak{m}S$. □

Proof of Theorem 7.114 Put $I = \mathfrak{m}S$. Let $g_1, g_2, \ldots, g_r \in S$ be the homogeneous elements, which become a basis of S/I over $R/\mathfrak{m}$. Since S is a finitely generated R-module, $g_1, g_2, \ldots, g_r$ generates S by (NAK). We show that S is a *free* R-module generated by these elements.

Let $f_1, f_2, \ldots, f_r \in R$ and assume that $f_1g_1 + f_2g_2 + \ldots + f_rg_r = 0$. Since $g_1 \notin I$, by Lemma 7.115, $f_1 \in (f_2, \ldots, f_r)R$. Putting $f_1 = a_2f_2 + \ldots + a_rf_r$, $a_2, \ldots, a_r \in R$, we have $f_2(a_2g_1 + g_2) + \ldots + f_r(a_rg_1 + g_r) = 0$. Since $a_2g_1 + g_2 \notin I$, we have $f_2 \in (f_3, \ldots, f_r)R$. Repeating this process we get $f_1 = f_2 = \ldots = f_r = 0$. We have proved that S is a free R-module. □

Proof of Theorem 7.113 Here we assume that $\mathrm{char}\, k = 0$. Refer to [KW] if $\mathrm{char}\, k = p > 0$.

Assume that $S^G = k[f_1, f_2, \ldots, f_n]$ is a polynomial ring, that is, generated by algebraically independent elements $f_1, f_2, \ldots, f_n$. Here, we assume that each f_i is homogeneous with $\deg f_i = d_i$. By Theorem 7.108(4), we have $r(G) = \sum_{i=1}^n (d_i - 1)$. On the other hand, let H be the subgroup of G generated by all pseudo-reflections of G. Since we have shown (2) ⇒(1), $S^H = k[g_1, g_2, \ldots, g_n]$ is a polynomial ring. Putting $\deg g_i = e_i$, then we have $r(H) = \sum_{i=1}^n (e_i - 1)$. Since $S^H \supseteq S^G$, for every i, we have $e_i \le d_i$. But since H contains every pseudo-reflection of G, $\sum_{i=1}^n (d_i - 1) = r(G) = r(H) = \sum_{i=1}^n (e_i - 1)$. Hence we have $d_i = e_i$ for every i. Hence by Theorem 7.108(1), we have $|G| = |H| = \prod_{i=1}^n d_i$ and $G = H$. Namely, G is generated by pseudo-reflection. □

For a finite group G, to give an embedding $\rho : G \hookrightarrow \mathrm{GL}(n, k)$ is the same as to give a faithful n-dimensional kG-module V with a fixed k-basis $v_1, \ldots, v_n$. If such a V and a basis are given, then we get ρ by $\rho(g) = (c_{ij}(g))$, where $gv_j = \sum_{i=1}^n c_{ij}(g)v_i$. If $\mathrm{char}(k) = p > 0$ and $|G| \neq 0$ in k, given a (faithful) n-dimensional G-module V, there exists a (faithful) $\mathbb{C}G$-module $V_{\mathbb{C}}$ which is deeply connected with V via the modulo p reduction. Namely:

Proposition 7.116 *Assume that* $\mathrm{char}(k) = p > 0$ *and* $|G| \neq 0$ *in* k. *If* V *is a finite-dimensional* kG*-module, then there exist a subring* R *of* $\mathbb{C}$ *which is a complete DVR whose residue field is a finite subfield of* k*, and an* RG*-module* V_R *which is finite free as an* R*-module and* $k \otimes_R V_R \cong V$ *as* kG*-modules. When we define* $V_{\mathbb{C}} = \mathbb{C} \otimes_R V_R$*, we have* $P(S_{\mathbb{C}}^G, t) = P(S, t)$*, where* $S = \mathrm{Sym}\, V^*$ *and* $S_{\mathbb{C}} = \mathrm{Sym}_{\mathbb{C}}\, V_{\mathbb{C}}^*$.

Corollary 7.117 *In the assumption of Proposition 7.116, S^G is a polynomial ring (resp. Gorenstein) if and only if so is $\mathbb{C}[X_1, X_2, \ldots, X_n]^{G'}$. Hence Theorems 7.113, 7.110 are also true if* char $(k) = p > 0$ *under the general assumption* $|G| \neq 0$ *in* k.

Proof of Corollary 7.117 The Gorenstein property of S^G is determined by Theorem 7.56. We have seen that (2) $\Rightarrow$(1) of Theorem 7.113 is valid in any characteristic. So is the proof of (1) $\Rightarrow$(2), noting that $r(G) = r(G')$. □

Proof of Proposition 7.116 By Maschke's theorem [Pie, (3.6)], $\mathbb{F}_pG$ is semisimple (that is, the $\mathbb{F}_pG$-module $\mathbb{F}_pG$ is the sum of its simple submodules). By Wedderburn's structure theorem [Pie, (3.5)], there exists an isomorphism

$$\mathbb{F}_pG \cong M_{n_1}(\mathbb{F}_{q_1}) \times \cdots \times M_{n_r}(\mathbb{F}_{q_r})$$

for some r and some $q_1, \ldots, q_r$ where $q_i = p^{e_i}$ is a power of p for each i. (Note that a finite division ring is a field by Wedderburn's little theorem [Pie, (13.6)].) Set $q' = q_1q_2 \cdots q_r$, and let q be the prime power such that $\mathbb{F}_q = \mathbb{F}_{q'} \cap k$. We set $k_0 = \mathbb{F}_q \subset k$. Note that

$$k \otimes_{\mathbb{F}_p} M_n(\mathbb{F}_{q_i}) = M_n(k\mathbb{F}_{q_i})^{s_i} \cong k \otimes_{\mathbb{F}_q} M_n(\mathbb{F}_q\mathbb{F}_{q_i})^{s_i},$$

where $s_i = [(\mathbb{F}_{q_i} \cap k) : \mathbb{F}_p]$. So every kG-module is a base change of a (unique) k_0G-module. (Note that this is true for simple modules, and every k_0G-module is a direct sum of simple modules.) Let V_0 be the k_0G-module such that $k \otimes_{k_0} V_0 \cong V$.

We can take an algebraic number field K (that is, a finite extension of $\mathbb{Q}$) such that there exists a maximal ideal $\mathfrak{m}$ of S such that $S/\mathfrak{m} \cong k_0$, where S is the ring of integers of K (that is, the integral closure of $\mathbb{Z}$ in K). Let R be the completion of $S_\mathfrak{m}$, and let L be the field of fractions of R. Note that $\mathrm{tr.deg}_\mathbb{Q} L \leq \#L \leq 2^{\aleph_0} = \mathrm{tr.deg}_\mathbb{Q}\mathbb{C}$, where $\aleph_0$ is the countably infinite cardinal. Since $\mathbb{C}$ is algebraically closed (cf. [Nag2, Theorem 2.12.4]), there is a K-algebra map $L \to \mathbb{C}$. This map is injective, and we consider that $S \subset R \subset L \subset \mathbb{C}$.

For each simple k_0G-module W, there is an indecomposable projective RG-module P and a surjective RG-linear map $P \to W$. Since R is Henselian by Hensel's lemma (Theorem 4.30), any idempotent of $\Lambda/t\Lambda$ lifts to $\Lambda = \mathrm{End}_{RG} P$, where t is a prime element of R. It follows that P/tP is an indecomposable k_0G-module. Since k_0P is semisimple, P/tP is simple, and hence $P/tP \cong W$. (It is easy to see that such a P is uniquely determined from W.) So any finite-dimensional k_0G-module is a base change of a (unique) finitely generated projective RG-module. Let V_R be the finite projective RG-module such that $k_0 \otimes_R V_R = V_R/tV_R \cong V_0$. Note that $k \otimes_R V_R \cong V$. We set $V_\mathbb{C} = \mathbb{C} \otimes_R V_R$.

Since $1/|G| \in R$, the map $\rho : RG \to R$ given by $\rho(\sum_g c_g g) = |G|^{-1} \sum_g c_g$ (the Reynolds operator) is the left inverse of the RG-linear map $\iota : R \to RG$ given by $\iota(1) = \sum_g g$. So the trivial RG-module R is a direct summand of RG, and is RG-projective. Note that for a finite projective RG-module F, an RG-module M, and an R-algebra T, the canonical map

$$T \otimes_R \operatorname{Hom}_{RG}(F, M) \to \operatorname{Hom}_{RG}(F, T \otimes_R M) \cong \operatorname{Hom}_{TG}(T \otimes_R F, T \otimes_R M)$$

is an isomorphism. This is obvious for $F = RG$, and it is also true for $F = (RG)^n$, and hence is also true for its direct summand. Hence the map is an isomorphism for any finite projective RG-module F. Letting $F = R$, the trivial RG-module, we have that the G-invariance and the base change are compatible for every T:

$$T \otimes_R M^G \cong (T \otimes_R M)^G.$$

Considering $M = \operatorname{Sym}_j V_R^*$ and letting $T = k$ and $T = \mathbb{C}$, we get:

$$\dim_{\mathbb{C}}(\operatorname{Sym}_j V_{\mathbb{C}}^*)^G = \operatorname{rank}_R(\operatorname{Sym}_j V_R^*)^G = \dim_k(\operatorname{Sym}_j V^*)^G$$

for each $j \geq 0$. (Note that Sym_j is compatible with the base change, see Appendix D, Lemma D.7.) We get the desired result immediately. □

7.118 We introduce the generalization of Theorem 7.110 to the modular case (that is, the case that the order $|G|$ of G is divisible by the characteristic of k).

Let k be a field, G a finite group, and B a k-algebra on which G acts. We define the *twisted group algebra* $B * G$ as the tensor product $B \otimes kG$, where kG is the group algebra of G over k, as a k-vector space, and the product of $B * G$ is given by $(b \otimes g)(b' \otimes g') = b(gb') \otimes gg'$ for $b, b' \in B$ and $g, g' \in G$. Note that $B * G$ is an associative k-algebra. If M is a $B * G$-module, then it is a G-module B-module in an obvious way, and is a (B, G)-module (see Definition 7.100), since

$$g(bm) = (1 \otimes g)(b \otimes 1)m = (gb \otimes g)m = (gb \otimes 1)(1 \otimes g)m = (gb)(gm).$$

Conversely, a (B, G)-module yields a $B * G$-module, and the category of (B, G)-modules and maps of (B, G)-modules is equivalent to the category $\operatorname{Mod}(B * G)$.

Proposition 7.119 *Let k be a field, and $G \subseteq \mathrm{GL}_n(k)$ be a finite subgroup without pseudo-reflections. Let $S = k[x_1, \ldots, x_n]$ on which G acts canonically, and $A = S^G$. Let* $\operatorname{Ref}(A)$ *be the full subcategory of* $\operatorname{Mod} A$ *consisting of reflexive finite A-modules. Let* $\operatorname{Ref}(S, G)$ *be the full subcategory of* $\operatorname{Mod}(S*G)$ *consisting of reflexive finite $(S * G)$-modules. Then $\mathcal{G} : \operatorname{Ref}(S, G) \to \operatorname{Ref}(A)$ given by $\mathcal{G}(M) = M^G$ is an equivalence, and $\mathcal{F} : \operatorname{Ref}(A) \to \operatorname{Ref}(S, G)$ given by $\mathcal{F}(N) = (S \otimes_A N)^{**}$ is its quasi-inverse, where $(-)^* = \operatorname{Hom}_S(-, S)$.*

Proof (sketch) Assume that $G \subseteq \mathrm{GL}_n(V)$ does not have a pseudo-reflection. If $M \in \operatorname{Ref}(G, S)$, then taking the S-dual $(-)^*$ of a presentation $(S * G)^n \xrightarrow{\varphi} (S * G)^m \xrightarrow{\psi} M^* \to 0$, we get an exact sequence

$$0 \to M \xrightarrow{\psi^*} (S * G)^m \xrightarrow{\varphi^*} (S * G)^n.$$

Applying the G-invariance functor, we get

$$0 \to M^G \xrightarrow{\psi^*} S^m \to S^n,$$

and M^G is (S_2), since S satisfies (S_2) as an A-module. By Proposition 3.51, we have that M^G is reflexive. Thus, $\mathcal{G} = (-)^G : \mathrm{Ref}(G, S) \to \mathrm{Ref}(A)$ is well-defined. Also, it is easy to see that the S-dual of any S-finite (G, S)-module lies in $\mathrm{Ref}(G, S)$. So $\mathcal{F}: (S \otimes_A -)^{**} : \mathrm{Ref}(A) \to \mathrm{Ref}(G, S)$ is also well-defined.

Now we define $\varepsilon : \mathcal{F}\mathcal{G} = (S \otimes_A (-)^G)^{**} \to \mathrm{Id}$ by

$$(S \otimes_A M^G)^{**} \xrightarrow{\Phi^{**}} M^{**} \cong M,$$

where $\Phi : S \otimes_A M^G \to M$ is given by $\Phi(s \otimes m) = sm$. We also define $u : \mathrm{Id} \to ((S \otimes_A (-))^{**})^G = \mathcal{G}\mathcal{F}$ by

$$N \xrightarrow{\rho} (S \otimes_A N)^G \xrightarrow{\delta^G} ((S \otimes_A N)^{**})^G,$$

where $\rho(n) = 1 \otimes n$, and $\delta : S \otimes_A N \to (S \otimes_A N)^{**}$ is the canonical map to the double dual. To prove that $\varepsilon : \mathcal{F}\mathcal{G}M \to M$ is an isomorphism, it suffices to show that $(S_\mathfrak{p} \otimes_{A_\mathfrak{p}} M_\mathfrak{p}^G)^{**} = S_\mathfrak{p} \otimes_{A_\mathfrak{p}} M_\mathfrak{p}^G \to M_\mathfrak{p}$ is an isomorphism for every height one prime ideal $\mathfrak{p}$ of $A = S^G$. Now $M \mapsto S_\mathfrak{p} \otimes_{A_\mathfrak{p}} M_\mathfrak{p}^G$ and $M \mapsto M_\mathfrak{p}$ are both left exact, and hence considering an exact sequence $0 \to M \to (S * G)^m \to (S * G)^n$, the problem is reduced to the case that $M = S * G$. To prove this, it suffices to show that $S \otimes_A S \to S \otimes kG$ given by $s \otimes t \mapsto \sum_{h \in G} s(ht) \otimes h$ is an isomorphism in codimension one. This follows from the fact that the map $G \times X \to X \times_Y X$ given by $(g, x) \mapsto (gx, x)$ is an isomorphism outside of a closed subset codimension two or more, from the assumption that G does not have a pseudo-reflection. Similarly, to prove that $u : N \to \mathcal{G}\mathcal{F}N$ is an isomorphism, we may assume that $N = A$, and this case is trivial. So, $\mathcal{F}$ and $\mathcal{G}$ are quasi-inverses of each other, and $\mathcal{F}$ is an equivalence of categories. □

The class group $\mathrm{Cl}(A)$ is isomorphic to the group of the isomorphism classes of rank-one reflexive A-modules. The sum $[L] + [N]$ is given by $[(L \otimes_A N)^{**}]$, where $(-)^{**}$ denotes the double dual. We define $\mathrm{Cl}(G, S)$ to be the group of the isomorphism classes of rank-one reflexive (G, S)-modules. As the equivalence $\mathcal{F}$: $\mathrm{Ref}(A) \to \mathrm{Ref}(G, S)$ preserves the rank of a module and the double dual of the tensor product, we get an isomorphism $\rho : \mathrm{Cl}(A) \to \mathrm{Cl}(G, S)$ with $\rho([N]) = [(S \otimes_A N)^{**}]$. Every rank-one reflexive S-module is free, and hence every rank-one reflexive (G, S)-module is of the form $\chi \otimes_k S$ with $\chi \in X(G)$ a character (i.e., one-dimensional representation) of G, where $X(G)$ is the character group of G, that is, the group of isomorphism classes of one-dimensional representations of G (the sum is given by the tensor product again). It is easy to see that $X(G) \to \mathrm{Cl}(G, S)$ given by $\chi \mapsto \chi \otimes_k S$ is an isomorphism. So $\mathrm{Cl}(A) \cong X(G)$. With this isomorphism, it is known that the class of the canonical module $[K_A] \in \mathrm{Cl}(A)$ corresponds to $\det_V$, the determinant representation $g \mapsto \det g$. Thus we have shown that the following theorem is also true for the modular case, that is, the case that the characteristic p of k is positive and divides the order $|G|$ of G.

Theorem 7.120 *Let $V = k^n$, and G be a finite subgroup of $\mathrm{GL}(V)$ without pseudo-reflections. Then we have:*

(1) *The class group $\mathrm{Cl}(A)$ is isomorphic to the character group $X(G)$. In particular, if $G = [G, G]$, then A is a UFD.*
(2) ([FlWo], [Bra]) *The following are equivalent:*

  (a) *A is quasi-Gorenstein. That is, K_A is a projective A-module.*
  (b) $[K_A] = [A]$ *in* $\mathrm{Cl}(A)$.
  (c) $\det_V = k$ *in* $X(G)$.
  (d) $G \subseteq \mathrm{SL}(V)$.

For the details, see [Has]. In the non-modular case (that is, the case that $|G|$ is not divisible by $p = \mathrm{char}(k)$), a finite subgroup $G \subseteq \mathrm{SL}(V)$ cannot have a pseudo-reflection. Moreover, A is Cohen–Macaulay in this case. So Theorem 7.120 is a generalization of Theorem 7.110.

Chapter 8
Multiplicities

In this chapter, we study the multiplicity $\mathrm{e}_0(\mathfrak{m}, A)$, which is one of the most important numerical invariants of a Noetherian local ring $(A, \mathfrak{m})$. In general, the multiplicity $\mathrm{e}_0(I, M)$ is defined for a pair of an ideal I and a finite module M such that $\ell_A(M/IM) < \infty$. We study the notion of reduction of ideals, superficial elements, and d-sequences, which are important tools to study multiplicities of rings and modules. We prove the celebrated theorem of Serre and Auslander–Buchsbaum which says that the multiplicity of a module with respect to a parameter ideal is expressed in terms of the Euler characteristic of the Koszul homology (Theorem 8.24). We also prove the associativity formula (Theorem 8.26) of the multiplicity which says that the calculation of the multiplicity of a module is reduced to the calculation of the multiplicities of integral domains. We also prove the inequality between the length and the multiplicity (Proposition 8.28). The equality holds if and only if the module is Cohen–Macaulay (Theorem 8.32). The theorem of Hayasaka–Hyry, generalizing this well-known fact on parameter ideals to powers of parameter ideals, is proved (Theorem 8.33). Finally, we prove the theorem of Nagata which says that an unmixed Noetherian local ring is a regular local ring if and only if its multiplicity is one (Theorem 8.36).

8.1 Multiplicities and Reductions

In this section, A denotes a Noetherian local ring, and $\mathfrak{m}$ its maximal ideal. M denotes a finitely generated A-module, and I denotes a proper ideal of A. A *parameter ideal* means an ideal generated by a system of parameters.

Let M be a finitely generated A-module of dimension d. Then we say that an ideal $Q \neq A$ of A is a *parameter ideal of M* if Q is generated by d elements and $\ell_A(M/QM) < \infty$.

S. Goto, K. Watanabe, *Commutative Algebra*, Graduate Texts in Mathematics 309,
https://doi.org/10.1007/978-981-95-4227-7_8

If $\ell_A(M/IM) < \infty$, then $\ell_A(M/I^{n+1}M) < \infty$ for all integers $n \geq 0$, and the function $\ell_A(M/I^{n+1}M)$ on n is a polynomial with coefficients in $\mathbb{Q}$ for sufficiently large n. If $M \neq 0$ and $d = \dim_A M$, then the degree of this polynomial is d, and there exists a unique sequence $\{\mathrm{e}_i(I, M)\}_{0 \leq i \leq d}$ of integers such that

$$\ell_A(M/I^{n+1}M) = \mathrm{e}_0(I, M)\binom{n+d}{d} - \mathrm{e}_1(I, M)\binom{n+d-1}{d-1} + \cdots + (-1)^d \mathrm{e}_d(I, M)$$

holds for all $n \gg 0$ (see Chap. 7, Sect. 7.4 and [AM, Corollary 11.2]). If $d = 0$, then $\mathrm{e}_0(I, M) = \ell_A(M)$. The integer $\mathrm{e}_i(I, M)$ $(0 \leq i \leq d)$ is called the ith *Hilbert coefficient* of M with respect to I. The integer $\mathrm{e}_0(I, M)$ is called the *multiplicity* of M with respect to I. It is an important invariant of M. If $M = A$, then $\mathrm{e}_i(I, A)$ is simply denoted by $\mathrm{e}_i(I)$.

Example 8.1

(1) Let M be a Cohen–Macaulay A-module, and assume that $d = \dim_A M \geq 1$. If $Q = (a_1, a_2, \ldots, a_d)$ is a parameter ideal of M, then the equality

$$\ell_A(M/Q^{n+1}M) = \ell_A(M/QM)\binom{n+d}{d}$$

holds for any integer $n \geq 0$. Therefore, we have $\mathrm{e}_0(Q, M) = \ell_A(M/QM)$ and $\mathrm{e}_i(Q, M) = 0\,(1 \leq i \leq d)$.

(2) (Cf. Theorem 8.36) If A is a regular local ring, then $\mathrm{e}_0(\mathfrak{m}) = 1$.

(3) Let $R = k[[X, Y, Z, W]]$ be the formal power series ring over a field k, and

$$A = R/[(X, Y) \cap (Z, W)].$$

Then for any integer $n \geq 0$, we have

$$\ell_A(A/\mathfrak{m}^{n+1}) = 2\binom{n+2}{2} - 1.$$

Therefore, $\mathrm{e}_0(\mathfrak{m}) = 2$, $\mathrm{e}_1(\mathfrak{m}) = 0$, and $\mathrm{e}_2(\mathfrak{m}) = -1$.

(4) Let $(R, \mathfrak{n})$ be a regular local ring, and $0 \neq f \in \mathfrak{n}$. When we set $A = R/(f)$ and $\mathfrak{m} = \mathfrak{n}/(f)$, then the equality

$$\mathrm{e}_0(\mathfrak{m}) = \max\{\ell \in \mathbb{Z} \mid f \in \mathfrak{n}^\ell\}$$

holds.

Proof

(1) We prove this by induction on the dimension d of M. As the case $d = 1$ is easy, we assume that $d > 1$ and the assertion is true up to $d - 1$. As the sequence $a_1, a_2, \ldots, a_d$ is M-regular, we have that

$$a_1 M \cap Q^{k+1}M = a_1 Q^k M$$

for any integer $k \geq 0$ (Proposition 8.2). Therefore, when we set $\overline{M} = M/a_1M$, then we have an exact sequence

$$0 \to M/Q^kM \xrightarrow{a_1} M/Q^{k+1}M \to \overline{M}/Q^{k+1}\overline{M} \to 0.$$

Hence we have

$$\begin{aligned}\ell_A(Q^kM/Q^{k+1}M) &= \ell_A(M/Q^{k+1}M) - \ell_A(M/Q^kM)\\ &= \ell_A(\overline{M}/Q^{k+1}\overline{M})\\ &= \ell_A(M/QM)\binom{k+d-1}{d-1},\end{aligned}$$

and the equalities

$$\begin{aligned}\ell_A(M/Q^{n+1}M) &= \sum_{k=0}^{n} \ell_A(Q^kM/Q^{k+1}M)\\ &= \ell_A(M/QM)\sum_{k=0}^{n}\binom{k+d-1}{d-1}\\ &= \ell_A(M/QM)\binom{n+d}{d}\end{aligned}$$

follow.

(2) The maximal ideal $\mathfrak{m}$ is generated by a regular sequence of length d. Hence $\ell_A(A/\mathfrak{m}^{n+1}) = \binom{n+d}{d}(n \geq 0)$, and it follows that $\mathrm{e}_0(\mathfrak{m}) = 1$.

(3) Look at the exact sequence of A-modules

$$0 \to A \to R/(X,Y) \oplus R/(Z,W) \to R/(X,Y,Z,W) \to 0.$$

When we set $B = R/(X,Y) \oplus R/(Z,W)$, then for any integer $n \geq 0$, we have $\mathfrak{m}^{n+1} = \mathfrak{m}^{n+1}B$. So we get the exact sequence

$$0 \to A/\mathfrak{m}^{n+1} \to B/\mathfrak{m}^{n+1}B \to A/\mathfrak{m} \to 0.$$

As the rings $R/(X,Y)$ and $R/(Z,W)$ are regular, we get the equality

$$\ell_A(A/\mathfrak{m}^{n+1}) = 2\binom{n+2}{2} - 1.$$

(4) We set $\dim R = d+1$, and $\ell = \max\{\ell \in \mathbb{Z} \mid f \in \mathfrak{n}^\ell\}$. As $f \in \mathfrak{n}^\ell \setminus \mathfrak{n}^{\ell+1}$, we have that $fR \cap \mathfrak{n}^{n+1} = f\mathfrak{n}^{n+1-\ell}$ for all $n \in \mathbb{Z}$. Hence we get the exact sequences

$$0 \to R/\mathfrak{n}^{n-\ell+1} \to R/\mathfrak{n}^{n+1} \to A/\mathfrak{m}^{n+1} \to 0,$$

and the equalities

$$\ell_A(A/\mathfrak{m}^{n+1}) = \binom{n+d+1}{d+1} - \binom{n-\ell+d+1}{d+1} \quad (n \geq \ell)$$

follow. As $\binom{n-\ell+d+1}{d+1} = \binom{n+d+1}{d+1} - \ell\binom{n+d}{d} + (\text{lower terms})$, we have $\mathrm{e}_0(\mathfrak{m}) = \ell$. □

Proposition 8.2 *If a sequence of elements $a_1, a_2, \ldots, a_d$ of A is M-regular, then*

$$(a_1, a_2, \ldots, a_i)M \cap (a_1, a_2, \ldots, a_d)^{n+1}M = [(a_1, a_2, \ldots, a_i)\cdot(a_1, a_2, \ldots, a_d)^n]M$$

holds for any integers $1 \leq i \leq d$ and $n \geq 0$.

Proof Set $Q = (a_1, a_2, \ldots, a_d)$ and $Q_i = (a_1, a_2, \ldots, a_i)$ for $1 \leq i \leq d$. We prove the assertion by descending induction on i. If $i = d$, then the assertion is obvious. Assume that $i < d$, and the assertion is true up to $i+1$. We show $Q_iM \cap Q^{n+1}M = (Q_iQ^n)M$. If $n = 0$, then the assertion is true. So assume that $n > 0$ and the assertion is true up to $n-1$. Then we have

$$\begin{aligned} Q_iM \cap Q^{n+1}M &= (Q_{i+1}M \cap Q^{n+1}M) \cap Q_iM \\ &= (Q_{i+1}Q^n)M \cap Q_iM \\ &= ([Q_i + (a_{i+1})]Q^n)M \cap Q_iM \\ &= (Q_iQ^n)M + [(a_{i+1}Q^n)M \cap Q_iM]. \end{aligned}$$

Now we prove $(a_{i+1}Q^n)M \cap Q_iM \subseteq (Q_iQ^n)M$. We take $x \in (a_{i+1}Q^n)M \cap Q_iM$, and we write $x = a_{i+1}y (y \in Q^nM)$. Since $a_{i+1}y \in Q_iM$, we have $y \in Q_iM \cap Q^nM = (Q_iQ^{n-1})M$, we get $x = a_{i+1}y \in (Q_iQ^n)M$, and hence

$$(a_{i+1}Q^n)M \cap Q_iM \subseteq (Q_iQ^n)M.$$

So we have $Q_iM \cap Q^{n+1}M = (Q_iQ^n)M$. □

In the theory of multiplicities, the notion of *reduction* is very important.

Definition 8.3 (Reduction) Let $J \subseteq I$ be ideals of A.

We say that $a \in A$ is *integral* over J if there exists a monic polynomial $f(t) = t^n + c_1t^{n-1} + \cdots + c_n \in A[t]$ such that $c_i \in J^i$ for $i = 1, \ldots, n$ and $f(a) = 0$. If every element a of I is integral over J, then we say that I is *integral over* J. It is easy to see that I is integral over J if and only if the Rees algebra $\mathrm{R}_A(I)$ is an integral extension of the ring $\mathrm{R}_A(J)$.

We say that J is a *reduction* of I if there exists some integer $r \geq 0$ such that $I^{r+1} = I^rJ$.

If $I = (a_1, \dots, a_s)$ is integral over J, we have $a_i^{n_i+1} \in JI^{n_i}$ for some $n_i \geq 0$. Then it is easy to see that $I^{n_1+\dots+n_s+1} \subseteq JI^{n_1+\dots+n_s}$ and J is a reduction of I. Conversely, if $I^{r+1} = I^r J$, then the Rees algebra $\mathrm{R}_A(I)$ is generated by the elements of degree up to r as an $\mathrm{R}_A(J)$-module, and it is a finite $\mathrm{R}_A(J)$-module. So I is integral over J. Hence we have the following.

Lemma 8.4 *Let $J \subseteq I$ be ideals of A. Then the following are equivalent:*

(1) *I is integral over J.*
(2) *$\mathrm{R}_A(I)$ is integral over $\mathrm{R}_A(J)$.*
(3) *$\mathrm{R}_A(I)$ is module-finite over $\mathrm{R}_A(J)$.*
(4) *J is a reduction of I.*

A minimal element of the set of reductions of I is called a *minimal reduction* of I. If $A/\mathfrak{m}$ is an infinite field, then the number of generators of all minimal reductions J of I is the same. This number $\mu_A(J)$ is called the *analytic spread* of I, and we denote it by $\ell(I)$. We have inequalities

$$\mathrm{ht}_A(I) \leq \ell(I) \leq \dim A.$$

If I is $\mathfrak{m}$-primary, then a minimal reduction of I is a parameter ideal.

Proposition 8.5 *Let J be a reduction of I. Then we have* $\mathrm{e}_0(I, M) = \mathrm{e}_0(J, M)$.

Proof Take $r \geq 0$ so that $I^{r+1} = JI^r$. Then we have $I^{n+1} = J^{n-r+1}I^r$ for every $n \geq r$. Thus

$$\ell_A(M/J^{n+1}M) \geq \ell_A(M/I^{n+1}M) \geq \ell_A(M/J^{n-r+1}M)$$

for $n \geq r$. The proposition follows from this. □

Remark 8.1 Surprisingly, the converse of Proposition 8.5 is true in the following sense. If $\dim \widehat{A}/\mathfrak{p} = \dim A$ for each minimal prime ideal $\mathfrak{p}$ of $\widehat{A}$ and $J \subseteq I$, then $\mathrm{e}_0(J) = \mathrm{e}_0(I)$ if and only if J is a reduction of I [R2].

8.2 Superficial Elements and d-Sequence

We develop a theory of multiplicities based on the theory of superficial elements. The next lemma is the key.

Lemma 8.6 *Let $R = \sum_{n\geq 0} R_n$ be a Noetherian graded ring, and set $R_+ = \sum_{n\geq 1} R_n$. Let L be a finitely generated graded R-module, and $f \in R_1$. Then the following are equivalent:*

(1) *The number of the nonzero homogeneous components $\sharp\{n \in \mathbb{Z} \mid [0 :_L f]_n \neq 0\}$ of the graded submodule $0 :_L f$ of L is finite.*
(2) *If $\mathfrak{p} \in \mathrm{Ass}_R L$ and $R_+ \not\subseteq \mathfrak{p}$, then $f \notin \mathfrak{p}$.*

Proof Note that we have

$$\mathrm{Ass}_R([0 :_L f]) = \mathrm{Ass}_R(\mathrm{Hom}_R(R/(f), L)) = \mathrm{V}(f) \cap \mathrm{Ass}_R L.$$

We have that (1) holds if and only if $R_+^\ell{\cdot}[0 :_L f] = 0$ for $\ell \gg 0$. This is equivalent to $\mathrm{Supp}_R[0 :_L f] = \mathrm{V}(\mathrm{Ann}([0 :_L f])) \subseteq \mathrm{V}(R_+)$. This is equivalent to say $\mathrm{V}(f) \cap \mathrm{Ass}_R L = \mathrm{Ass}_R[0 :_L f] \subseteq \mathrm{V}(R_+)$, which is precicely condition (2). □

Unless otherwise specified, let A be a Noetherian local ring with the maximal ideal $\mathfrak{m}$, M be a finite A-module, and I a proper ideal of A.

We apply Lemma 8.6 to the Rees algebra $R = \mathrm{R}_A(I) = A[It]$ and the associated graded module $L = \mathrm{G}_I(M)$ of M with respect to I. When we define $\varphi : I \to R$ by $\varphi(a) = at$, then $I \subseteq \varphi^{-1}(\mathfrak{p})$ and $R_+ \subseteq \mathfrak{p}$ are equivalent for $\mathfrak{p} \in \mathrm{Spec}\, R$. So if the residue field $A/\mathfrak{m}$ of A is infinite, then for any finite subset $\mathcal{F}$ of $\mathrm{Proj}\, R$, we have $I \not\subseteq \bigcup_{\mathfrak{p}\in\mathcal{F}} \varphi^{-1}(\mathfrak{p})$, where

$$\mathrm{Proj}\, R := \{\mathfrak{p} \in \mathrm{Spec}\, R \mid \mathfrak{p} \text{ is homogeneous and } \mathfrak{p} \not\supseteq R_+\}.$$

When we set $\mathcal{F} = \{\mathfrak{p} \in \mathrm{Ass}_R L \mid \mathfrak{p} \not\supseteq R_+\}$, then there exists some $a \in I$ such that $f = at$ satisfies condition (1) of Lemma 8.6 for $L = \mathrm{G}_I(M)$. If, moreover, $I \neq 0$, then we can find such an a from $I \setminus \mathfrak{m}I$.

Definition 8.7 Let $a \in I$. We assume $\mathrm{Proj}(R) \neq \emptyset$ and $L_n \neq 0$ for $n > 0$. We say that a is a *superficial element* of M with respect to the ideal I, if the conditions of Lemma 8.6 are satisfied by $f = at \in R_1$, where $R = \mathrm{R}_A(I)$ and $L = \mathrm{G}_I(M)$.

Proposition 8.8 *If the residue field $A/\mathfrak{m}$ is infinite, then the ideal I contains a superficial element of M with respect to I.*

In the rest of this section, let $a \in I$ be a superficial element of M with respect to the ideal I.

Proposition 8.9 *There exists an integer $N \geq 0$ such that*

$$[I^{n+1}M : a] \cap I^N M = I^n M$$

holds for any integer $n \geq N$.

Proof We take $N \geq 0$ so that $[0 :_{\mathrm{G}_I(M)} at]_n = 0$ holds for any integer $n \geq N$ by Lemma 8.6 (1). Indeed, for $x \in [I^{n+1}M : a] \cap I^N M$, if $x \notin I^n M$, then letting $\ell = \max\{\ell \in \mathbb{Z} \mid x \in I^\ell M\}$, we have $N \leq \ell < n$. Since $ax \in I^{n+1}M$, we have $at{\cdot}\overline{xt^\ell} = 0$ in $\mathrm{G}_I(M)$, and so $0 \neq \overline{xt^\ell} \in [0 :_{\mathrm{G}_I(M)} at]_\ell$. This is absurd. □

Corollary 8.10 *There exists some integer $N \geq 0$ such that*

$$I^{n+1}M : a = I^n M + [0 :_M a], \quad I^n M \cap [0 :_M a] = 0$$

for any $n \geq N$.

Proof Let $N_1 \geq 0$ be as in Proposition 8.9. By the Artin–Rees lemma, there exists some integer $N_2 \geq 0$ such that

$$I^{n+N_2}M \cap aM = I^n[I^{N_2}M \cap aM]$$

holds for any $n \geq 0$. Let $n \geq N_1$ be an integer, and $x \in I^{n+N_2+1}M : a$. As we have $ax \in aI^{n+1}M$, there exists some $y \in I^{n+1}M$ such that $x - y \in 0 :_M a$. As we have $n \geq N_1$, it follows that

$$y \in [I^{n+N_2+1}M : a] \cap I^{N_1}M = I^{n+N_2}M.$$

Hence we have $x \in I^{n+N_2}M + [0 :_M a]$, and it follows that

$$I^{n+1}M : a = I^nM + [0 :_M a]$$

for any $n \geq N := N_1 + N_2$. On the other hand, when $n \geq N$,

$$\begin{aligned} I^nM \cap [0 :_M a] &\subseteq [I^{\ell+1}M : a] \cap I^nM \\ &\subseteq [I^{\ell+1}M : a] \cap I^{N_1}M \\ &= I^{\ell}M \end{aligned}$$

for any integer $\ell \geq N$. Hence we have $I^nM \cap [0 :_M a] \subseteq \bigcap_{\ell \geq N} I^{\ell}M = 0$, and $I^nM \cap [0 :_M a] = 0$ holds for any integer $n \geq N$. □

Proposition 8.11 *The following hold:*

(1) *Let $M \neq 0$. If $I \not\subseteq \bigcup_{\mathfrak{p} \in \mathrm{Assh}_A M} \mathfrak{p}$, then a is a part of a system of parameters of M.*
(2) *If $\ell_A(M/IM) < \infty$, then $\ell_A(0 :_M a) < \infty$.*

Proof By Corollary 8.10, we have $I^n{\cdot}[0 :_M a] = 0$ for $n \gg 0$. Hence we have (2). We prove (1). Assume that there exists some $\mathfrak{p} \in \mathrm{Assh}_A M$ such that $a \in \mathfrak{p}$. Then $I \not\subseteq \mathfrak{p}$, and hence we have $[0 :_M a]_{\mathfrak{p}} = 0$. □

On the other hand, since $\mathrm{Ass}_A([0 :_M a]) = V(a) \cap \mathrm{Ass}_A(M)$, $\mathfrak{p} \in \mathrm{Supp}_A([0 :_M a])$, which is a contradiction.

Theorem 8.12 *Let $a \in I$ be a superficial element of M with respect to I. If $M \neq 0$, $d = \dim_A M \geq 1$, put $\overline{M} = M/aM$. If $\ell_A(M/IM) < \infty$, then we have $\ell_A(\overline{M}/I\overline{M}) < \infty$, and*

$$\mathrm{e}_i(I, \overline{M}) = \begin{cases} \mathrm{e}_i(I, M) & (0 \leq i \leq d-2), \\ \mathrm{e}_{d-1}(I, M) + (-1)^{d-1}\ell_A(0 :_M a) & (i = d-1) \end{cases}$$

Proof By Proposition 8.11, we have $\dim_A \overline{M} = d - 1$. When we take an integer $N \geq 0$ as in Corollary 8.10, then we get an exact sequence

$$0 \to 0 :_M a \to M/I^nM \xrightarrow{a} M/I^{n+1}M \to \overline{M}/I^{n+1}\overline{M} \to 0$$

of A-modules and it follows that

$$\ell_A(I^nM/I^{n+1}M) = \ell_A(\overline{M}/I^{n+1}\overline{M}) - \ell_A(0 :_M a)$$

for any integer $n \geq N$. Hence

$$\sum_{i=0}^{d-1}(-1)^i \, \mathrm{e}_i(I, M)\binom{n+d-i-1}{d-i-1}$$
$$= \sum_{i=0}^{d-1}(-1)^i \, \mathrm{e}_i(I, \overline{M})\binom{n+d-i-1}{d-i-1} - \ell_A(0 :_M a)$$

for any integer $n \geq N$. Therefore, we have $\mathrm{e}_i(I, M) = \mathrm{e}_i(I, \overline{M})$ for $0 \leq i \leq d-2$, and $\mathrm{e}_{d-1}(I, M) = \mathrm{e}_{d-1}(I, \overline{M}) + (-1)^d \ell_A(0 :_M a)$ if $i = d-1$. $\square$

Now we define a d-sequence. Leaving the settings immediately after Lemma 8.6, let A be a general commutative ring, $\underline{a} = a_1, a_2, \ldots, a_s (s \geq 1)$ a sequence of elements of A, and E an A-module.

Definition 8.13 (C. Huneke[1] [Hu1])

(1) We say that $\underline{a}$ is a *d-sequence* on E if the equalities

$$(a_1, \ldots, a_{i-1})E :_E a_i a_j = (a_1, \ldots, a_{i-1})E :_E a_j$$

hold for all integers i, j such that $1 \leq i \leq j \leq s$.

(2) We say that $\underline{a}$ is a *d^+-sequence* if $a_{\sigma(1)}^{n_1}, a_{\sigma(2)}^{n_2}, \ldots, a_{\sigma(s)}^{n_s}$ is a d-sequence for any natural numbers $n_1, n_2, \ldots, n_s \geq 1$ and any permutation σ of $\{1, \ldots, s\}$.

Note that a weak regular sequence on E is a d-sequence on E. **In what follows, let $\underline{a}$ be a d-sequence on E. We set $Q_i = (a_1, \ldots, a_i)$ for $0 \leq i \leq s$, and $Q = Q_s$**. We set $a_{s+1} = 1$.

Proposition 8.14 *Let i and n be integers such that $0 \leq i \leq s$ and $n \geq 0$. Then we have*

$$[Q_iE :_E a_{i+1}] \cap Q^{n+1}E = (Q_iQ^n)E.$$

Proof We show $[Q_iE :_E a_{i+1}] \cap QE = Q_iE$. As $a_{s+1} = 1$, we may assume that $i < s$. As $a_{i+1}, \ldots, a_s$ is a d-sequence on E/Q_iE, we may assume that $i = 0$. Namely, we are to check that $[0 :_E a_1] \cap QE = 0$. We prove this by induction on s.

[1] Craig Huneke, 1951–.

Let $s = 1$. If $x = a_1 y \in [0 :_E a_1] \cap QE$. Then as $y \in 0 :_E a_1^2 = 0 :_E a_1$, we have that $x = a_1 y = 0$. Now assume that $s > 1$ and that the assertion is true up to $s - 1$. Let $x \in [0 :_E a_1] \cap QE$, and write $x = \sum_{i=1}^{s} a_i x_i (x_i \in E)$. Then since

$$\sum_{i=1}^{s-1}(a_1 a_i)x_i + (a_1 a_s)x_s = a_1 x = 0,$$

it follows that $(a_1 a_s)x_s \in Q_{s-1}E$. By the definition of a d-sequence, we have that $a_s x_s \in Q_{s-1}E$. Hence $x \in [0 :_E a_1] \cap Q_{s-1}E$. As $a_1, \ldots, a_{s-1}$ is a d-sequence on E, we have $x = 0$ by induction. Hence $[0 :_E a_1] \cap QE = 0$ has been proved.

Next, we show that $[Q_i E :_E a_{i+1}] \cap Q^{n+1}E = (Q_i Q^n)E$. We prove this by descending induction on i. As the case $i = s$ is trivial, we may assume that $i < s$ and the assertion is true up to $i + 1$. We prove the equality by induction on n. The case $n = 0$ is already proved, and so we assume that $n > 0$ and the assertion is proved up to $n - 1$. Note that

$$[Q_i E :_E a_{i+1}] \cap Q^{n+1}E \subseteq [Q_i E :_E a_{i+1}] \cap QE = Q_i E.$$

On the other hand, we have

$$\begin{aligned} Q_i E \cap Q^{n+1}E &= (Q_{i+1}E \cap Q^{n+1}E) \cap Q_i E = (Q_{i+1}Q^n)E \cap Q_i E \\ &= ([Q_i + (a_{i+1})]Q^n)E \cap Q_i E = (Q_i Q^n)E + [(a_{i+1}Q^n)E \cap Q_i E]. \end{aligned}$$

So it remains to show that $a_{i+1}Q^n E \cap Q_i E \subseteq Q_i Q^n E$. Let $x \in a_{i+1}Q^n E \cap Q_i E$, and write $x = a_{i+1}y$ with $y \in Q^n E$. Then $y \in Q_i E :_E a_{i+1}$. By the induction assumption, we have

$$y \in [Q_i E :_E a_{i+1}] \cap Q^n E = (Q_i Q^{n-1})E,$$

and so $x = a_{i+1}y \in (Q_i Q^n)E$, and the assertion has been proved. Hence $Q_i E \cap Q^{n+1}E = (Q_i Q^n)E$, as desired. □

Corollary 8.15 *Let i and n be integers such that $0 \le i \le s$ and $n \ge 1$, we have*

$$Q_i E \cap Q^n E = (Q_i Q^{n-1})E.$$

Let $\mathrm{R}_A(Q) = A[Qt]$ be the Rees algebra of the ideal Q, and $L = \mathrm{G}_Q(E)$ be the associated graded module of E with respect to Q. For each $1 \le i \le s + 1$, we set $f_i = a_i t \in \mathrm{R}_A(Q)$.

Proposition 8.16 *The following hold:*

(1) *$f_1, f_2, \ldots, f_s$ is a d-sequence on L.*
(2) *For $1 \le i \le s$, there is an isomorphism*

$$[(f_1, \ldots, f_{i-1})L :_L f_i]/(f_1, \ldots, f_{i-1})L$$
$$= [[(f_1, \ldots, f_{i-1})L :_L f_i]/(f_1, \ldots, f_{i-1})L]_0 \cong [Q_{i-1}E :_E a_i]/Q_{i-1}E.$$

(3) *If* $0 \leq i \leq s$, *then there is an isomorphism of the associated graded modules*

$$L/(f_1, \ldots, f_i)L \cong \mathrm{G}_{\overline{Q}}(\overline{E}),$$

where $\overline{E} = E/Q_iE$ *and* $\overline{Q} = Q/Q_i$.

Proof Let $1 \leq i \leq j \leq s$, and $\varphi = \overline{xt^n} \in L_n$, where $n \geq 0$ and $x \in Q^nE$. If

$$(f_i f_j)\varphi \in (f_1, \ldots, f_{i-1})L,$$

then $(a_i a_j)x \in Q_{i-1}E + Q^{n+3}E$. Hence we can choose $y \in Q_{i-1}E$ such that

$$(a_i a_j)x - y \in Q^{n+3}E \cap Q_iE = (Q_iQ^{n+2})E.$$

When we write $(a_i a_j)x - y = \sum_{k=1}^{i} a_k y_k$ $(y_k \in Q^{n+2}E)$, as we have

$$a_i(a_jx - y_i) = y + \sum_{k=1}^{i-1} a_k y_k \in Q_{i-1}E,$$

we have that

$$a_jx - y_i \in [Q_{i-1}E :_E a_i] \cap Q^{n+1}E = (Q_{i-1}Q^n)E.$$

As $y_i \in Q^{n+2}E$, it follows that $f_j\varphi \in (f_1, \ldots, f_{i-1})L$. Hence $f_1, f_2, \ldots, f_s$ is a d-sequence on L.

Letting $j = i$ and $z = y_i - a_ix$ in the argument above, we have

$$z + a_ix = y_i \in Q_iE \cap Q^{n+2}E = (Q_iQ^{n+1})E.$$

So when we write $z + a_ix = z' + a_ix'$ $(z' \in (Q_{i-1}Q^{n+1})E, x' \in Q^{n+1}E)$, then we have $a_i(x - x') = z' - z \in Q_{i-1}E$. Namely, $x - x' \in [Q_{i-1}E :_E a_i] \cap Q^nE$. If $n > 0$, then we have $[Q_{i-1}E :_E a_i] \cap Q^nE = (Q_{i-1}Q^{n-1})E$, and it follows that

$$\varphi = \overline{xt^n} \in (f_1, \ldots, f_{i-1})L.$$

Hence the graded module $[(f_1, \ldots, f_{i-1})L :_L f_i]/(f_1, \ldots, f_{i-1})L$ is concentrated in degree zero.

If $n = 0$, then as we have $x \in [Q_{i-1}E :_E a_i] + QE$, the isomorphism

$$[[(f_1, \ldots, f_{i-1})L :_L f_i]/(f_1, \ldots, f_{i-1})L]_0 \cong [(Q_{i-1}E :_E a_i) + QE]/QE$$

follows. As we have $[Q_{i-1}E :_E a_i] \cap QE = Q_{i-1}E$, we get the isomorphism

$$\begin{aligned}&[(f_1, \ldots, f_{i-1})L :_L f_i]/(f_1, \ldots, f_{i-1})L \\ &= [[(f_1, \ldots, f_{i-1})L :_L f_i]/(f_1, \ldots, f_{i-1})L]_0 \\ &\cong [Q_{i-1}E :_E a_i]/Q_{i-1}E.\end{aligned}$$

Thus (1) and (2) have been proved. As

$$[Q^nE + Q_iE]/[Q^{n+1}E + Q_iE] \cong Q^nE/[Q^{n+1}E + (Q_iE \cap Q^nE)]$$

and $Q_iE \cap Q^nE = (Q_iQ^{n-1})E$ for $n \geq 0$, assertion (3) follows. □

Now let A be a Noetherian local ring, and M a finite A-module.

Corollary 8.17 *Let $s = \dim_A M \geq 1$, and $a_1, \ldots, a_s$ be a system of parameters of M which is a d-sequence on M. Then a_1 is a superficial element of M with respect to $Q = (a_1, a_2, \ldots, a_s)$.*

Proof As $0 :_L f_1 = [0 :_L f_1]_0$, the assertion follows from the definition of a superficial element. □

Example 8.18 Let $a_1, a_2, \ldots, a_d$ be a system of parameters of A which is also a d-sequence. When we set $Q = (a_1, a_2, \ldots, a_d)$, then the element a_1 is a superficial element of A with respect to Q. For example, let $R = k[[X, Y, Z, W]]$ be a formal power series ring over a field k, and $A = R/[(X^n, Y^n) \cap (Z, W)] (n \geq 2)$, then the sequence $a = xy^{n-1} - z$, $b = x^n + y^n - w$ (where x, y, z, w respectively denote the images of X, Y, Z, W in A) is a system of parameters of A which is also a d-sequence. Hence a is a superficial element of A with respect to Q. However, b is not a superficial element of A with respect to Q (check it). So the sequence b, a is not a d-sequence of A. Thus being a d-sequence depends on the ordering. This also shows that there is a d-sequence which is not a d^+-sequence.

If A is a Noetherian local ring and E is a Buchsbaum A-module (Definition 9.11), then each system of parameters of E is a d-sequence on E. Hence any system of parameters of a Buchsbaum module is a d^+-sequence. As there are many non-Cohen–Macaulay Buchsbaum rings, there are many d-sequences which are not regular sequences.

8.3 Euler Characteristics and Multiplicities

In this section, let A be a Noetherian ring, not necessarily local. Let $a_1, a_2, \ldots, a_d (d \geq 1)$ be a sequence of elements of A, and $Q = (a_1, a_2, \ldots, a_d)$. Let E be a finitely generated A-module, and assume that $\ell_A(E/QE) < \infty$. Then all the Koszul homology modules $\mathrm{H}_i(a_1, a_2, \ldots, a_d; E)$ have finite length. We define

$$\chi(a_1, a_2, \ldots, a_d; E) = \sum_{i \in \mathbb{Z}} (-1)^i \ell_A(\mathrm{H}_i(a_1, a_2, \ldots, a_d; E)),$$

and we call it the *Euler characteristic* of E with respect to $a_1, a_2, \ldots, a_d$.

Lemma 8.19 *The following hold:*

(1) *If* $0 \to E' \to E \to E'' \to 0$ *is a short exact sequence of A-module, then*

$$\chi(a_1, a_2, \ldots, a_d; E) = \chi(a_1, a_2, \ldots, a_d; E') + \chi(a_1, a_2, \ldots, a_d; E'').$$

(2) *If there exists an integer* $n \geq 1$ *such that* $a_1^n E = 0$*, then* $\chi(a_1, a_2, \ldots, a_d; E) = 0$.

(3) *Let* $d \geq 2$*, and let* a_1 *be a non-zerodivisor on* E*. Then we have*

$$\chi(a_1, a_2, \ldots, a_d; E) = \chi(a_2, \ldots, a_d; E/a_1E).$$

Proof

(1) This is because there is a short exact sequence of Koszul complexes

$$0 \to \mathrm{K}_\bullet(a_1, a_2, \ldots, a_d; E') \to \mathrm{K}_\bullet(a_1, a_2, \ldots, a_d; E) \to \mathrm{K}_\bullet(a_1, a_2, \ldots, a_d; E'') \to 0.$$

(2) Let $d = 1$. As we have $\ell_A(E) < \infty$, it follows that $\ell_A(0 :_E a_1) = \ell_A(E/a_1E)$ by the exact sequence

$$0 \to 0 :_E a_1 \to E \xrightarrow{a_1} E \to E/a_1E \to 0.$$

Hence $\chi(a_1; E) = 0$.

Let $d > 1$. As we have $\ell_A(E/(a_2, \ldots, a_d)E) < \infty$, the long exact sequence

$$\cdots \to \mathrm{H}_i(a_2, \ldots, a_d; E) \xrightarrow{a_1} \mathrm{H}_i(a_2, \ldots, a_d; E) \to \mathrm{H}_i(a_1, a_2, \ldots, a_d; E) \\ \to \mathrm{H}_{i-1}(a_2, \ldots, a_d; E) \xrightarrow{a_1} \mathrm{H}_{i-1}(a_2, \ldots, a_d; E) \to \cdots \tag{8.1}$$

yields $\chi(a_1, a_2, \ldots, a_d; E) = 0$.

(3) Comparing the long exact sequence of Koszul homology

$$0 \to \mathrm{H}_i(a_2, \ldots, a_d; E)/a_1\mathrm{H}_i(a_2, \ldots, a_d; E) \to \mathrm{H}_i(a_2, \ldots, a_d; E/a_1E) \\ \to 0 :_{\mathrm{H}_{i-1}(a_2, \ldots, a_d; E)} a_1 \to 0$$

with respect to the sequence $a_2, \ldots, a_d$ coming from the short exact sequence $0 \to E \xrightarrow{a_1} E \to E/a_1E \to 0$ with the long exact sequence (8.1) above, we

get the equality

$$\ell_A(\mathrm{H}_i(a_1, a_2, \ldots, a_d; E)) = \ell_A(\mathrm{H}_i(a_2, \ldots, a_d; E/a_1E))$$

for any integer i, and hence the equality

$$\chi(a_1, a_2, \ldots, a_d; E) = \chi(a_2, \ldots, a_d; E/a_1E)$$

follows.

□

Theorem 8.20 *Let $d \geq 2$. Then we have*

$$\chi(a_1, a_2, \ldots, a_d; E) = \chi(a_2, \ldots, a_d; E/a_1E) - \chi(a_2, \ldots, a_d; 0 :_E a_1)$$

Proof Letting $a = a_1$, and we take $n \gg 0$ so that $0 :_E a^n = 0 :_E a^{n+1}$. As we have

$$a^{n+1}E :_E a = a^nE + [0 :_E a], \quad a^nE \cap [0 :_E a] = 0,$$

there is an exact sequence

$$0 \to 0 :_E a \to E/a^nE \xrightarrow{a} E/a^{n+1}E \to E/aE \to 0. \tag{8.2}$$

Considering also the exact sequences

$$0 \to 0 :_E a^n \to E \to a^nE \to 0, \tag{8.3}$$

$$0 \to a^nE/a^{n+1}E \to E/a^{n+1}E \to E/a^nE \to 0, \tag{8.4}$$

we get:

$$\begin{aligned}
&\chi(a_1, a_2, \ldots, a_d; E)\\
&= \chi(a_1, a_2, \ldots, a_d; a^nE) \quad \text{(by (8.3))}\\
&= \chi(a_2, \ldots, a_d; a^nE/a^{n+1}E) \quad (a \text{ is } a^nE\text{-regular})\\
&= \chi(a_2, \ldots, a_d; E/a^{n+1}E) - \chi(a_2, \ldots, a_d; E/a^nE) \quad \text{(by (8.4))}\\
&= \chi(a_2, \ldots, a_d; E/aE) - \chi(a_2, \ldots, a_d; 0 :_E a) \quad \text{(by (8.2))}.
\end{aligned}$$

□

Corollary 8.21 *Let A be a Noetherian local ring. If $\dim_A E < d$, then we have*

$$\chi(a_1, a_2, \ldots, a_d; E) = 0.$$

Proof We may assume that $Q = (a_1, a_2, \ldots, a_d) \subsetneq A$. If $\ell_A(E) < \infty$, then we can take $n \gg 0$ so that $a_1^n E = 0$, and hence $\chi(a_1, a_2, \ldots, a_d; E) = 0$ by Lemma 8.19, (2). So we may assume that $E \neq 0$ and $s = \dim_A E > 0$. By induction, we may assume that the assertion is true up to $s - 1$. When we set $a = a_1$ and take $n \gg 0$ as in the proof of Theorem 8.20, we get

$$\begin{aligned}\chi(a_1, a_2, \ldots, a_d; E) &= \chi(a_1, a_2, \ldots, a_d; a^n E)\\ &= \chi(a_2, \ldots, a_d; a^n E/a^{n+1} E).\end{aligned}$$

We may assume that $a^n E \neq 0$. As a is $a^n E$-regular, we get $\dim_A a^n E/a^{n+1} E = \dim_A a^n E - 1 \le s - 1 < d - 1$. By the induction assumption on s, we get

$$\chi(a_2, \ldots, a_d; a^n E/a^{n+1} E) = 0,$$

and $\chi(a_1, a_2, \ldots, a_d; E) = 0$ follows. □

Corollary 8.22 *Let $Q_i = (a_1, a_2, \ldots, a_i)(0 \le i \le d)$. Then we have*

$$\begin{aligned}\chi(a_1, a_2, \ldots, a_d; E) &= \ell_A(E/QE) - \ell_A([Q_{d-1}E :_E a_d]/Q_{d-1}E)\\ &\quad - \sum_{0<k<d} \chi(a_{k+1}, \ldots, a_d; [Q_{k-1}E :_E a_k]/Q_{k-1}E).\end{aligned}$$

Proof Induction on d. Use Theorem 8.20. □

Theorem 8.23 *Let $a, b \in A$, and assume that $\ell_A(E/(a_1, \ldots, a_{d-1}, a)E) < \infty$ and $\ell_A(E/(a_1, \ldots, a_{d-1}, b)E) < \infty$. Then we have $\ell_A(E/(a_1, \ldots, a_{d-1}, ab)E) < \infty$, and the equality*

$$\chi(a_1, \ldots, a_{d-1}, ab; E) = \chi(a_1, \ldots, a_{d-1}, a; E) + \chi(a_1, \ldots, a_{d-1}, b; E)$$

holds.

Proof By induction on d and Theorem 8.20, we may assume that $d = 1$.

Let $R = \mathbb{Z}[\alpha, \beta]$ be the polynomial ring in two variables over the ring of integers $\mathbb{Z}$. Then A can be viewed as an R-algebra by the ring homomorphism $R \to A$ given by $\alpha \mapsto a$ and $\beta \mapsto b$. It is easy to see that $\alpha\beta R :_E \beta = \alpha R$, and hence

$$0 \to R/\alpha R \xrightarrow{\beta} R/\alpha\beta R \to R/\beta R \to 0 \qquad (\natural)$$

is exact. Note that $0 \to R \xrightarrow{\beta} R \to R/\beta R \to 0$ is a projective resolution of $R/\beta R$. So $\mathrm{Tor}_i^R(R/\beta R, E)$ is isomorphic to E/bE if $i = 0$, $0 :_E b$ if $i = 1$, and 0 otherwise. Note also that $R/\alpha\beta R$ and $R/\alpha R$ have similar resolutions, and similar descriptions of $\mathrm{Tor}_i^R(-, E)$. So the long exact sequence of Tor for the short exact sequence ($\natural$) is the exact sequence of A-modules of finite length

$$0 \to 0 :_E a \to 0 :_E ab \to 0 :_E b \to E/aE \to E/abE \to E/bE \to 0.$$

So we get $\chi(ab; E) = \chi(a; E) + \chi(b; E)$. This is what we wanted to prove. □

Now let A be a Noetherian local ring with $d = \dim A \geq 1$, and I be its $\mathfrak{m}$-primary ideal. Let M be a finitely generated A-module. Then the Hilbert function $\ell_A(M/I^{n+1}M)$ is a polynomial of n with coefficients in $\mathbb{Q}$ for $n \gg 0$, and its degree is at most d.

Now we define the sequence of integers $\{\mathrm{e}_I^i(M)\}_{0 \leq i \leq d}$ by

$$\ell_A(M/I^{n+1}M) = \mathrm{e}_I^0(M)\binom{n+d}{d} - \mathrm{e}_I^1(M)\binom{n+d-1}{d-1} + \cdots + (-1)^d \mathrm{e}_I^d(M)$$

($n \gg 0$). Do not confuse $\mathrm{e}_I^i(M)$ with $\mathrm{e}_i(I, M)$. By definition, $\mathrm{e}_I^0(M) = \mathrm{e}_0(I, M) > 0$ if $\dim_A M = d = \dim A$, while $\mathrm{e}_I^0(M) = 0$ if $\dim_A M < d$.

Let $Q = (a_1, a_2, \ldots, a_d)$ be a parameter ideal of the local ring A. Then the following holds.

Theorem 8.24 (Serre [Se], Auslander–Buchsbaum [AB]) *Let M be a finitely generated A-module. Then we have*

$$\chi(a_1, a_2, \ldots, a_d; M) = \mathrm{e}_Q^0(M) \geq 0.$$

Proof Induction on d. By Corollary 8.21, we may assume that $M \neq 0$ and $\dim_A M = d$. Replacing A by $B = A[X]_{\mathfrak{m}A[X]}$, where X is an indeterminate, we may assume that the residue field $A/\mathfrak{m}$ of A is infinite. Note that the Koszul complex is determined up to isomorphisms only by Q and is independent of the set of minimal generators. So without loss of generality, we may assume that $a_1 \in Q$ is a superficial element of M with respect to Q.

If $d = 1$, then as we have $\dim_A M = 1$,

$$\begin{aligned}
\mathrm{e}_Q^0(M) &= \mathrm{e}_0(Q, M) \\
&= \mathrm{e}_0(Q, M/a_1M) - \ell_A(0 :_M a_1) \\
&= \ell_A(M/a_1M) - \ell_A(0 :_M a_1) \\
&= \chi(a_1; M)
\end{aligned}$$

by Theorem 8.12.

Assume that $d \geq 2$ and the theorem is true up to $d-1$. Let $\overline{a_i}$ denote the image of a_i in the ring $\overline{A} = A/(a_1)$. If N is a finitely generated A-module and $a_1N = 0$, then we have $\chi(a_2, \ldots, a_d; N) = \chi(\overline{a_2}, \ldots, \overline{a_d}; N)$. Since we have $\ell_A(0 :_M a_1) < \infty$,

$$\begin{aligned}
\chi(a_1, a_2, \ldots, a_d; M) &= \chi(a_2, \ldots, a_d; M/a_1M) - \chi(a_2, \ldots, a_d; 0 :_M a_1) \\
&= \chi(a_2, \ldots, a_d; M/a_1M) \\
&= \chi(\overline{a_2}, \ldots, \overline{a_d}; M/a_1M)
\end{aligned}$$

by Theorem 8.20 and Lemma 8.19 (2). Therefore, as we have $\dim_A M/a_1M = d-1$,

$$\chi(a_1, a_2, \dots, a_d; M) = \mathrm{e}^0_{\overline{Q}}(M/a_1M)$$

by the induction assumption on d. On the other hand, as we have

$$\begin{aligned}\mathrm{e}^0_{\overline{Q}}(M/a_1M) &= \mathrm{e}_0(Q, M/a_1M)\\ &= \mathrm{e}_0(Q, M)\\ &= \mathrm{e}^0_Q(M)\end{aligned}$$

by Theorem 8.12, the equality $\chi(a_1, a_2, \dots, a_d; M) = \mathrm{e}^0_Q(M)$ follows. □

Corollary 8.25 *Let I be an $\mathfrak{m}$-primary ideal of A, and $0 \to L \to M \to N \to 0$ be a short exact sequence of finitely generated A-modules. Then we have $\mathrm{e}^0_I(M) = \mathrm{e}^0_I(L) + \mathrm{e}^0_I(N)$.*

Proof Replacing A by $B = A[X]_{\mathfrak{m}A[X]}$, where X is an indeterminate, we may assume that the residue field $A/\mathfrak{m}$ of A is infinite. Let Q be a parameter ideal of A which is a reduction of I. Then $\mathrm{e}^0_I(M) = \mathrm{e}^0_Q(M)$ holds for any finitely generated A-module M. So by Lemma 8.19 and Theorem 8.24, the equality $\mathrm{e}^0_I(M) = \mathrm{e}^0_I(L) + \mathrm{e}^0_I(N)$ follows. □

8.4 Associativity Formula for Multiplicities

Let A be a Noetherian local ring, M be a nonzero finitely generated A-module, and I be a proper ideal of A such that $\ell_A(M/IM) < \infty$.

The following is called the *associativity formula* for multiplicities.

Theorem 8.26 *Under the setting above, we have*

$$\mathrm{e}^0_I(M) = \sum_{\mathfrak{p} \in \operatorname{Assh}_A A} \ell_{A_\mathfrak{p}}(M_\mathfrak{p})\, \mathrm{e}^0_I(A/\mathfrak{p}).$$

In particular, we have

$$\mathrm{e}_0(I, M) = \sum_{\mathfrak{p} \in \operatorname{Assh}_A M} \ell_{A_\mathfrak{p}}(M_\mathfrak{p})\, \mathrm{e}_0(I, A/\mathfrak{p}).$$

Proof We prove the first equality. We denote the right-hand side of the first equation above by $e'_I(M)$. We know that both e^0_I and e'_I are additive. Namely, if

$$0 \to L \to M \to N \to 0$$

is a short exact sequence, then $\ell_A(L/IL) < \infty$ and $\ell_A(N/IL) < \infty$, and we have that $\mathrm{e}_I^0(M) = \mathrm{e}_I^0(L) + \mathrm{e}_I^0(N)$ and $\boldsymbol{e}'_I(M) = \boldsymbol{e}'_I(L) + \boldsymbol{e}'_I(N)$.

We have a Bourbaki filtration

$$(0) = M_0 \subsetneq M_1 \subsetneq M_2 \subsetneq \cdots \subsetneq M_n = M$$

of M, that is, there is a filtration of A-modules, such that

$$M_i/M_{i-1} \cong A/P_i$$

for some $P_i \in \operatorname{Spec} A$ for each i, see Lemma 2.18. So we may assume that $M = A/P$ for some $P \in \operatorname{Spec} A$ by additivity. For each $\mathfrak{p} \in \operatorname{Assh}_A A$, we have $\ell_{A_\mathfrak{p}}((A/P)_\mathfrak{p}) = 1$ if $P = \mathfrak{p}$ and 0 otherwise. So $\mathrm{e}_I^0(A/P) = \boldsymbol{e}'_I(A/P) = 0$ if $P \notin \operatorname{Assh}_A A$, and $\mathrm{e}_I^0(A/P) = \boldsymbol{e}'_I(A/P)$ if $P \in \operatorname{Assh}_A A$, and the first equality follows.

For the second equality, replacing A by $A/\operatorname{Ann}_A M$ and I by its image, we may assume that M is faithful. Note that $\operatorname{Supp}_A M = \operatorname{Spec} A$, and hence $\dim M = \dim A$ and $\operatorname{Assh}_A M = \operatorname{Assh}_A A$. So the left-hand side and the right-hand side respectively equal those of the first equality. Hence the second equality follows from the first, which has already been proved. □

Example 8.27 (see Chap. 7, Sect. 7.6) Let $n \geq 1$ and set $V = \{1, 2, \ldots, n\}$. Let Δ be a simplicial complex on V. Let $S = k[\Delta]$ be the Stanley–Reisner ring of Δ over k, and we set $A = S_P(P = S_+)$. Then we have

$$\mathrm{e}_0(\mathfrak{m}) = \sharp\{F \in \Delta \mid \dim F = \dim \Delta\}.$$

Note that for each $\mathfrak{p} \in \operatorname{Assh} S$, the ring $S/\mathfrak{p}$ is the polynomial ring over k. As the ring S is reduced and $\operatorname{Assh} S$ is in one-to-one correspondence with the facets of Δ whose dimension is $\dim \Delta$, the assertion follows.

8.5 Multiplicities with Respect to Parameter Ideals and Cohen–Macaulay Property

Let A be a Noetherian local ring, and M be a nonzero finitely generated A-module with $d = \dim_A M \geq 0$. Let $Q = (a_1, a_2, \ldots, a_d)$ be a parameter ideal of M. The aim of this section is to analyze the Cohen–Macaulay property of M using the behavior of the multiplicity $\mathrm{e}_0(Q, M)$.

Proposition 8.28 (cf. Example 8.1 (1)) *We have $\ell_A(M/QM) \geq \mathrm{e}_0(Q, M)$. If M is Cohen–Macaulay, then the equality holds.*

Proof We may assume that $d \geq 1$. Replacing A by $A/\mathrm{Ann}_A(M)$ if necessary, we may assume that M is faithful. Then Q is a parameter ideal of the ring A, and we have that

$$\mathrm{e}_0(Q, M) = \mathrm{e}_Q^0(M) = \chi(a_1, a_2, \ldots, a_d; M).$$

By Corollary 8.22 and Theorem 8.24, we have

$$\begin{aligned}\ell_A(M/QM) - \mathrm{e}_0(Q, M) &= \ell_A([Q_{d-1}M :_M a_d]/Q_{d-1}M) \\ &\quad + \sum_{0<k<d} \chi(a_{k+1}, \ldots, a_d; [Q_{k-1}M :_M a_k]/Q_{k-1}M) \\ &= \ell_A([Q_{d-1}M :_M a_d]/Q_{d-1}M) \\ &\quad + \sum_{0<k<d} \mathrm{e}^0_{Q/Q_k}([Q_{k-1}M :_M a_k]/Q_{k-1}M) \\ &\geq 0,\end{aligned}$$

where $Q_i = (a_1, a_2, \ldots, a_i)$. If M is Cohen–Macaulay, then the last ≥ 0 is $= 0$, and we have $\ell_A(M/QM) = \mathrm{e}_0(Q, M)$. □

Lemma 8.29 *Let E be an A-module and $a \in A$. If $\mathrm{Supp}_A[0 :_E a] \subseteq \{\mathfrak{m}\}$, then there is a long exact sequence of local cohomology modules*

$$\begin{aligned}0 \to 0 :_E a \to \mathrm{H}^0_{\mathfrak{m}}(E) &\xrightarrow{a} \mathrm{H}^0_{\mathfrak{m}}(E) \to \mathrm{H}^0_{\mathfrak{m}}(E/aE) \\ \to \mathrm{H}^1_{\mathfrak{m}}(E) &\xrightarrow{a} \mathrm{H}^1_{\mathfrak{m}}(E) \to \mathrm{H}^1_{\mathfrak{m}}(E/aE) \\ \cdots \to \mathrm{H}^i_{\mathfrak{m}}(E) &\xrightarrow{a} \mathrm{H}^i_{\mathfrak{m}}(E) \to \mathrm{H}^i_{\mathfrak{m}}(E/aE) \to \cdots.\end{aligned}$$

Proof When we break the exact sequence $0 \to 0 :_E a \to E \xrightarrow{a} E \to E/aE \to 0$ into the two short exact sequences:

$$0 \to 0 :_E a \to E \to aE \to 0; \tag{8.5}$$

$$0 \to aE \to E \to E/aE \to 0, \tag{8.6}$$

then by (8.5), we get the exact sequence

$$0 \to 0 :_E a \to \mathrm{H}^0_{\mathfrak{m}}(E) \to \mathrm{H}^0_{\mathfrak{m}}(aE) \to 0$$

and isomorphisms

$$\mathrm{H}^i_{\mathfrak{m}}(E) \xrightarrow{\sim} \mathrm{H}^i_{\mathfrak{m}}(aE) \quad (i \geq 1).$$

Then replacing all the $\mathrm{H}^i_{\mathfrak{m}}(aE)(i \geq 1)$ by $\mathrm{H}^i_{\mathfrak{m}}(E)$ in the long exact sequence

$$\begin{aligned} 0 \to \mathrm{H}^0_{\mathfrak{m}}(aE) &\to \mathrm{H}^0_{\mathfrak{m}}(E) \to \mathrm{H}^0_{\mathfrak{m}}(E/aE) \\ \to \mathrm{H}^1_{\mathfrak{m}}(aE) &\to \mathrm{H}^1_{\mathfrak{m}}(E) \to \mathrm{H}^1_{\mathfrak{m}}(E/aE) \\ \cdots \to \mathrm{H}^i_{\mathfrak{m}}(aE) &\to \mathrm{H}^i_{\mathfrak{m}}(E) \to \mathrm{H}^i_{\mathfrak{m}}(E/aE) \to \cdots \end{aligned}$$

associated with the short exact sequence (8.6), we get the desired long exact sequence. □

Lemma 8.30 *Assume that $d \geq 2$, and let $a \in \mathfrak{m}$. If $\ell_A(0 :_M a) < \infty$ and M/aM is Cohen–Macaulay, then M is also Cohen–Macaulay.*

Proof As $\ell_A(0 :_M a) < \infty$, we have that $\dim_A M/aM = d - 1$, and we get an exact sequence of the local cohomology modules

$$0 \to 0 :_M a \to \mathrm{H}^0_{\mathfrak{m}}(M) \xrightarrow{a} \mathrm{H}^0_{\mathfrak{m}}(M) \to \mathrm{H}^0_{\mathfrak{m}}(M/aM).$$

As we have $\mathrm{H}^0_{\mathfrak{m}}(M/aM) = 0$, we have that $\mathrm{H}^0_{\mathfrak{m}}(M) = a\,\mathrm{H}^0_{\mathfrak{m}}(M)$, and hence $\mathrm{H}^0_{\mathfrak{m}}(M) = 0$ by NAK. Hence we have $0 :_M a = 0$. As M/aM is Cohen–Macaulay, M is Cohen–Macaulay. □

Corollary 8.31 *Assume that $d = \dim_A M \geq 2$.*

(1) *Let $I \subsetneq A$ be an ideal of the ring A such that $\ell_A(M/IM) < \infty$. If $a \in I$ is a superficial element of M with respect to I and M/aM is Cohen–Macaulay, then M is Cohen–Macaulay.*
(2) *If M/aM is Cohen–Macaulay for any $a \in \mathfrak{m}$ which is a part of a system of parameters, then M is Cohen–Macaulay.*

Proof (1) follows from Lemma 8.30 immediately. For (2), note that $\ell_A(0 :_M a) < \infty$ for any $a \in \mathfrak{m} \setminus \bigcup_{\mathfrak{p} \in \mathrm{Ass}_A M, \mathfrak{p} \neq \mathfrak{m}} \mathfrak{p}$. □

Now we state a criterion for the Cohen–Macaulay property.

Theorem 8.32 *The following are equivalent:*

(1) *M is Cohen–Macaulay.*
(2) *$\ell_A(M/QM) = \mathrm{e}_0(Q, M)$.*

Proof We only have to prove (2) $\Rightarrow$ (1). We may assume that $d > 0$, and the theorem is true up to $d - 1$. We may assume that $A/\mathfrak{m}$ is infinite. Let $a \in Q \setminus \mathfrak{m}Q$ be a superficial element of M with respect to Q, and set $\overline{A} = A/(a)$, $\overline{M} = M/aM$, and $\overline{Q} = Q/(a)$.

If $d = 1$, then as we have

$$\begin{aligned} \mathrm{e}_0(Q, M) &= \mathrm{e}_0(Q, \overline{M}) - \ell_A(0 :_M a) \\ &= \ell_A(M/QM) - \ell_A(0 :_M a), \end{aligned}$$

a is M-regular, and hence M is Cohen–Macaulay.

Now let $d \geq 2$. Then since we have $\ell_A(M/QM) = \ell_{\overline{A}}(\overline{M}/\overline{QM})$ and $\mathrm{e}_0(Q, M) = \mathrm{e}_0(Q, \overline{M}) = \mathrm{e}_0(\overline{Q}, \overline{M})$, we get $\ell_{\overline{A}}(\overline{M}/\overline{QM}) = \mathrm{e}_0(\overline{Q}, \overline{M})$, and hence $\overline{M}$ is Cohen–Macaulay by induction. By Corollary 8.31, we conclude that M is Cohen–Macaulay. □

In what follows, we consider the case $M = A$. Let $d = \dim A$. The assertion $\mathrm{e}_1(Q) \leq 0$ in the next theorem is due to M. Mandal, B. Singh, and J. K. Verma [MSV].

Theorem 8.33 ([HaHy]) *Let $d = \dim A \geq 1$ and $Q = (a_1, a_2, \ldots, a_d)$ be a parameter ideal of the ring A. Then for any integer $n \geq 0$, we have*

$$\ell_A(A/Q^{n+1}) \geq \mathrm{e}_0(Q)\binom{n+d}{d}.$$

Hence in particular, $\mathrm{e}_1(Q) \leq 0$. If the equality

$$\ell_A(A/Q^{n+1}) = \mathrm{e}_0(Q)\binom{n+d}{d}$$

holds for some $n \geq 0$, then A is Cohen–Macaulay, and for all $n \geq 0$, we have $\ell_A(A/Q^{n+1}) = \mathrm{e}_0(Q)\binom{n+d}{d}$.

Proof Let $R = A[X_1, X_2, \ldots, X_d]$ be the polynomial ring, and set $\mathfrak{M} = \mathfrak{m}R + (X_1, X_2, \ldots, X_d)$ and $B = R_{\mathfrak{M}}$. When we set $f_i = X_i - a_i$, then as we have $R = A[f_1, f_2, \ldots, f_d]$, $f_1, f_2, \ldots, f_d$ are algebraically independent over A, and hence they form an R-sequence. Let $\mathfrak{q} = (f_1, f_2, \ldots, f_d)$. Then since we have

$$B/[\mathfrak{q}^{n+1} + (X_1, X_2, \ldots, X_d)B] \cong A/Q^{n+1}$$

for $n \geq 0$, we have that $X_1, X_2, \ldots, X_d$ is a system of parameters of the B-module $B/\mathfrak{q}^{n+1}$. By Theorem 8.26, we have

$$\begin{aligned}\ell_A(A/Q^{n+1}) &= \ell_B(B/[\mathfrak{q}^{n+1} + (X_1, X_2, \ldots, X_d)B])\\ &\geq \mathrm{e}_0((X_1, X_2, \ldots, X_d)B, B/\mathfrak{q}^{n+1}) \quad \text{(Proposition 8.28)}\\ &= \sum_{\mathfrak{p} \in \mathrm{Assh}_B B/\mathfrak{q}^{n+1}} \ell_{B_\mathfrak{p}}(B_\mathfrak{p}/\mathfrak{q}^{n+1}B_\mathfrak{p})\, \mathrm{e}_0((X_1, X_2, \ldots, X_d)B, B/\mathfrak{p}).\end{aligned}$$

Since we have $\mathrm{Assh}_B B/\mathfrak{q}^{n+1} = \mathrm{Assh}_B B/\mathfrak{q}$ and $B/\mathfrak{q} \cong A$, it follows that $\dim B/\mathfrak{p} = d$ for any $\mathfrak{p} \in \mathrm{Assh}_B B/\mathfrak{q}^{n+1}$. Hence as $f_1, f_2, \ldots, f_d \in \mathfrak{p}$ is a B-sequence, the ring $B_\mathfrak{p}$ is a d-dimensional Cohen–Macaulay local ring, and $f_1, f_2, \ldots, f_d$ is a system of parameters of $B_\mathfrak{p}$. Hence we have

$$\ell_{B_\mathfrak{p}}(B_\mathfrak{p}/\mathfrak{q}^{n+1}B_\mathfrak{p}) = \ell_{B_\mathfrak{p}}(B_\mathfrak{p}/\mathfrak{q}B_\mathfrak{p})\binom{n+d}{d}.$$

As we have $B/\mathfrak{q} \cong A$ and the ideal $[(X_1, X_2, \ldots, X_d)B + \mathfrak{q}]/\mathfrak{q}$ corresponds to the ideal Q, we have that

$$\begin{aligned}
\ell_A(A/Q^{n+1}) &\geq \sum_{\mathfrak{p}\in \operatorname{Assh}_B B/\mathfrak{q}} \ell_{B_\mathfrak{p}}(B_\mathfrak{p}/\mathfrak{q}^{n+1}B_\mathfrak{p})\, \mathrm{e}_0((X_1, X_2, \ldots, X_d)B, B/\mathfrak{p}) \\
&= \left[\sum_{\mathfrak{p}\in \operatorname{Assh}_B B/\mathfrak{q}} \ell_{B_\mathfrak{p}}(B_\mathfrak{p}/\mathfrak{q}B_\mathfrak{p})\, \mathrm{e}_0((X_1, X_2, \ldots, X_d)B, B/\mathfrak{p})\right]\binom{n+d}{d} \\
&= \mathrm{e}_0((X_1, X_2, \ldots, X_d)B, B/\mathfrak{q})\binom{n+d}{d} \\
&= \mathrm{e}_0(Q)\binom{n+d}{d}
\end{aligned}$$

for any $n \geq 0$. Hence if $\ell_A(A/Q^{n+1}) = \mathrm{e}_0(Q)\binom{n+d}{d}$, then the equality

$$\ell_B(B/[\mathfrak{q}^{n+1} + (X_1, X_2, \ldots, X_d)B]) = \mathrm{e}_0((X_1, X_2, \ldots, X_d)B, B/\mathfrak{q}^{n+1})$$

holds, and hence by Theorem 8.29, the ring $B/\mathfrak{q}^{n+1}$ is Cohen–Macaulay. As the ideal $\mathfrak{q}$ is generated by the B-sequence $f_1, f_2, \ldots, f_d$, the powers $\mathfrak{q}^{n+1}$ are perfect, see [EN]. So B is Cohen–Macaulay, and hence $A = B/\mathfrak{q}$ is also Cohen–Macaulay. □

Exercise 8.1 Prove that if $B/\mathfrak{q}^{n+1}$ is Cohen–Macaulay (for some $n \geq 0$), then B is Cohen–Macaulay, and complete the proof of Theorem 8.33 (see [EN]).

In what follows, we set $v(A) = \mu_A(\mathfrak{m})$, $\mathrm{e}(A) = \mathrm{e}_0(\mathfrak{m})$, and $d = \dim A$.

Theorem 8.34 ([Sal]) *Assume that A is Cohen–Macaulay. Then we have $v(A) \leq \mathrm{e}(A) + d - 1$. Assume that A contains a parameter ideal Q which is a reduction of $\mathfrak{m}$. Then the equality holds in the formula above if and only if $\mathfrak{m}^2 = Q\mathfrak{m}$.*

Proof To prove that $v(A) \leq \mathrm{e}(A) + d - 1$, we may assume that $A/\mathfrak{m}$ is infinite. So we may assume that there is a parameter ideal $Q = (a_1, a_2, \ldots, a_d)$ of A which is a reduction of $\mathfrak{m}$. Then since $a_1, a_2, \ldots, a_d$ are a part of minimal generators of $\mathfrak{m}$, $Q \cap \mathfrak{m}^2 = Q\mathfrak{m}$. So by the exact sequence $0 \to Q \to \mathfrak{m} \to \mathfrak{m}/Q \to 0$, we get a short exact sequence

$$0 \to Q/\mathfrak{m}Q \to \mathfrak{m}/\mathfrak{m}^2 \to L/\mathfrak{m}L \to 0$$

of vector spaces over the residue field $A/\mathfrak{m}$, where $L = \mathfrak{m}/Q$. Hence we have

$$\begin{aligned}
v(A) &= \mu_A(L) + d \\
&\leq \ell_A(L) + d \\
&= [\ell_A(A/Q) - \ell_A(A/\mathfrak{m})] + d
\end{aligned}$$

$$= \mathrm{e}_0(Q) + d - 1$$
$$= \mathrm{e}(A) + d - 1.$$

In order to have that $v(A) = \mathrm{e}(A) + d - 1$, $\mathfrak{m}L = 0$, that is, $\mathfrak{m}^2 = Q\mathfrak{m}$ is necessary and sufficient. □

Exercise 8.2 (**[Nor]**[2]) Assume that A is Cohen–Macaulay and $d \geq 1$. Let I be an $\mathfrak{m}$-primary ideal of A, and let Q be a parameter ideal of A which is also a reduction of I. Prove that the following assertions are true, reducing the problem to the case $d = 1$.

(1) The inequalities $\mu_A(I/Q) = \mu_A(I) - d \leq \mathrm{e}_0(I) - \ell_A(A/I) \leq \mathrm{e}_1(I)$ hold. In particular, $\mathrm{e}_1(I) \geq 0$.
(2) We have the equality $\mu_A(I/Q) = \mathrm{e}_0(I) - \ell_A(A/I)$ if and only if $\mathfrak{m}I = \mathfrak{m}Q$.
(3) $\mathrm{e}_0(I) - \ell_A(A/I) = \mathrm{e}_1(I)$ if and only if $I^2 = QI$.

Lemma 8.35 *Let M be a nonzero finitely generated A-module, and assume that* $\dim_A M = \dim A \geq 2$. *Assume also that the residue field of A is infinite, and A has a canonical module K_A. Assume that* $\dim A/\mathfrak{p} = \dim_A M$ *for any* $\mathfrak{p} \in \mathrm{Ass}_A M$. *Then we have the following:*

(1) $\mathrm{H}^1_{\mathfrak{m}}(M)$ *is finitely generated.*
(2) *If I is an $\mathfrak{m}$-primary ideal of A, then there exists $a \in I \setminus \mathfrak{m}I$ such that a is a superficial element of M with respect to I, and that*

$$\mathrm{Ass}_A M/aM \setminus \{\mathfrak{m}\} \subseteq \mathrm{Assh}_A M/aM.$$

Proof Let $\varphi : M \to L := \mathrm{Hom}_A(\mathrm{Hom}_A(M, K_A), K_A)$ be the canonical morphism, and set $C = \mathrm{Coker}\ \varphi$. Since we have $\dim A/\mathfrak{p} = \dim_A M$ for any $\mathfrak{p} \in \mathrm{Ass}_A M$, the map φ is injective, and we get the short exact sequence of A-modules

$$0 \to M \xrightarrow{\varphi} L \to C \to 0. \tag{8.7}$$

As we have $\mathrm{depth}_A L \geq 2$, the isomorphism

$$\mathrm{H}^1_{\mathfrak{m}}(M) \cong \mathrm{H}^0_{\mathfrak{m}}(C)$$

is obtained from this short exact sequence, and we know that $\mathrm{H}^1_{\mathfrak{m}}(M)$ is finitely generated. We take $a \in I \setminus \mathfrak{m}I$ so that a is a superficial element of all of A, M, L, and C, with respect to I. As the element a is a part of a system of parameters of A, it is K_A-regular, and hence is also L-regular. Now look at the exact sequence of A-modules

[2] Douglas Geoffrey Northcott, 1916–2005.

$$0 \to 0 :_C a \to M/aM \to L/aL \to C/aC \to 0$$

obtained from (8.7). As we have $[0 :_C a]_{\mathfrak{p}} = 0$ for any $\mathfrak{p} \in \operatorname{Ass}_A M/aM \setminus \{\mathfrak{m}\}$, we get an embedding

$$0 \to M_{\mathfrak{p}}/aM_{\mathfrak{p}} \to L_{\mathfrak{p}}/aL_{\mathfrak{p}}.$$

Hence $\mathfrak{p} \in \operatorname{Ass}_A L/aL$. As L satisfies Serre's (S_2) condition, we have that $\dim A/\mathfrak{p} = d - 1$, and hence $\mathfrak{p} \in \operatorname{Assh}_A M/aM$. □

The following is due to Masayoshi Nagata [Nag].

Theorem 8.36 *Let A be an unmixed Noetherian local ring. Namely, assume that $\dim \widehat{A}/\mathfrak{p} = d$ for any $\mathfrak{p} \in \operatorname{Ass}_{\widehat{A}} \widehat{A}$. If $\mathrm{e}(A) = 1$, then A is regular.*

Proof We may assume that A is complete and the residue field of A is infinite. Set $d = \dim A$. We may assume that $d > 0$, and that the theorem is true up to $d - 1$. By Lemma 8.35, (2), we can take $a \in \mathfrak{m} \setminus \mathfrak{m}^2$ such that a is a superficial element of A with respect to $\mathfrak{m}$, and that $\operatorname{Ass}_A A/(a) \setminus \{\mathfrak{m}\} \subseteq \operatorname{Assh}_A A/(a)$. Since A is unmixed, a is A-regular. When we set $\overline{A} = A/(a)$, then we have $\mathrm{e}(\overline{A}) = \mathrm{e}(A) = 1$ by Theorem 8.12, $\overline{A} = A/(a)$ is a field if $d = 1$, and hence A is a discrete valuation ring. So we may assume that $d \geq 2$. We set $W = \mathrm{H}^0_{\mathfrak{m}}(\overline{A})$, and $B = \overline{A}/W$. Then since $\operatorname{Ass}_A \overline{A}\setminus\{\mathfrak{m}\} \subseteq \operatorname{Assh}_A \overline{A}$, the ring B is unmixed, and as we have $\ell_A(W) < \infty$, it follows $\mathrm{e}(B) = \mathrm{e}(\overline{A}) = 1$. Hence B is regular by induction. If $d \geq 3$, then by the exact sequence of the local cohomology modules

$$0 \to \mathrm{H}^0_{\mathfrak{m}}(\overline{A}) \to \mathrm{H}^1_{\mathfrak{m}}(A) \xrightarrow{a} \mathrm{H}^1_{\mathfrak{m}}(A) \to \mathrm{H}^1_{\mathfrak{m}}(\overline{A}) = 0$$

obtained from the short exact sequence $0 \to A \xrightarrow{a} A \to \overline{A} \to 0$, it follows that $\mathrm{H}^1_{\mathfrak{m}}(A) = 0$, as we have $\mathrm{H}^1_{\mathfrak{m}}(A)$ is finitely generated (Lemma 8.35, (1)) and $\mathrm{H}^1_{\mathfrak{m}}(A) = a\,\mathrm{H}^1_{\mathfrak{m}}(A)$. So $W = \mathrm{H}^0_{\mathfrak{m}}(\overline{A}) = 0$, and hence $B = \overline{A}$ is regular. It follows that A is also regular.

Now assume that $d = 2$. Letting $\widetilde{A} = \operatorname{Hom}_A(K_A, K_A)$, we get an exact sequence

$$0 \to A \xrightarrow{\varphi} \widetilde{A} \to C \to 0,$$

where $\widetilde{A}$ is commutative, and for each $a \in A$, $\varphi(a)$ is the map of multiplication by a. As we have $\operatorname{depth}_A \widetilde{A} = 2$, we have that $\widetilde{A}$ is a Cohen–Macaulay A-module. As the ring A is unmixed, we have that $\ell_A(C) < \infty$. Hence, if we take a parameter ideal Q of A which is also a reduction of $\mathfrak{m}$, then the equality

$$\mathrm{e}_0(Q, \widetilde{A}) = \mathrm{e}_0(Q) = \mathrm{e}(A) = 1$$

holds (Corollary 8.25). Hence $\ell_A(\widetilde{A}/Q\widetilde{A}) = 1$. As A is complete local and $\widetilde{A}$ is finitely generated as an A-module, the A-algebra $\widetilde{A}$ is a direct product of finitely many local rings. But since $\ell_A(\widetilde{A}/Q\widetilde{A}) = 1$, the ring $\widetilde{A}$ is a local ring in fact, and letting $\widetilde{\mathfrak{m}}$ be the maximal ideal of $\widetilde{A}$, we immediately have that

$$\widetilde{\mathfrak{m}} = Q\widetilde{A}, \quad A/\mathfrak{m} \cong \widetilde{A}/\widetilde{\mathfrak{m}}.$$

Hence $\widetilde{\mathfrak{m}} = \mathfrak{m}\widetilde{A}$, and by NAK, we have that $\widetilde{A} = A$. Namely, we have $\mathfrak{m} = Q$, and the ring A is regular. □

Exercise 8.3 (see [GhGHOPV]) Prove the following assertion using Lemma 8.35.

Let A be a Noetherian local ring, M be a nonzero finitely generated A-module, and $d = \dim_A M \geq 1$. Assume that M is unmixed, that is, $\dim \widehat{A}/\mathfrak{p} = d$ for any $\mathfrak{p} \in \mathrm{Ass}_{\widehat{A}} \widehat{M}$. If there exists a parameter ideal Q of M in A such that $\mathrm{e}_1(Q, M) \geq 0$, then M is Cohen–Macaulay.

Chapter 9
Modules with Finite Local Cohomology and Buchsbaum Modules

In this chapter, let $(A, \mathfrak{m})$ be a Noetherian local ring, and M be a nonzero finitely generated A-module of dimension d. We study A-modules which are "near" to Cohen–Macaulay modules.

We say that an A-module M of dimension d has finite local cohomology (FLC for short) if $H^i_{\mathfrak{m}}(M)$ is a finitely generated A-module for every $i < d$. If M has FLC, then for any $\mathfrak{p} \in \mathrm{Supp}_A M \setminus \{\mathfrak{m}\}$, $M_{\mathfrak{p}}$ is a Cohen–Macaulay $A_{\mathfrak{p}}$-module, and $d = \dim_A M_{\mathfrak{p}} + \dim A/\mathfrak{p}$. The converse is also true if A is a homomorphic image of a Cohen–Macaulay ring.

In Sect. 9.2, we study Buchsbaum A-modules, which are "nearest" to Cohen–Macaulay A-modules. There is a very beautiful theory regarding them, and we will give some criteria for an A-module M to be a Buchsbaum A-module.

9.1 Modules with Finite Local Cohomology

The purpose of this section is to develop some basics on the modules with FLC. We start with the definition.

Definition 9.1 ([GO], [STC]) We say that M has finitely generated local cohomology modules (*finite local cohomology*, *FLC* for short) if $\mathrm{H}^i_{\mathfrak{m}}(M)$ is a finitely generated A-module for each $i \neq d$.

If $d \leq 1$ or M is Cohen–Macaulay, then M has FLC. M has FLC if and only if the $\widehat{A}$-module $\widehat{M}$ does so.

Lemma 9.2 *Let $d \geq 1$ and assume that M has FLC. Assume that $a \in \mathfrak{m}$. If a is a part of a system of parameters of M, then $\ell_A(0 :_M a) < \infty$.*

S. Goto, K. Watanabe, *Commutative Algebra*, Graduate Texts in Mathematics 309,
https://doi.org/10.1007/978-981-95-4227-7_9

Proof Let $a \in \mathfrak{m}$ and assume that $\dim_A M/aM = d-1$. We set $W = \mathrm{H}^0_\mathfrak{m}(M)$, and $L = M/W$. If $d = 1$, then L is Cohen–Macaulay, and a is L-regular; so $0 :_M a \subseteq W$. Hence $\ell_A(0 :_M a) < \infty$. Assume that $d \geq 2$ and the assertion of the lemma is true up to $d-1$. We take $b \in \mathfrak{m}$ so that b is L-regular and $\dim_A M/(a,b)M = d-2$. Since $\mathrm{H}^i_\mathfrak{m}(M)$ is finitely generated, we can replace b by its sufficiently large power so that $b\mathrm{H}^i_\mathfrak{m}(M) = 0$ $(i \neq d)$. Now let $n \geq 1$ be an integer, and we set $c = b^n$. Then since c is L-regular, we have $0 :_M c \subseteq W$ and hence $\ell_A(0 :_M c) < \infty$. So by Lemma 8.29, we have the long exact sequence of local cohomology modules

$$
\begin{aligned}
0 \to 0 :_M c \to \mathrm{H}^0_\mathfrak{m}(M) &\xrightarrow{c} \mathrm{H}^0_\mathfrak{m}(M) \to \mathrm{H}^0_\mathfrak{m}(M/cM) \\
\to \mathrm{H}^1_\mathfrak{m}(M) &\xrightarrow{c} \mathrm{H}^1_\mathfrak{m}(M) \to \mathrm{H}^1_\mathfrak{m}(M/cM) \\
\cdots \to \mathrm{H}^i_\mathfrak{m}(M) &\xrightarrow{c} \mathrm{H}^i_\mathfrak{m}(M) \to \mathrm{H}^i_\mathfrak{m}(M/cM) \to \cdots.
\end{aligned}
$$

Since $c\mathrm{H}^i_\mathfrak{m}(M) = 0$ $(i \neq d)$, this long exact sequence splits into the short exact sequences

$$0 \to \mathrm{H}^i_\mathfrak{m}(M) \to \mathrm{H}^i_\mathfrak{m}(M/cM) \to \mathrm{H}^{i+1}_\mathfrak{m}(M) \to 0 \qquad (\#)$$

for $0 \leq i \leq d-2$. Hence M/cM has FLC. As $\dim_A M/cM = d-1$, we have that $\ell_A([cM :_M a]/cM) < \infty$ by the induction assumption. Hence we have

$$[cM :_M a]/cM \subseteq \mathrm{H}^0_\mathfrak{m}(M/cM).$$

Now let $I = \mathrm{Ann}_A \mathrm{H}^0_\mathfrak{m}(M) \cap \mathrm{Ann}_A \mathrm{H}^1_\mathfrak{m}(M)$. Then by the short exact sequence (#), we have that

$$I^2 \mathrm{H}^0_\mathfrak{m}(M/cM) = 0,$$

and hence we have

$$I^2 \cdot [b^n M :_M a] \subseteq b^n M$$

for all $n \geq 1$. Then since obviously we have that $0 :_M a \subseteq b^n M :_M a$, we have that

$$I^2 \cdot [0 :_M a] \subseteq \bigcap_{n \geq 1} b^n M,$$

and hence $I^2 \cdot [0 :_M a] = 0$. Therefore, we have $\ell_A(0 :_M a) < \infty$. □

Theorem 9.3 *Assume that M has FLC and $d \geq 1$. We set*

$$I = \bigcap_{i=0}^{d-1} \mathrm{Ann}_A \mathrm{H}^i_\mathfrak{m}(M).$$

Let $a_1, a_2, \ldots, a_d$ be a system of parameters of M. Then taking integers i, j such that $0 \le j \le d-1$ and $0 \le i \le d-j-1$, and letting $Q_j = (a_1, a_2, \ldots, a_j)$ and $n_j = 2^j$, we have that

$$I^{n_j} \cdot \mathrm{H}^i_{\mathfrak{m}}(M/Q_jM) = 0.$$

Hence the A-module M/Q_jM has FLC.

Proof We prove the theorem by induction on j. Assume that $j \ge 1$ and that the theorem is true up to $j-1$. Set $N = M/Q_{j-1}M$. Then N has FLC. Since we have $\ell_A(0 :_N a_j) < \infty$ by Lemma 9.2, we get a long exact sequence of the local cohomology modules

$$\cdots \to \mathrm{H}^i_{\mathfrak{m}}(N) \to \mathrm{H}^i_{\mathfrak{m}}(M/Q_jM) \to \mathrm{H}^{i+1}_{\mathfrak{m}}(N) \to \cdots$$

by Lemma 8.29. When we set $m = 2^{j-1}$ and $J = I^m$, as we have that $J \cdot \mathrm{H}^i_{\mathfrak{m}}(N) = 0$ and $J \cdot \mathrm{H}^{i+1}_{\mathfrak{m}}(N) = 0$ for $0 \le i \le d-j-1$,

$$J^2 \cdot \mathrm{H}^i_{\mathfrak{m}}(M/Q_jM) = 0.$$

by the assumptions on j. As we have $I^{n_j} = J^2$, it holds that $I^{n_j} \cdot \mathrm{H}^i_{\mathfrak{m}}(M/Q_jM) = 0$ for any $0 \le i \le d-j-1$. □

Corollary 9.4 *Assume that $d \ge 1$ and M has FLC. Let $I = \bigcap_{i=0}^{d-1} \mathrm{Ann}_A \mathrm{H}^i_{\mathfrak{m}}(M)$. Then for any system of parameters $a_1, a_2, \ldots, a_d$ of M and any integer $1 \le i \le d$, we have that*

$$I^n \cdot [Q_{i-1}M :_M a_i] \subseteq Q_{i-1}M,$$

where $Q_j = (a_1, a_2, \ldots, a_j)$ $(0 \le j \le d)$ and $n = 2^{d-1}$.

Proof By Theorem 9.3, $M/Q_{i-1}M$ has FLC. Hence we have

$$[Q_{i-1}M :_M a_i]/Q_{i-1}M \subseteq \mathrm{H}^0_{\mathfrak{m}}(M/Q_{i-1}M)$$

by Lemma 9.2. By Theorem 9.3, we have that

$$I^{n_i} \cdot [Q_{i-1}M :_M a_i] \subseteq Q_{i-1}M$$

for $n_i = 2^{i-1} (\le n = 2^{d-1})$. Therefore,

$$I^n \cdot [Q_{i-1}M :_M a_i] \subseteq Q_{i-1}M,$$

as desired. □

Proposition 9.5 *Assume that $d \ge 1$ and M has FLC. If $Q = (a_1, a_2, \ldots, a_d)$ is a parameter ideal of M, then we have*

$$\ell_A(M/QM) - e_0(Q, M) = \ell_A([(a_1, a_2, \ldots, a_{d-1})M :_M a_d]/(a_1, a_2, \ldots, a_{d-1})M).$$

Hence if $d \geq 2$ in addition, then we have

$$e_0(Q, M) = e_0((a_2, a_3, \ldots, a_d), M/a_1M).$$

Proof Passing to $A/\mathrm{Ann}_A M$, we may assume that $\mathrm{Ann}_A M = 0$. As Q is also a parameter ideal of A, we have that

$$\begin{aligned}\ell_A(M/QM) - e_0(Q, M) &= \ell_A(M/QM) - \chi(a_1, a_2, \ldots, a_d; M)\\ &= \ell_A([Q_{d-1}M :_M a_d]/Q_{d-1}M)\\ &\quad + \sum_{0<k<d} \chi(a_{k+1}, \ldots, a_d; [Q_{k-1}M :_M a_k]/Q_{k-1}M)\end{aligned}$$

by Corollary 8.22, where $Q_i = (a_1, a_2, \ldots, a_i)$. As we have

$$\ell_A([Q_{k-1}M :_M a_k]/Q_{k-1}M) < \infty$$

for any $0 < k < d$ by Corollary 9.4, we also have

$$\chi(a_{k+1}, \ldots, a_d; [Q_{k-1}M :_M a_k]/Q_{k-1}M) = 0,$$

and we get $\ell_A(M/QM) - e_0(Q, M) = \ell_A([Q_{d-1}M :_M a_d]/Q_{d-1}M)$. □

Theorem 9.6 *Assume that $d \geq 1$ and M has FLC. Then for $\mathfrak{p} \in \mathrm{Supp}_A M \setminus \{\mathfrak{m}\}$, the $A_\mathfrak{p}$-module $M_\mathfrak{p}$ is Cohen–Macaulay, and the equality on the dimensions*

$$d = \dim_{A_\mathfrak{p}} M_\mathfrak{p} + \dim A/\mathfrak{p}$$

holds. In particular, as we have $\mathrm{Ass}_A M \setminus \{\mathfrak{m}\} = \mathrm{Assh}_A M$, if $\mathrm{depth}_A M > 0$ in addition, then M is unmixed.

Proof We may assume that $d > 1$ and the assertion is true for the rings of dimension less than or equal to $d - 1$.

Let $\mathfrak{p} \in \mathrm{Supp}_A M \setminus \{\mathfrak{m}\}$. First, we prove that if $\mathfrak{p} \in \mathrm{Ass}_A M$, then $\mathfrak{p} \in \mathrm{Assh}_A M$. Assume that $\mathfrak{p} \notin \mathrm{Assh}_A M$. Then as we have $\mathfrak{p} \not\subseteq \bigcup_{P \in \mathrm{Assh}_A M} P$, we can take $a \in \mathfrak{p}$ as a part of system of parameters on M. Then since $\mathfrak{p} \neq \mathfrak{m}$ and $\ell_A(0 :_M a) < \infty$, we have $[0 :_{M_\mathfrak{p}} a] = [0 :_M a]_\mathfrak{p} = 0$. But this is impossible, since $\mathrm{depth}_{A_\mathfrak{p}} M_\mathfrak{p} = 0$. Thus we have

$$\mathrm{Ass}_A M \setminus \{\mathfrak{m}\} = \mathrm{Assh}_A M.$$

Next, assume that $\mathfrak{p} \notin \mathrm{Ass}_A M$. Taking $a \in \mathfrak{p}$ so that a is a part of a system of parameters on M and letting $\overline{M} = M/aM$, we have that $\overline{M}$ has FLC and $\dim_A \overline{M} = d - 1$. Hence $\overline{M}_\mathfrak{p} = M_\mathfrak{p}/aM_\mathfrak{p}$ is Cohen–Macaulay, and we have that $\dim_{A_\mathfrak{p}} \overline{M}_\mathfrak{p} +$

$\dim A/\mathfrak{p} = d - 1$. As we have $\mathfrak{p} \neq \mathfrak{m}$ and $\ell_A(0 :_M a) < \infty$, it follows that a is $M_\mathfrak{p}$-regular. So $M_\mathfrak{p}$ is Cohen–Macaulay, and hence $\dim_{A_\mathfrak{p}} \overline{M}_\mathfrak{p} = \dim_{A_\mathfrak{p}} M_\mathfrak{p} - 1$. Therefore, we have

$$\dim_A M_\mathfrak{p} + \dim A/\mathfrak{p} = d,$$

and the proof is completed. □

Theorem 9.7 *Assume that A is a homomorphic image of a Cohen–Macaulay ring, and $d \geq 1$. If the $A_\mathfrak{p}$-module $M_\mathfrak{p}$ is Cohen–Macaulay and the equality*

$$d = \dim_{A_\mathfrak{p}} M_\mathfrak{p} + \dim A/\mathfrak{p}$$

holds for any $\mathfrak{p} \in \operatorname{Supp}_A M \setminus \{\mathfrak{m}\}$, then the A-module M has FLC.

Proof We may assume that A is Cohen–Macaulay and $\dim A = d$. Take any $P \in \operatorname{Supp}_{\widehat{A}} \widehat{M} \setminus \{\widehat{\mathfrak{m}}\}$, and we set $\mathfrak{p} = P \cap A$. Then $\mathfrak{p} \in \operatorname{Supp}_A M \setminus \{\mathfrak{m}\}$. As the morphism $A_\mathfrak{p} \to \widehat{A}_P$ is flat and $\widehat{A}_P/\mathfrak{p}\widehat{A}_P$ is Cohen–Macaulay, we have that $\widehat{M}_P$ is a Cohen–Macaulay $\widehat{A}_P$-module, and we have

$$\begin{aligned}\dim_{\widehat{A}_P} \widehat{M}_P &= \dim \widehat{A}_P/\mathfrak{p}\widehat{A}_P + \dim_{A_\mathfrak{p}} M_\mathfrak{p}\\ &= (\dim \widehat{A}_P - \dim A_\mathfrak{p}) + \dim_{A_\mathfrak{p}} M_\mathfrak{p}\\ &= \dim \widehat{A}_P.\end{aligned}$$

So we have $\dim_{\widehat{A}_P} \widehat{M}_P + \dim \widehat{A}/P = \dim \widehat{A}_P + \dim \widehat{A}/P = d$. So replacing A by $\widehat{A}$ if necessary, we may assume that $A = \widehat{A}$. Let K_A be the canonical module of A, and take $i \neq 0$. Then for any $\mathfrak{p} \in \operatorname{Supp}_A M \setminus \{\mathfrak{m}\}$, we have

$$\operatorname{Ext}^i_A(M, K_A)_\mathfrak{p} \cong \operatorname{Ext}^i_{A_\mathfrak{p}}(M_\mathfrak{p}, K_{A_\mathfrak{p}}).$$

As we have $i \neq 0$ and $\dim_{A_\mathfrak{p}} M_\mathfrak{p} = d - \dim A/\mathfrak{p} = \dim A_\mathfrak{p}$, it holds that

$$\operatorname{Ext}^i_{A_\mathfrak{p}}(M_\mathfrak{p}, K_{A_\mathfrak{p}}) = 0.$$

In other words, we have $\ell_A(\operatorname{Ext}^i_A(M, K_A)) < \infty$, and hence $\ell_A(\mathrm{H}^{d-i}_\mathfrak{m}(M)) < \infty$ for $i \neq 0$, and thus M has FLC. □

Remark 9.1 Nagata [Nag, Bad examples] has shown an example of a two-dimensional Noetherian local domain $(A, \mathfrak{m})$ such that $\mathrm{e}_0(\mathfrak{m}) = 1$ but A is not regular. By Theorem 8.36, A is not unmixed. As $\dim A = 2$, for any $\mathfrak{p} \in \operatorname{Spec} A \setminus \{\mathfrak{m}\}$, $A_\mathfrak{p}$ is a Cohen–Macaulay local ring, and $2 = \dim A_\mathfrak{p} + \dim A/\mathfrak{p}$. However, A does not have FLC. Indeed, since depth $A > 0$, if A had FLC, then A were unmixed (see Theorem 9.6). By Theorem 9.8 (6) below, we also have that

$$\sup_{a,b \text{ is an s.o.p. of } A} \ell_A([(a) :_A b]/(a)) = \infty.$$

The fundamental theorem that characterizes modules with FLC is the following.

Theorem 9.8 (**[STC]**) *Assume that $d \geq 1$. Then the following conditions on M are equivalent:*

(1) *M has FLC.*
(2) *There exists an $\mathfrak{m}$-primary ideal I of A such that*

$$I\cdot[(a_1, a_2, \ldots, a_{i-1})M :_M a_i] \subseteq (a_1, a_2, \ldots, a_{i-1})M$$

for any system of parameters $a_1, a_2, \ldots, a_d$ of M and any integer i with $1 \leq i \leq d$.
(3) *There exists some $\mathfrak{m}$-primary ideal I of A such that for any system of parameters $a_1, a_2, \ldots, a_d$ of M*

$$I\cdot[(a_1, a_2, \ldots, a_{d-1})M :_M a_d] \subseteq (a_1, a_2, \ldots, a_{d-1})M.$$

(4) *There exists some integer $\ell \gg 0$ such that any system of parameters $a_1, a_2, \ldots, a_d$ on M that is contained in $\mathfrak{m}^\ell$ acts as a d-sequence on M.*
(5) *There exists some system of parameters $a_1, a_2, \ldots, a_d$ of M which acts as a d^+-sequence (see Definition 8.13) on M.*
(6) *$\sup_Q[\ell_A(M/QM) - \mathrm{e}_0(Q, M)] < \infty$, where Q varies all the parameter ideals of M.*

If these equivalent conditions hold, then letting $\mathbb{I}(M) = \displaystyle\sum_{i=0}^{d-1} \binom{d-1}{i} \ell_A(\mathrm{H}^i_{\mathfrak{m}}(M))$, the equality

$$\sup_Q[\ell_A(M/QM) - \mathrm{e}_0(Q, M)] = \mathbb{I}(M)$$

holds. ($\mathbb{I}(M)$ is called the Buchsbaum invariant of M.)

Proof (1) $\Rightarrow$(2). See Corollary 9.4.

(2) $\Rightarrow$(3). This is trivial.

(3) $\Rightarrow$(2). Let $a_1, a_2, \ldots, a_d$ be a system of parameters of M. Then for any $1 \leq i \leq d$ and $n \geq 1$, as we have

$$I\cdot[(a_1, \ldots, a_{i-1}, a_{i+1}^n, \ldots, a_d^n)M :_M a_i] \subseteq (a_1, \ldots, a_{i-1})M + (a_{i+1}^n, \ldots, a_d^n)M,$$

it follows that $I\cdot[(a_1, \ldots, a_{i-1})M :_M a_i] \subseteq (a_1, \ldots, a_{i-1})M$.

(2) $\Rightarrow$(4). If $a_1, a_2, \ldots, a_d$ is a system of parameters of M and $a_1, a_2, \ldots, a_d \in I$, then for any integer $1 \leq i \leq d$, we have

$$(a_1, \ldots, a_{i-1})M :_M I = (a_1, \ldots, a_{i-1})M :_M a_i.$$

Hence if $1 \le i \le j \le d$, then

$$(a_1, \ldots, a_{i-1})M :_M a_i a_j = (a_1, \ldots, a_{i-1})M :_M I = (a_1, \ldots, a_{i-1})M :_M a_j.$$

(4) $\Rightarrow$(5). If $a_1, a_2, \ldots, a_d \in \mathfrak{m}^\ell$ is a system of parameters of M, then $a_1^{n_1}, a_2^{n_2}, \ldots, a_d^{n_d}$ ($n_i \ge 1$) in any order is a d-sequence.

(5) $\Rightarrow$(1). Assume that $a_1, a_2, \ldots, a_d$ is a system of parameters on M forming a d^+-sequence. We set $Q = (a_1, a_2, \ldots, a_d)$. We prove that $Q\mathrm{H}^i_\mathfrak{m}(M) = 0$ ($i \ne d$) by induction on d. As $0 :_M a_1 = 0 :_M a_1^2$, we have that $a_1 \mathrm{H}^0_\mathfrak{m}(M) = 0$. Hence we may assume that $d > 1$ and the assertion is true up to $d-1$. Take an integer $n \ge 1$ and set $\overline{M} = M/a_1^n M$. Then $a_2, \ldots, a_d$ is a system of parameters on $\overline{M}$, and the sequence acts as a d^+-sequence on $\overline{M}$. So $\overline{M}$ has FLC, and $(a_2, \ldots, a_d)\mathrm{H}^i_\mathfrak{m}(\overline{M}) = 0$ ($i \ne d-1$). As $a_1^n, a_2, \ldots, a_d$ is a d-sequence on M, $(a_2, \ldots, a_d)[0 :_M a_1^n] = 0$. As $\ell_A(0 :_M a_1^n) < \infty$, we get the long exact sequence of local cohomology modules

$$\begin{aligned} 0 \to 0 :_M a_1^n \to \mathrm{H}^0_\mathfrak{m}(M) &\xrightarrow{a_1^n} \mathrm{H}^0_\mathfrak{m}(M) \to \mathrm{H}^0_\mathfrak{m}(\overline{M}) \\ \to \mathrm{H}^1_\mathfrak{m}(M) &\xrightarrow{a_1^n} \mathrm{H}^1_\mathfrak{m}(M) \to \mathrm{H}^1_\mathfrak{m}(\overline{M}) \qquad (\sharp) \\ \cdots \to \mathrm{H}^i_\mathfrak{m}(M) &\xrightarrow{a_1^n} \mathrm{H}^i_\mathfrak{m}(M) \to \mathrm{H}^i_\mathfrak{m}(\overline{M}) \\ \to \mathrm{H}^{i+1}_\mathfrak{m}(M) &\xrightarrow{a_1^n} \mathrm{H}^{i+1}_\mathfrak{m}(M) \to \mathrm{H}^{i+1}_\mathfrak{m}(\overline{M}) \to \cdots, \end{aligned}$$

and we have

$$(a_2, \ldots, a_d)[0 :_{\mathrm{H}^{i+1}_\mathfrak{m}(M)} a_1^n] = 0$$

for $0 \le i \le d-2$. As the integer $n \ge 1$ is arbitrary, we have $(a_2, \ldots, a_d)\mathrm{H}^{i+1}_\mathfrak{m}(M) = 0$. Hence changing the order of $a_1, a_2, \ldots, a_d$, we get $Q\mathrm{H}^{i+1}_\mathfrak{m}(M) = 0$. On the other hand, when we take $n \ge 1$ such that $a_1^n \mathrm{H}^0_\mathfrak{m}(M) = 0$, then we can check that $\mathrm{H}^0_\mathfrak{m}(M) \subseteq \mathrm{H}^0_\mathfrak{m}(\overline{M})$, using the exact sequence ($\sharp$), and this proves $Q\mathrm{H}^0_\mathfrak{m}(M) = 0$.

(1) $\Rightarrow$(6). We set $\mathbb{I}(M) = \sum_{i=0}^{d-1} \binom{d-1}{i} \ell_A(\mathrm{H}^i_\mathfrak{m}(M))$. First, we show that

$$\sup_Q [\ell_A(M/QM) - \mathrm{e}_0(Q, M)] \le \mathbb{I}(M).$$

Let $Q = (a_1, a_2, \ldots, a_d)$ be a parameter ideal of M. Then we have

$$\ell_A(M/QM) - \mathrm{e}_0(Q, M) = \ell_A([Q_{d-1}M :_M a_d]/Q_{d-1}M),$$

where $Q_{d-1} = (a_1, a_2, \ldots, a_{d-1})$ (see Proposition 9.5). Therefore, if $d = 1$, as we have

$$0 :_M a_1 \subseteq \mathrm{H}^0_\mathfrak{m}(M),$$

we get

$$\ell_A(M/QM) - e_0(Q, M) \leq \ell_A(H^0_{\mathfrak{m}}(M)) = \mathbb{I}(M).$$

Assume that $d > 1$ and that the assertion is true up to $d - 1$. Set $\overline{M} = M/a_1M$. Then by the long exact sequence

$$\cdots H^i_{\mathfrak{m}}(M) \xrightarrow{a_1} H^i_{\mathfrak{m}}(M) \to H^i_{\mathfrak{m}}(\overline{M}) \to H^{i+1}_{\mathfrak{m}}(M) \xrightarrow{a_1} H^{i+1}_{\mathfrak{m}}(M) \to \cdots$$

obtained from Lemma 8.29, we have

$$\begin{aligned}\ell_A(H^i_{\mathfrak{m}}(\overline{M})) &\leq \ell_A(0 :_{H^i_{\mathfrak{m}}(M)} a_1) + \ell_A(0 :_{H^{i+1}_{\mathfrak{m}}(M)} a_1)\\ &\leq \ell_A(H^i_{\mathfrak{m}}(M)) + \ell_A(H^{i+1}_{\mathfrak{m}}(M))\end{aligned}$$

for $0 \leq i \leq d - 2$. Hence we have

$$\begin{aligned}\mathbb{I}(\overline{M}) &\leq \sum_{i=0}^{d-2} \binom{d-2}{i} [\ell_A(0 :_{H^i_{\mathfrak{m}}(M)} a_1) + \ell_A(0 :_{H^{i+1}_{\mathfrak{m}}(M)} a_1)]\\ &\leq \sum_{i=0}^{d-2} \binom{d-2}{i} [\ell_A(H^i_{\mathfrak{m}}(M)) + \ell_A(H^{i+1}_{\mathfrak{m}}(M))] \qquad (\sharp)\\ &= \sum_{i=0}^{d-1} \binom{d-1}{i} \ell_A(H^i_{\mathfrak{m}}(M))\\ &= \mathbb{I}(M).\end{aligned}$$

Hence by Proposition 9.5, we have

$$\begin{aligned}\ell_A(M/QM) - e_0(Q, M) &= \ell_A(\overline{M}/(a_2, \ldots, a_d)\overline{M}) - e_0((a_2, \ldots, a_d), \overline{M})\\ &\leq \mathbb{I}(\overline{M})\\ &\leq \mathbb{I}(M).\end{aligned}$$

Next, we show that if $a_1, a_2, \ldots, a_d$ form a d^+-sequence on M, then

$$\ell_A(M/QM) - e_0(Q, M) = \mathbb{I}(M).$$

Consider the case $d = 1$. Then we have $0 :_M a_1^2 = 0 :_M a_1$, and the assertion is true since $0 :_M a_1 = H^0_{\mathfrak{m}}(M)$. If $d \geq 2$ and the assertion is true up to $d - 1$, then since the sequence $a_2, \ldots, a_d$ acts as a d^+-sequence on $\overline{M} = M/a_1M$, we have

$$\ell_A(M/QM) - e_0(Q, M) = \ell_A(\overline{M}/(a_2, \ldots, a_d)\overline{M}) - e_0((a_2, \ldots, a_d), \overline{M})$$

$$= \mathbb{I}(\overline{M}).$$

On the other hand, as we have $Q\mathrm{H}^i_{\mathfrak{m}}(M) = 0$ $(i \neq d)$ (see the proof of (5) ⇒(1)), we get the short exact sequence of local cohomology

$$0 \to \mathrm{H}^i_{\mathfrak{m}}(M) \to \mathrm{H}^i_{\mathfrak{m}}(\overline{M}) \to \mathrm{H}^{i+1}_{\mathfrak{m}}(M) \to 0$$

for $0 \leq i \leq d-2$, and we get $\ell_A(\mathrm{H}^i_{\mathfrak{m}}(\overline{M})) = \ell_A(\mathrm{H}^i_{\mathfrak{m}}(M)) + \ell_A(\mathrm{H}^{i+1}_{\mathfrak{m}}(M))$. Hence by (♯), the equality $\mathbb{I}(\overline{M}) = \mathbb{I}(M)$ follows, and

$$\ell_A(M/QM) - \mathrm{e}_0(Q, M) = \mathbb{I}(M)$$

follows.

(6) ⇒(3). Replacing A by $A/\mathrm{Ann}_A M$, we may assume that $\mathrm{Ann}_A M = 0$. If $Q = (a_1, a_2, \ldots, a_d)$ is a parameter ideal on M, then by Corollary 8.22, we have

$$\begin{aligned}\ell_A(M/QM) - \mathrm{e}_0(Q, M) &= \ell_A([Q_{d-1}M :_M a_d]/Q_{d-1}M) \\ &\quad + \sum_{0<k<d} \chi(a_{k+1}, \ldots, a_d; [Q_{k-1}M :_M a_k]/Q_{k-1}M),\end{aligned}$$

where $Q_i = (a_1, a_2, \ldots, a_i)$. So when we set

$$q = \sup_Q [\ell_A(M/QM) - \mathrm{e}_0(Q, M)],$$

we have $q \geq \ell_A([Q_{d-1}M :_M a_d]/Q_{d-1}M)$, and hence

$$\mathfrak{m}^q \cdot ([Q_{d-1}M :_M a_d]/Q_{d-1}M) = 0$$

holds for any parameter ideal $Q = (a_1, a_2, \ldots, a_d)$ of M.

So Theorem 9.8 has been proved. □

Corollary 9.9 (Corollary to the Proof of Theorem 9.8) *Let $d \geq 1$, and $Q = (a_1, a_2, \ldots, a_d)$ be a parameter ideal of M. The following are equivalent:*

(1) *The sequence $a_1, a_2, \ldots, a_d$ acts on M as a d^+-sequence.*
(2) *M has FLC, and $\ell_A(M/QM) - \mathrm{e}_0(Q, M) = \mathbb{I}(M)$.*

If the conditions above are satisfied, then $Q\mathrm{H}^i_{\mathfrak{m}}(M) = 0$ $(i \neq d)$.

Proof By the proof of Theorem 9.8 ((5) ⇒ (1), (1) ⇒ (6)), the assertion (1) ⇒ (2) and the last assertion $Q\mathrm{H}^i_{\mathfrak{m}}(M) = 0$ $(i \neq d)$ has already been proved. Here we briefly sketch the proof of (2) ⇒ (1). Let $Q = (a_1, a_2, \ldots, a_d)$ be a parameter ideal on M such that $\ell_A(M/QM) - \mathrm{e}_0(Q, M) = \mathbb{I}(M)$. Then when we set $Q_{d-1} = (a_1, a_2, \ldots, a_{d-1})$, for any integer $n \geq 1$, we have

$$\begin{aligned}\ell_A(M/QM) - e_0(Q, M) &= \ell_A([Q_{d-1}M :_M a_d]/Q_{d-1}M)\\ &\le \ell_A([Q_{d-1}M :_M a_d^n]/Q_{d-1}M)\\ &= \ell_A(M/(a_1, \ldots, a_{d-1}, a_d^n)M) - e_0((a_1, \ldots, a_{d-1}, a_d^n), M) \le \mathbb{I}(M).\end{aligned}$$

So once we have

$$\ell_A(M/QM) - e_0(Q, M) = \mathbb{I}(M),$$

we have that $Q_{d-1}M :_M a_d = Q_{d-1}M :_M a_d^n$, and

$$\ell_A(M/(a_1, \ldots, a_{d-1}, a_d^n)M) - e_0((a_1, \ldots, a_{d-1}, a_d^n), M) = \mathbb{I}(M)$$

holds for any integer $n \ge 1$. On the other hand, as we have that whether

$$\ell_A(M/QM) - e_0(Q, M) = \mathbb{I}(M)$$

holds or not is independent of the ordering of the system of parameters $a_1, a_2, \ldots, a_d$, the argument shows that for any $n_1, n_2, \ldots, n_d \ge 1$ and any permutation $\sigma \in S_d$, the equality

$$\ell_A(M/(a_{\sigma(1)}^{n_1}, a_{\sigma(2)}^{n_2}, \ldots, a_{\sigma(d)}^{n_d})M) - e_0((a_{\sigma(1)}^{n_1}, a_{\sigma(2)}^{n_2}, \ldots, a_{\sigma(d)}^{n_d}), M) = \mathbb{I}(M)$$

holds. Hence it suffices to show that $a_1, a_2, \ldots, a_d$ acts on M as a d-sequence.

Assume that $d = 1$. Then as we have $0 :_M a_1 = 0 :_M a_1^2$, the assertion is obvious. Let $d > 1$ and assume that the assertion is true up to $d - 1$. When we set $\overline{M} = M/a_1M$, then

$$\begin{aligned}\mathbb{I}(\overline{M}) &\ge \ell_A(\overline{M}/(a_2, \ldots, a_d)\overline{M}) - e_0((a_2, \ldots, a_d), \overline{M})\\ &= \ell_A(M/QM) - e_0(Q, M) = \mathbb{I}(M).\end{aligned}$$

As is shown in the proof of (1) $\Rightarrow$ (6) of Theorem 9.8, we have $\mathbb{I}(\overline{M}) \le \mathbb{I}(M)$, and so in fact

$$\ell_A(\overline{M}/(a_2, \ldots, a_d)\overline{M}) - e_0((a_2, \ldots, a_d), \overline{M}) = \mathbb{I}(\overline{M}) = \mathbb{I}(M),$$

and thus $a_1\mathrm{H}_{\mathfrak{m}}^i(M) = 0$ $(i \ne d)$. Hence reordering $a_1, a_2, \ldots, a_d$, we have $Q\mathrm{H}_{\mathfrak{m}}^i(M) = 0$ $(i \ne d)$. Hence by induction on d, the sequence $a_2, \ldots, a_d$ acts on M as a d-sequence. So letting $1 \le i \le j \le d$ and taking an element $x \in (a_1, \ldots, a_{i-1})M :_M a_ia_j$, we get $a_jx \in (a_1, \ldots, a_{i-1})M$ if $i \ge 2$, considering the module $\overline{M} = M/a_1M$. Consider the case $i = 1$. As we have $(a_1a_j)x = 0$, it follows that $a_jx \in 0 :_M a_1 \subseteq \mathrm{H}_{\mathfrak{m}}^0(M)$. Since $a_j\mathrm{H}_{\mathfrak{m}}^0(M) = 0$ and hence $a_j^2x = 0$, we have that

$$x \in 0 :_M a_j^2 \subseteq \mathrm{H}^0_{\mathfrak{m}}(M),$$

so $a_j x = 0$ follows. Hence $a_1, a_2, \ldots, a_d$ acts on M as a d-sequence. □

Definition 9.10 Let $d \geq 1$ and M has FLC. We say that a parameter ideal Q on M is *standard* if $\ell_A(M/QM) - \mathrm{e}_0(Q, M) = \mathbb{I}(M)$ is satisfied.

9.2 Buchsbaum Module

Let A be a Noetherian local ring. Let M be a nonzero finitely generated A-module of dimension d.

Definition 9.11 ([SV2]) We say that M is a *Buchsbaum module* if for any parameter ideal Q on M, the difference $\ell_A(M/QM) - \mathrm{e}_0(Q, M)$ has a constant value and is independent of the choice of Q. We say that A is a Buchsbaum local ring if the A-module A is Buchsbaum.

By Proposition 8.28, a Cohen–Macaulay A-module is Buchsbaum, while a Buchsbaum module has FLC by Theorem 9.8.

Proposition 9.12 *Let $d \geq 1$. The following are equivalent:*

(1) *M is Buchsbaum.*
(2) *A has FLC, and any parameter ideal is standard.*
(3) *Any system of parameters of M is a d-sequence.*

Proof See Corollary 9.9. □

Remark 9.2 Even if any parameter ideal on M is generated by a d-sequence, M may not be a Buchsbaum module. Let $(R, \mathfrak{n})$ be a 2-dimensional regular local ring, $\mathfrak{n} = (X, Y)$, $D = R/(X)$, and let $A = R \ltimes D$ be the idealization. Then we have $\dim A = 2$ and $\mathrm{H}^1_{\mathfrak{m}}(A) \cong \mathrm{H}^1_{\mathfrak{n}}(D)$. As $\mathrm{H}^1_{\mathfrak{m}}(A)$ is not a finitely generated A-module, A does not have FLC. In particular, it is not a Buchsbaum ring. However, if a, b is a system of parameters of A, then either a, b or b, a is a d-sequence. We briefly sketch the proof. We regard an R-module as an A-module through the morphism $p : A \to R$. Now consider the exact sequence

$$0 \to D \xrightarrow{j} A \xrightarrow{p} R \to 0,$$

where $j(x) = (0, x)$. Then a, b is a system of parameters on the A-module R. On the other hand, as D is a DVR, we have $aD \subseteq bD$ or $bD \subseteq aD$. We show that if $bD \subseteq aD$, then a, b in this order forms a d-sequence. As $aD \neq 0$, a is a non-zerodivisor on D, and hence is a non-zerodivisor on A. Hence we have the short exact sequence

$$0 \to D/aD \xrightarrow{\overline{j}} A/(a) \xrightarrow{\overline{p}} R/aR \to 0.$$

If $\varphi \in [(a) :_A b^2]/(a)$, then b is a non-zerodivisor on R/aR, and we have $\overline{p}(\varphi) = 0$. Hence we can express $\varphi = \overline{j}(\psi)$ $(\psi \in D/aD)$. As we have $b \cdot (D/aD) = 0$, it follows that $b\psi = 0$, that is, $b\varphi = 0$. As a is a non-zerodivisor on A, a, b forms a d-sequence of A.

Corollary 9.13 *Let $d \geq 1$. If M is Buchsbaum, then for any $a \in \mathfrak{m}$ which is a part of a system of parameters of M, M/aM is also Buchsbaum. If $d \geq 2$, then $\mathbb{I}(M/aM) = \mathbb{I}(M)$.*

Proof The fact that $\overline{M} = M/aM$ is Buchsbaum follows immediately from Proposition 9.12. We show that $\mathbb{I}(M/aM) = \mathbb{I}(M)$. If $a_2, \ldots, a_d$ is a system of parameters on $\overline{M}$, then $a_1 = a, a_2, \ldots, a_d$ is a system of parameters on M, and

$$\ell_A(M/(a_1, a_2, \ldots, a_d)M) = \ell_A(\overline{M}/(a_2, \ldots, a_d)\overline{M}).$$

On the other hand, by Corollary 9.5, we have

$$\mathrm{e}_0((a_1, a_2, \ldots, a_d), M) = \mathrm{e}_0((a_2, \ldots, a_d), \overline{M}).$$

Hence we have $\mathbb{I}(M/aM) = \mathbb{I}(M)$. □

Theorem 9.14 *If $d \geq 1$, then the following are equivalent:*

(1) *M is Buchsbaum.*
(2) *If $a_1, a_2, \ldots, a_d$ is a system of parameters of M, then for any integers $1 \leq i \leq d$, we have*

$$\mathfrak{m} \cdot [[(a_1, \ldots, a_{i-1})M :_M a_i]/(a_1, \ldots, a_{i-1})M] = 0.$$

If these conditions are satisfied, then we have $\mathfrak{m} \cdot \mathrm{H}^i_{\mathfrak{m}}(M) = 0$ $(i \neq d)$.

Proof (2) $\Rightarrow$(1) As we have $(a_1, \ldots a_{i-1})M :_M a_i = (a_1, \ldots, a_{i-1})M : \mathfrak{m}$ for $1 \leq i \leq j \leq d$, we have

$$(a_1, \ldots, a_{i-1})M : a_i a_j = (a_1, \ldots, a_{i-1})M : \mathfrak{m} = (a_1, \ldots, a_{i-1})M : a_j,$$

and every system of parameters on M forms a d-sequence.

(1) $\Rightarrow$(2). By Corollary 9.9, we have $Q\mathrm{H}^i_{\mathfrak{m}}(M) = 0$ $(i \neq d)$ for any system of parameters Q of M. Taking a set of generators $\{x_1, \ldots, x_v\}$ of the maximal ideal $\mathfrak{m}$ such that each x_i is a part of a system of parameters on M, $\mathfrak{m} \cdot \mathrm{H}^i_{\mathfrak{m}}(M) = 0$ $(i \neq d)$ follows immediately. Next, let $a_1, a_2, \ldots, a_d$ be a system of parameters on M. Then, as we have seen, $M/(a_1, \ldots, a_{i-1})M$ is Buchsbaum for $1 \leq i \leq d$, and we have

$$[(a_1, \ldots, a_{i-1})M :_M a_i]/(a_1, \ldots, a_{i-1})M \subseteq \mathrm{H}^0_{\mathfrak{m}}(M/(a_1, \ldots, a_{i-1})M)$$

(see Lemma 9.2). Hence

$$\mathfrak{m}\cdot [[(a_1, \ldots, a_{i-1})M :_M a_i]/(a_1, \ldots, a_{i-1})M] = 0$$

follows. □

Corollary 9.15 (Cf. Corollary 8.31) *If $d \geq 2$, then the converse of Corollary 9.13 holds.*

Proof Let $a_1, a_2, \ldots, a_d$ be a system of parameters of M. We show that

$$\mathfrak{m}\cdot [[(a_1, \ldots, a_{i-1})M :_M a_i]/(a_1, \ldots, a_{i-1})M] = 0$$

holds for any $1 \leq i \leq d$. As M/a_1M is Buchsbaum, we may assume that $i = 1$. Let $x \in 0 :_M a_1$, and $n \geq 1$ be an integer. Letting $\overline{x}$ be the image of the element x in M/a_2^nM, we have that $a_1\overline{x} = 0$, and M/a_2^nM is Buchsbaum, and hence $\mathfrak{m}\overline{x} = 0$. Namely, we have $\mathfrak{m}x \subseteq a_2^nM$ for any $n \geq 1$, and hence $\mathfrak{m}x = 0$. It follows that $\mathfrak{m}\cdot[0 :_M a_1] = 0$. By Theorem 9.14, M is a Buchsbaum module. □

As the next example shows, even if $\mathfrak{m}\cdot \mathrm{H}^i_{\mathfrak{m}}(M) = 0$ $(i \neq d)$, M may not be Buchsbaum, although M has FLC.

Example 9.16 Let k be a field, and $R = k[[X, Y, Z, W]]$ be the formal power series ring. Let $n \geq 1$ be an integer, and set $A = R/[(X, Y)^n \cap (Z, W)]$. Then we have:

(1) $\dim A = 2$ and $\operatorname{depth} A = 1$, A has FLC, and $\mathbb{I}(A) = \binom{n+1}{2}$.
(2) A is a Buchsbaum ring if and only if $n = 1$.
(3) When we set $B = R/[(X, Y) \cap (Z, W) \cap (X^2, Y, Z^2, W)]$, then $\dim B = 2$ and $\mathfrak{n}\cdot \mathrm{H}^i_{\mathfrak{n}}(B) = 0$ $(i = 0, 1)$. But B is not a Buchsbaum ring, where $\mathfrak{n}$ is the maximal ideal of R.

Proof (1), (2) By the exact sequence of R-modules

$$0 \to A \to R/(X, Y)^n \oplus R/(Z, W) \to R/[(X, Y)^n + (Z, W)] \to 0, \qquad (\sharp)$$

we have $\mathrm{H}^0_{\mathfrak{n}}(A) = 0$, $\mathrm{H}^1_{\mathfrak{n}}(A) \cong R/[(X, Y)^n + (Z, W)]$. $\dim A = 2$ is readily proved, and $\operatorname{depth} A = 1$ follows immediately. As $\ell_A(\mathrm{H}^1_{\mathfrak{n}}(A)) = \binom{n+1}{2}$, we have $\mathbb{I}(A) = \binom{n+1}{2}$. If A is Buchsbaum, then $\mathfrak{n}\cdot \mathrm{H}^1_{\mathfrak{n}}(A) = 0$, and hence $n = 1$ follows. We check that if $n = 1$ then A is Buchsbaum. Let α, β be a system of parameters of A. By Davis's lemma, we can take $a, b \in R$ such that α and β are respectively the images of a and b in A, and a, b is a regular sequence in R. When we set $\mathfrak{q} = (a, b)R$ and $Q = (\alpha, \beta)A$, as a, b is a system of parameters on $R/(X, Y)$ and $R/(Z, W)$, by the exact sequence ($\sharp$), we get the exact sequence of $\operatorname{Tor}^R(R/(a, b), -)$

$$0 \to (R/\mathfrak{n})^2 \to A/Q \to R/[(X, Y) + \mathfrak{q}] \oplus R/[(Z, W) + \mathfrak{q}] \to R/\mathfrak{n} \to 0.$$

Hence we have

$$\begin{aligned}\ell_A(A/Q) &= \mathrm{e}_0(\mathfrak{q}, R/(X, Y)) + \mathrm{e}_0(\mathfrak{q}, R/(Z, W)) + 1 \\ &= \mathrm{e}_0(Q, A) + 1 \qquad \text{(see Theorem 8.26)},\end{aligned}$$

and hence A is Buchsbaum and $\mathbb{I}(A) = 1$.

(3) We set $I = (X, Y) \cap (Z, W)$ and $J = (X^2, Y, Z^2, W)$. Then we have

$$I + J = (X^2, Y, Z^2, XZ, W), \quad I \cap J = (XW, YZ, YW, X^2Z, XZ^2).$$

We look at the exact sequence of R-modules

$$0 \to B \to R/I \oplus R/J \to R/[I + J] \to 0. \tag{$\sharp$}$$

As we have seen in (2), we have depth $R/I = 1$ and hence by the exact sequence ($\sharp$), we get the exact sequence

$$0 \to \mathrm{H}^0_{\mathfrak{n}}(B) \to \mathrm{H}^0_{\mathfrak{n}}(R/J) \to \mathrm{H}^0_{\mathfrak{n}}(R/[I + J]) \to 0.$$

So we have $\mathrm{H}^0_{\mathfrak{n}}(B) \cong [I + J]/J$, and it follows that $\ell_A(\mathrm{H}^0_{\mathfrak{n}}(B)) = 1$. On the other hand, we get an isomorphism

$$\mathrm{H}^1_{\mathfrak{n}}(B) \xrightarrow{\sim} \mathrm{H}^1_{\mathfrak{n}}(R/I)$$

by the exact sequence ($\sharp$), we have $\ell_A(\mathrm{H}^1_{\mathfrak{n}}(B)) = 1$ by the proof of (1), (2). Hence we have $\mathfrak{n}{\cdot}\mathrm{H}^i_{\mathfrak{n}}(B) = 0$ $(i = 0, 1)$. Now when we set $a = X + Z$ and $b = Y + W$, a, b is a system of parameters on the R-module B, where we have

$$B/(X + Z)B \cong k[[X, Y, W]]/(XW, XY, YW, X^3),$$

and $Y + W$ is a parameter on $C = k[[X, Y, W]]/(XW, XY, YW, X^3)$. However,

$$(Y + W)X \equiv 0 \mod (XW, XY, YW, X^3),$$

but

$$X^2 \not\equiv 0 \mod (XW, XY, YW, X^3).$$

Hence C is not Buchsbaum. So B is not Buchsbaum, either, by Corollary 9.13. □

We give some applications.

Proposition 9.17 *Let A be a Buchsbaum local ring, and assume that a finite group G acts on A as ring automorphisms. If $|G| \in \mathrm{U}(A)$, then the ring of invariants A^G is Buchsbaum.*

Proof Note that A is integral over A^G and A^G is a direct summand of the A^G-module A. We may assume that $d = \dim A \geq 1$. If $a_1, a_2, \ldots, a_d$ is a system of parameters of the Noetherian local ring A^G, then it is also a system of parameters of A, and hence we have

$$(a_1, \ldots, a_{j-1})A :_A a_i a_j = (a_1, \ldots, a_{i-1})A :_A a_j$$

for any $1 \leq i \leq j \leq d$. Hence if an element $x \in A^G$ satisfies

$$(a_i a_j)x \in (a_1, \ldots, a_{i-1})A^G,$$

then $(a_i a_j)x \in (a_1, \ldots, a_{i-1})A$, and hence we have

$$a_j x \in (a_1, \ldots, a_{i-1})A \cap A^G = (a_1, \ldots, a_{i-1})A^G,$$

and we have

$$(a_1, \ldots, a_{i-1})A^G :_{A^G} a_i a_j = (a_1, \ldots a_{i-1})A^G :_{A^G} a_j.$$

Hence A^G is a Buchsbaum ring. □

Proposition 9.18 ([GO]) *Assume that A contains a field of positive characteristic p. Assume that the Frobenius map* $F : A \to {}^1A$, $F(a) = a^p$ *is a split monomorphism (see Definition 10.26). Then A has FLC if and only if A is Buchsbaum.*

Proof We show that if A has FLC, then A is a Buchsbaum ring. We may assume that $d \geq 1$. As A has FLC, when we can take $e \gg 0$ such that any system of parameters $b_1, b_2, \ldots, b_d$ contained in $\mathfrak{m}^q$ forms a d-sequence in A, where $q = p^e$. Hence for any system of parameters $a_1, a_2, \ldots, a_d$ of A, the sequence $a_1^q, a_2^q, \ldots, a_d^q$ is a d-sequence in A. Let $1 \leq i \leq j \leq d$ and let $x \in A$ satisfy $(a_i a_j)x \in (a_1, a_2, \ldots, a_{i-1})$. Then as we have

$$(a_i^q a_j^q)x^q \in (a_1^q, a_2^q, \ldots, a_{i-1}^q),$$

it follows that $a_j^q x^q \in (a_1^q, a_2^q, \ldots, a_{i-1}^q)$. As $F^e : A \to {}^eA$, $F^e(a) = a^q$ is a split monomorphism, we have that $a_j x \in (a_1, a_2, \ldots, a_{i-1})$. Hence $a_1, a_2, \ldots, a_d$ forms a d-sequence, and by Proposition 9.12, A is a Buchsbaum ring. □

We state a criterion for Buchsbaum property of a ring and an existence theorem of a Buchsbaum ring. Let M be a nonzero finitely generated A-module of dimension d. As we have $\mathrm{H}_{\mathfrak{m}}^i(M) = \lim_{n\to\infty} \mathrm{Ext}_A^i(A/\mathfrak{m}^n, M)$ $(i \in \mathbb{Z})$, there is a canonical map from $\mathrm{Ext}_A^i(A/\mathfrak{m}, M)$ to the inductive limit $\mathrm{H}_{\mathfrak{m}}^i(M)$,

$$\mathrm{Ext}_A^i(A/\mathfrak{m}, M) \to \mathrm{H}_{\mathfrak{m}}^i(M).$$

The following theorem is a criterion for Buchsbaum property of an A-module using this map.

Theorem 9.19 ([SV1]) *Let M be a nonzero finitely generated module over a Noetherian local ring A with $\dim_A M = d$. If the canonical map*

$$\mathrm{Ext}^i_A(A/\mathfrak{m}, M) \to \mathrm{H}^i_{\mathfrak{m}}(M)$$

is surjective for $i \neq d$, then M is a Buchsbaum module. If A is a regular local ring, then the converse is also true.

Corollary 9.20 ([SV1]) *Let M be a nonzero finitely generated module over a Noetherian local ring A with $\dim_A M = d$. Assume that $d > t = \mathrm{depth}_A M$ and $\mathrm{H}^i_{\mathfrak{m}}(M) = 0$ $(\forall i \neq t, d)$. Then M is a Buchsbaum module if and only if $\mathfrak{m}\mathrm{H}^t_{\mathfrak{m}}(M) = 0$.*

Proof As $t = \mathrm{depth}_A M$, it is easy to see that the map

$$\mathrm{Ext}^t_A(A/\mathfrak{m}, M) \to \mathrm{H}^t_{\mathfrak{m}}(M)$$

is surjective if $\mathfrak{m}\mathrm{H}^t_{\mathfrak{m}}(M) = 0$. □

Note that in Example 9.16, the Buchsbaum property can be checked also by using Corollary 9.20. Both Theorem 9.19 and Corollary 9.20 are powerful criteria, and many applications are known. The following is one of these.

Theorem 9.21 (Goto[1], [G]) *For given integers $d \geq 1$ and $h_0, h_1, \ldots, h_{d-1} \geq 0$, there exists a d-dimensional Buchsbaum local ring A such that*

$$\ell_A(\mathrm{H}^i_{\mathfrak{m}}(A)) = h_i, \quad 0 \leq \forall i \leq d-1.$$

If $h_0 = 0$, then the Buchsbaum ring A can be chosen to be a domain. If, moreover, $d \geq 2$ and $h_0 = h_1 = 0$, then it can be chosen to be a normal domain.

[1] Shiro Goto, 1946–2022.

Chapter 10
Characteristic p Methods

In 1985, M. Hochster[1] and C. Huneke[2] found the notion of *tight closure* of ideals in rings of characteristic p. This method enables us to make many things clearer and sometimes to generalize in broader settings. In this chapter, we explain the concept of tight closures and applications.

10.1 Tight Closure of Ideals, F-Regular Rings, and Theorem of Briançon–Skoda

In this section, we fix a prime number p and assume all rings are of characteristic p. If the characteristic of a ring A is p and $a, b \in A$, then we have

$$(a + b)^p = a^p + b^b$$

in such rings. Hence the *Frobenius map*

$$F : A \to A, \quad F(a) = a^p$$

is a ring homomorphism. In the following, the letter q will denote always a power of p ($q = p^e$ for some e).

Let us introduce the definition of tight closure.

Definition 10.1 Let A be a Noetherian ring of characteristic p and I be an ideal of A.

(1) We denote by A° the set of elements of A which are not in any minimal prime ideal of A. For example, if A is an integral domain, then $A^\circ = A \setminus \{0\}$.

---
[1] Melvin Hochster, 1943–.

[2] Craig Huneke, 1971–.

S. Goto, K. Watanabe, *Commutative Algebra*, Graduate Texts in Mathematics 309,
https://doi.org/10.1007/978-981-95-4227-7_10

(2) We put $I^{[q]} = (a^q \mid a \in I)$.
(3) We put $I^* = \{x \in A \mid \exists c \in A^\circ, cx^q \in I^{[q]} (\forall q = p^n \gg 1)\}$. We call I^* the *tight closure* of I.

Let us list some fundamental properties of tight closure. In the following, we always assume that A is a Noetherian ring of characteristic p.

Proposition 10.2 *Let $I \subseteq J$ be ideals of A and $x \in A$.*

(1) *I^* is an ideal of A containing I and we have $(I^*)^* = I^*$.*
(2) *If $I \subseteq J$, then $I^* \subseteq J^*$.*
(3) *$x \in I^*$ if and only if for every minimal prime $\mathfrak{p}$ of A we have $x+\mathfrak{p} \in ((I+\mathfrak{p})/\mathfrak{p})^*$ in $A/\mathfrak{p}$.*

Proof (1) Let $a, b \in I^*$. It suffices to show that $a + b \in I^*$. If $c, c' \in A^\circ$ and $ca^q, c'b^q \in I^{[q]}$ then $(cc')(a + b)^q \in I^{[q]}$ for all $q \gg 1$. Note that for finite elements $a_1, \ldots, a_n \in I^*$, we can take $c \in A^\circ$ such that $ca_i^q \in I^{[q]}$ for all a_i. Since A is Noetherian, we can take $c \in A^\circ$ so that $c(I^*)^{[q]} \subseteq I^{[q]}$ for $q \gg 1$. Now if $b \in (I^*)^*$ and $c'b^q \in (I^*)^{[q]}$, then $(cc')b^q \in I^{[q]}$ and $b \in I^*$. (2) is obvious.

(3) If $x \in I^*$, from $cx^q \in I^{[q]}$, we have $(c + \mathfrak{p})(x + \mathfrak{p})^q \in ((I + \mathfrak{p})/\mathfrak{p})^{[q]}$ and $x + \mathfrak{p} \in ((I + \mathfrak{p})/\mathfrak{p})^*$.

Conversely, assume that for every minimal prime ideal $\mathfrak{p}$ of A, $x + \mathfrak{p} \in ((I + \mathfrak{p})/\mathfrak{p})^*$. Let $\{\mathfrak{p}_1, \ldots, \mathfrak{p}_r\}$ be the set of minimal primes of A. Then for every i, we can take $c_i \in A^\circ$ so that $c_i x^q \in I^{[q]} + \mathfrak{p}_i$. Then by "prime avoidance," we can take $t_i \in A$ so that $t_i \in \cap_{j \neq i} \mathfrak{p}_j$, $t_i \notin \mathfrak{p}_i$. Put $c = \sum c_i t_i$, then for every $q \gg 1$, we have $cx^q \in I^{[q]} + \cap_{i=1}^r \mathfrak{p}_i = I^{[q]} + \mathfrak{N}(A)$. Take q' so that $\mathfrak{N}(A)^{q'} = 0$ (not depending on x, I), then since $c_\mathfrak{p} x^q \in I^{[q]} + \mathfrak{p}$, putting $c = \prod_\mathfrak{p} c_\mathfrak{p}$, we have $c^{q'} x^{qq'} \in I^{[qq']}$, hence $x \in I^*$. □

Remark 10.1 In the definition of the tight closure; "$\exists c \in A^\circ, cx^q \in I^{[q]}$ $(\forall q = p^n \gg 1)$", it is natural to ask "Can we change "$\forall q = p^n \gg 1$" to "$\forall q = p^n \geq 1$"? Actually, if A is reduced or $\text{ht}(I) > 0$, we can change the definition to "$\exists c \in A^\circ, cx^q \in I^{[q]}$ $(\forall q = p^n)$". In fact, if for finite q, $cx^q \notin I^{[q]}$, let q_0 be the maximal of such q and if $\text{ht}(I) > 0$, take $c' \in I \cap A^\circ$. Then we have $c(c')^{q_0} x^q \in I^{[q]}$ for all q.

On the other hand, if A has nonzero nilpotent elements and if $\text{Ass}(A) = \text{Assh}(A)$, then since the elements of A° are non-zerodivisors, a nilpotent element $x \neq 0$ does not satisfy the condition "$\exists c \in A^\circ$ such that $cx^q = 0$" for $q = 1$. But certainly a nilpotent element x is in $(0)^*$ by Definition 10.1.

Let us compute some examples.

Example 10.3 Let k be a field of characteristic p.

(1) Let $A = k[X, Y]$ and $I = (X^2, Y^2)$. Then $XY \notin I^*$ and since (X^2, XY, Y^2) is the unique minimal ideal strictly containing I, we have $I^* = I$.
(2) Let $A = k[X, Y, Z]/(X^n + Y^n + Z^n)$ with $(n, p) = 1$. If $I = (y, z)$, then $x^2 \in I^*$, where x, y, z are the images of X, Y, Z, respectively.

Proof

(1) If $c(XY)^q = cX^qY^q \in I^{[q]} = (X^{2q}, Y^{2q})$, then we must have $c \in (X^q, Y^q)$. Since this must hold for all $q \gg 1$, we have $c = 0$ and hence $XY \notin I^*$.
(2) Let $q \gg 1$ and put $q = ns + r,\ 0 \le r < n$. Then $(x^2)^q = x^{2r}(y^n + z^n)^{2s}$. In the expansion of $(y^n + z^n)^{2s}$, every term except $(yz)^{ns}$ is divisible by either y^q or z^q. Hence putting $c = y^n$, we have $cx^q \in I^{[q]}\,\forall q \gg 0$. Hence we have $x^2 \in I^*$.

□

If $(A, \mathfrak{m})$ is a regular local ring, then for every ideal I of A we have $I^* = I$. To prove that, the following theorem by E. Kunz is essential.

Theorem 10.4 (E. Kunz) *For a Noetherian local ring $(A, \mathfrak{m})$ of dimension d, the following conditions are equivalent:*

(1) *A is regular.*
(2) *$F : A \to A$ is a flat map.*
(3) $\forall q,\ \ell_A(A/\mathfrak{m}^{[q]}) = q^d$.
(4) $\exists q,\ \ell_A(A/\mathfrak{m}^{[q]}) = q^d$.

Proof We will outline the proof. Since the statement is equivalent if we replace A with $\widehat{A}$, we assume that A is complete. Also, we can take a faithfully flat extension $(A, \mathfrak{m}) \to (B, \mathfrak{n})$ with $\mathfrak{n} = \mathfrak{m}B$ and $B/\mathfrak{n}$ is a perfect field by Proposition 5.79. Since the equivalence in the theorem holds for A if and only for B, we can assume that $A/\mathfrak{m}$ is perfect. So, we assume that A is complete and $A/\mathfrak{m}$ is a perfect field. Since $A/\mathfrak{m}$ is perfect, by $F : A \to A$, A is a finite A module.

Now, the implication (1) ⇒(3) ⇒(4) is obvious. If A is an integral domain, it is easy to see that $\operatorname{rank}_A A^{1/q} = q^d$. Since $\ell_A(A/\mathfrak{m}^{[q]})$ is the number of minimal generators of $A^{1/q}$ over A, (4) implies that $A^{1/q}$ is a free A module, hence we have (4) ⇒(2).

Next, assume (2) and let $(x_1, \ldots, x_n)$ be a set of minimal generators of $\mathfrak{m}$. Then for every $q = p^e$, we can show that $(x_1^q, \ldots, x_n^q)$ is "independent" in the sense of Exercise 10.1 below, and by multiplicity theory, we can show $n = d$ and hence A is regular. □

Exercise 10.1 Let $(A, \mathfrak{m})$ be a Noetherian local ring. We say a set $(f_1, \ldots, f_s) \subseteq A$ is *independent* if it satisfies the condition (#):

(#) If $(a_1, \ldots, a_s) \subseteq A$ and $\sum_{i=1}^{s} a_i f_i = 0$, then $a_1, \ldots, a_s \in (f_1, \ldots, f_s)$.

(1) Let $(x_1, \ldots, x_n)$ be a minimal generator system of $\mathfrak{m}$. If $A \to A^{1/q}$ is flat, then show that $(x_1^q, \ldots, x_n^q)$ is independent.
(2) If $(f_1, \ldots, f_s)$ is independent and if $f_1 \in (g_1)$, show that $(g_1, f_2, \ldots, f_s)$ is independent and show also that $(f_2, \ldots, f_s) : g_1 \subseteq (f_1, f_2, \ldots, f_s)$.
(3) Show that if $(x_1, \ldots, x_n)$ is a set of minimal generators of $\mathfrak{m}$ and if $(x_1, \ldots, x_n)$ is independent, show that $\ell_A(A/(x_1^q, \ldots, x_n^q)) = q^n$.

The following fundamental fact of tight closure theory follows from Theorem 10.4.

Theorem 10.5 *If $(A, \mathfrak{m})$ is a regular local ring, then $I^* = I$ for every ideal I of A.*

Proof Let $x \in I^*$ and assume $cx^q \in I^{[q]}$. Then since $F : A \to A$ is flat, we have $[I^{[q]} : x^q] = [I : x]^{[q]}$. If $x \notin I$, then $I : x \subseteq \mathfrak{m}$ and $[I^{[q]} : x^q] \subseteq \mathfrak{m}^{[q]}$. Hence if $cx^q \in I^{[q]}$ for $q \gg 1$, then $c \in \bigcap_q \mathfrak{m}^{[q]} = 0$. This shows that $x \in I$. □

From this theorem we define F-regular rings, which behave like regular rings with respect to tight closure.

Definition 10.6 A Noetherian ring A of characteristic $p > 0$ is called a *weakly F-regular* ring if $I^* = I$ holds for every ideal I of A. We say A is *F-regular* if every localization $S^{-1}A$ of A is weakly F-regular.

Compatibility of tight closure with localization is a subtle subject and has been a big issue of the theory.

Remark 10.2 Let A be a Noetherian ring and $S \subseteq A$ a multiplicatively closed set.

(1) For every ideal I of A it is easy to show that $S^{-1}I^* \subseteq (S^{-1}I)^*$.
(2) The converse implication of (1) was a celebrated open problem of tight closure theory for over 20 years. Then H. Brenner and P. Monsky gave a counterexample in the case $p = 2$ in 2007 [BM]. The counterexample given there is a hypersurface of dimension 3 given by $k[X, Y, Z, T]/(Z^4 + XYZ^2 + X^3Z + Y^3Z + TX^2Y^2)$, where k is the algebraic closure of $\mathbb{F}_2$. But it seems that no counterexamples appeared for $p \geq 3$.
(3) If $\sqrt{I} = \mathfrak{m}$ is a maximal ideal, we can show $I^*A_\mathfrak{m} = (IA_\mathfrak{m})^*$.
(4) We can show that A is weakly F-regular if and only if for every maximal ideal $\mathfrak{m}$ and for every $\mathfrak{m}$ primary ideal I we have $I^* = I$. Hence A is weakly F-regular if and only if so is $A_\mathfrak{m}$ for every maximal ideal $\mathfrak{m}$ of A.
(5) If A is regular, then A is F-regular.

Proof (1) If $x \in I^*$ and if $cx^q \in I^{[q]}$, for some $c \in A^\circ$, then $(c/1)(x/1)^q \in I^{[q]}(S^{-1}A)$ and thus $x/1 \in (S^{-1}I)^*$.

(3) If I is $\mathfrak{m}$ primary, so is $I^{[q]}$. Hence if $(c/s)(x/1)^q \in (S^{-1}A)I^{[q]}$, $\exists s' \notin \mathfrak{m}$, $s'cx^q \in I^{[q]}$. Hence $cx^q \in I^{[q]}$ and we have $x \in I^*$.

(4) Since every ideal in a Noetherian ring is a possibly infinite intersection of $\mathfrak{m}$ primary ideals for maximal ideals $\mathfrak{m}$, (4) follows from (3). (5) follows from 10.5. □

Next, we will see the relationship between tight closure and integral closure.

Definition–Proposition 10.7 Let A be a Noetherian ring and I be an ideal of A with $\mathrm{ht}(I) > 0$. We say that $x \in A$ is *integral over I* if there exist an integer $n > 0$ and $c_i \in I^i$ $(1 \leq i \leq n)$ such that

$$x^n + c_1x^{n-1} + \cdots + c_n = 0.$$

The set of all elements of A that are integral over I forms an ideal of A, which we denote by

$$\overline{I} = \{ x \in A \mid x \text{ is integral over } I \}$$

and call the *integral closure* of I. Then the following statements hold:

(1) $x \in \overline{I}$ if and only if there exist $c \in A^\circ$ and an infinite subset Λ of $\mathbb{N}$ such that $\forall N \in \Lambda, cx^N \in I^N$.
(2) $I^* \subseteq \overline{I}$.

Proof

(1) By the definition of $\overline{I}$, $x \in \overline{I} \Leftrightarrow \exists r, \forall n \geq r, (I + xA)^n = (I + xA)^r I^{n-r}$. Hence, if we take $c \in I^r \cap A^\circ$, then we have $cx^n \in I^n$ $(\forall n \geq r)$. To show the converse, we may assume A is an integral domain considering $I(A/\mathfrak{p})$ in $A/\mathfrak{p}$ for a minimal prime $\mathfrak{p}$ of A. If $cx^n \in I^n$ $(\forall n \gg 0)$, then for every discrete valuation v of the field of fractions of A we have $v(x) \geq v(I)$. Then by 6.8.2 of [SwHu] we get $x \in \overline{I}$.
(2) If $x \in I^*$, then for some $c \in A^\circ$ and for $q = p^e \gg 1, cx^q \in I^{[q]} \subseteq I^q$. This is just the condition in (1), taking $\Lambda = \{q = p^e \mid q \gg 1\}$.

□

In many cases, I^* is much smaller than $\overline{I}$ (although, in some cases, I^* is much bigger than I). This fact is the reason for the name "tight" closure.

The next theorem is called "Briançon–Skoda theorem" and is very important explaining the behavior of integral closure. The proof of this theorem, which used to be very difficult before, became surprisingly easier by introducing the notion of tight closure.

Theorem 10.8 (Theorem of Briançon–Skoda) *Let A be a Noetherian ring and I be an ideal of A, generated by n elements. Then for every integer $w \geq 1$ we have*

$$\overline{I^{n+w-1}} \subseteq (I^w)^*.$$

The proof reduces to the following simple lemma.

Lemma 10.9 *Let I be an ideal generated by n elements. Then for every q and integer $w \geq 1$, we have*

$$I^{q(w-1)+(q-1)n+1} \subseteq (I^{[q]})^w = (I^w)^{[q]}.$$

Proof Let $I = (a_1, \ldots, a_n)$. Then $I^{q(w-1)+(q-1)n+1}$ is generated by the products of $q(w-1) + (q-1)n + 1$ elements of $a_1, \ldots, a_n$. But if $i_1 + i_2 + \ldots + i_n = q(w-1) + (q-1)n + 1$, then we have $\sum_{j=1}^n [i_j/q] \geq w$, which shows our result.

□

Proof of Theorem 10.8 Take an integer r so that

$$\overline{I^{n+w-1}}^{\,n+r} = \overline{I^{n+w-1}}^{\,r} (I^{n+w-1})^n \quad (\forall n \geq 0).$$

If $c \in \overline{I^{n+w-1}}^{\,r} I \cap A^\circ$, then for every q and for any $x \in \overline{I^{n+w-1}}^{\,r}$ we have

$$cx^q \in I^{(n+w-1)q+1}.$$

Since $(n+w-1)q+1 > (w-1)q+(q-1)n+1$, we can assert $cx^q \in (I^w)^{[q]}$ by Lemma 10.9. This shows $x \in (I^w)^*$. □

In particular, if $n = w = 1$, since $\overline{I} \subseteq I^*$, we have the following.

Proposition 10.10 *If $x \in A^\circ$ and if $I = (x)$, we have $(x)^* = \overline{(x)}$.*

As a corollary of this proposition we get the following important result.

Corollary 10.11 *F-regular rings are normal.*

Proof Assume that A is F-regular. If $x \in A$ is a non-zerodivisor of A and if for $y \in A$, y/x is integral over A, then since $y \in \overline{(x)} = (x)$, we have $y/x \in A$. Hence A is normal. □

The next theorem is called "Boutot's theorem," which is very important for the theory of rational singularities in algebraic geometry over a field of characteristic 0. Again, in the following form, the proof becomes almost trivial.

Theorem 10.12 *Let A be a pure subring of an F-regular ring B and assume that $A^\circ \subseteq B^\circ$. Then A is also F-regular.*

Proof Let I be any ideal of A. By our assumption, we have $I^* \subseteq (IB)^* = IB$ since B is F-regular. Then $I^* \subseteq (IB)^* \cap A = IB \cap A = I$, since $A \subseteq B$ is pure. □

In the next section, we will show that F-regular rings are Cohen–Macaulay. Then we get a simple and beautiful proof of the fact "a pure subring of a regular ring is Cohen–Macaulay."

10.2 Colon Capturing and F-Rational Rings

Let $(A, \mathfrak{m})$ be a Noetherian local ring and $(x_1, \ldots, x_d)$ be a system of parameters of A. Then it is very important to know how much bigger $(x_1, \ldots, x_i) : x_{i+1}$ is compared to $(x_1, \ldots, x_i)$. For example, A is Cohen–Macaulay if and only if $(x_1, \ldots, x_i) : x_{i+1} = (x_1, \ldots, x_i)$ for every (some) system of parameters $(x_1, \ldots, x_d)$.

We will show that $(x_1, \ldots, x_i) : x_{i+1} \subseteq (x_1, \ldots, x_i)^*$. We call this property of a system of parameters the *colon capturing* property. Then we define an *F-rational* ring to be a ring whose parameter ideals are tightly closed. Then we show that F-rational rings are normal and Cohen–Macaulay. Also, we will show later that F-rational rings are closely related to the notion of *rational singularities* in characteristic 0. Let us begin with "colon capturing" of a system of parameters.

Theorem 10.13 *Let $R \subseteq A$ be an extension of rings and assume that R is a regular domain and A is finitely generated as an R-module. Then for all ideals I, J of R, we have*

$$IA : JA \subseteq ((I : J)A)^*.$$

In particular, if $(x_1, \ldots, x_d)$ *is a system of parameters of* R*, then* $(x_1, \ldots, x_i)A : x_{i+1} \subseteq ((x_1, \ldots, x_i)A)^*$ *for every* i.

Proof Let K be the field of fractions of R and assume $A \otimes_R K \cong K^r$. Then take a basis $\{a_1, \ldots, a_r\}$ of $A \otimes_R K$ from A.

Then let L be the free R submodule of A generated by $\{a_1, \ldots, a_r\}$. Then, since A is finitely generated as an R-module, we can take $c \in R, c \neq 0$ such that $cA \subseteq L$. Also, since the Frobenius map on R is flat, we have $I^{[q]} : J^{[q]} = [I : J]^{[q]}$.

Now, let $x \in A, x \in IA : JA$. Since $xJ \subseteq IA$, we have $x^q J^{[q]} \subseteq I^{[q]}A$. If we multiply both sides by c, we get $cx^q J^{[q]} \subseteq I^{[q]}L$. Since L is a free R-module, we have $cx^q \in (I^{[q]}L : J^{[q]}) = (I : J)^{[q]}L \subseteq ((I : J)A)^{[q]}$ and we get the claim of the theorem. □

We will give a proof of "colon capturing" in the case where A is a homomorphic image of a Cohen–Macaulay local ring.

Theorem 10.14 *Assume that a Noetherian local ring* A *is a homomorphic image of a Cohen–Macaulay local ring* $(B, \mathfrak{n})$ *and if* $\dim A = \dim A/\mathfrak{p}$ *holds for every minimal prime ideal* $\mathfrak{p}$ *of* A*, then for every system of parameters* $(x_1, \ldots, x_d)$ *of* A *and for every* i, $0 \leq i \leq d-1$*, we have* $(x_1, \ldots, x_i) : x_{i+1} \subseteq (x_1, \ldots, x_i)^*$.

Proof Let $A = B/\mathfrak{a}$ and assume that $\mathrm{ht}(\mathfrak{a}) = r$. Take $y_1, \ldots, y_d \in B$ so that $x_i = y_i \bmod \mathfrak{a}$. Also, let $\mathfrak{a} = \mathfrak{q}_1 \cap \ldots \cap \mathfrak{q}_s$ be the primary decomposition of $\mathfrak{a}$ in B. We put $\sqrt{\mathfrak{q}_i} = \mathfrak{p}_i$ $(i = 1, \ldots, s)$. Since $(y_1, \ldots, y_d) + \mathfrak{a}$ is an $\mathfrak{n}$ primary ideal, we can take $(z_1, \ldots, z_r) \subseteq \mathfrak{a}$ so that $(y_1, \ldots, y_d, z_1, \ldots, z_r)$ is a system of parameters of B.

Since B is Cohen–Macaulay, $(y_1, \ldots, y_d, z_1, \ldots, z_r)$ forms a regular sequence.

Let $(z_1, \ldots, z_r) = \mathfrak{b} \cap \mathfrak{q}'_1 \cap \ldots \cap \mathfrak{q}'_t$ be the primary decomposition of $(z_1, \ldots, z_r)$. Here we assume that $\mathrm{Ass}_B(B/\mathfrak{b}) = \{\mathfrak{p}_1, \ldots, \mathfrak{p}_s\}$ and $\mathfrak{q}'_i$ $(1 \leq i \leq t)$ are the primary ideals belonging to the other prime ideals. Then take $d \in \mathfrak{q}'_1 \cap \ldots \cap \mathfrak{q}'_t$ with $d \notin \mathfrak{p}_i$ $(1 \leq i \leq s)$. Then the image $c \in A$ of d is in A°. Also note that there exists n such that $c\mathfrak{a}^n \subseteq (z_1, \ldots, z_r)$.

Now, let $u \in (x_1, \ldots, x_i) : x_{i+1}$ and put $ux_{i+1} = \sum_{j=1}^{i} a_j x_j$. Choose $v, b_j \in B$ $(1 \leq j \leq i)$ so that the image of v (resp. b_j) in A is u (resp. a_j). Then we have

$$vy_{i+1} - \sum_{j=1}^{i} b_j y_j \in \mathfrak{a}.$$

Then for a fixed $q' = p^{e'}$ we have $c\mathfrak{a}^{q'} \in (z_1, \ldots, z_r)$ and then for all $q = p^e, e \geq e'$, we have

$$d(vy_{i+1})^q \in (y_1^q, \ldots, y_i^q, z_1, \ldots, z_r).$$

Since $(y_1^q, \ldots, y_d^q, z_1, \ldots, z_r)$ is a regular sequence in B, we have

$$dv^q \in (y_1^q, \dots, y_i^q, z_1, \dots, z_r),$$

which means $cu^q \in (x_1, \dots, x_i)^{[q]}$ and this shows that $u \in (x_1, \dots, x_i)^*$. □

Remark 10.3 The condition "$\dim A = \dim A/\mathfrak{p}$ for every minimal prime ideal $\mathfrak{p}$ of A" in Theorem 10.14 is necessary.

For example, let k be a field and let $A = k[[X, Y, Z]]/(XY, XZ)$. We let x, y, z be the images of X, Y, Z in A, respectively. Then $(y, x-z)$ is a system of parameters of A. In this case we have $x \in 0 : y$. But since A is reduced, we have $(0)^* = (0)$ and $0 : y \subseteq (0)^*$ does not hold. In this case, the minimal primes of A are (x) and (y, z) with $\dim A/(x) = 2 \neq \dim A/(y, z) = 1$.

In Theorem 10.14, we can assert a slightly stronger result $(x_1, \dots, x_i)^* : x_{i+1} \subseteq (x_1, \dots, x_i)^*$, which will be necessary later.

Theorem 10.15 *Let $(B, \mathfrak{n})$ be a Cohen–Macaulay local ring and $A = B/\mathfrak{a}$ as in Theorem 10.14. We also assume that $\dim A = \dim A/\mathfrak{p}$ holds for every minimal prime ideal $\mathfrak{p}$ of A. Then for every system of parameters $(x_1, \dots, x_d)$ and for every i $(0 \leq i \leq d-1)$, we have $(x_1, \dots, x_i)^* : x_{i+1} \subseteq (x_1, \dots, x_i)^*$.*

Proof We use the same notation as in the proof of Theorem 10.14. If $u \in (x_1, \dots, x_i)^* : x_{i+1}$, then for some $c' \in A^\circ$ and for every $q = p^e \gg 1$, we have $c'u^q \in [(x_1, \dots, x_i) : x_{i+1}]^{[q]}$. Hence we have $c'u^q x_{i+1}^q \in (x_1, \dots, x_i)^{[q]}$. Take $d' \in B$ whose image in A is c', then we can take $b_j \in B$ $(1 \leq j \leq i)$ so that

$$d'v^q y_{i+1}^q - \sum_{j=1}^{i} b_j y_j^q \in \mathfrak{a}.$$

Since $d\mathfrak{a}^{[q']} \subseteq (z_1, \dots, z_r)$ for a fixed $q' = p^{e'}$ *by the proof of Theorem 10.14*, we have

$$dd'^{q'} v^{qq'} y_{i+1}^{qq'} - d\sum_{j=1}^{i} b_j^{q'} y_j^{qq'} \in (z_1, \dots z_r)$$

and we have $dd'^{q'} v^{qq'} \in (y_1^{qq'}, \dots, y_i^{qq'}, z_1, \dots, z_r)$. Taking the image of this equation in A, we have

$$cc'^{q'} u^{qq'} \in (x_1^{qq'}, \dots, x_i^{qq'})$$

and hence we have $u \in (x_1, \dots, x_i)^*$. □

Let us give the definition of *F-rational* rings. The name "F-rational" means "characteristic p version of rational singularities in characteristic 0" and actually, this "equivalence" was proved by K. Smith[3] [Sm], N. Hara[4] [Hara] and V. Mehta[5]–V. Srinivas[6] [MSV].

Definition 10.16 (1) Let $(A, \mathfrak{m})$ be a Noetherian local ring containing a field of characteristic $p > 0$ and we also assume that A is a homomorphic image of a Cohen–Macaulay local ring. We say A is F-rational if A is equidimensional and for every system of parameters $(x_1, \ldots, x_d)$, we have $(x_1, \ldots, x_d)^* = (x_1, \ldots, x_d)$.

(2) If A is a Noetherian ring which is a homomorphic image of a Cohen–Macaulay ring B, we call A F-rational if for every prime ideal $\mathfrak{p}$ of A, $A_{\mathfrak{p}}$ is F-rational in the sense of (1).

F-rational rings have many nice properties.

Theorem 10.17 *Let $(A, \mathfrak{m})$ be an equidimensional local ring which is a homomorphic image of a Cohen–Macaulay local ring. Let $(x_1, \ldots, x_d)$ be a system of parameters of A, and assume that $(x_1, \ldots, x_d)^* = (x_1, \ldots, x_d)$. Then we have the following properties:*

(1) *For every $i, 0 \leq i \leq d - 1$, we have $(x_1, \ldots, x_i)^* = (x_1, \ldots, x_i)$.*
(2) *A is Cohen–Macaulay.*
(3) *$(y_1, \ldots, y_d)^* = (y_1, \ldots, y_d)$ holds for every system of parameters $(y_1, \ldots, y_d)$ of A.*
(4) *A is an integrally closed integral domain.*
(5) *(Y. Nakamura, Hochster–Huneke) For every prime ideal $\mathfrak{p}$ of A, $A_{\mathfrak{p}}$ is also an F-rational ring.*
(6) *If A is Gorenstein and if A is F-rational, then A is F-regular. Namely, for Gorenstein rings, "F-rational" and "F-regular" are equivalent conditions.*

Proof

(1) We show that $(x_1, \ldots, x_i)^* = (x_1, \ldots, x_i)$ by descending induction on i. We have the equality for $i = d$ by our assumption. Assume that $(x_1, \ldots, x_i, x_{i+1})^* = (x_1, \ldots, x_i, x_{i+1})$ and assume $u \in (x_1, \ldots, x_i)^*$. Since

$$(x_1, \ldots, x_i)^* \subseteq (x_1, \ldots, x_i, x_{i+1})^* = (x_1, \ldots, x_i, x_{i+1}),$$

we have $u \in (x_1, \ldots, x_i, x_{i+1})$. Put $u = y + vx_{i+1}, y \in (x_1, \ldots, x_i)$. Then $v \in (x_1, \ldots, x_i)^* : x_{i+1} = (x_1, \ldots, x_i)^*$ by Theorem 10.15. Hence we have

$$(x_1, \ldots, x_i)^* \subseteq (x_1, \ldots, x_i) + (x_1, \ldots, x_i)^* x_{i+1}$$

and we have $(x_1, \ldots, x_i)^* = (x_1, \ldots, x_i)$ by NAK.

[3] Karen E. Smith, 1965–.
[4] Nobuo Hara, 1961–.
[5] Vikram Bhagvandas Mehta, 1946–2014.
[6] Vasudevan Srinivas, 1958–.

(2) By (1) and "colon capturing" (Theorem 10.14), $(x_1, \ldots, x_d)$ is a regular sequence and hence A is Cohen–Macaulay.

(3) Let $(y_1, \ldots, y_d)$ be any system of parameters of A. To show $(y_1, \ldots, y_d)^* = (y_1, \ldots, y_d)$, it suffices to show that

$$(y_1, \ldots, y_d)^* \cap [(y_1, \ldots, y_d) : \mathfrak{m}] = (y_1, \ldots, y_d).$$

First, we will show that if $(x_1, \ldots, x_d)^* = (x_1, \ldots, x_d)$, then for every positive integer n, we have $(x_1^n, \ldots, x_d^n)^* = (x_1^n, \ldots, x_d^n)$. Since A is Cohen–Macaulay by (2), the natural homomorphism

$$\phi : [(x_1, \ldots, x_d) : \mathfrak{m}]/(x_1, \ldots, x_d) \to [(x_1^n, \ldots, x_d^n) : \mathfrak{m}]/(x_1^n, \ldots, x_d^n)$$

given multiplication of $(x_1 x_2 \cdots x_d)^{n-1}$ is bijective (5.92). If $u(x_1 x_2 \cdots x_d)^{n-1} \in (x_1^n, \ldots, x_d^n)^*$, then since we have $u \in (x_1, \ldots, x_d)^* = (x_1, \ldots, x_d)$, we also have that $(x_1^n, \ldots, x_d^n)^* = (x_1^n, \ldots, x_d^n)$.

Now, let $(y_1, \ldots, y_d)$ be any system of parameters. Replacing $(x_1, \ldots, x_d)$ by $(x_1^n, \ldots, x_d^n)$, we may assume $(y_1, \ldots, y_d) \supseteq (x_1, \ldots, x_d)$. Take $z \in A$ so that the multiplication by z

$$[(y_1, \ldots, y_d) : \mathfrak{m}]/(y_1, \ldots, y_d) \xrightarrow{z} [(x_1, \ldots, x_d) : \mathfrak{m}]/(x_1, \ldots, x_d)$$

is bijective. Then note that

$$[(y_1, \ldots, y_d)^{[q]} : \mathfrak{m}]/(y_1, \ldots, y_d)^{[q]} \xrightarrow{z^q} [(x_1, \ldots, x_d)^{[q]} : \mathfrak{m}]/(x_1, \ldots, x_d)^{[q]}$$

is also a bijection.

Now take $u \in [(y_1, \ldots, y_d) : \mathfrak{m}] \cap (y_1, \ldots, y_d)^*$ and assume that $cu^q \in (y_1, \ldots, y_d)^{[q]}$ for some $c \in A^\circ$. Then since $c(zu)^q \in (x_1, \ldots, x_d)^{[q]}$, we have $zu \in (x_1, \ldots, x_d)^* = (x_1, \ldots, x_d)$. This shows that $u \in (y_1, \ldots, y_d)$ as desired.

(4) For every $x \in A^\circ$, we can find a system of parameters $(x_1, \ldots, x_d)$ with $x = x_1$. Then by (1) and (3), we show that $(x)^* = (x)$. Then A is a normal domain in the same manner as Corollary 10.11 since $\overline{(x)} = (x)^*$ by Proposition 10.10.

(5) By descending induction on $\operatorname{ht} \mathfrak{p}$, it suffices to assume that $\operatorname{ht} \mathfrak{p} = d - 1$. Then for some element $f \in A \setminus \mathfrak{p}$, $\mathfrak{p}$ is a maximal ideal of A_f. If $(f_1, \ldots, f_{d-1})$ is a system of parameters of $A_\mathfrak{p}$, then $(f_1, \ldots, f_{d-1}, f)$ is a system of parameters of A.

It suffices to show $((f_1, \ldots, f_{d-1})A_f)^* = (f_1, \ldots, f_{d-1})^* A_f$. If $x \in A$ and $x/1 \in ((f_1, \ldots, f_{d-1})A_f)^*$, then there exists $c \neq 0$ such that $cf^n x^q \in (f_1, \ldots, f_{d-1})^{[q]}$ ($\forall q \gg 1$). Since $(f_1, \ldots, f_{d-1}, f)$ is a system of parameters of A and since A is Cohen–Macaulay, we have $\forall q \gg 1, cx^q \in (f_1, \ldots, f_{d-1})^{[q]}$ and then $x \in (f_1, \ldots, f_{d-1})^* = (f_1, \ldots, f_{d-1})$.

(6) It suffices to show that $I^* = I$ for every $\mathfrak{m}$ primary ideal I. Since we have $I = J : (J : I)$ for every system of parameters $J \subseteq I$ (Proposition 6.10), we have $I^* = I$ by Lemma 10.18 below.

□

Lemma 10.18 *Let I be an ideal of A. If $I^* = I$, then for every ideal $\mathfrak{a}$ of A, we have $[I : \mathfrak{a}]^* = [I : \mathfrak{a}]$.*

Proof If $\mathfrak{a} = (a_1, \ldots, a_n)$, then $[I : \mathfrak{a}] = \cap_{i=1}^{n}[I : (a_i)]$. Hence we can assume that $\mathfrak{a} = (a)$. If $u \in [I : (a)]^*$ and if $cu^q \in (I : (a))^{[q]}$, then $cu^q = bv^q$ for some $v \in [I : (a)]$ and $b \in A$. Then we have $c(ua)^q \in I^{[q]}$ and we have $ua \in I^* = I$. □

Remark 10.4 Let A be F-rational and *not* F-regular. We will construct an ideal for which $I^* \neq I$.

The notion of F-injective rings is a weaker version of F-rational rings.

Definition 10.19 Let $(A, \mathfrak{m})$ be a local ring of characteristic $p > 0$ and let $J = (x_1, \ldots, x_d)$ be a system of parameters of A. We say that A is F-injective if the multiplication by $(x_1 \cdots x_d)^{q-1} : A/J \to A/J^{[q]}$ is injective. (This condition is equivalent to saying that the Frobenius morphism on $\mathrm{H}^d_{\mathfrak{m}}(A)$ is injective (see Sect. 10.3).)

Exercise 10.2 (Fedder) Let $(A, \mathfrak{m})$ be a local ring of characteristic $p > 0$ and let $x \in \mathfrak{m}$ be a non-zerodivisor of A. Assume that A/xA is Cohen–Macaulay, normal, F- injective and that A_x is F-rational. Then show that A is F-rational.

In Exercises 10.3 and 10.4, assume that k is a field of characteristic $p > 0$.

Exercise 10.3 (Huneke) Let x, T be variables over k and put

$$R = k[T, xT^4, x^{-1}T^4, (x-1)^{-1}T^4].$$

Then show the following statements:

(1) R is F-rational.
(2) R is not F -regular.
(3) Put $I = (xT^4, x^{-1}T^4)$. Then show that $T^7 \notin I$ and $T^7 \in (xT^4, x^{-1}T^4)^*$.

Exercise 10.4 ([Wa6]) Let $A = k[[X, Y, Z]]/(X^2 + Y^3 + Z^4)$ and $\mathfrak{m}$ be the maximal ideal of A. Let $R = A[\mathfrak{m}T] \subseteq A[T]$ be the Rees algebra of $\mathfrak{m}$ over A. Then show the following statements:

(1) Show that for every prime p, R is F-rational.
(2) Show that A is F-rational for $p \geq 5$ and not F-rational for $p = 2, 3$.
(3) Show that A is a pure subring of R. Consequently, the statement "If A is a pure subring of B and if B is F-rational, then so is A" (the "F-rational version" of Boutot's theorem for rational singularities) does not hold true.

10.3 Tight Closure of Modules, F-Pure Rings

Let A be a ring of characteristic $p > 0$. We define Frobenius functors F^e on the category of A-modules, and in a fixed A-module M, we define tight closure N_M^* of a submodule N of M. In particular, the tight closure of submodules of local cohomology module $\mathrm{H}_{\mathfrak{m}}^i(A)$ and injective envelope $E_A(A/\mathfrak{m})$ are very important.

This concept is applied to theory of complexes over the ring like the "phantom homology" theorem.

In this section, the rings are Noetherian rings of characteristic $p > 0$ and we put $q = p^e$.

Definition 10.20

(1) We write ${}^eA$ the A-module A whose A-module structure is defined by

$$a.x = a^q x$$

for $a \in A$ and $x \in {}^eA$.

(2) For an A-module M, we define $F^e(M) = M \otimes_A {}^eA$. For an A-module homomorphism $u : M \to M'$, we define $F^e(u) : F^e(M) \to F^e(M')$ by $F^e(u)(x \otimes 1) = u(x) \otimes 1$.

(3) For an A submodule N of M, we define

$$N^{[q]} = \mathrm{Im}(F^e(i) : F^e(N) \to F^e(M)),$$

where $i : N \to M$ is the inclusion map of N in M. Since F^e is not an exact functor, note that $F^e(i)$ is not injective in general. In particular, if $N = I$ is an ideal of $A = M$, $N^{[q]} = I^{[q]}$ coincides with Definition 10.1.

Example 10.21 Let M be a finitely generated A-module.

(1) Suppose M is represented by the following exact sequence ((a_{ij}) is an $m \times n$ matrix):

$$A^m \xrightarrow{(a_{ij})} A^n \to M \to 0.$$

Tensoring ${}^eA$ to this exact sequence (put $q = p^e$) we have the exact sequence

$$A^m \otimes_A {}^eA \xrightarrow{(a_{ij}^q)} A^n \otimes_A {}^eA \to F^e(M) \to 0.$$

(2) In particular, for an ideal I of A, $F^e(A/I) \cong A/I^{[q]}$.

Let us note that the following theorem follows from the theorem of Buchsbaum and Eisenbud, Theorem 5.115.

Theorem 10.22 *Let*

$$G_\bullet : 0 \to G_n \overset{\varphi_n}{\to} G_{n-1} \overset{\varphi_{n-1}}{\to} G_{n-2} \to \cdots \to G_1 \overset{\varphi_1}{\to} G_0$$

be a complex of finitely generated free A-modules. Then G is acyclic if and only if so is $F^e(G_\bullet)$. In particular, if $\mathrm{hd}_A\, I < \infty$ if and only if $\mathrm{hd}_A\, I^{[q]} < \infty$ for every $q = p^e$.

Proof Condition (2) of Theorem 5.115 for $G_\bullet$ and $F^e(G_\bullet)$ are clearly equivalent. □

Let us define the tight closure of a submodule.

Definition 10.23 Let M be an A-module and N be a submodule of M.

(1) Let $x \in M$. We denote by x^q the image of x by the Frobenius morphism $F^e(x) = x \otimes 1$. Then we have equalities

$$(x + y)^q = x^q + y^q, \text{ and } (ax)^q = a^q x^q$$

for $x, y \in M$ and $a \in A$.

(2) We define the *tight closure* N^*_M of N in M by

$$x \in N^*_M \iff \exists c \in A^\circ, \forall q \gg 1, cx^q \in N^{[q]}.$$

N^*_M is an A submodule of M containing N.

Remark 10.5 For $x \in M$, $x \in N^*_M$ if and only if the image of x in M/N is an element of $(0)^*_{M/N}$. Namely, we have $N^*_M/N \cong (0)^*_{M/N}$.

In the applications of tight closure of modules, the tight closures in local cohomology modules and injective envelopes are most typical.

Example 10.24 Let $(A, \mathfrak{m})$ be a local ring with $\dim A = d$ and let $\underline{x} = (x_1, \ldots, x_d)$ be a system of parameters of A.

(1) By Theorem 5.105, recall that $\mathrm{H}^d_\mathfrak{m}(A) = \mathrm{H}^d_{\underline{x}}(A) \cong A_{x_1 x_2 \cdots x_d}/\mathrm{Im}[C^{d-1} \to C^d]$. Hence we have $F^e(\mathrm{H}^d_\mathfrak{m}(A)) = \mathrm{H}^d_{\underline{x^q}}(A) \cong \mathrm{H}^d_\mathfrak{m}(A)$.

If we write by $[\frac{s}{(x_1 x_2 \cdots x_d)}]$ the image of $\frac{s}{(x_1 x_2 \cdots x_d)} \in C^d$ in $\mathrm{H}^d_{\underline{x}}(A)$, then for $\xi = [\frac{s}{(x_1 x_2 \cdots x_d)}]$, we have $\xi^q = [\frac{s^q}{(x_1 x_2 \cdots x_d)^q}]$.

(2) If A is Cohen–Macaulay, A is F rational if and only if $(0)^*_{\mathrm{H}^d_\mathfrak{m}(A)} = 0$. In fact, for $\xi = [\frac{s}{(x_1 x_2 \cdots x_d)}]$ and $c \in A^\circ$, $c\xi^q = 0$ in $\mathrm{H}^d_\mathfrak{m}(A)$ is equivalent to saying that $cs^q \in (\underline{x})^{[q]}$. Hence $\xi \in (0)^*_{\mathrm{H}^d_\mathfrak{m}(A)} \Leftrightarrow s \in (\underline{x})^*$.

(3) (K. Smith) Let $(A, \mathfrak{m})$ be a complete normal local ring. Then A is F-rational if and only if $\mathrm{H}^d_\mathfrak{m}(A)$ has no proper submodule closed under the Frobenius

action. (We say that an A submodule N of $\mathrm{H}^d_{\mathfrak{m}}(A)$ is closed under the action of Frobenius if for every $x \in N$ we have $x^q \in N$.) Actually, if N is a proper A submodule of $\mathrm{H}^d_{\mathfrak{m}}(A)$, the Matlis dual $N^\vee$ of N is a proper quotient of K_A. Then we can take $c \in A^\circ$ such that $cN^\vee = 0$, which implies that $\forall x \in N, cx^q = 0$ and then $N \subseteq (0)^*_{\mathrm{H}^d_{\mathfrak{m}}(A)}$.

The action of Frobenius endomorphism characterizes F-regular rings.

Theorem 10.25 *Assume that a Noetherian local ring* $(A, \mathfrak{m})$, *satisfies the condition* $(*)$ *For every* $n > 0$, *there exists an irreducible* $\mathfrak{m}$*-primary ideal* $J \subseteq \mathfrak{m}^n$.
Then the following conditions on A are equivalent:

(1) *A is weakly F-regular.*
(2) *For every finitely generated A module M and for any submodule N of M, we have* $N^*_M = N$.
(3) *Let* $E = E_A(A/\mathfrak{m})$ *be the injective envelope of* $A/\mathfrak{m}$ *as A module. Then we have* $(0)^{*fg}_E = 0$, *where* $(0)^{*fg}_E$ *is the sum of* $(0)^*_M$ *for all finitely generated A submodule M of E.*

Proof We have shown in Remark 10.2 that A is weakly F-regular if for every $\mathfrak{m}$-primary ideal I, we have $I^* = I$. Since any $\mathfrak{m}$ primary ideal is an intersection of irreducible $\mathfrak{m}$-primary ideals, it suffices to assume that I is irreducible. Then there is an injection $A/I \to E_A(A/\mathfrak{m})$ and we have (3) ⇒(1). Next, we show (1) ⇒(2). Assume that $x \in N^*_M, x \notin N$. Then assuming that N is a maximal submodule with $x \notin N$, we may assume that $\mathfrak{m}x \subseteq N$ and $M/N \subseteq E$. Put $I = \mathrm{Ann}_A(M/N)$ and take an irreducible $\mathfrak{m}$ primary ideal $J \subseteq I$. Since $A/J \to A/I$ is surjective, taking the Matlis dual, $M/N \cong (A/I)^\vee \subseteq (A/J)^\vee \cong A/J$. Then we have $x \in J^*/J$, which implies A is not weakly F-regular. Thus we show (1) ⇒(2). □

Next, we will introduce the notion of F-pure rings. This concept is historically the oldest one among the notions that appear in this chapter.

Definition 10.26 A is called *F-pure*, if A is a pure subring of ${}^1A$ by $F : A \to {}^1A$.

Exercise 10.5 For a local ring $(A, \mathfrak{m})$, let $E = E_A(A/\mathfrak{m})$ be the injective envelope of $A/\mathfrak{m}$. Show that A is F-pure if and only if $F : E \to E \otimes_A {}^1A$ is injective.

Remark 10.6 By Exercise 10.5, if A is F-regular, then A is F-pure. There is no implication between "F-pure" and "F-rational."

Exercise 10.6 Let R be a graded ring over a field $k = R_0$ of characteristic $p > 0$.

(1) Show that a face ring $k[\Delta]$ is F-pure.
(2) Show that if R is F-pure (resp. F-rational), then $a(R) \leq 0$ (resp. $a(R) < 0$).
(3) Moreover, assume that k is algebraically closed. If $\dim R = 1$ and R is F-pure, then R is isomorphic to a face ring (see also [GW2]).

Exercise 10.7 (Fedder's criterion) Let $(S, \mathfrak{n})$ be a regular local ring and $A = S/I$.

(1) If we put $E = E_S(S/\mathfrak{n})$, then show that $E_A(A/\mathfrak{m}) = [0 :_E I]$ and $E_A(A/\mathfrak{m}) \otimes {}^1A \cong [0 :_E I^{[p]}]/I[0 :_E I^{[p]}]$.
(2) (Fedder's criterion) Show that A is F-pure if and only if $[I^{[p]} : I] \not\subseteq \mathfrak{n}^{[p]}$.
(3) In particular, if I is generated by a regular sequence $(f_1, \ldots, f_r)$, then A is f-pure if and only if $(f_1 f_2 \cdots f_r)^{p-1} \notin \mathfrak{n}^{[p]}$.
(4) Let $A = k[X, Y, Z]/(X^3 + Y^3 + Z^3)$ $(p \neq 3)$. Then A is F-pure if and only if $p \equiv 1 \pmod 3$. Also, show that $A = k[X, Y, Z]/(X^2 + Y^4 + Z^4)$ $(p \neq 2)$ is F-pure if and only if $p \equiv 1 \pmod 4$.

Let us introduce the notion of "phantom homology."

Definition 10.27 Let G be a complex of finitely generated A modules. Then we say G has *phantom homology* if $\forall i$, $\mathrm{Ker}(G_i \to G_{i-1}) \subseteq [\mathrm{Im}(G_{i+1} \to G_i)]^*$.

The following theorem is a generalization of Theorem 5.115 (Buchsbaum–Eisenbud) in terms of tight closure.

Theorem 10.28 (Phantom Homology) *Let* $G_\bullet : 0 \to G_n \overset{\varphi_n}{\to} G_{n-1} \overset{\varphi_{n-1}}{\to} G_{n-2} \to \cdots \to G_1 \overset{\varphi_1}{\to} G_0$ *be a complex of finitely generated free A-modules with* $G_i \neq 0$ $(1 \leq i \leq n)$. *Then the following conditions for* $G_\bullet$ *are equivalent:*

(1) $G_\bullet$ *has phantom homology.*
(2) *For every i, $1 \leq i \leq n$, the following two conditions are satisfied:*

 (a) $\mathrm{ht}(\mathrm{I}(\varphi_i)) \geq i$.
 (b) $\mathrm{rank}(\varphi_i, M) + \mathrm{rank}(\varphi_{i+1}, M) = \mathrm{rank}_A F_i$.

 Here, we put $\mathrm{rank}(\varphi_{n+1}, M) = 0$.

We have seen that the theory of positive characteristic yields arguments very quickly. If you are interested to know more of the theory, refer to the lecture notes of C. Huneke [Hu2].

Also this theory is known to be closely related to the geometric properties of singularities of algebraic varieties defined over a field of characteristic 0. For example, a rational singularity A over a field of characteristic 0 is characterized by the property "for infinitely many p, A_p, the reduction modulo p of A is F-rational" (K. Smith, N. Hara and V. Mehta–V. Srinivas). (Later, it turned out that it suffices to show that A_p is F-rational for a single p [MaSch].) Also, "log-terminal singularities" are characterized by "$\mathbb{Q}$-Gorenstein and A_p is F-regular for infinitely many p". Also, the multiplier ideals in characteristic 0 are known to be computable by a characteristic p method by N. Hara–K.-i. Yoshida and S. Takagi.

Appendix A
Direct Sums and Direct Products

Let A be a ring, and $\{M_\lambda\}_{\lambda\in\Lambda}$ be a family of A-modules. The *direct product*

$$\prod_{\lambda\in\Lambda} M_\lambda = \{(x_\lambda)_{\lambda\in\Lambda} \mid x_\lambda \in M_\lambda\}$$

of $\{M_\lambda\}_{\lambda\in\Lambda}$ is an A-module with the addition given by $(x_\lambda) + (y_\lambda) = (x_\lambda + y_\lambda)$ and the scalar multiplication given by $a(x_\lambda) = (ax_\lambda)$. We define the *direct sum* $\bigoplus_{\lambda\in\Lambda} M_\lambda$ of $\{M_\lambda\}$ by

$$\bigoplus_{\lambda\in\Lambda} M_\lambda = \{(x_\lambda)_{\lambda\in\Lambda} \in \prod_{\lambda\in\Lambda} M_\lambda \mid \#\{\lambda \in \Lambda \mid x_\lambda \neq 0\} < \infty\}.$$

Then the direct sum $\bigoplus\limits_{\lambda\in\Lambda} M_\lambda$ is an A-submodule of $\prod\limits_{\lambda\in\Lambda} M_\lambda$. If the index set Λ is finite, then $\prod\limits_{\lambda\in\Lambda} M_\lambda = \bigoplus\limits_{\lambda\in\Lambda} M_\lambda$, but they do not agree in general.

If the index set Λ is empty, then we consider that there is a unique family $\mathbb{M}_\emptyset = \{M_\lambda\}_{\lambda\in\emptyset}$ of A-modules indexed by the empty set, and $\bigoplus_{\lambda\in\emptyset} M_\lambda = \prod_{\lambda\in\emptyset} M_\lambda = 0$, as can be seen easily from the definition.

Note that for each $\lambda \in \Lambda$, $p_\lambda : \prod_{\lambda\in\Lambda} M_\lambda \to M_\lambda$ given by $(x_\lambda)_{\lambda\in\Lambda} \mapsto x_\lambda$ is a surjective A-linear map. Also, $\iota_\lambda : M_\lambda \to \bigoplus_{\lambda\in\Lambda} M_\lambda$ given by $\iota_\lambda(x) = (y_\lambda)$, where $y_\lambda = x$, and $y_\mu = 0$ for $\mu \neq \lambda$, is an injective A-linear map. Direct products and direct sums have the following *universality*.

Proposition A.1 *The following hold:*

(1) *For an A-module X, the map*

$$\Phi : \mathrm{Hom}_A(\bigoplus_{\lambda\in\Lambda} M_\lambda, X) \to \prod_{\lambda\in\Lambda} \mathrm{Hom}_A(M_\lambda, X)$$

S. Goto, K. Watanabe, *Commutative Algebra*, Graduate Texts in Mathematics 309,
https://doi.org/10.1007/978-981-95-4227-7

given by $\Phi(\psi) = (\psi \circ \iota_\lambda)_{\lambda \in \Lambda}$ *is an isomorphism of A-modules.*

(2) *For an A-module X, the map*

$$\Psi : \mathrm{Hom}_A(X, \prod_{\lambda \in \Lambda} M_\lambda) \to \prod_{\lambda \in \Lambda} \mathrm{Hom}_A(X, M_\lambda)$$

given by $\Psi(\varphi) = (p_\lambda \circ \varphi)_{\lambda \in \Lambda}$ *is an isomorphism of A-modules.*

Proof

(1) The inverse Φ^{-1} is given by $\Phi^{-1}(\psi_\lambda)((x_\lambda)) = \sum_{\lambda \in \Lambda} \psi_\lambda(x_\lambda)$ (the sum is a finite sum, since $(x_\lambda) \in \bigoplus_\lambda M_\lambda$), and Φ is an isomorphism.
(2) Ψ^{-1} is given by $\Psi^{-1}((\varphi_\lambda))(x) = ((\varphi_\lambda(x)))$, and Ψ is an isomorphism.

□

A.2 Let M be an A-module and $\{N_\lambda\}_{\lambda \in \Lambda}$ be its family of A-submodules. Then the A-linear map $\varphi = \Phi^{-1}((j_\lambda)) : \bigoplus_{\lambda \in \Lambda} N_\lambda \to M$ given by $\varphi((x_\lambda)) = \sum_\lambda x_\lambda \in M$, where $j_\lambda : N_\lambda \to M$ is the inclusion, and Φ is as in the proposition. Then we have $\mathrm{Im}\,\varphi = \sum_\lambda N_\lambda$. So if φ is injective, then $\varphi : \bigoplus_{\lambda \in \Lambda} N_\lambda \to \sum_{\lambda \in \Lambda} N_\lambda$ is an isomorphism. With this reasoning, we write the submodule $\sum_{\lambda \in \Lambda} N_\lambda$ by the symbol $\bigoplus_{\lambda \in \Lambda} N_\lambda$, if φ is injective. Note that φ is injective if and only if $\sum_{\lambda \in \Lambda} x_\lambda = 0$ in M $(x_\lambda \in N_\lambda)$ implies $x_\lambda = 0$ for all λ. This is equivalent to saying that $N_\lambda \cap \sum_{\mu \neq \lambda} N_\mu = 0$ for any $\lambda \in \Lambda$. If Λ is totally ordered, then yet another equivalent condition is, $N_\lambda \cap \sum_{\mu < \lambda} N_\mu = 0$ for any $\lambda \in \Lambda$. If φ is an isomorphism, then $M = \bigoplus_{\lambda \in \Lambda} N_\lambda$, and then we say that M is the direct sum of the submodules N_λ. In order to emphasize that the direct sum is taken in a module, we say that the sum is an internal direct sum. For example, if $\Lambda = \{1, 2\}$, then the sum $N_1 + N_2 \subseteq M$ is an internal direct sum if and only if $N_1 \cap N_2 = 0$. Indeed, the sequence

$$0 \to N_1 \cap N_2 \xrightarrow{\psi} N_1 \oplus N_2 \xrightarrow{\varphi} N_1 + N_2 \to 0$$

is exact, where $\psi(n) = (n, -n)$, and $\varphi(n_1, n_2) = n_1 + n_2$.

Let $\{M_\lambda\}_{\lambda \in \Lambda}$, and $\{N_\lambda\}_{\lambda \in \Lambda}$ be families of A-modules indexed by the same index set Λ, and let $\{\varphi_\lambda : M_\lambda \to N_\lambda\}$ be a family of A-linear maps. Then we define $\prod_\lambda \varphi_\lambda : \prod_\lambda M_\lambda \to \prod_{\lambda \in \Lambda} N_\lambda$ by $(\prod_\lambda \varphi_\lambda)((x_\lambda)_{\lambda \in \Lambda}) = (\varphi_\lambda(x_\lambda))$. Its restriction $\bigoplus_{\lambda \in \Lambda} M_\lambda \to \bigoplus_{\lambda \in \Lambda} N_\lambda$ is denoted by $\bigoplus_{\lambda \in \Lambda} \varphi_\lambda$.

Theorem A.3 *Let* $\{L_\lambda\}_{\lambda \in \Lambda}$, $\{M_\lambda\}_{\lambda \in \Lambda}$, *and* $\{N_\lambda\}_{\lambda \in \Lambda}$ *be families of A-modules indexed by the same index set* Λ. *Let* $\{\varphi_\lambda : L_\lambda \to M_\lambda\}$ *and* $\{\psi_\lambda : M_\lambda \to N_\lambda\}$ *be families of A-linear maps such that*

$$L_\lambda \xrightarrow{\varphi_\lambda} M_\lambda \xrightarrow{\psi_\lambda} N_\lambda$$

is exact for any $\lambda \in \Lambda$. *Then the sequences*

$$\prod_{\lambda\in\Lambda} L_\lambda \xrightarrow{\prod_\lambda \varphi_\lambda} \prod_{\lambda\in\Lambda} M_\lambda \xrightarrow{\prod_\lambda \psi_\lambda} \prod_{\lambda\in\Lambda} N_\lambda \tag{A.1}$$

and

$$\bigoplus_{\lambda\in\Lambda} L_\lambda \xrightarrow{\bigoplus_\lambda \varphi_\lambda} \bigoplus_{\lambda\in\Lambda} M_\lambda \xrightarrow{\bigoplus_\lambda \psi_\lambda} \bigoplus_{\lambda\in\Lambda} N_\lambda \tag{A.2}$$

are exact.

Proof Let (x_λ) be an element of $\operatorname{Ker}(\prod_\lambda \psi_\lambda)$. Then $\psi_\lambda(x_\lambda) = 0$ for all λ. So for each λ, there exists some $y_\lambda \in L_\lambda$ such that $\varphi_\lambda(y_\lambda) = x_\lambda$. Then $(x_\lambda) = (\prod_\lambda \varphi_\lambda)((y_\lambda))$, and hence (A.1) is exact. If, moreover, $S = \{\lambda \in \Lambda \mid x_\lambda \neq 0\}$ is finite and hence $(x_\lambda) \in \bigoplus_\lambda M_\lambda$, then replacing y_λ by 0 if $\lambda \in \Lambda \setminus S$, and so we can take (y_λ) from $\bigoplus_{\lambda\in\Lambda} L_\lambda$ this case, and the exactness of (A.2) has also been proved. □

Appendix B
Inductive and Projective Limits

B.1 Inductive Systems and Inductive Limits

Let A be a commutative ring, and I be an ordered set. We set $T = \{(i, j) \mid i, j \in I, i \leq j\}$. A pair $(\{M_i\}_{i \in I}, \{f_{ji} : M_i \to M_j\}_{(i,j) \in T})$ is called an *inductive system* of A-modules on I if M_i is an A-module for each $i \in I$, $f_{ji} \in \operatorname{Hom}_A(M_i, M_j)$ for each $(i, j) \in T$, and the conditions

(1) for each $i \in I$, we have $f_{ii} = 1_{M_i}$;
(2) for each $i, j, k \in I$ such that $i \leq j \leq k$, we have $f_{ki} = f_{kj} \circ f_{ji}$

are satisfied. Sometimes $\{f_{ji}\}$ is omitted and we say that $\{M_i\}_{i \in I}$ is an inductive system.

In what follows, let $\mathbb{M} = \{M_i\}_{i \in I}$ be an inductive system of A-modules on I, and we set $L = \bigoplus_{i \in I} M_i$. Let $\xi_i : M_i \to L$ be the canonical inclusion for each $i \in I$, and we denote the A-submodule of L generated by the set

$$\{\xi_j(f_{ji}(x)) - \xi_i(x) \mid (i, j) \in T\} \subseteq L$$

by N. We set $M = L/N$. For each $i \in I$, we set $f_i : M_i \xrightarrow{\xi_i} L \xrightarrow{\varepsilon} M$, where $\varepsilon : L \to M$ is the canonical projection.

Proposition B.1 *If $(i, j) \in T$, then $f_i = f_j \circ f_{ji}$. Or equivalently, the diagram*

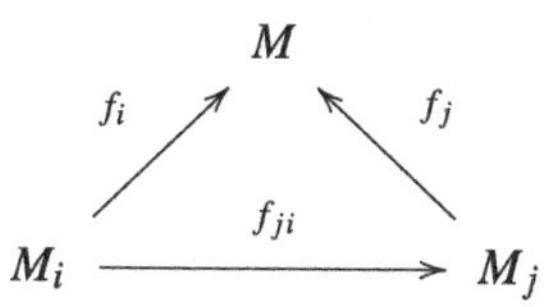

is commutative.

S. Goto, K. Watanabe, *Commutative Algebra*, Graduate Texts in Mathematics 309,
https://doi.org/10.1007/978-981-95-4227-7

Proof For $(i, j) \in T$ and $x \in M_i$, we have that $\xi_j(f_{ji}(x)) - \xi_i(x) \in N$. Applying ε to it, we get the result. □

Theorem B.2 *Let P be an A-module, and $\{g_i : M_i \to P\}_{i \in I}$ be a family of A-linear maps such that the diagram*

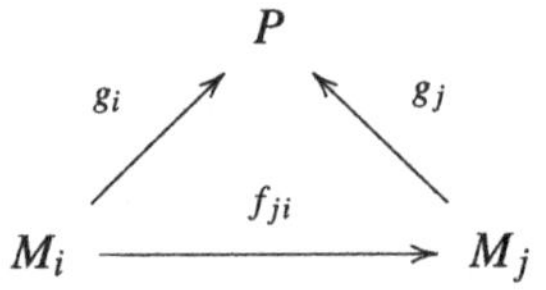

is commutative for each $(i, j) \in T$. Then there exists a unique $h \in \mathrm{Hom}_A(M, P)$ such that $g_i = hf_i$ for each i.

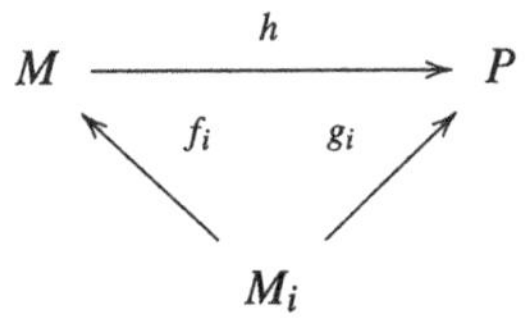

Proof Let $\tilde{h} : L = \bigoplus_{i \in I} M_i \to P$ be the A-linear map such that $\tilde{h}\xi_i = g_i$ for each i (it uniquely exists by Proposition A.1 (as the map $\Phi^{-1}((g_i)_{i \in I})$)). If $(i, j) \in T$ and $x \in M_i$, then $\tilde{h}(\xi_j(f_{ji}(x)) - \xi_i(x)) = g_j f_{ji} x - g_i x = 0$ by assumption. Hence $\tilde{h}(N) = 0$. By the homomorphism theorem, there exists an unique $h : M = L/N \to P$ such that $h\varepsilon = \tilde{h}$. So $g_i = \tilde{h}\xi_i = h\varepsilon\xi_i = hf_i$ for each $i \in I$. Conversely, if $h : M \to P$ satisfies $hf_i = g_i$ for each $i \in I$, then defining $\tilde{h} = h\varepsilon$, we have $g_i = hf_i = h\varepsilon\xi_i = \tilde{h}\xi_i$ for each i. Such an $\tilde{h}$ is unique as above. Then h is also unique, and we are done. □

Exercise B.1 *Prove that the module M is uniquely determined by the property as stated in Theorem B.2 up to isomorphisms.*

We call M the *inductive limit* of $\{M_i\}_{i \in I}$ and denote it by $\varinjlim M_i$.

Definition B.3 We say that the ordered set I is a *directed set* if $I \neq \emptyset$, and for any $i, j \in I$, there exists some $k \in I$ such that $i \leq k$ and $j \leq k$. Note that an ordered set I is directed if and only if any finite subset of I has an upper bound.

Proposition B.4 *Let I be a directed set. Then the following hold:*

(1) $M = \bigcup_{i \in I} \mathrm{Im}\, f_i$.
(2) *If $i \in I$, $x \in M_i$, and $f_i(x) = 0$, then there exists some $j \in I$ such that $i \leq j$ and $f_{ji}(x) = 0$.*

Proof

(1) For any $x \in M$, there exists some finite subset J of I and an element $\{x_j\}_{j \in J} \in \bigoplus_{j \in J} x_j$ such that $x = \sum_{j \in J} f_j(x_j)$. Then we can take an $i \in I$ such that $j \leq i$ for each $j \in J$. Since

$$x = \sum_{j \in J} f_i(f_{ij}(x_j)) = f_i(\sum_{j \in J} f_{ij}(x_j)),$$

we have $x \in \operatorname{Im} f_i$.

(2) Since $f_i(x) = 0$, we can express $x = \sum_{s=1}^{r}(x_s - f_{j_s i_s}(x_s))$, where $(i_s, j_s) \in T$ and $x_s \in M_{i_s}$ for $s = 1, \ldots, r$. Take $j \in I$ such that $j \geq i$, $j \geq j_s$ ($s = 1, \ldots, r$). Then we define $h : \bigoplus_{u \in J} M_u \to M_j$ by $h((m_u)) = \sum_u f_{ju}(m_u)$, where $J = \{u \in I \mid u \leq j\}$. So

$$f_{ji}(x) = h(x) = \sum_s h(x_s - f_{j_s i_s}(x_s)) = \sum_s (f_{ji_s}(x_s) - f_{jj_s} f_{j_s i_s}(x_s)) = 0.$$

□

We say that $J \subseteq I$ is *cofinal* (or *final*) subset of I if for any $i \in I$, there exists some $j \in J$ such that $i \leq j$.

Proposition B.5 *Let I be a directed set, and J be its cofinal subset. Then $\rho : \varinjlim_{j \in J} M_j \to \varinjlim_{i \in I} M_i$ is an isomorphism, where $f'_j : M_j \to \varinjlim_{j \in J} M_j$ is the canonical map, and ρ is the unique map coming from the universality (the map such that $\rho f'_j = f_j$ for any $j \in J$).*

Proof If $\alpha \in \varinjlim_{i \in I} M_i$, then we can find some $i \in I$ and $x \in M_i$ such that $\alpha = f_i(x)$ by Proposition B.4. Then we can find $j \in J$ such that $j \geq i$. Hence $\alpha = \rho(f'_j f_{ji}(x))$, and ρ is surjective.

Let $\beta \in \operatorname{Ker} \rho$. We can find some $j \in J$ and $x \in M_j$ such that $f'_j(x) = \beta$. Then $f_j(x) = \rho(\beta) = 0$, and hence by Proposition B.4, we can find some $i \in I$ such that $f_{ij}(x) = 0$. Taking $j' \in J$ such that $j' \geq i$, we have $f_{j'j}(x) = f_{j'i}(f_{ij}(x)) = 0$. So $\beta = f'_j(x) = f'_{j'}(f_{j'j}(x)) = f'_{j'}(0) = 0$, and hence ρ is injective. □

Let $\mathbb{L} = \{L_i\}_{i \in I}$ and $\mathbb{M} = \{M_i\}_{i \in I}$ be inductive systems of A-modules on an ordered set I. We say that a family of A-linear maps $\boldsymbol{\alpha} = \{\alpha_i : L_i \to M_i\}_{i \in I}$ is a *morphism of inductive systems* $\boldsymbol{\alpha} : \mathbb{L} \to \mathbb{M}$ if

$$\begin{array}{ccc} L_i & \xrightarrow{\alpha_i} & M_i \\ {\scriptstyle f^L_{ji}}\downarrow & & \downarrow{\scriptstyle f^M_{ji}} \\ L_j & \xrightarrow{\alpha_j} & M_j \end{array}$$

is commutative for any $(i, j) \in T$. If $\{\alpha_i : L_i \to M_i\}_{i \in I} : \mathbb{L} \to \mathbb{M}$ is a morphism of inductive systems, then by Theorem B.2, There exists a unique A-linear map $\alpha : \varinjlim L_i \to \varinjlim M_i$ such that the diagram

$$\begin{array}{ccc} L_i & \xrightarrow{\alpha_i} & M_i \\ {\scriptstyle f_i^L}\downarrow & & \downarrow{\scriptstyle f_i^M} \\ \varinjlim L_i & \xrightarrow{\alpha} & \varinjlim M_i \end{array}$$

is commutative for any $i \in I$. We denote this map α by $\varinjlim \alpha_i$.

Theorem B.6 *Let I be a directed set, and $\mathbb{L} = \{L_i\}_{i\in I}$, $\mathbb{M} = \{M_i\}_{i\in I}$, and $\mathbb{N} = \{N_i\}_{i\in I}$ be inductive systems of A-modules on I. Let $\mathbb{L} \xrightarrow{\{\alpha_i\}_{i\in I}} \mathbb{M} \xrightarrow{\{\beta_i\}_{i\in I}} \mathbb{N}$ be morphisms of inductive systems such that $0 \to L_i \to M_i \to N_i \to 0$ is exact for any $i \in I$. Then the sequence of A-modules*

$$0 \to \varinjlim L_i \xrightarrow{\varinjlim \alpha_i} \varinjlim M_i \xrightarrow{\varinjlim \beta_i} \varinjlim N_i \to 0$$

is exact.

Proof The diagram

$$\begin{array}{ccccccc}
\bigoplus_{(i,j)\in T} L_i & \xrightarrow{q} & \bigoplus_{i\in I} L_i & \xrightarrow{\varepsilon} & \varinjlim L_i & \longrightarrow & 0 \\
\downarrow{\scriptstyle \bigoplus_{(i,j)\in T}\varphi_i} & & \downarrow{\scriptstyle \bigoplus_{i\in I}\varphi_i} & & \downarrow{\scriptstyle \varinjlim \varphi_i} & & \\
\bigoplus_{(i,j)\in T} M_i & \xrightarrow{q} & \bigoplus_{i\in I} M_i & \xrightarrow{\varepsilon} & \varinjlim M_i & \longrightarrow & 0 \\
\downarrow{\scriptstyle \bigoplus_{(i,j)\in T}\psi_i} & & \downarrow{\scriptstyle \bigoplus_{i\in I}\varphi_i} & & \downarrow{\scriptstyle \varinjlim \varphi_i} & & \\
\bigoplus_{(i,j)\in T} N_i & \xrightarrow{q} & \bigoplus_{i\in I} N_i & \xrightarrow{\varepsilon} & \varinjlim N_i & \longrightarrow & 0 \\
\downarrow & & \downarrow & & \downarrow & & \\
0 & & 0 & & 0 & &
\end{array}$$

is commutative with exact rows, where ε is the canonical projection, and q is given by $q(\xi_{(i,j)}(x)) = \xi_i(x) - \xi_j(f_{ji}(x))$. The first and the second columns (from the left) are exact, and hence so is the third (proved easily by diagram chasing or the snake lemma). So far, we have not used the assumption that I is directed.

It only remains to show that $\varinjlim \varphi_i : \varinjlim L_i \to \varinjlim M_i$ is injective, provided each φ_i is injective. Let $\alpha \in \mathrm{Ker}(\varinjlim \varphi_i)$. Then there exists some $i \in I$ and $x \in L_i$ such that $f_i^L(x) = \alpha$. Since $f_i^M(\varphi_i(x)) = (\varinjlim \varphi_i) f_i^L(x) = (\varinjlim \varphi_i)(\alpha) = 0$, there exists some $j \geq i$ such that $f_{ji}^M(\varphi_i(x)) = 0$. So $\varphi_j(f_{ji}^L(x)) = 0$. As φ_j is injective, we have that $f_{ji}^L(x) = 0$. Hence $\alpha = f_i^L(x) = f_j^L f_{ji}^L(x) = f_j^L(0) = 0$. Hence $\varinjlim \varphi_i$ is injective, as desired. □

Exercise B.2 *Instead of A-modules, we can consider an inductive system of A-algebras $\mathbb{A} = \{A_i\}_{i\in I}$. It consists of a family of A-algebras $\{A_i\}_{i\in I}$ and a family of A-algebra maps $\{f_{ij} : A_i \to A_j\}_{i,j\in I, i\leq j}$ among them. Prove that if I is a directed set, then $\varinjlim A_i$ has a unique A-algebra structure such that $f_i : A_i \to \varinjlim A_i$ is an A-algebra map for each $i \in I$.*

Lemma B.7 *Let I be an arbitrary ordered set, $\{(M_i)\}$ an inductive system on I, and N an A-module. Then the canonical map $f_i : M_i \to \varinjlim M_i$ induces an isomorphism $\rho : \varinjlim (M_i \otimes_A N) \to \varinjlim M_i \otimes_A N$.*

Proof Note that $f_i \otimes 1_N : M_i \otimes_A N \to \varinjlim M_i \otimes_A N$ induces ρ. The diagram

$$\begin{array}{ccccccc}
\bigoplus_{(i,j)\in T}(M_i \otimes_A N) & \longrightarrow & \bigoplus_{i\in I}(M_i \otimes_A N) & \longrightarrow & \varinjlim(M_i \otimes_A N) & \longrightarrow & 0 \\
\downarrow & & \downarrow & & \downarrow \rho & & \\
(\bigoplus_{(i,j)\in T} M_i) \otimes_A N & \longrightarrow & (\bigoplus_{i\in I} M_i) \otimes_A N & \longrightarrow & (\varinjlim M_i) \otimes_A N & \longrightarrow & 0
\end{array}$$

is commutative with exact rows. By the five lemma and Proposition 1.112, (5), ρ is an isomorphism. □

B.2 Projective Limits

Let I be an ordered set. A pair $(\{M_i\}_{i\in I}, \{f_{ji} : M_j \to M_i\}_{(i,j)\in T})$ such that $\{M_i\}_{i\in I}$ is a family of A-modules and $\{f_{ij} : M_j \to M_i\}_{(i,j)\in T}$ is a family of A-linear maps is called a *projective system* of A-modules on I if the conditions

(1) for each $i \in I$, we have $f_{ii} = 1_{M_i}$;
(2) if $i, j, k \in I$ and $i \leq j \leq k$, then $f_{ik} = f_{ij} \circ f_{jk}$

are satisfied, where $T = \{(i, j) \in I^2 \mid i \leq j\}$. Note that $(\{M_i\}_{i\in I}, \{f_{ij} : M_j \to M_i\}_{(i,j)\in T})$ is a projective system on I if and only if it is an inductive system on the opposite I^{op} of I. Then letting $L = \prod_{i\in I} M_i$, the subset

$$M = \{(x_i)_{i\in I} \in L \mid \forall (i, j) \in T \ f_{ij}(x_j) = x_i\}$$

is an A-submodule of L.

Proposition B.8 *For each $i \in I$, we set $f_i : M \xrightarrow{\iota} L \xrightarrow{p_i} M_i$, where p_i is the canonical projection, and ι is the inclusion. Then for $(i, j) \in T$, we have $f_i = f_{ij} \circ f_j$.*

Theorem B.9 *Let P be an A-module, and $\{g_i : P \to M_i\}_{i\in I}$ be a family of A-linear maps such that the diagram*

$$\begin{array}{ccc} & P & \\ {\scriptstyle g_j}\swarrow & & \searrow{\scriptstyle g_i} \\ M_j & \xrightarrow{\ f_{ij}\ } & M_i \end{array}$$

is commutative. Then there exists a unique A-linear map $h : P \to M$ such that $g_i = f_i \circ h$ for each $i \in I$.

Proof There exists a unique A-linear map $\tilde{h} : P \to L := \prod_i M_i$ such that $g_i = p_i \tilde{h}$ for any $i \in I$. Consider the exact sequence

$$0 \to M \xrightarrow{\iota} \prod_{i \in I} M_i \xrightarrow{Q} \prod_{(i,j) \in T} M_i,$$

where ι is the inclusion, and Q is given by $Q((x_i)_{i\in I}) = (x_i - f_{ij}(x_j))_{(i,j)\in T}$. Let $\varphi : P \to L = \prod_{i\in I} M_i$ be the unique map such that $\pi_i \varphi = g_i$ for each i. So $\varphi(x) = (g_i(x))_{i \in I}$. Then $Q\varphi(x) = (g_i(x) - f_{ij}(g_j(x)))_{(i,j)} = 0$, and hence $\varphi(P) \subseteq M$, and this defines $h : P \to M$ such that $g_i = f_i \circ h$ uniquely. □

The module M defined by Theorem B.9, which is uniquely determined by the projective system $\mathbb{M} = \{M_i\}_{i \in I}$, is called the *projective limit* or limit of $\mathbb{M}$, and is denoted by $\varprojlim M_i$.

Let $\mathbb{L}$ and $\mathbb{M}$ be the projective systems. A morphism of projective systems $(\alpha_i : L_i \to M_i)_{i\in I}$ is an element of $\prod_i \mathrm{Hom}_A(L_i, M_i)$ such that the diagram

$$\begin{CD} L_j @>{\alpha_j}>> M_j \\ @V{f_{ij}}VV @VV{f_{ij}}V \\ L_i @>{\alpha_i}>> M_i \end{CD}$$

is commutative for any $(i, j) \in T$ A morphism of projective systems induces a map $\varprojlim \alpha_i : \varprojlim L_i \to \varprojlim M_i$ given by $(\varprojlim \alpha_i)((x_i)_{i\in I}) = (\alpha_i(x_i))_{i \in I}$.

Exercise B.3 *Let $\mathbb{A} = (\{A_i\}_{i\in I}, \{f_{ij}\}_{(i,j)\in T})$ be a projective system such that each A_i is an A-algebra and each f_{ij} is an A-algebra map. Prove that $\varprojlim A_i$ has a structure of an A-algebra such that each $f_i : \varprojlim A_i \to A_i$ is an A-algebra map.*

Here we give a lemma necessary in constructing the completion. Set $I = \{1, 2, 3, \ldots\}$. Let $\mathbb{L} = \{L_i\}_{i\in I}$, $\mathbb{M} = \{M_i\}_{i \in I}$, and $\mathbb{N} = \{N_i\}_{i\in I}$ be projective systems, and consider morphisms

$$\mathbb{L} \xrightarrow{\{\alpha_n\}_{n\in I}} \mathbb{M} \xrightarrow{\{\beta_n\}_{n\in I}} \mathbb{N}.$$

Lemma B.10 *Assume that the sequence of A-modules* $0 \to L_n \overset{\alpha_n}{\to} M_n \overset{\beta_n}{\to} N_n \to 0$ *is exact for any* $n \in I$. *Then:*

(1) *The sequence of A-modules* $0 \to \varprojlim L_n \xrightarrow{\varprojlim \alpha_n} \varprojlim M_n \xrightarrow{\varprojlim \beta_n} \varprojlim N_n$ *is exact.*
(2) *If* $f_{n,n+1} : L_{n+1} \to L_n$ *is surjective for any* $n \in I$, *then*

$$0 \to \varprojlim L_n \xrightarrow{\varprojlim \alpha_n} \varprojlim M_n \xrightarrow{\varprojlim \beta_n} \varprojlim N_n \to 0$$

is exact.

Proof Left to the readers. □

Exercise B.4 *Let* $\{N_n\}_{n \geq 0}$ *be a family of A-modules. Set* $M_n = \bigoplus_{0 \leq i \leq n} N_i$. *For* $0 \leq i \leq j$, *define the A-linear map* $f_{ij} : M_j \to M_i$ *by* $f_{ij}(m_0, \ldots, m_j) = (m_0, \ldots, m_i)$. *Let* $M = \prod_{n \geq 0} N_n$ *be the direct product, and let* $g_n : M \to M_n = \bigoplus_{0 \leq i \leq n} N_i$ *be the map given by* $g_n((m_i)) = (m_0, \ldots, m_n)$. *Prove that* $((M_i), (f_{ij}))$ *forms a projective system of A-modules, and* $g_i : M \to M_i$ *induces an isomorphism of A-modules* $g : M \to \varprojlim M_i$.

Lemma B.11 *Let X be an A-module, and* $(\{M_i\}_{i \in I}, \{f_{ji} : M_i \to M_j\}_{(i,j) \in T})$ *be an inductive system of A-modules, where* $T = \{(i, j) \in I^2 \mid i \leq j\}$. *Then* $(\{\mathrm{Hom}_A(M_i, X)\}_{i \in I}, \{f_{ji}^* : \mathrm{Hom}_A(M_j, X) \to \mathrm{Hom}_A(M_i, X)\}_{(i,j) \in T})$ *is a projective system of A-modules indexed by T. The isomorphism*

$$\Phi : \mathrm{Hom}_A(\bigoplus_{i \in I} M_i, X) \to \prod_{i \in I} \mathrm{Hom}_A(M_i, X)$$

in Proposition A.1 induces an isomorphism

$$\overline{\Phi} : \mathrm{Hom}_A(\varinjlim M_i, X) \to \varprojlim \mathrm{Hom}_A(M_i, X).$$

This map is natural with respect to both X and $\{M_i\}$ *in the sense that for each A-linear map* $f : X \to X'$ *and each map of inductive systems* $\alpha = \{\alpha_i : M_i' \to M_i\}$, *the diagram*

$$\begin{array}{ccc} \mathrm{Hom}_A(M_i, X) & \xrightarrow{\overline{\Phi}} & \varprojlim \mathrm{Hom}_A(M_i, X) \\ \downarrow {\scriptstyle \mathrm{Hom}_A(\alpha_i, f)} & & \downarrow {\scriptstyle \varprojlim \mathrm{Hom}_A(\alpha_i, f)} \\ \mathrm{Hom}_A(M_i', X') & \xrightarrow{\overline{\Phi}} & \varprojlim \mathrm{Hom}_A(M_i', X') \end{array}$$

Proof The lemma says that $\psi : \bigoplus_i M_i \to X$ factors through $q : \bigoplus_i M_i \to \varinjlim M_i$ if and only if $(\psi_i)_{i \in I} \in \prod_i \mathrm{Hom}_A(M_i, X)$ lies in $\varprojlim \mathrm{Hom}_A(M_i, X)$, where $\psi_i = \psi|_{M_i}$. This is equivalent to saying that $f_{ji}^*(\psi_j) = \psi_j f_{ji} = \psi_i$ for every $(i, j) \in T$, and this is nothing but the definition of the inductive limit and the homomorphism theorem. □

The following is also proved similarly.

Lemma B.12 *Let X be an A-module, and $(\{M_i\}_{i\in I}, \{f_{ij} : M_i \to M_j\}_{(i,j)\in T})$ be a projective system of A-modules, where $T = \{(i,j) \in I^2 \mid \leq j\}$. Then $(\{\mathrm{Hom}_A(X, M_i)\}_{i\in I}, \{(f_{ij})_* : \mathrm{Hom}_A(X, M_j) \to \mathrm{Hom}_A(X, M_i)\}_{(i,j)\in T})$ is a projective system of A-modules indexed by T. The isomorphism*

$$\Phi : \mathrm{Hom}_A(X, \prod_{i\in I} M_i) \to \prod_{i\in I} \mathrm{Hom}_A(X, M_i)$$

in Proposition A.1 induces an isomorphism

$$\overline{\Phi} : \mathrm{Hom}_A(X, \varprojlim M_i) \to \varprojlim \mathrm{Hom}_A(X, M_i).$$

Let X be an A-module, and $(\{M_i\}_{i\in I}, \{f_{ji} : M_i \to M_j\}_{(i,j)\in T})$ be an inductive system of A-modules, where $T = \{(i,j) \in I^2 \mid \leq j\}$. Then $(\iota_i)_* : \mathrm{Hom}_A(X, M_i) \to \mathrm{Hom}_A(X, \varinjlim M_i)$ induces a map $\Phi' : \varinjlim \mathrm{Hom}_A(X, M_i) \to \mathrm{Hom}_A(X, \varinjlim M_i)$.

Lemma B.13 *If I is a directed set, then*

$$\Phi' : \varinjlim \mathrm{Hom}_A(X, M_i) \to \mathrm{Hom}_A(X, \varinjlim M_i)$$

is injective. If, moreover, X is finitely generated, then Φ' is an isomorphism.

Proof Let h be an element of $\mathrm{Ker}\,\Phi' \subseteq \varinjlim \mathrm{Hom}_A(X, M_i)$. Then it is in the image of $h_i \in \mathrm{Hom}_A(X, M_i) \to \varinjlim \mathrm{Hom}_A(X, M_i)$ for some i, since I is directed. Since the composition

$$X \xrightarrow{h_i} M_i \xrightarrow{f_i} \varinjlim M_i$$

is zero, $h_j = f_{ji}h_i = 0$ for some $j \geq i$. Since h equals the image of $h_j = 0$ in $\varinjlim \mathrm{Hom}_A(X, M_i)$, we have that $\mathrm{Ker}\,\Phi' = 0$. So the first assertion has been proved.

Now assume that X is finitely generated, and let

$$Y \xrightarrow{\varphi} F_0 \xrightarrow{\psi} X \to 0$$

be an exact sequence where F_0 is a finitely generated A-free module. Then we have a commutative diagram

$$\begin{array}{ccccccc}
0 \longrightarrow & \varinjlim \mathrm{Hom}_A(X, M_i) & \xrightarrow{\varinjlim \psi_X^*} & \varinjlim \mathrm{Hom}_A(F_0, M_i) & \xrightarrow{\varinjlim \varphi_{F_0}^*} & \varinjlim \mathrm{Hom}_A(Y, M_i) \\
& \downarrow \Phi' & & \downarrow \Phi' & & \downarrow \Phi'_Y \\
0 \longrightarrow & \mathrm{Hom}_A(X, \varinjlim M_i) & \xrightarrow{\psi^*} & \mathrm{Hom}_A(F_0, \varinjlim M_i) & \xrightarrow{\varphi^*} & \mathrm{Hom}_A(Y, \varinjlim M_i)
\end{array}$$

with exact rows. By the five lemma, it suffices to show the lemma for the case that $X = A^n$ is a finitely generated free module, since we know that $\Phi'_Y : \varinjlim \mathrm{Hom}_A(Y, M_i) \to \mathrm{Hom}_A(Y, \varinjlim M_i)$ is injective. Then we have that Φ' for $X = A^n$ is the composition of the isomorphisms

$$\varinjlim \mathrm{Hom}_A(A^n, M_i) \cong \varinjlim \mathrm{Hom}_A(A, M_i)^n \xrightarrow{(\Phi')^n} \mathrm{Hom}_A(A, \varinjlim M_i)^n \cong \mathrm{Hom}_A(A^n, \varinjlim M_i),$$

and we are done. □

The following is proved similarly.

Corollary B.14 *If X is a finitely generated A-module and $\{M_\lambda\}$ a family of A-modules, then the canonical map $h : \bigoplus_\lambda \mathrm{Hom}_A(X, M_\lambda) \to \mathrm{Hom}_A(X, \bigoplus_\lambda M_\lambda)$ is an isomorphism.*

Proof Let I be the set of finite subsets of Λ with the incidence relation as an order. For each $i \in I$, $\bigoplus_{\lambda\in i} \mathrm{Hom}_A(X, M_\lambda) \to \mathrm{Hom}_A(X, \bigoplus_{\lambda\in i} M_\lambda)$ is an isomorphism by Proposition A.1, (2). So we have a sequence of isomorphisms

$$\begin{aligned}
\bigoplus_{\lambda\in\Lambda} \mathrm{Hom}_A(X, M_\lambda) &\cong \varinjlim \bigoplus_{\lambda\in i} \mathrm{Hom}_A(X, M_\lambda) \cong \varinjlim \mathrm{Hom}_A(X, \bigoplus_{\lambda\in i} M_\lambda) \\
&\cong \mathrm{Hom}_A(X, \varinjlim \bigoplus_{\lambda\in i} M_\lambda) \cong \mathrm{Hom}_A(X, \bigoplus_{\lambda\in\Lambda} M_\lambda),
\end{aligned}$$

whose composition is the map in question. □

Appendix C
Ext and Tor

C.1 Ext

Let A be a ring. For a set I, we denote the direct sum of the family $\{M_i = A\}_{i \in I}$ of A-modules by $A^{(I)}$. Namely, we define:

$$A^{(I)} = \{(a_i)_{i \in I} \mid a_i \in A, \text{ for almost all } i \in I, \ a_i = 0\}.$$

We define the element $\mathbf{e}_i$ of $A^{(I)}$ by $\boldsymbol{e}_i = (\delta_{ij})_{j \in J} \in A^{(I)}$, where δ_{ij} is *Kronecker's delta*. That is, $\delta_{ij} = 1$ if $i = j$ and $\delta_{ij} = 0$ otherwise. If we give a family $\{m_i\}_{i \in I}$ of elements of M, then there exists a unique A-linear map $\varphi : A^{(I)} \to M$ such that $\varphi(\boldsymbol{e}_i) = m_i$. Namely, $\varphi((a_i)) = \sum_i a_i m_i$. The family $\{m_i\}$ is said to be *linearly independent* if φ is injective. We say that $\{m_i\}$ *generates* M if φ is surjective. We say that $\{m_i\}$ is a *free basis* or a *basis* of M if φ is bijective. We say that an A-module F is an A-free module if F has a free basis. Note that for any A-module M, there is a surjective A-linear map $\varphi : F \to M$ with F being A-free. Indeed, letting $F = A^{(M)}$ and defining $\varphi : F \to M$ by $\varphi((a_m)_{m \in M}) = \sum_{m \in M} a_m m$, φ is a surjective A-linear map.

Let F and L be A-modules. We say that $\alpha \in \operatorname{Hom}_A(L, F)$ is a *split monomorphism* (or *split injective* A-linear map) if there exists some $\beta \in \operatorname{Hom}_A(F, L)$ such that $\beta\alpha = 1_L$. We say that $\beta \in \operatorname{Hom}_A(F, L)$ is a *split epimorphism* (or *split surjective* A-linear map) if there exists some $\alpha \in \operatorname{Hom}_A(L, F)$ such that $\beta\alpha = 1_L$. We say that $e : F \to F$ is a *projector* if $e^2 = e$.

Lemma C.1 *Let F and L be A-modules. The following are equivalent:*

(1) *There exists some split monomorphism $\alpha : L \to F$.*
(2) *There exists some split epimorphism $\beta : F \to L$.*
(3) *There exist some $\alpha : L \to F$ and $\beta : F \to L$ such that $\beta\alpha = 1_F$.*

S. Goto, K. Watanabe, *Commutative Algebra*, Graduate Texts in Mathematics 309,
https://doi.org/10.1007/978-981-95-4227-7

(3)′ *There exist some homomorphisms* $\alpha : L \to F$ *and* $\beta : F \to L$ *such that* $\alpha\beta$ *is a projector of* F, β *is surjective, and* α *is injective.*
(4) *There exist some* A*-submodules* F_1 *and* F_2 *of* F *such that* $F = F_1 \oplus F_2$ *and an isomorphism* $\psi : L \to F_1$.

More precisely, if (1) (*resp.* (2)) *is assumed, then there exists some* β (*resp.* α) *such that* (3) *is satisfied by the same* α (*resp.* β). (3) *is satisfied by* α *and* β, *then* (3)′ *is satisfied by the same* α *and* β, *and vice versa. For* α *and* β *satisfying* (3) *or* (3)′, *the condition in* (4) *is satisfied by* $F_1 = \operatorname{Im}\alpha$ *and* $F_2 = \operatorname{Ker}\beta$. *Conversely, if* (4) *is assumed, then defining* $\alpha : L \to F$ *by* $\alpha(\ell) = \psi(\ell)$, *and* $\beta : F \to L$ *by* $\beta(f_1 + f_2) = \psi^{-1}(f_1)$ *for* $\ell \in L$, $f_1 \in F_1$, *and* $f_2 \in F_2$, (3) *is satisfied, and* $\operatorname{Im}\alpha = F_1$ *and* $\operatorname{Ker}\beta = F_2$.

Proof (1) ⇔(2) ⇔(3) is obvious.

(3) ⇒(3)′. Clearly, β is surjective and α is injective. Moreover, since $\alpha\beta\alpha\beta = \alpha 1_L \beta = \alpha\beta$, we have that $\alpha\beta$ is a projector.

(3)′ ⇒(3). Since $\alpha(1_L - \beta\alpha)\beta = \alpha\beta - \alpha\beta\alpha\beta = \alpha\beta - \alpha\beta = 0$, β is surjective, and α is injective, we have that $1_L - \beta\alpha = 0$.

(3) ⇒(4). We set $F_1 = \operatorname{Im}\alpha$ and $F_2 = \operatorname{Ker}\beta$. We define $\psi : L \to F_1$ by $\psi(x) = \alpha(x)$ for $x \in L$. As α is injective, ψ is an isomorphism. If $m \in F$, then $\beta(1_F - \psi\beta)(m) = \beta m - \beta\alpha\beta m = \beta m - \beta m = 0$. Hence $(1_F - \psi\beta)(m) \in \operatorname{Ker}\beta = F_2$. So $m = \psi\beta m + (1_F - \psi\beta)m \in F_1 + F_2$. If $m = \alpha(x) \in F_1 \cap F_2$, then $x = \beta\alpha x = \beta m = 0$. Hence $F = F_1 \oplus F_2$ is a direct sum.

(4) ⇒(1). We define $\alpha : L \to F$ by $\alpha(x) = \psi(x)$, and $\beta : F \to L$ by $\beta(m_1 + m_2) = \psi^{-1}(m_1)$. Then $\beta\alpha(x) = x$, and $\operatorname{Im}\alpha = F_1$ and $\operatorname{Ker}\beta = F_2$.

The rest of the assertions are clear from the proof above. □

We say that L is a *direct summand* of F if the equivalent conditions in Lemma C.1 are satisfied. A direct summand of an A-free module is called a projective A-module.

Proposition C.2 *For an* A*-module* P, *the following are equivalent:*

(1) P *is a projective* A*-module.*
(2) $\operatorname{Hom}_A(P, *)$ *is an exact functor.*
(3) *If* $\varphi : X \to Y$ *is a surjective* A*-linear map and* $f : P \to Y$ *is an* A*-linear map, then there exists some* A*-linear map* $g : P \to X$ *such that* $\varphi g = f$.

Proof (1) ⇒(3). First, we assume that P is an A-free module with a free basis Z. For each $z \in Z$, we can find $x_z \in X$ such that $\varphi(x_z) = f(z)$, since φ is surjective. Defining $g : P \to X$ by $g(\sum_z c_z z) = \sum_z c_z x_z$ for $(c_z) \in A^{(Z)}$, we have $\varphi g = f$. Next, we consider the general case. Let F be an A-free module, and $\alpha : P \to F$ and $\beta : F \to P$ be A-linear maps such that $\beta\alpha = 1_P$. As the assertion is true for free modules, there exists some $\tilde{g} : F \to X$ such that $\varphi\tilde{g} = f\beta$. Then defining $g = \tilde{g}\alpha$, we get $\varphi g = \varphi\tilde{g}\alpha = f\beta\alpha = f$.

(3) ⇒(2). Since $\operatorname{Hom}_R(P, *)$ is left exact in general, it suffices to show that $\varphi_* : \operatorname{Hom}_R(P, X) \to \operatorname{Hom}_R(P, Y)$ is surjective for any surjective A-linear map $\varphi : X \to Y$, where $\varphi_*(g) = \varphi g$. But this is nothing but assertion (3) itself.

(2) $\Rightarrow$(1). Let $\varphi : X = F = A^{(P)} \to P = Y$ be a surjective A-linear map, which exists. Since $f = 1_P : P \to P = Y$ can be lift to $g : P \to F$ (that is, $\varphi g = f = 1_P$), P is a direct summand of F. □

Let M be an A-module. We can take a surjective A-linear map $\varphi_0 : F_0 \to M$ with F_0 a free A-module. Then $M_1 := \mathrm{Ker}(\varphi_0)$ admits a surjective A-linear map $\varphi_1 : F_1 \to M_1$ with F_1 a free A-module. Continuing this way, we get an exact sequence

$$0 \to M_{i+1} \xrightarrow{\psi_i} F_i \xrightarrow{\varphi_i} M_i \to 0$$

such that F_i is a free module. Then setting $\partial_i : F_i \to F_{i-1}$ to be the map $\psi_{i-1}\varphi_i$, we get an exact sequence

$$\cdots \xrightarrow{\partial_{i+1}} F_i \xrightarrow{\partial_i} F_{i-1} \xrightarrow{\partial_{i-1}} \cdots \xrightarrow{\partial_2} F_1 \xrightarrow{\partial_1} F_0 \to M \to 0$$

such that all the F_i are free A-modules. We call such an exact sequence a *free resolution* of the A-module M. In particular, we have an exact sequence

$$\cdots \to P_i \xrightarrow{\partial_i} P_{i-1} \to \cdots \to P_1 \xrightarrow{\partial_1} P_0 \xrightarrow{\varepsilon} M \to 0$$

such that all the P_i are projective A-modules, since a free A-module is projective. We call such a sequence a *projective resolution* of M.

For two chain complexes of A-modules

$$P_\bullet : \cdots \to P_{n+1} \xrightarrow{\partial^P_{n+1}} P_n \xrightarrow{\partial^P_n} P_{n-1} \to \cdots$$

and

$$Q_\bullet : \cdots \to Q_{n+1} \xrightarrow{\partial^Q_{n+1}} Q_n \xrightarrow{\partial^Q_n} Q_{n-1} \to \cdots,$$

a collection of A-linear maps $f_\bullet = \{f_n\}_{n\in\mathbb{Z}} \in \prod_{n\in\mathbb{Z}} \mathrm{Hom}_A(P_n, Q_n)$ is called a *chain map* $f_\bullet : P_\bullet \to Q_\bullet$ if $\partial_n f_n = f_{n-1}\partial_n$ for all $n \in \mathbb{Z}$. The set of chain maps $C(P_\bullet, Q_\bullet)$ from $P_\bullet$ to $Q_\bullet$ forms an A-submodule of $\prod_{n\in\mathbb{Z}} \mathrm{Hom}_A(P_n, Q_n)$. If $g_\bullet : Q_\bullet \to R_\bullet$ is a chain map, we have that the composite $g_\bullet f_\bullet = \{g_n f_n\}_{n\in\mathbb{Z}}$ is a chain map again. The identity map $1_{P_\bullet} = \{1_{P_n}\}_{n\in\mathbb{Z}}$ is a chain map. Thus we get the category of chain complexes of A-modules and chain maps $\mathcal{C}(A)$. It is an A-linear abelian category.

We define $Z_n(P_\bullet) = \mathrm{Ker}\,\partial_n$ and $B_n(P_\bullet) = \mathrm{Im}\,\partial_{n+1}$, so that $\mathrm{H}_n(P_\bullet) = Z_n(P_\bullet)/B_n(P_\bullet)$, the nth homology group of $P_\bullet$. If $f_\bullet : P_\bullet \to Q_\bullet$ is a chain map, then $f_n(Z_n(P_\bullet)) \subseteq Z_n(Q_\bullet)$ and $f_n(B_n(P_\bullet)) \subseteq B_n(Q_\bullet)$. Hence f_n induces an A-linear map $\mathrm{H}_n(f_\bullet) : \mathrm{H}_n(P_\bullet) \to \mathrm{H}_n(Q_\bullet)$. Obviously, we have $\mathrm{H}_n(g_\bullet)\,\mathrm{H}_n(f_\bullet) =$

$\mathrm{H}_n(g_\bullet f_\bullet)$, and $\mathrm{H}_n(1_{P_\bullet}) = 1_{\mathrm{H}_n(P_\bullet)}$. Thus we get an additive functor $\mathrm{H}_n : C(A) \to \operatorname{Mod} A$ for each n.

We say that two chain maps $f_\bullet, g_\bullet : P_\bullet \to Q_\bullet$ are *homotopic* if there exists some collection of A-linear maps $s_\bullet = \{s_n : P_n \to Q_{n+1}\}_{n\in\mathbb{Z}}$ such that $f_n - g_n = \partial^Q_{n+1} s_n + s_{n-1}\partial^P_n$. If so,

$$(\mathrm{H}_n(f_\bullet) - \mathrm{H}_n(g_\bullet))(\alpha + B_n(P_\bullet)) = (\partial^Q_{n+1} s_n \alpha + s_{n-1}\partial^P_n \alpha) + B_n(Q_\bullet) = 0$$

for $\alpha \in Z_n(P_\bullet)$. Namely, $\mathrm{H}_n(f_\bullet) = \mathrm{H}_n(g_\bullet)$. We say that a chain map $f_\bullet : P_\bullet \to Q_\bullet$ is *null-homotopic* if $f_\bullet$ and the zero map 0 are homotopic. If $f_\bullet : P_\bullet \to Q_\bullet$ and $g_\bullet : Q_\bullet \to R_\bullet$ are chain maps and either $f_\bullet$ or $g_\bullet$ is null-homotopic, then $g_\bullet f_\bullet$ is null-homotopic. We can define the homotopy category $K(A)$ of A whose object set is the same as that of $C(A)$, while the hom-set from $P_\bullet$ to $Q_\bullet$ is $K(A)(P_\bullet, Q_\bullet) = C(P_\bullet, Q_\bullet)/\mathrm{NH}(P_\bullet, Q_\bullet)$, where $\mathrm{NH}(P_\bullet, Q_\bullet)$ is the set of null-homotopic chain maps from $P_\bullet$ to $Q_\bullet$. A chain complex $P_\bullet \in C(A)$ is said to be *split exact* or *homotopically trivial* if the identity map $1_{P_\bullet} : P_\bullet \to P_\bullet$ is null-homotopic, or equivalently, $P_\bullet$ is isomorphic to the zero complex 0 in $K(A)$.

Lemma C.3 *Let M and N be A-modules,*

$$\cdots \to P_i \xrightarrow{\partial_i} P_{i-1} \to \cdots \xrightarrow{\partial_2} P_1 \xrightarrow{\partial_1} P_0 \xrightarrow{\varepsilon} M \to 0$$

be a complex of A-modules where each P_i is A-projective,

$$\cdots \to Q_i \xrightarrow{\partial'_i} Q_{i-1} \to \cdots \xrightarrow{\partial'_2} Q_1 \xrightarrow{\partial'_1} Q_0 \xrightarrow{\varepsilon'} N \to 0$$

be an exact sequence of A-modules, and $f : M \to N$ be an A-linear map. Then we can take a chain map $\{f_\bullet : P_\bullet \to Q_\bullet\}_{i\geq 0}$ such that $f\varepsilon = \varepsilon' f_0$. The chain map $f_\bullet$ is called the lift *of the A-linear map f. The lift $f_\bullet$ is unique up to homotopy. That is, the image of $f_\bullet$ in $K(A)(P_\bullet, Q_\bullet) = C(P_\bullet, Q_\bullet)/\mathrm{NH}(P_\bullet, Q_\bullet)$ is unique.*

Let $\mathcal{P}_+$ be the full subcategory of $K(A)$ whose objects are complexes of A-modules

$$\cdots \to P_i \xrightarrow{\partial_i} P_{i-1} \to \cdots$$

such that each P_i is projective for $i \in \mathbb{Z}$, and $P_i = 0$ for $i \ll 0$. The set of morphisms $\mathcal{P}_+(P_\bullet, Q_\bullet)$ is $K(A)(P_\bullet, Q_\bullet)$. We define a functor $\boldsymbol{P} : \operatorname{Mod} A \to \mathcal{P}_+$ as follows. For $M \in \operatorname{Mod} A$, we choose a projective resolution $\boldsymbol{P}(M)$ of M. For $f \in \operatorname{Hom}_A(M, N)$, let $\boldsymbol{P}(f) : \boldsymbol{P}(M) \to \boldsymbol{P}(N)$ be the lift of f, which is unique up to homotopy, and is well-defined in $K(A)(\boldsymbol{P}(M), \boldsymbol{P}(N))$. From the uniqueness, $\boldsymbol{P}(\mathrm{id}_M) = \mathrm{id}_{\boldsymbol{P}(M)}$ and $\boldsymbol{P}(g)\circ\boldsymbol{P}(f) = \boldsymbol{P}(g\circ f)$ follow, and $\boldsymbol{P}$ is a functor. We define a functor $\operatorname{Ext}^i_A(-,-) : (\operatorname{Mod} A)_{\mathrm{op}} \times \operatorname{Mod} A \to \operatorname{Mod} A$ as follows, where $(-)_{\mathrm{op}}$ denotes the opposite category. We define $\operatorname{Ext}^i_A(M, N) = \mathrm{H}^i(\operatorname{Hom}_A(\boldsymbol{P}(M), N))$.

For morphisms $f : M' \to M$ and $g : N \to N'$ in Mod A, we define $\mathrm{Ext}^i_A(f, g) : \mathrm{Ext}^i_A(M, N) \to \mathrm{Ext}^i_A(M', N')$ to be $\mathrm{H}^i(\mathrm{Hom}_A(\boldsymbol{P}(f), g))$.

If

$$\boldsymbol{P}_\bullet : \cdots \to P_i \xrightarrow{\partial_i} P_{i-1} \xrightarrow{\partial_{i-1}} \cdots \xrightarrow{\partial_2} P_1 \xrightarrow{\partial_1} P_0 \xrightarrow{\varepsilon} M \to 0$$

is a projective resolution of M, then $\boldsymbol{P}_\bullet$ and $\boldsymbol{P}(M)$ are isomorphic in $\mathcal{P}_+$, and hence the complex

$$\mathrm{Hom}_A(\boldsymbol{P}_\bullet, N) : \cdots \to 0 \to \mathrm{Hom}_R(P_0, N) \xrightarrow{\partial_1^*} \mathrm{Hom}_A(P_1, N)$$
$$\xrightarrow{\partial_2^*} \cdots \xrightarrow{\partial_i^*} \mathrm{Hom}_A(P_i, N) \to \cdots$$

is isomorphic to $\mathrm{Hom}_A(\boldsymbol{P}(M), N)$ in $K(A)$. So $\mathrm{H}^i(\mathrm{Hom}_A(\boldsymbol{P}_\bullet, N))$ is isomorphic to $\mathrm{Ext}^i_A(M, N)$.

Proposition C.4 $\mathrm{Ext}^i_A(-, -)$ *is a functor from the category* $(\mathrm{Mod}\,A)_{\mathrm{op}} \times \mathrm{Mod}\,A$ *to* Mod A.

Note that the functor $\mathrm{Ext}^i_A(-, -)$ is independent of the choice of $\boldsymbol{P}$, up to isomorphisms of functors.

Definition C.5 A category C is said to be an A-category if $C(c, c')$ is an A-module for $c, c' \in C$, and the composition $C(c', c'') \times C(c, c') \to C(c, c'')$ is A-bilinear. For $a \in A$ and $c \in C$, the morphism $a1_c : c \to c$ is called the *homothety* of a. A functor $F : C \to C'$ is called A-*linear* if $F : C(c_1, c_2) \to C'(Fc_1, Fc_2)$ is A-linear for each $c_1, c_2 \in C$. This is equivalent to saying that for each $a \in A$ and $c \in C$, $F(a1_c) = a1_{Fc}$ holds.

It is easy to see that Mod A and its opposite $(\mathrm{Mod}\,A)_{\mathrm{op}}$ are A-categories, and $\mathrm{Ext}^i_A(-, N)$ and $\mathrm{Ext}^i_A(M, -)$ are A-linear functors.

C.2 Long Exact Sequences of Ext

Let M be an A-module, and

$$\mathbb{X} : 0 \to X \xrightarrow{\alpha} Y \xrightarrow{\beta} Z \to 0$$

be a short exact sequence of A-modules. Since each term of the projective resolution $\boldsymbol{P}(M)$ of M is A-projective,

$\mathrm{Hom}_A(\boldsymbol{P}(M), \mathbb{X})$:

$$0 \to \mathrm{Hom}_A(\boldsymbol{P}(M), X) \xrightarrow{\alpha_*} \mathrm{Hom}_A(\boldsymbol{P}(M), Y) \xrightarrow{\beta_*} \mathrm{Hom}_A(\boldsymbol{P}(M), Z) \to 0$$

is a short exact sequence of complexes of A-modules. This short exact sequence of complexes yields a long exact sequence of Ext groups

$$\cdots \to \mathrm{Ext}_A^{i-1}(M, Y) \to \mathrm{Ext}_A^{i-1}(M, Z) \xrightarrow{\Delta} \mathrm{Ext}_A^i(M, X)$$
$$\to \mathrm{Ext}_A^i(M, Y) \to \mathrm{Ext}_A^i(M, Z) \xrightarrow{\Delta} \mathrm{Ext}_A^{i+1}(M, X) \to \cdots .$$

Moreover, if

$$\begin{array}{ccccccccc} 0 & \longrightarrow & X & \xrightarrow{\alpha} & Y & \xrightarrow{\beta} & Z & \longrightarrow & 0 \\ & & {\scriptstyle f}\downarrow & & \downarrow{\scriptstyle g} & & \downarrow{\scriptstyle h} & & \\ 0 & \longrightarrow & X' & \xrightarrow{\alpha'} & Y' & \xrightarrow{\beta'} & Z' & \longrightarrow & 0 \end{array}$$

is a commutative diagram of A-modules with exact rows, then the diagram

$$\begin{array}{ccc} \mathrm{Ext}_A^i(M, Z) & \xrightarrow{\Delta} & \mathrm{Ext}_A^i(M, X) \\ {\scriptstyle h_*}\downarrow & & \downarrow{\scriptstyle f_*} \\ \mathrm{Ext}_A^i(M, Z') & \xrightarrow{\Delta'} & \mathrm{Ext}_A^i(M, X') \end{array}$$

is commutative, where Δ and Δ' are connecting homomorphisms, see Section 1.6 of the text.

Lemma C.6 *Let*

$$0 \to X \xrightarrow{\alpha} Y \xrightarrow{\beta} Z \to 0 \tag{C.1}$$

be a short exact sequence of A-modules, and $\boldsymbol{P}_\bullet$ and $\boldsymbol{Q}_\bullet$ be projective resolutions of X and Z, respectively. Then there exists some projective resolution $\boldsymbol{R}_\bullet$ of Y and an exact sequence of complexes

$$0 \to \boldsymbol{P}_\bullet \xrightarrow{f_\bullet} \boldsymbol{R}_\bullet \xrightarrow{g_\bullet} \boldsymbol{Q}_\bullet \to 0 \tag{C.2}$$

such that $f_\bullet$ and $g_\bullet$ are lifts of α and β, respectively.

Proof First, consider the diagram

$$\begin{array}{ccccccccc} & & 0 & & & & 0 & & \\ & & \uparrow & & & & \uparrow & & \\ 0 & \longrightarrow & X & \xrightarrow{\alpha} & Y & \xrightarrow{\beta} & Z & \longrightarrow & 0 \\ & & {\scriptstyle\varepsilon}\uparrow & & & & {\scriptstyle\tau}\uparrow & & \\ & & P_0 & & & & Q_0 & , & \end{array}$$

where ε and τ are augmentations of the respective resolutions. As Q_0 is projective and β is surjective, we can take $\gamma : Q_0 \to Y$ such that $\beta\gamma = \tau$. Then the diagram

$$\begin{array}{ccccccccc} 0 & \longrightarrow & X & \xrightarrow{\alpha} & Y & \xrightarrow{\beta} & Z & \longrightarrow & 0 \\ & & {\scriptstyle\varepsilon}\uparrow & & {\scriptstyle\rho}\uparrow & & {\scriptstyle\tau}\uparrow & & \\ 0 & \longrightarrow & P_0 & \underset{\binom{1}{0}}{\xrightarrow{i}} & P_0 \oplus Q_0 & \underset{(0\ 1)}{\xrightarrow{p}} & Q_0 & \longrightarrow & 0 \end{array}$$

is commutative with exact rows, where $\rho(x, y) = \alpha\varepsilon(x) + \gamma(y)$. We set $R_0 = P_0 \oplus Q_0$. By the five lemma, $\rho : R_0 \to Y$ is surjective. So by the snake lemma, $0 \to \operatorname{Ker} \varepsilon \to \operatorname{Ker} \rho \to \operatorname{Ker} \tau \to 0$ is exact. Letting this exact sequence be

$$0 \to X_1 \xrightarrow{\alpha_1} Y_1 \xrightarrow{\beta_1} Z_1 \to 0,$$

we can form a diagram with exact rows and columns

$$\begin{array}{ccccccccc} & & 0 & & & & 0 & & \\ & & \uparrow & & & & \uparrow & & \\ 0 & \longrightarrow & X_1 & \xrightarrow{\alpha_1} & Y_1 & \xrightarrow{\beta_1} & Z_1 & \longrightarrow & 0 \\ & & {\scriptstyle\partial}\uparrow & & & & {\scriptstyle\partial}\uparrow & & \\ & & P_1 & & & & Q_1 & . & \end{array}$$

Then we can repeat the same argument, and continuing this way, the lemma is proved. □

If N is an A-module, then letting (C.1) and (C.2) be as in Lemma C.6,

$$0 \to \operatorname{Hom}_A(Q_\bullet, N) \to \operatorname{Hom}_A(R_\bullet, N) \to \operatorname{Hom}_A(P_\bullet, N) \to 0$$

is a short exact sequence of complexes. Taking the cohomology, we get a long exact sequence

$$\cdots \xrightarrow{\Delta} \mathrm{Ext}^i_A(Z, N) \to \mathrm{Ext}^i_A(Y, N) \to \mathrm{Ext}^i_A(X, N) \xrightarrow{\Delta} \mathrm{Ext}^{i+1}_A(Z, N) \to \cdots.$$

The connecting homomorphism Δ in this long exact sequence is also compatible with a morphism between short exact sequences.

We remark that $\mathrm{Ext}^i_A(M, N)$ can also be calculated using the injective resolution of N. As a matter of fact, if $0 \to N \to I^\bullet$ is an *injective resolution* of N, that is, if the sequence is exact and each term of $I^\bullet$ is an injective A-module, then we have $\mathrm{Ext}^i_A(M, N) \cong \mathrm{H}^i(\mathrm{Hom}_A(M, I^\bullet))$, see [Wei, Theorem 2.7.6].

Lemma C.7 *Let A be a Noetherian ring, and M a finitely generated A-module. Then there is a free resolution $\mathbb{F}$ of M whose terms are finite free A-modules.*

Proof Take a surjective map $\varepsilon : F_0 \to M$ such that F_0 is finite free, and set $M_1 = \mathrm{Ker}\,\varepsilon$. Then take a surjective map $\varepsilon_1 : F_1 \to M_1$ such that F_1 is finite free, and set $M_2 = \mathrm{Ker}\,\varepsilon$. Continuing like this, we get a desired resolution

$$\cdots \to F_n \xrightarrow{d_n} F_{n-1} \to \cdots \to F_1 \xrightarrow{d_1} F_0 \xrightarrow{\varepsilon} M \to 0,$$

where d_n is the composition of the surjective map $\varepsilon_n : F_n \to M_n$ and the inclusion $M_n \hookrightarrow F_{n-1}$. □

Proposition C.8 *Let A be a Noetherian ring, M a finitely generated A-module, I a directed set, and $\{N_i\}$ an inductive system on I. Then the canonical map $\varinjlim \mathrm{Ext}^n_A(M, N_i) \to \mathrm{Ext}^n_A(M, \varinjlim N_i)$ is an isomorphism for every n.*

Proof By Lemma C.7, we can take a free resolution $\mathbb{F}$ of M whose terms are finite free. Then $\varinjlim \mathrm{Ext}^n_A(M, N_i) \cong \varinjlim \mathrm{H}^n(\mathrm{Hom}_A(\mathbb{F}, N_i)) \cong \mathrm{H}^n(\varinjlim \mathrm{Hom}_A(\mathbb{F}, N_i)) \cong \mathrm{H}^n(\mathrm{Hom}_A(\mathbb{F}, \varinjlim N_i)) \cong \mathrm{Ext}^i_A(M, \varinjlim N_i)$. □

C.3 Tor

We define the functor $\mathrm{Tor}^A_i(-,-) : \mathrm{Mod}\,A \times \mathrm{Mod}\,A \to \mathrm{Mod}\,A$ by $\mathrm{Tor}^A_i(M, N) = \mathrm{H}_i(\boldsymbol{P}(M) \otimes_A N)$. Note that $\mathrm{Tor}^A_i(-,-)$ is obtained by the composition of the functors

$$\mathrm{Mod}\,A \times \mathrm{Mod}\,A \xrightarrow{\boldsymbol{P}\times \mathrm{Id}} \mathcal{P}_+ \times \mathrm{Mod}\,A \xrightarrow{-\otimes_A -} K(A) \xrightarrow{\mathrm{H}_i(-)} \mathrm{Mod}\,A.$$

As a matter of fact, we have a functorial isomorphism

$$\mathrm{Tor}^A_i(M, N) = \mathrm{H}_i(\boldsymbol{P}(M) \otimes_A N) \cong \mathrm{H}_i(M \otimes_A \boldsymbol{P}(N)),$$

and hence $\mathrm{Tor}^A_i(M, N) \cong \mathrm{Tor}^A_i(N, M)$ as functors on $\mathrm{Mod}\,A \times \mathrm{Mod}\,A$, see [Wei, Theorem 2.7.2]. The definition of $\mathrm{Tor}^A_i(M, N)$ is independent of the choice of the projective resolution $\boldsymbol{P}$. Indeed, if $Q_\bullet$ is another projective resolution of M, then we can take a lift $f_\bullet : \boldsymbol{P}(M) \to Q_\bullet$ of the identity map 1_M of M. Then $f_\bullet$ is a homotopy equivalence (that is, an isomorphism in $K(A)$), and hence so is $f_\bullet \otimes 1_N : \boldsymbol{P}(M) \otimes_A N \to Q \otimes_A N$, which induces isomorphisms of homology groups.

Note that for each $M \in \operatorname{Mod} A$ and $i \geq 0$, $\operatorname{Tor}_i^A(M, -)$ and $\operatorname{Tor}_i^A(-, M)$ are A-linear functors from $\operatorname{Mod} A$ to itself.

If

$$0 \to X \xrightarrow{f} Y \xrightarrow{g} Z \to 0$$

is a short exact sequence of A-modules and M is an A-module, then

$$0 \to \boldsymbol{P}(M) \otimes_A X \xrightarrow{1 \otimes_A f} \boldsymbol{P}(M) \otimes_A Y \xrightarrow{1 \otimes_A g} \boldsymbol{P}(M) \otimes_A Z \to 0$$

is a short exact sequence of complexes of A-modules, which yields a long exact sequence

$$\cdots \to \operatorname{Tor}_{i+1}^A(M, Z) \xrightarrow{\Delta} \operatorname{Tor}_i^A(M, X) \to \operatorname{Tor}_i^A(M, Y) \to \operatorname{Tor}_i^A(M, Z) \xrightarrow{\Delta} \cdots.$$

The connecting homomorphism Δ is compatible with morphisms between short exact sequences. Similarly, For an A-module N a short exact sequence $0 \to X \to Y \to Z \to 0$, we get a long exact sequence

$$\cdots \to \operatorname{Tor}_{i+1}^A(Z, N) \xrightarrow{\Delta} \operatorname{Tor}_i^A(X, N) \to \operatorname{Tor}_i^A(Y, N) \to \operatorname{Tor}_i^A(Z, N) \xrightarrow{\Delta} \cdots$$

whose connecting homomorphism Δ is compatible with morphisms between short exact sequences.

Proposition C.9 *For an A-module N, the following are equivalent:*

(1) *N is a flat A-module.*
(2) *For any A-module M and an integer $i \geq 1$, we have* $\operatorname{Tor}_i^A(M, N) = 0$.
(3) *For any finitely generated A-module M, we have* $\operatorname{Tor}_1^A(M, N) = 0$.
(4) *For any finitely generated ideal I of A, we have* $\operatorname{Tor}_1^A(A/I, N) = 0$.
(5) *For any finitely generated ideal I of A, we have that $I \otimes_A N \to N$ ($a \otimes n \mapsto an$) is injective.*

Proof (1) $\Rightarrow$(2). Let

$$P_\bullet : \cdots \to P_n \to \cdots \to P_1 \to P_0 \to M \to 0$$

is a projective resolution. Then

$$P_\bullet \otimes_A N : \cdots \to P_n \otimes_A N \to \cdots \to P_1 \otimes_A N \to P_0 \otimes_A N \to M \otimes_A N \to 0$$

is still exact, since N is A-flat. Hence $\mathrm{Tor}_i^A(M, N) \cong \mathrm{H}_i(P_\bullet \otimes_A N) = 0$ for $i > 0$.

(2) ⇒(3) ⇒(4) is obvious.

(4) ⇔(5). Since $0 \to I \to A \to A/I \to 0$ is a short exact sequence, it induces a long exact sequence of Tor:

$$\mathrm{Tor}_1^A(A, N) = 0 \to \mathrm{Tor}_1^A(A/I, N) \to I \otimes_A N \to A \otimes_A N = N,$$

which is exact. Hence $\mathrm{Tor}_1^A(A/I, N) = 0$ if and only if $I \otimes_A N \to N$ is injective.

(5) ⇒(1). Let I be an ideal of A. Then $I = \varinjlim J$, where J runs through the set of finitely generated ideals contained in I, which forms a directed set with respect to the incidence relation. Since $J \otimes_A N \to A \otimes_A N = N$ is injective, we have that $I \otimes_A N \cong \varinjlim(J \otimes_A N) \to \varinjlim N = N$ is also injective. It follows that $\mathrm{Tor}_1^A(A/I, N) = 0$ for each ideal I of A, as can be seen as in the proof of (4) ⇔(5) above.

Let M be an A-module and M_0 be a submodule of M. Let Ω be the set of A-submodules M_1 of M, containing M_0, such that $M_0 \otimes_A N \to M_1 \otimes_A N$ is injective. Using Zorn's lemma and Theorem B.6, we have that Ω has a maximal element M_1. If $M_1 \neq M$, then there exists some $m \in M \setminus M_1$. Let $M_2 = M_1 + Am \subseteq M$. Then

$$0 \to M_1 \to M_2 \to A/I \to 0$$

is exact, where $I = [M_1 :_A m] = \mathrm{Ann}_A(M_2/M_1)$. Then $\mathrm{Tor}_1^A(A/I, N) = 0 \to M_1 \otimes_A N \to M_2 \otimes_A N$ is injective. So $M_0 \otimes_A N \to M_2 \otimes_A N$ is also injective, and this contradicts the maximality of M_1. Hence we have $M_1 = M$, and we have that N is A-flat. □

Proposition C.10 *Let I be a directed set, $\{M_i\}$ an inductive system of A-modules on I, and N an A-module. Then for every i, there is an isomorphism*

$$\varinjlim \mathrm{Tor}_i^A(M_i, N) \cong \mathrm{Tor}_i^A(\varinjlim M_i, N).$$

Proof Let $\mathbb{P}$ be a projective resolution of N. Then

$$\begin{aligned}\varinjlim \mathrm{Tor}_i^A(M_i, N) &\cong \varinjlim \mathrm{H}_i(M_i \otimes_A \mathbb{P}) \cong \mathrm{H}_i(\varinjlim(M_i \otimes_A \mathbb{P})) \\ &\cong \mathrm{H}_i((\varinjlim M_i) \otimes_A \mathbb{P}) \cong \mathrm{Tor}_i^A(\varinjlim M_i, N).\end{aligned}$$

□

Corollary C.11 *Let I be a directed set, and $\{M_i\}$ an inductive system of flat A-modules on I. Then $\varinjlim M_i$ is also A-flat.*

Proof Since $\mathrm{Tor}_1^A(M_i, N) = 0$ for each $i \in I$, we have that $\mathrm{Tor}_1^A(\varinjlim M_i, N) \cong \varinjlim \mathrm{Tor}_1^A(M_i, N) = 0$. □

Lemma C.12 *Let M be a finitely presented A-module, and $\{N_\lambda\}$ a family of A-modules. Then the map*

$$\rho_M : M \otimes_A \prod_\lambda N_\lambda \to \prod_\lambda (M \otimes_A N_\lambda) \quad (m \otimes (n_\lambda) \mapsto (m \otimes n_\lambda)) \quad (m \in M,\ n_\lambda \in N_\lambda)$$

is an isomorphism.

Proof The case that $M = A$ is trivial, and the case that $M = A^r$ also follows without difficulty.

We consider the general case. Since M is finitely presented, there exists an exact sequence $F_1 \to F_0 \to M \to 0$ with F_1 and F_0 finite free. Then we have a commutative diagram with exact rows

$$\begin{array}{ccccccc}
F_1 \otimes_A \prod_\lambda N_\lambda & \longrightarrow & F_0 \otimes_A \prod_\lambda N_\lambda & \longrightarrow & M \otimes_A \prod_\lambda N_\lambda & \longrightarrow & 0 \\
\downarrow \rho_{F_1} & & \downarrow \rho_{F_0} & & \downarrow \rho_M & & \\
\prod_\lambda (F_1 \otimes_A N_\lambda) & \longrightarrow & \prod_\lambda (F_0 \otimes_A N_\lambda) & \longrightarrow & \prod_\lambda (M \otimes_A N_\lambda) & \longrightarrow & 0\ .
\end{array}$$

Since ρ_{F_1} and ρ_{F_0} are isomorphisms, ρ_M is also an isomorphism by the five lemma. □

Theorem C.13 (Chase's Theorem) *Let A be a Noetherian ring, and $\{F_\lambda\}$ a family of flat A-modules. Then the direct product $\prod_\lambda F_\lambda$ is A-flat.*

Proof Let M be a finitely generated A-module. Then we can take a free resolution $\mathbb{P}$ of M whose terms are finitely generated. Then by Lemma C.12 and Theorem A.3, we have

$$\operatorname{Tor}_1^A(M, \prod_\lambda F_\lambda) = \mathrm{H}_1(\mathbb{P} \otimes_A \prod_\lambda F_\lambda) \cong \mathrm{H}_1\left(\prod_\lambda (\mathbb{P} \otimes_A F_\lambda)\right) \cong$$
$$\prod_\lambda \mathrm{H}_1(\mathbb{P} \otimes_A F_\lambda) \cong \prod_\lambda \operatorname{Tor}_1^A(M, F_\lambda) = 0,$$

and hence M is A-flat by Proposition C.9. □

C.14 Let $(A, \mathfrak{m})$ be a Noetherian local ring, and M be a finitely generated A-module. If $x_1, \ldots, x_r$ form a minimal set of generators of M and $\varepsilon_0 = \varepsilon : F_0 = A^r \to M$ be the map given by $\varepsilon(e_i) = x_i$, where $e_1, \ldots, e_r$ is the standard basis of A^r, we can form a short exact sequence

$$0 \to M_1 \xrightarrow{\iota_1} F_0 \xrightarrow{\varepsilon} M \to 0,$$

where $M_1 = \operatorname{Ker} \varepsilon$, and $\iota_1 : M_1 \to F_1$ is the inclusion. Then we have

$$M_1 \otimes_A A/\mathfrak{m} \xrightarrow{\iota_1 \otimes A/\mathfrak{m}} F_0 \otimes_A A/\mathfrak{m} \xrightarrow{\varepsilon \otimes A/\mathfrak{m}} M \otimes_A A/\mathfrak{m} \to 0$$

is exact. As $\varepsilon \otimes A/\mathfrak{m} : F_0/\mathfrak{m}F_0 = F_0 \otimes_A A/\mathfrak{m} \to M \otimes_A A/\mathfrak{m} = M/\mathfrak{m}M$ maps the basis $\overline{e}_1, \ldots, \overline{e}_r$ of $F_0/\mathfrak{m}F_0$ to the basis $\overline{x}_1, \ldots, \overline{x}_r$ of $M/\mathfrak{m}M$ bijectively, we have that $\varepsilon \otimes A/\mathfrak{m}$ is an isomorphism. Hence $\iota_1 \otimes A/\mathfrak{m} : M_1/\mathfrak{m}M_1 \to F_0/\mathfrak{m}F_0$ is a zero map. In other words, we have that $M_1 = \operatorname{Ker} \varepsilon \subseteq \mathfrak{m}F_0$. The same argument shows that we can take a finite free A-module F_1, a surjective map $\varepsilon_1 : F_1 \to M_1$ such that $M_2 := \operatorname{Ker} \varepsilon_1 \subseteq \mathfrak{m}F_1$. Continuing this way, we can construct an exact sequence

$$F_\bullet : \cdots \xrightarrow{\partial_{i+1}} F_i \xrightarrow{\partial_i} F_{i-1} \xrightarrow{\partial_{i-1}} \cdots \xrightarrow{\partial_2} F_1 \xrightarrow{\partial_1} F_0 \xrightarrow{\varepsilon} M \to 0,$$

where each F_i is a finite free A-module, and $\partial_i(F_i) \subseteq \mathfrak{m}F_{i-1}$ for $i = 1, 2, \ldots$. We call this complex $P_\bullet$ the *minimal free resolution* of M.

Let $L_\bullet$ be another projective resolution of M, and let $f_\bullet : F_\bullet \to L_\bullet$ and $g_\bullet : L_\bullet \to F_\bullet$ be chain maps that cover $1_M : M \to M$. Since both $1_{F_\bullet}$ and $g_\bullet f_\bullet$ cover 1_M, there exists some $\{s_i : F_i \to F_{i+1}\}_{i \in \mathbb{Z}}$ such that $1_{F_i} - g_i f_i = \partial_{i+1}s_i + s_{i-1}\partial_i$. Since $\partial_{i+1}(F_{i+1}) \subseteq \mathfrak{m}F_i$ and $\partial_i(F_i) \subseteq \mathfrak{m}F_{i-1}$, we have that $\operatorname{Im}(1_{F_i} - g_i f_i) \subseteq \mathfrak{m}F_i$. Hence $\operatorname{Im}(g_i f_i) + \mathfrak{m}F_i = F_i$. By NAK, we have that $g_i f_i : F_i \to F_i$ is surjective, and hence is an isomorphism by Corollary 3.4. So $g_\bullet f_\bullet : F_\bullet \to F_\bullet$ is an automorphism of complexes. Replacing $g_\bullet$ by $g'_\bullet : (g_\bullet f_\bullet)^{-1} g_\bullet$, we have that $g'_\bullet \circ f_\bullet = \mathrm{id}_{F_\bullet}$. Hence $f_\bullet$ is a split mono, and any projective resolution $L_\bullet$ of M is of the form $L_\bullet \cong F_\bullet \oplus L'_\bullet$. Note that $L'_\bullet$ is a projective resolution of the zero module, and hence is split exact. If $L_\bullet$ is also a minimal free resolution, then $g_\bullet$ is both a split mono and a split epi, and hence is an isomorphism. This shows that a minimal free resolution is unique up to isomorphism of complexes covering the identity map 1_M of M. It follows that the minimal free resolution $F_\bullet$ is the shortest projective resolution, whose length gives the value of $\operatorname{hd}_A M$. Note also that the boundary maps of $F_\bullet \otimes_A A/\mathfrak{m}$ are all zero. So we have

$$\operatorname{Tor}_i^A(M, A/\mathfrak{m}) = \mathrm{H}_i(F_\bullet \otimes_A A/\mathfrak{m}) = F_i \otimes_A A/\mathfrak{m}.$$

The dimension $\dim_{A/\mathfrak{m}} \operatorname{Tor}_i^A(M, A/\mathfrak{m}) = \operatorname{rank}_A F_i$ is called the ith *Betti number* of M, and is denoted by $\beta_i(M)$. Combining these, we get the following theorem.

Theorem C.15 *Let $(A, \mathfrak{m})$ be a Noetherian local ring, M be a finitely generated A-module, and $F_\bullet$ be the minimal free resolution of M. Then we have*

$$\operatorname{hd}_A M = \sup\{i \geq 0 | F_i \neq 0\} = \sup\{i \geq 0 | \operatorname{Tor}^i_A(M, A/\mathfrak{m}) \neq 0\} \leq \operatorname{hd}_A A/\mathfrak{m}.$$

If $L_\bullet$ is a projective resolution of M, then we have $L_\bullet \cong F_\bullet \oplus L'_\bullet$ for some split exact projective complex $L'_\bullet$.

Appendix D
Multilinear Algebra

This appendix gives a brief survey of multilinear algebra without proof. Readers are kindly invited to refer to [Nor] for details.

Let A be a commutative ring, and E an A-module. In this appendix, an A-algebra need not be commutative, unless it is explicitly called commutative.

We define the *tensor algebra* $TE = \bigoplus_{n \geq 0} E^{\otimes n}$, where $E^{\otimes n} = E \otimes_A E \otimes_A \cdots \otimes_A E$ (E appears n times). When we need to emphasize A, we also write $T_A E$. We define the product of TE as

$$(e_1 \otimes \cdots \otimes e_r) \cdot (f_1 \otimes \cdots \otimes f_s) = e_1 \otimes \cdots \otimes e_r \otimes f_1 \otimes \cdots \otimes f_s \in E^{\otimes(r+s)} \subseteq TE$$

for $e_1, \ldots, e_r, f_1, \ldots, f_s \in E$, and this makes TE an A-algebra. It is a graded algebra in the sense that $(E^{\otimes r}) \cdot (E^{\otimes s}) \subseteq E^{\otimes(r+s)}$.

D.1 Let I be the (two-sided) ideal of TE generated by $\{e \otimes e \mid e \in E\}$. Since the generators are homogeneous of degree two, I is a homogeneous ideal. So the quotient $\bigwedge E = TE/I = \bigoplus_{n \geq 0} \bigwedge^n E$ is a graded A-algebra in a natural way, where $\bigwedge^n E$ is the image of $E^{\otimes n}$ in $\bigwedge^n E$. We also write $\bigwedge_A E$, if necessary. The image of $e_1 \otimes \cdots \otimes e_n$ in $\bigwedge^n E$ is denoted by $e_1 \wedge \cdots \wedge e_n$ for $e_1, \ldots, e_n \in E$. We call $\bigwedge E$ the *exterior algebra* of E, and $\bigwedge^n E$ the nth *exterior power* of E. In $\bigwedge^2 E$, we have

$$e \wedge e' + e' \wedge e = (e + e') \wedge (e + e') - e \wedge e - e' \wedge e' = 0,$$

and hence $e \wedge e' = -e' \wedge e$ for $e, e' \in E$.

D.2 Let J be the ideal of TE generated by $\{e \otimes e' - e' \otimes e \mid e, e' \in E\}$. The quotient $\operatorname{Sym} E = TE/J = \bigoplus_{n \geq 0} \operatorname{Sym}_n E$ is a commutative graded A-algebra. We also write $\operatorname{Sym}_A E$, if necessary. The A-algebra $\operatorname{Sym} E$ is called the *symmetric*

S. Goto, K. Watanabe, *Commutative Algebra*, Graduate Texts in Mathematics 309,
https://doi.org/10.1007/978-981-95-4227-7

algebra of E. The homogeneous component $\mathrm{Sym}_n E$ of degree n is called the nth *symmetric power* of E.

D.3 Let C be an A-algebra, and $\phi : E \to C$ be an A-linear map. Then the A-linear map $\phi_n : E^{\otimes n} \to C$ given by $\phi_n(e_1 \otimes \cdots \otimes e_n) = \phi(e_1) \cdots \phi(e_n)$ is well-defined, since $(e_1, \ldots, e_n) \to e_1 \cdots e_n$ is multilinear. So $\Phi : TE \to C$ is induced, and it is easy to see that Φ is the unique A-algebra map whose restriction to $E = E^{\otimes 1}$ is ϕ. This universality characterizes the A-algebra TE. Note that Φ is degree-preserving if C is graded and $\phi(E) \subseteq C_1$. Note also that Φ induces the unique A-algebra map $\overline{\Phi} : \mathrm{Sym}\, E \to C$ whose restriction on E is ϕ, if C is commutative. We say that an $\mathbb{N}$-graded A-algebra $C = \bigoplus_{n \geq 0} C_n$ is *skew commutative* if $a^2 = 0$ for $a \in C_{2n+1}$ $(n \geq 0)$ and $ab = (-1)^{st} ba$ for $a \in C_s$ and $b \in C_t$ $(s, t \geq 0)$. Note that, if $1/2 \in A$, then $a^2 = (1/2)(ab + ba) = (1/2)(ab + (-1)^{(2n+1)^2} ab) = 0$ by the second condition, and so the first condition $a^2 = 0$ for $a \in C_{2n+1}$ follows from the second condition, where $b = a$. Note that $\bigwedge E$ is skew commutative. If C is an $\mathbb{N}$-graded skew commutative A-algebra and $\phi : E \to C_1$ is an A-linear map, then $\Phi : TE \to C$ induces a unique homomorphism of graded algebras $\overline{\Phi} : \bigwedge E \to C$.

D.4 Let $f : E \to E'$ be an A-linear map between A-modules. Then $f : E \to TE'$ is extended to a unique A-algebra map $Tf : TE \to TE'$, which is degree-preserving. Similarly, f induces unique A-algebra maps $\bigwedge f : \bigwedge E \to \bigwedge E'$ and $\mathrm{Sym}\, f : \mathrm{Sym}\, E \to \mathrm{Sym}\, E'$. These are checked easily using D.3. It is easy to see that T, $\bigwedge$, and Sym are functors from the category of A-modules to that of $\mathbb{N}$-graded A-algebras. Their homogeneous components $(-)^{\otimes n}$, $\bigwedge^n$, and Sym_n are functors from the category of A-modules to itself.

D.5 Let E and E' be A-modules. Then $T(E \oplus E') \to TE \otimes_A TE'$ given by $e + e' \mapsto e \otimes 1 + 1 \otimes e'$ is an isomorphism. It induces an isomorphism $\mathrm{Sym}(E \oplus E') \cong \mathrm{Sym}\, E \otimes_A \mathrm{Sym}\, E'$. Taking the homogeneous components, we have $\mathrm{Sym}_n(E \oplus E') \cong \bigoplus_{i+j=n} \mathrm{Sym}_i E \otimes_A \mathrm{Sym}_j E$.

Let C and D be $\mathbb{N}$-graded A-algebras. Then we define the graded A-algebra $C \mathbin{\underline{\otimes}}_A D$ as follows. As a graded A-module, $C \mathbin{\underline{\otimes}}_A D$ is the same as $C \otimes_A D$. So the homogeneous component $(C \mathbin{\underline{\otimes}}_A D)_n$ of degree n is $\bigoplus_{i+j=n} C_i \otimes_A D_j$. The product is given by $(c \otimes d)(c' \otimes d') = (-1)^{s't} cc' \otimes dd'$ for $c \in C$, $c' \in C_{s'}$, $d \in D_t$, and $d' \in D$. We call $C \mathbin{\underline{\otimes}}_A D$ the *skew tensor product* of C and D. It is easy to see that $C \mathbin{\underline{\otimes}}_A D$ is a graded A-algebra. If C and D are skew commutative, then so is $C \mathbin{\underline{\otimes}}_A D$. The A-algebra map $T(E \oplus E') \to TE \mathbin{\underline{\otimes}}_A TE'$ given by $e + e' \mapsto e \otimes 1 + 1 \otimes e'$ is also an isomorphism, and it induces an isomorphism $\bigwedge(E \oplus E') \cong \bigwedge E \mathbin{\underline{\otimes}}_A \bigwedge E'$ of graded A-algebras. So we have an isomorphism of A-modules $\bigwedge^n(E \oplus E') \cong \bigoplus_{i+j=n} \bigwedge^i E \otimes \bigwedge^j E'$.

D.6 If F is a finite free A-module of rank r with a free basis $x_1, \ldots, x_r$, then $F^{\otimes n}$ has a free basis

$$\{x_{i_1} \otimes \cdots \otimes x_{i_n} \mid 1 \leq i_1, i_2, \ldots, i_n \leq r\}.$$

Note that $\bigwedge^n F$ has a free basis

$$\{x_{i_1} \wedge \cdots \wedge x_{i_n} \mid 1 \le i_1 < i_2 < \cdots < i_n \le r\}.$$

In particular, $\bigwedge^n F$ is A-free of rank $\binom{r}{n}$ if $0 \le n \le r$, and $\bigwedge^n F = 0$ if $n < 0$ or $n > r$. Note also that $\varphi : A[x_1, \dots, x_n] \to \operatorname{Sym} F$ is an isomorphism of graded A-algebras, where $\varphi(x_i) = x_i$. Thus we often identify $\operatorname{Sym} F$ with $A[x_1, \dots, x_n]$. In particular, the set of monomials of degree n

$$\{x_{i_1} \cdots x_{i_n} \mid 1 \le i_1 \le i_2 \le \cdots \le i_n \le r\}$$

is a free basis of $\operatorname{Sym}_n F$. So $\operatorname{rank} \operatorname{Sym}_n F = \binom{r+n-1}{n}$. If P is a direct summand of F and $\iota : P \to F$ and $\pi : F \to P$ satisfy $\pi\iota = 1_P$, then $\bigwedge^n \pi \bigwedge^n \iota = \bigwedge^n (\pi\iota) = 1_{\bigwedge^n P}$, and hence $\bigwedge^n P$ is a direct summand of the finitely generated free module $\bigwedge^n F$. Hence if P is a finitely generated projective A-module, then so is $\bigwedge^n P$. Similarly, $\operatorname{Sym}_n P$ and $P^{\otimes n}$ are also finitely generated projective A-modules if P is so.

The following base change property holds.

Lemma D.7 *If B is a commutative A-algebra, then $\theta : B \otimes_A T_A E \to T_B(B \otimes_A E)$ given by*

$$\theta(s \otimes (e_1 \otimes e_n)) = s(e_1 \otimes 1) \otimes \cdots \otimes (e_n \otimes 1) \qquad (s \in B,\ e_1, \dots, e_n \in E)$$

is an isomorphism, and it induces isomorphisms $\overline{\theta} : B \otimes_A \bigwedge_A E \to \bigwedge_B (B \otimes_A E)$ and $\overline{\theta} : B \otimes_A \operatorname{Sym}_A E \to \operatorname{Sym}_A (B \otimes_A E)$. Since θ and $\overline{\theta}$ preserve gradings, we have isomorphisms of B-modules $B \otimes E^{\otimes n} \cong (B \otimes_A E)^{\otimes n}$, $B \otimes_A \bigwedge^n E \cong \bigwedge^n (B \otimes_A E)$, and $B \otimes_A \operatorname{Sym}_n E \cong \operatorname{Sym}_n (B \otimes_A E)$.

D.8 Let P be a finitely generated projective A-module. We define the *rank function* of P on $\operatorname{Spec} A$ by $\operatorname{rank}_P(\mathfrak{p}) = \operatorname{rank}_{A_\mathfrak{p}} P_\mathfrak{p} = \dim_{\kappa(\mathfrak{p})} P_\mathfrak{p}/\mathfrak{p} P_\mathfrak{p}$, where $\kappa(\mathfrak{p}) = A_\mathfrak{p}/\mathfrak{p} A_\mathfrak{p}$. It is a locally constant function on $\operatorname{Spec} A$, hence it is constant if $\operatorname{Spec} A$ is connected. If rank_P is constant, then we say that P has a *constant rank*, and its constant value is denoted by $\operatorname{rank}_A P$ or $\operatorname{rank} P$. If P has a constant rank and $r = \operatorname{rank}_A P$, then the finitely generated projective A-modules $P^{\otimes n}$, $\bigwedge^n P$, and $\operatorname{Sym}_n P$ have constant ranks r^n, $\binom{r}{n}$, and $\binom{r+n-1}{n}$, respectively. This is checked locally using Lemma D.7, and is reduced to the case that P is free, see D.6.

If the total quotient ring $Q = \mathrm{Q}(A)$ is semi-local (e.g., A is Noetherian), then P has a constant rank if and only if P has a rank in the sense of 1.134, that is, $P \otimes_A Q$ is a free Q-module, as can be seen easily by Proposition 5.110.

D.9 If P is a finitely generated projective A-module which has a constant rank and $r = \operatorname{rank}_A P$, the exterior power $\bigwedge^r P$ is called the *top exterior power* of P. It is finitely generated projective of constant rank one. It is easy to see that $\delta : A \to$

$\mathrm{Hom}_A(\bigwedge^r P, \bigwedge^r P)$ $(a \mapsto (x \mapsto ax))$ is an isomorphism. (We can check this after localizing at a prime, and then $\bigwedge^r P$ is a free module of rank one.) For each $\phi \in \mathrm{Hom}_A(P, P)$, we call $\delta^{-1}(\bigwedge^r \phi) \in A$ the *determinant* of ϕ, and we denote it by $\det \phi$. If $P = A^r$ is free, then $\det \phi$ is the determinant of the representation matrix of ϕ, which is independent of the choice of free basis of P.

Appendix E
From Field Theory

This appendix gives a brief survey of field theory without proof. Readers are kindly invited to refer to [Nag2].

E.1 Let k be a field, A be a nonzero ring, and $f : k \to A$ be a homomorphism. Unless $A = 0$, we have $f(1) = 1 \neq 0$. Hence $\operatorname{Ker} f \neq k$ and $\operatorname{Ker} f = 0$. So f must be injective. In particular, any homomorphism $k \to K$ between fields is injective. When we consider only one homomorphism from k to K, it is often considered as an extension K/k.

E.2 Let K/k be a field extension, and $\alpha \in K$. Then α is algebraic over k (see Definition 1.33), if and only if it is integral (see Definition 3.1) over k. When any $\alpha \in K$ is algebraic over k, then we say that the extension K/k is *algebraic*. This is equivalent to saying that the ring extension K/k is integral. Otherwise we say that the extension K/k is *transcendental*. For $\alpha \in K$, there exists a unique k-algebra map $\phi_\alpha : k[X] \to K$ such that $\phi_\alpha(X) = \alpha$.

E.3 By definition, α is algebraic over k if and only if ϕ_α is not injective. If so, since $k[X]$ is a PID, $\operatorname{Ker} \phi_\alpha$ is a principal ideal, and is generated by a nonzero polynomial of the lowest degree among $f \in k[X] \setminus \{0\}$ such that $f(\alpha) = 0$. The monic generator of $\operatorname{Ker} \phi_\alpha$ is called the *minimal polynomial* of α over k, and we denote it by f_α. For example, the minimal polynomial of $i = \sqrt{-1}$ over $\mathbb{R}$ is $X^2 + 1$. Since $k[\alpha]$ is integral over k, it is a field by Lemma 3.11. Since $k[X]/(f_\alpha) \cong k[\alpha]$, we have that f_α is an irreducible polynomial. Conversely, if $f \in k[X]$ is a monic irreducible polynomial and $f(\alpha) = 0$, then $(f) \subseteq (f_\alpha) \neq k[X]$ and (f) is maximal, so $f = cf_\alpha$ for some $c \in k[X]^\times = k^\times$. Since both f and f_α are monic, we have $f = f_\alpha$. A monic irreducible polynomial $f \in k[X]$ is the minimal polynomial of X in the field $K = k[X]/(f)$. So a polynomial $f \in k[X]$ is a minimal polynomial of some algebraic element α of some field extension K/k if and only if f is monic and irreducible.

S. Goto, K. Watanabe, *Commutative Algebra*, Graduate Texts in Mathematics 309,
https://doi.org/10.1007/978-981-95-4227-7

E.4 We say that a field extension K/k is *finite* or *finite algebraic* if $\dim_k K < \infty$. We often write $\dim_k K = [K : k]$, and call it the *degree* of the field extension K/k. For $f \in k[X] \setminus k$, the ring $k[X]/(f)$ has $1, X, \ldots, X^{n-1}$ as its k-basis, where $n = \deg f$. So if $K = k(\alpha)$ and α is algebraic over k, then K/k is finite and $[K : k] = \deg f_\alpha$. A field extension is finite if and only if it is algebraic and finitely generated.

Let k be a field and $f \in k[X] \setminus k$. We say that f *splits* (over k) if $f(X) = c(X - \alpha_1)(X - \alpha_2) \cdots (X - \alpha_n)$ for some $c \in k \setminus \{0\}$ and $\alpha_1, \ldots, \alpha_n \in k$, where $n = \deg f$. If K/k is a field extension and $f \in k[X] \setminus k$ splits over K, then we say that K is a *splitting field* of k. If, moreover, $K = k(\alpha_1, \ldots, \alpha_n)$ and $f(X) = c(X-\alpha_1)(X-\alpha_2) \cdots (X-\alpha_n)$ for some $c \in k \setminus \{0\}$, then we say that K is a *minimal splitting field* of f. If $\alpha \in K$, $f \in k[X] \setminus k$, and $f(\alpha) = 0$, then we say that α is a *root* of f in K. Obviously, there are at most n roots in K if $n = \deg f$.

Lemma E.5 *For a field k, the following are equivalent:*

(1) *Every $f \in k[X] \setminus k$ splits over k.*
(2) *Every $f \in k[X] \setminus k$ has a root in k. That is, there exists $\alpha \in k$ such that $f(\alpha) = 0$.*
(3) *If K is an algebraic extension field of k, then $K = k$.*

If the equivalent conditions in the lemma are satisfied, we say that k is algebraically closed. It is known that the field of complex numbers $\mathbb{C}$ is algebraically closed (*the fundamental theorem of algebra*, see [Nag2, Theorem 2.12.4]). We say that an extension field K of k is the algebraic closure of k if K is an algebraic extension of k, and K is *algebraically closed*. It is known that for every field k, there exists an algebraic closure of K uniquely up to isomorphisms of k-algebras, see [Nag2, Theorem 2.12.2]. Usually, for $f \in k[X] \setminus k$, a root of f means a root in the algebraic closure of k, if no extension field K of k is specified.

E.6 Let $f \in k[X] \setminus k$ be a polynomial, and consider the decomposition $f = c(X - \alpha_1)(X - \alpha_2) \cdots (X - \alpha_n)$ in $\overline{k}[X]$, where $\overline{k}$ is the algebraic closure of k. Let $\alpha \in \overline{k}$ be a root of f. The number of i such that $\alpha = \alpha_i$ is called the *multiplicity* of the root α of f. A root is called a *simple root* if its multiplicity is one, and is called a *multiple root* otherwise. If α is a root of f, then $\alpha_1, \ldots, \alpha_n$ are called the *conjugates* of α. If f does not have a multiple root, or equivalently, f has $n = \deg f$ roots in $\overline{k}$, we say that f is *separable*. A root α of f is a multiple root if and only if it is also a root of the derivative f' of f. In characteristic zero, if $\deg f \geq 1$, then $f' \neq 0$ and $\deg f' = \deg f - 1$. If f is irreducible and α is a root of f, then $f = cf_\alpha$ ($c \in k \setminus \{0\}$), and $f'(\alpha) \neq 0$, otherwise f divides f' and $\deg f' < \deg f$. So an irreducible polynomial does not have a multiple root in characteristic zero. On the other hand, if the characteristic of k is $p > 0$, then $f(X) = X^p - t^p \in k[t^p]$ is an irreducible polynomial, but it has a multiple root t, since $f(X) = (X - t)^p$.

E.7 Let K/k be a field extension. We say that α is *separable algebraic*, or often simply called separable, if α is algebraic over k and its minimal polynomial f_α is a separable polynomial. A field extension K/k is called *separable algebraic* (sometimes simply called separable, if there is no danger of confusion) if K/k is algebraic, and every $\alpha \in K$ is separable over k. If K/k and L/K are field extensions, then L/k is separable algebraic if and only if both K/k and L/K are so. Every algebraic extension is separable in characteristic zero. A field k is said to be *perfect* if every algebraic extension field K of k is separable. A field of characteristic zero is perfect. A finite field is also perfect. If k is a field of characteristic $p > 0$, then $k(x)/k(x^p)$ is not separable. An algebraic field extension that is not separable is called an *inseparable extension*.

E.8 An algebraic extension of fields K/k is called a *normal extension* if for any $\alpha \in K$, the minimal polynomial f_α of α splits over K. Or equivalently, K/k is normal if for every $\alpha \in K$, all the conjugates of α also lie in K. For example, $\mathbb{Q}(\sqrt[3]{2})/\mathbb{Q}$ is not a normal extension, since the conjugates of $\sqrt[3]{2}$ are $\sqrt[3]{2}, \sqrt[3]{2}\omega, \sqrt[3]{2}\omega^2$, where $\omega = (-1+\sqrt{-3})/2$, but $\sqrt[3]{2}\omega \notin \mathbb{Q}(\sqrt[3]{2})$. If K/k is an algebraic extension, adding all the conjugates of elements of K, we get the smallest normal extension L of k containing K. We call L the *normal closure* of K/k. If K/k is finite, then so is L.

E.9 We say that a field extension K/k is *Galois* if it is separable algebraic and normal. If K/k is a separable algebraic field extension, then so is its normal closure L. So L/k is Galois, and L is also called the *Galois closure* of K/k. For a field extension K/k, the group of k-algebra automorphisms of K, equipped with the group structure by the composition, is denoted by $\operatorname{Aut}_k K$. A finite field extension K/k is Galois if and only if $\#\operatorname{Aut}_k K = [K:k]$. If K/k is a Galois extension, then $\operatorname{Aut}_k K$ is often denoted by $\operatorname{Gal}_k K$, and called the *Galois group* of K/k. If K/k_0 is a field extension, G a finite subgroup of $\operatorname{Aut}_{k_0} K$, and $k = K^G = \{a \in K \mid \forall g \in G\ ga = a\}$, then k is an intermediate field of K/k_0. Moreover, K/k is Galois of the Galois group G. In particular, we have $[K:k] = \#G < \infty$.

The following theorem is very important in Galois theory.

Theorem E.10 (Fundamental Theorem of Galois Theory) *Let K/k be a finite Galois extension, and $G = \operatorname{Gal}_k(K)$. Let $\mathcal{G}$ be the set of subgroups of G, and $\mathcal{K}$ be the set of intermediate fields of k and K. Let $\Phi : \mathcal{G} \to \mathcal{K}$ be the map given by $\Phi(H) = K^H$, and $\Psi : \mathcal{K} \to \mathcal{G}$ be the map given by $\Psi(L) = \operatorname{Gal}_L(K) \subseteq \operatorname{Gal}_k(K) = G$. Then:*

(1) *Φ and Ψ are order-reversing bijections, and $\Psi = \Phi^{-1}$.*
(2) *$N \in \mathcal{G}$ is a normal subgroup of G if and only if $\Phi(N) = K^N$ is a Galois extension of k. If this is the case, the restriction $\rho = (-)|_{K^N} : \operatorname{Gal}_k(K) \to \operatorname{Gal}_k(K^N)$ is well-defined and surjective, and $N = \operatorname{Ker}\rho$.*
(3) *For $H, H' \in \mathcal{G}$, $\Phi(H \cap H')$ is nothing but the composite field $\Phi(H)\Phi(H')$, that is, the subfield of K generated by $\Phi(H) \cup \Phi(H')$.*

E.11 Let K/k be a field extension. A subset $S \subseteq K$ is said to be *algebraically independent* if for every sequence $s_1, \dots, s_r$ of distinct finitely many elements in S and every $f(X_1, \dots, X_r) \in k[X_1, \dots, X_r] \setminus \{0\}$, $f(s_1, \dots, s_r) \neq 0$. We say that

$S \subseteq K$ is a *transcendence basis* of K/k if S is algebraically independent over k and $K/k(S)$ is an algebraic extension.

Lemma E.12 *Let K/k be a field extension, and S and T be subsets of K such that $S \subseteq T$, S is algebraically independent over k, and $K/k(T)$ is algebraic. Then there exists a transcendence basis U of K/k such that $S \subseteq U \subseteq T$.*

A transcendence basis is not unique, but its cardinality is independent of the choice of transcendence basis. We denote it by $\mathrm{tr.deg}_k K$, and call it the *transcendence degree* of K/k. If k is a field and $X_1, \ldots, X_n$ are indeterminates, then $X_1, \ldots, X_n$ is a transcendence basis of $K = k(X_1, \ldots, X_n)$ over k, and we have $\mathrm{tr.deg}_k K = n$. If K/k and L/K are extensions of fields, then we have $\mathrm{tr.deg}_k L = \mathrm{tr.deg}_k K + \mathrm{tr.deg}_K L$.

E.13 A transcendence basis S of K/k is called *separating* if $K/k(S)$ is separable algebraic. A field extension K/k is said to be *separably generated* if it has a separating transcendence basis. A field extension K/k is said to be *separable* if $\overline{k} \otimes_k K$ is reduced, where $\overline{k}$ is the algebraic closure of k. An algebraic extension K/k is separable in this sense if and only if it is separable algebraic. So it is easy to see that a separably generated field extension is separable. A finitely generated extension field $K = k(\alpha_1, \ldots, \alpha_n)$ is separable over k if and only if it is separably generated. Any field extension in characteristic zero is separably generated. However, if k is a field of characteristic $p > 0$, then $K = k(X, X^{1/p}, X^{1/p^2}, X^{1/p^3}, \ldots) = \varinjlim k(X^{1/p^e})$ is separable over k, but is not separably generated.

Appendix F
Theorems of Hochster–Ratliff and Matijevic–Roberts Type

For a commutative ring A and an A-algebra B, we say that B is *essentially of finite type* over A if there exists some finitely generated A-algebra C and a multiplicatively closed subset S of C such that B is isomorphic to C_S as an A-algebra. A homomorphism $f : A \to B$ between Noetherian rings is said to be a *regular homomorphism* if f is flat, and for any prime ideal $\mathfrak{p}$ of A and any finite extension field K of $\kappa(\mathfrak{p}) = A_\mathfrak{p}/\mathfrak{p}A_\mathfrak{p}$, the base change $B \otimes_A K$ is regular. A homomorphism between Noetherian rings is called *smooth* if it is essentially of finite type and regular.

Let Γ be a multiplicative abelian group which is isomorphic to the additive group $\mathbb{Z}^n$, and $R = \bigoplus_{\gamma \in \Gamma} R_\gamma$ be a Noetherian Γ-graded ring. We denote the set of H-maximal ideals (see Definition 7.20) of R by $\operatorname{H-Max} R$.

For an ideal I of R, we define I^* to be the ideal generated by all the homogeneous elements of I. I^* is the largest homogeneous ideal contained in I. Let $R\Gamma = \bigoplus_{\gamma \in \Gamma} R\gamma$ be the group ring, which is isomorphic to the Laurent polynomial ring $R[t_1^{\pm 1}, \ldots, t_n^{\pm 1}]$. Let $\iota : R \hookrightarrow R\Gamma$ be the inclusion, and $\omega : R \to R\Gamma$ be the R_0-linear map given by $\omega(\sum_\gamma a_\gamma) = \sum_\gamma a_\gamma \gamma$, where $a_\gamma \in R_\gamma$ and the sum is finite. It is easy to see that ω is a homomorphism of graded rings, where the grading of $R\Gamma$ is given by $[R\Gamma]_\gamma = R\gamma$ for $\gamma \in \Gamma$. For each ideal I of R, we have $I^* = \omega^{-1}(\iota(I)R\Gamma)$. If $P \in \operatorname{Spec} R$, then $PR\Gamma$ is a prime ideal, and hence $P^* = \omega^{-1}(PR\Gamma)$ is also a prime ideal. We also define R_0-linear maps $\Box : R\Gamma \to R\Gamma$ and $\Box^- : R\Gamma \to R\Gamma$ by $\Box(a\gamma) = a\delta\gamma$ and $\Box^-(a\gamma) = a\delta^{-1}\gamma$ for $a \in R_\delta$ and $\gamma \in \Gamma$, respectively. Note that $\Box$ and $\Box^-$ are R_0-algebra maps, and they are inverses of each other. Being an isomorphism, $\Box$ is faithfully flat and smooth. Since $\omega = \Box \circ \iota$ and both $\Box$ and ι are faithfully flat and smooth, we have that ω is also faithfully flat and smooth.

Let $\mathcal{N}$ denote the class of Noetherian rings, and $\mathcal{A}$ be its subclass. We assume that if $A \in \mathcal{A}$, then for any finitely generated A-algebra B and any prime ideal P of B, the local ring B_P lies in $\mathcal{A}$. Let C be a subclass of $\mathcal{A}$.

S. Goto, K. Watanabe, *Commutative Algebra*, Graduate Texts in Mathematics 309,
https://doi.org/10.1007/978-981-95-4227-7

Theorem F.1 *Assume the following:*

(1) *If $A \in C$, then for any $\mathfrak{p} \in \operatorname{Spec} A$, we have $A_{\mathfrak{p}} \in C$.*
(2) *If $(A, \mathfrak{m}) \in \mathcal{A}$ and $(A, \mathfrak{m}) \to (B, \mathfrak{n})$ is a smooth local homomorphism, then $A \in C$ if and only if $B \in C$.*

Let Γ be a multiplicative abelian group which is isomorphic to the additive group $\mathbb{Z}^n$, and $R = \bigoplus_{\gamma \in \Gamma} R_\gamma$ be a Noetherian Γ-graded ring. Assume that all the local rings of R lie in $\mathcal{A}$. Then the following hold:

(i) *For $P \in \operatorname{Spec} R$, we have $R_P \in C$ if and only if $R_{P^*} \in C$.*
(ii) *If $R_{\mathfrak{m}} \in C$ for any $\mathfrak{m} \in \operatorname{H-Max} R$, then $R_P \in C$ for any $P \in \operatorname{Spec} R$.*

Proof

(i) Assume that $R_{P^*} \in C$. Since $\omega : R \to R\Gamma$ is smooth and the prime ideal $PR\Gamma$ lies over P^*, we have a smooth local homomorphism $R_{P^*} \to (R\Gamma)_{PR\Gamma}$. Since $R_{P^*} \in C$, we have that $(R\Gamma)_{PR\Gamma} \in C$. On the other hand, $\iota : R \to R\Gamma$ is smooth, and the prime ideal $PR\Gamma$ lies over P. Hence we get a smooth local homomorphism $R_P \to (R\Gamma)_{PR\Gamma}$, and hence $R_P \in C$. Conversely, if $R_P \in C$, then $R_{P^*} \in C$, since R_{P^*} is a localization of R_P.
(ii) Take an H-maximal ideal $\mathfrak{m}$ of R containing P^*. Since $R_{\mathfrak{m}} \in C$, we have $R_{P^*} \in C$. By (i), we have $R_P \in C$.

□

The corollary below follows immediately by the theorem. The corollary for the case that $\mathbb{P}$ is Cohen–Macaulay was proposed by M. Nagata, and proved by J. Matijevic and P. Roberts [MR] and M. Hochster and L. J. Ratliff, Jr. [HR] independently.

Corollary F.2 *Let $\mathbb{P}$ be one of the following properties of Noetherian local rings:* (a) *Cohen–Macaulay;* (b) *Cohen–Macaulay of type $\leq r$ $(r \geq 1)$;* (c) *Gorenstein;* (d) *local complete intersection;* (e) *regular;* (f) *smooth over a fixed field k;* (g) (S_n) *$(n \geq 1)$;* (h) (R_n) *$(n \geq 0)$;* (i) *reduced;* (j) *normal. Let R be a $\mathbb{Z}^n$-graded Noetherian ring. If $R_{\mathfrak{m}}$ satisfies $\mathbb{P}$ for any H-maximal ideal $\mathfrak{m}$ of R, then $R_{\mathfrak{p}}$ satisfies $\mathbb{P}$ for any $\mathfrak{p} \in \operatorname{Spec} R$.*

The approach to Theorem F.1 follows [HM, Section 7].

Appendix G
Purity and Hochster–Eagon Theorem

Let A be a ring and $\varphi : C \to D$ be an A-linear map between A-modules. We say that φ is *pure* or *A-pure* if $\varphi \otimes 1_M : C \otimes_A M \to D \otimes_A M$ is injective for any A-module M. Hence a ring homomorphism $\varphi : A \to B$ is pure if it is so as an A-linear map (cf. Definition 7.5). A composition of pure A-linear maps is again pure. If B is an A-algebra and $\varphi : C \to D$ is A-pure, then $1_B \otimes \varphi : B \otimes_A C \to B \otimes_A D$ is B-pure. An A-linear map $\varphi : C \to D$ is A-pure if and only if $\varphi_\mathfrak{m} : C_\mathfrak{m} \to D_\mathfrak{m}$ is $A_\mathfrak{m}$-pure for any maximal ideal $\mathfrak{m}$ of A.

Proposition G.1 *Let $(A, \mathfrak{m})$ be a Noetherian local ring, B an A-module, and $\varphi : A \to B$ be an A-linear map. Then the following are equivalent:*

(1) *φ is pure.*
(2) *$\varphi \otimes 1_E : A \otimes_A E \to B \otimes_A E$ is injective.*
(3) *The composition $\rho : A/\mathfrak{m} \hookrightarrow E \cong A \otimes_A E \xrightarrow{\varphi \otimes 1_E} B \otimes_A E$ is nonzero.*
(4) *For any A-module M of finite length, $\varphi \otimes 1_M : A \otimes_A M \to B \otimes_A M$ is injective.*
(5) *For any finitely generated A-module M, $\varphi \otimes 1_M : A \otimes_A M \to B \otimes_A M$ is injective,*

where E is the injective envelope of the residue field $A/\mathfrak{m}$.

Proof (1) $\Rightarrow$(2) is obvious.

(2) $\Rightarrow$(3). The composition map ρ is injective, and $A/\mathfrak{m}$ is nonzero. So ρ is nonzero.

(3) $\Rightarrow$(2). If $\operatorname{Ker}(\varphi \otimes 1_E)$ is nonzero, then it intersects the image of $A/\mathfrak{m}$, since $A/\mathfrak{m} \hookrightarrow E \cong A \otimes_A E$ is an essential mono. Since $A/\mathfrak{m}$ is a simple A-module, ρ must be zero then.

(2) $\Rightarrow$(4). Let $j : M \to E'$ be the injective envelope. Since E' is isomorphic to E^n for some $n \geq 0$, we have that $\varphi \otimes 1_{E'} : A \otimes_A E' \to B \otimes_A E'$ is injective. Hence $(\varphi \otimes 1_{E'})(1_A \otimes j) = \varphi \otimes j = (1_B \otimes j)(\varphi \otimes 1_M)$ is injective. Hence $\varphi \otimes 1_M$ is injective.

S. Goto, K. Watanabe, *Commutative Algebra*, Graduate Texts in Mathematics 309,
https://doi.org/10.1007/978-981-95-4227-7

(4) $\Rightarrow$(5). If $\varphi \otimes 1_M : A \otimes_A M \to B \otimes_A M$ is not injective, then we can take $m \in M$ such that $\varphi(1) \otimes m = 0$. We can take $n \geq 1$ such that $m \notin \mathfrak{m}^n M$. Then $M/\mathfrak{m}^n M$ is of finite length, and $\varphi \otimes 1_{M/\mathfrak{m}^n M} : A \otimes_A M/\mathfrak{m}^n M \to B \otimes_A M/\mathfrak{m}^n M$ is not injective.

(5) $\Rightarrow$(1). If M is an A-module and $\varphi \otimes 1_M : A \otimes_A M \to B \otimes_A M$ is not injective, then we can find $m \in M \setminus \{0\}$ such that $\varphi(1) \otimes m = 0$ in $B \otimes_A M$. Then there exists some finitely generated A-submodule N of M such that $m \in N$, and $\varphi(1) \otimes m$ is zero in $B \otimes_A N$. This means that $\varphi \otimes 1_N : A \otimes_A N \to B \otimes_A N$ is not injective for the finitely generated module N. □

Corollary G.2 *Let* $(A, \mathfrak{m})$ *be a Noetherian ring, and* $\varphi : A \to B$ *a pure ring homomorphism. Then there exists some maximal ideal* $\mathfrak{n}$ *of* B *such that* $A \to B_{\mathfrak{n}}$ *is pure.*

Proof By Proposition G.1, $\rho(1) \in B \otimes_A E$ is nonzero, where $\rho : A/\mathfrak{m} \to B \otimes_A E$ is as in condition (3) in the proposition. So there exists some maximal ideal $\mathfrak{n}$ of B such that $\rho(1) \in B_{\mathfrak{n}} \otimes_A E$ is nonzero. By Proposition G.1 again, $A \to B_{\mathfrak{n}}$ is pure. □

If $\varphi : A \to B$ is a pure ring homomorphism between local rings $(A, \mathfrak{m})$ and $(B, \mathfrak{n})$, then φ is a local homomorphism. Indeed, since $A/\mathfrak{m} = A \otimes_A A/\mathfrak{m} \to B \otimes_A A/\mathfrak{m} \cong B/\mathfrak{m}B$ is injective, $\mathfrak{m}B \neq B$.

Lemma G.3 *If* A *is a Noetherian ring and* $f : A \to B$ *is a pure ring homomorphism, then* $\dim A \leq \dim B$.

Proof Let $\mathfrak{p}_0 \supsetneq \mathfrak{p}_1 \supsetneq \mathfrak{p}_2 \supsetneq \cdots \supsetneq \mathfrak{p}_d$ be a chain of prime ideals of A. Note that $A_{\mathfrak{p}_0} \to B_{\mathfrak{p}_0}$ is pure. By Lemma G.2, there exists some prime ideal P_0 of B lying over $\mathfrak{p}_0$ such that $A_{\mathfrak{p}_0} \to (B_{\mathfrak{p}_0})_{P_0 B_{\mathfrak{p}_0}} = B_{P_0}$ is pure. Using this argument to the pure ring homomorphism $A_{\mathfrak{p}_0} \to B_{P_0}$, we know that there exists some prime ideal $P_1 \subseteq P_0$ such that $P_1 \cap A = \mathfrak{p}_1$ and $A_{\mathfrak{p}_1} \to B_{P_1}$ is pure. Continuing this, we can take a chain of prime ideals

$$P_0 \supsetneq P_1 \supsetneq P_2 \supsetneq \cdots \supsetneq P_d$$

such that $P_i \cap A = \mathfrak{p}_i$ and $A_{\mathfrak{p}_i} \to B_{P_i}$ is pure. This proves the lemma. □

Proposition G.4 *Let* $f : (A, \mathfrak{m}) \to (B, \mathfrak{n})$ *be a pure local ring homomorphism such that* B *is Noetherian and the fiber ring* $B/\mathfrak{m}B$ *is Artinian. Then* $\dim A = \dim B$ *and* $\operatorname{depth} A \geq \operatorname{depth} B$.

Proof We have $\dim A \leq \dim B \leq \dim A + \dim B/\mathfrak{m}B = \dim A$ by Lemma G.3 and the assumption that $\dim B/\mathfrak{m}B$ is Artinian.

Let $r = \operatorname{depth} A$ and $s = \operatorname{depth} B$. We prove that $r \geq s$ by induction on s. If $s = 0$, then the assertion is trivial. Assume that $s > 0$. If $r = 0$, then there is an injective A-linear map $h : A/\mathfrak{m} \to A$. Then $1_B \otimes h : B/\mathfrak{m}B \to B$ is nonzero, since $f(h(1)) \neq 0$. Since $B/\mathfrak{m}B$ is of finite length, the image of $1_B \otimes h$ is a nonzero finite-length module. Hence $s = 0$, and this is a contradiction. So $r > 0$. For each

$P \in \operatorname{Ass} B$, we have $P \cap A \neq \mathfrak{m}$, since $\mathfrak{n} \cap A = \mathfrak{m}$, $P \subsetneq \mathfrak{n}$, and $B/\mathfrak{m}B$ is zero-dimensional. By prime avoidance, there exists some

$$y \in \mathfrak{m} \setminus \left(\left(\bigcup_{P \in \operatorname{Ass} B} (P \cap A) \right) \cup \left(\bigcup_{\mathfrak{p} \in \operatorname{Ass} A} \mathfrak{p} \right) \right).$$

So y is both A-regular and B-regular. Hence $A/yA \to B/yB$ is pure, and $(B/yB)/\mathfrak{m}(B/yB)$ is Artinian. By the induction assumption, we have

$$r - 1 = \operatorname{depth} A/yA \geq \operatorname{depth} B/yB = s - 1$$

and hence $r \geq s$, as required. □

Theorem G.5 (Hochster–Eagon [HE]) *Let $f : A \to B$ be a pure ring homomorphism. Let B be Noetherian and Cohen–Macaulay. If the fiber ring $B \otimes_A \kappa(\mathfrak{m})$ is zero-dimensional for any maximal ideal $\mathfrak{m}$ of A, then A is Cohen–Macaulay.*

Proof It suffices to show that $A_{\mathfrak{m}}$ is a Cohen–Macaulay local ring for any maximal ideal $\mathfrak{m}$ of A. Replacing f by $f_{\mathfrak{m}} : A_{\mathfrak{m}} \to B_{\mathfrak{m}}$, we may assume that $(A, \mathfrak{m})$ is local. Then we can find a maximal ideal $\mathfrak{n}$ of B lying over $\mathfrak{m}$ such that $A \to B_{\mathfrak{n}}$ is pure. As a localization of a zero-dimensional ring is still zero-dimensional, we may assume that f is a local homomorphism between local rings. By the Cohen–Macaulay property of B and Proposition G.4, we have

$$\operatorname{depth} A \geq \operatorname{depth} B = \dim B = \dim A \geq \operatorname{depth} A,$$

and the assertion follows. □

Exercise G.5 (Cf. [Has2]) *Prove the following. Let $f : A \to B$ be a pure ring homomorphism. Let B be a Noetherian ring which satisfies Serre's (S_n) condition. If the fiber ring $B \otimes_A \kappa(\mathfrak{p})$ is zero-dimensional for any prime ideal $\mathfrak{p}$ of A, then A satisfies (S_n), too.*

A pure subring of an F-rational ring of prime characteristic p need not be F-rational in general, see Exercise 10.4. However, we have the following.

Theorem G.6 *Let B be an F-rational Noetherian ring of prime characteristic p, and A its pure subring. If the closed fiber ring $B/\mathfrak{m}B$ is zero-dimensional for any maximal ideal $\mathfrak{m}$ of A, then A is also an F-rational Noetherian ring of characteristic p.*

Proof Being a pure subring of a Noetherian ring of characteristic p, A is Noetherian and of characteristic p. Since B is Cohen–Macaulay by Theorem 10.17, (2), we have that A is also Cohen–Macaulay by Theorem G.5. So by Theorem 10.17, (5), it suffices to show that $A_{\mathfrak{m}}$ is F-rational local for each $\mathfrak{m} \in \operatorname{Max} A$. Localizing at $\mathfrak{m}$, we may assume that $(A, \mathfrak{m})$ is local. We can take a maximal ideal $\mathfrak{n}$ of B such that $A \to B_{\mathfrak{n}}$ is pure and local. So we may assume that B is also local, and $A \to B$ is a pure local homomorphism whose closed fiber ring $B/\mathfrak{m}B$ is zero-dimensional. Since A is Cohen–Macaulay, it is equi-dimensional. By Theorem 10.17, (3), it suffices to prove that a system of parameters $(x_1, \ldots, x_d)$ of A is tightly closed, where $d = \dim A$. Note that $\dim B = \dim A = d$ by Proposition G.4, and $x_1, \ldots, x_d$ is also an s.o.p. of B by assumption. Let $x \in (x_1, \ldots, x_d)^*$ and assume that $c \in A^\circ$ and $cx^q \in (x_1^q, \ldots, x_d^q)$ for $q \gg 1$. Since $\dim B/cB = \dim A/cA = d - 1$ and B is Cohen–Macaulay, we have that $c \in B^\circ$, and hence $x \in ((x_1, \ldots, x_d)B)^* = (x_1, \ldots, x_d)B$. So $x \in (x_1, \ldots, x_d)B \cap A = (x_1, \ldots, x_d)$, since pure implies cyclically pure. Hence $(x_1, \ldots, x_d)$ is tightly closed. This proves that A is F-rational, as desired. □

Corollary G.7 *Let $(A, \mathfrak{m})$ be a Noetherian local ring whose completion $\widehat{A}$ is F-rational. Then A is F-rational.*

The converse holds for excellent local rings.

Proposition G.8 (cf. [Vel, Proposition 0.10]) *Let $(A, \mathfrak{m})$ be an excellent F-rational Noetherian local ring of prime characteristic p. Then the completion $\widehat{A}$ is also F-rational.*

Bibliography

[AN] Y. Akizuki and M. Nagata, *Kindai Daisugaku* (Modern Algebra), reissue, Kyoritsu Shuppan, 2012 (originally published in 1957).

[Aoy] Y. Aoyama, Some basic results on canonical modules, *J. Math. Kyoto Univ.* **23** (1983), 85–94.

[Ar] M. Artin, On the joins of Hensel rings, *Adv. Math.* **7** (1971), 282–296.

[AM] M. F. Atiyah and I. G. Macdonald, *Introduction to Commutative Algebra*, Addison-Wesley Publishing Co., Reading, Mass.-London-Don Mills, Ont., 1969.

[AB] M. Auslander and D. A. Buchsbaum, Codimension and multiplicity, *Ann. of Math.* **68** (1958), 625–657.

[BFS] V. Barucci, R. Fröberg, M. Şahin, On free resolutions of some semigroup rings, *J. Pure Appl. Algebra* **218** (2014), 1107–1116.

[Ba] H. Bass, Big projective modules are free, *Illinois J. Math.* **7** (1963), 24–31.

[Bo] Bourbaki, *Commutative Algebra*, Chapters 1–7, Springer-Verlag, 1985.

[Bra] A. Braun, On the Gorenstein property for modular invariants, *J. Algebra* **345** (2011), 81–99.

[Bre] H. Bresinsky, Symmetric semigroups of integers generated by 4 elements, *Manuscripta Math.* **17** (1975), 205–219.

[BH] W. Bruns and J. Herzog, *Cohen–Macauly Rings*, revised ed., Cambridge University Press, 1997.

[BV] W. Bruns and U. Vetter, *Determinantal Rings*, Springer Lecture Notes in Math., **1327** (1988).

[BE1] D. A. Buchsbaum and D. Eisenbud, What makes a complex exact?, *J. Algebra* **25** (1973), 259–268.

[BE2] D. A. Buchsbaum and D. Eisenbud, Algebra structures for finite free resolutions, and some structure theorems for ideals of codimension 3, *Amer. J. Math.* **99** (1977), 447–485.

[BM] H. Brenner and P. Monsky, Tight closure does not commute with localization, *Ann. of Math.* (2) **171** (2010), 571–588.

[Che] C. Chevalley, Invariants of finite groups generated by reflections, *American J. Math.*, **88** (1955), 778–782.

[Coh] I. S. Cohen, On the structure and ideal theory of complete local rings, *Trans. Amer. Math. Soc.* **59** (1946), 54–106.

[DEP] C. De Cocini, D. Eisenbud and C. Procesi, Hodge algebras, *Asterisque*, **91**, Soc. Math. de France, (1982).

S. Goto, K. Watanabe, *Commutative Algebra*, Graduate Texts in Mathematics 309,
https://doi.org/10.1007/978-981-95-4227-7

[Del] C. Delorme, Sous-monoïdes d'intersection complète de ℕ, *Ann. Sci. 'Ecole Norm. Sup.* (4) **9** (1976), 145–154.

[EN] J. A. Eagon and D. G. Northcott, Ideals defined by matrices and a certain complex associated with them, *Proc. Roy. Soc.* **269** (1962), 188–204.

[Ei] D. Eisenbud, Introduction to algebras with straightening laws, in B. R. McDonald (ed.), *Ring Theory and Algebra, III*, Lect. Notes in Pure and Appl. Math. **55**, M. Dekker (1980), 243–268.

[Ei2] D. Eisenbud, *Commutative Algebra with a View Toward Algebraic Geometry*, Grad. Texts in Math. **150**, Springer-Verlag, 1995.

[FW] R. Fedder and K.-i. Watanabe, A characterization of F-regularity in terms of F-purity, in *Commutative Algebra* (Berkeley, CA, 1987), Math. Sci. Res. Inst. Publ. **15**, Springer, New York (1989), 227–245.

[FlWo] P. Fleischmann and C. Woodcock, Relative invariants, ideal classes and quasi-canonical modules of modular rings of invariants, *J. Algebra* **348** (2011), 110–134.

[Fo] R. Fossum, *The Divisor Class Group of Krull Domains*, Springer Verlag, 1973.

[GhGHOPV] L. Ghezzi, S. Goto, J. Hong, K. Ozeki, T. T. Phuong, and W. V. Vasconcelos, Cohen–Macaulayness versus the vanishing of the first Hilbert coefficient of parameter ideals, *J. London Math. Soc.* **81** (2010), 679–695.

[G] S. Goto, On Buchsbaum rings, *J. Algebra* **67** (1980), 272–279.

[GO] S. Goto and T. Ogawa, A note on rings with finite local cohomology. *Tokyo J. Math.* **6** (1983), 403–411.

[GW1] S. Goto and K.-i. Watanabe, On graded rings, I, *J. Math. Soc. Japan* **30** (1978), 179–213.

[GW2] S. Goto and K.-i. Watanabe, The structure of one-dimensional F-pure rings, *J. Algebra* **49** (1977), 415–421.

[EGAIII-1] A. Grothendieck, Éléments de géometrie algébrique. III. Étude cohomologique des faisceaux cohérents, Première partie, Publ. IHES **11** (1961).

[Gro] A. Grothendieck, *Local Cohomology*, Lecture Notes in Math. **41**, Springer Verlag, Heidelberg, 1967.

[Hara] N. Hara, A characterization of rational singularities in terms of injectivity of Frobenius map, *Amer. J. Math.* **120** (1998), 981–996.

[HY] N. Hara and K. Yoshida, A generalization of tight closure and multiplier ideals, *Trans. Amer. Math. Soc.* **355** (2003), 3143–3174.

[Has2] M. Hashimoto, *Hochster–Eagon type theorem for Serre's* (S_n) *condition*, *Proc. Amer. Math. Soc.* **153** (2025), 1041–1044.

[Has] M. Hashimoto, Equivariant class group. III. Almost principal fiber bundles, `arXiv:1503.02133`

[HM] M. Hashimoto and M. Miyazaki, G-prime and G-primary G-ideals on G-schemes, *Comm. Algebra* **41** (2013), 2254–2296.

[HaHy] F. Hayasaka and E. Hyry, On the Buchsbaum-Rim function of a parameter module, *J. Algebra* **327** (2011), 307–315.

[He] J. Herzog, Generators and relations of abelian semigroups and semigroup rings, *Manuscripta Math.* **3** (1970), 175–193.

[Hi] T. Hibi, Distributive lattices, affine semigroup rings, and algebras with straightening laws, in M. Nagata and H. Matsumura (eds.), *Commutative Algebra and Combinatorics*, Advanced studies in Pure Math. **11**, North Holland (1987), 93–109.

[Ho1] M. Hochster, Rings of invariants of tori, Cohen-Macaulay rings generated by monomials, and polytopes, *Ann. of Math.* **96** (1972), 318–337.

[Ho2] M. Hochster, Cyclic purity versus purity in excellent Noetherian rings, *Trans. Amer. Math. Soc.* **231** (1977), 463–488.

[HE] M. Hochster and J. A. Eagon, Cohen–Macaulay rings, invariant theory, and the generic perfection of determinantal loci, *Amer. J. Math.* **93** (1971), 1020–1058.

[HH1] M. Hochster and C. Huneke, Tight closure, invariant theory and the Briançon–Skoda theorem, *J. Amer. Math. Soc.* **3** (1990), 31–116.

[HR] M. Hochster and L. J. Ratliff, Jr., Five theorems on Macaulay rings, *Pacific J. Math.* **44** (1973), 147–172.

[Hu1] C. Huneke, The theory of d-sequences and powers of ideals, *Adv. in Math.* **46** (1982), 249–279.

[Hu2] C. Huneke, *Tight Closure and Its Applications*, Regional Conference Series in Math. **88**, Amer. Math. Soc., 1996.

[KW] V. Kac and K.-i. Watanabe, Finite linear groups whose ring of invariants is a complete intersection, *Bull. Amer. Math. Soc.* **6** (1982), 221–223.

[Ka] Y. Kamoi, Noetherian rings generated by an Abelian group, *Tokyo J. Math.*, **18** (1995), 31–48.

[LW] G. J. Leuschke and R. Wiegand, *Cohen–Macaulay Representations*, Math. Surveys Monogr. **181**, Amer. Math. Soc., 2012.

[MaSch] L. Ma and K. Schwede, Singularities in mixed characteristic via big Cohen-Macaulay algebras, *Duke Math. J.* **170** (2021), 2815–2890.

[Mac] F.S. Macaulay, *Algebraic Theory of Modular Systems*, Cambridge University Press, 1916.

[MSV] M. Mandal, B. Singh, and J. K. Verma, On some conjectures about the Chern numbers of filtrations, *J. Algebra* **325** (2011), 147–162.

[MV] M. Mandal and J. K. Verma, On the Chern number of an ideal, *Proc. Amer. Math. Soc.* **138** (2010), 1995–1999.

[MR] J. Matijevic and P. Roberts, A conjecture of Nagata on graded Cohen–Macaulay rings, *J. Math. Kyoto Univ.* **14** (1974), 125–128.

[Ma1] H. Matsumura, *Commutative Algebra. Second ed.*, Mathematics Lecture Note Series, **56**, Benjamin/Cummings Publishing Co., Inc., Reading, Mass., 1980.

[Ma2] H. Matsumura, *Kakan Kan Ron*, Kyoritsu Shuppan, 1980 (in Japanese, augmented English translation is available as [Ma3]).

[Ma3] H. Matsumura, *Commutative Ring Theory*. Paperback ed., Cambridge Studies in Advanced Mathematics **8**, Cambridge University Press, Cambridge, 1989.

[MS] V. B. Mehta and V. Srinivas, A characterization of rational singularities, *Asian J. Math.* **1** (1997), 249–278.

[Mil] J. S. Milne, *Fields and Galois Theory*, Kea Books, Ann Arbor, 2022.

[Mol] T. Molien, Uber der Invarianten der linearen Substitutionsgruppen, *Sitzungsber. König. Press. Akad. Wiss.* (1897), 291–302.

[Nag] M. Nagata, *Local Rings*, Corrected Reprints, Robert Krieger Publishing Company, New York, 1975.

[Nag2] M. Nagata, *Theory of Commutative Fields*, Translations of Mathematical Monographs **125**, AMS, Providence, 1991.

[NW] H. Nakajima and K.-i. Watanabe, The classification of quotient singularities which are complete intersections, in *Complete Intersections* (Acireale, 1983), Lect. Notes in Math. **1092**, Springer, (1984), 102–120.

[Nak] Y. Nakamura, Local Rings of Cohen-Macaulay F-rational rings are F-rational, *Tokyo J. Math.* **14** (1991), 41–44.

[Neu] J. Neukirch, *Algebraic Number Theory*, Springer, 1999.

[Nor] D. G. Northcott, A note on the coefficients of the abstract Hilbert function, *J. London Math. Soc.* **35** (1960), 209–214.

[Nor2] D. G. Northcott, *Multilinear Algebra*, Cambridge University Press, 1984.

[NR] D. G. Northcott, D. Rees, Reductions of ideals in local rings, *Proc. Cambridge Philos. Soc.* **50** (1954), 145–158.

[Og] T. Ogoma, Cohen Macaulay factorial domain is not necessarily Gorenstein, *Mem. Fac. Sci. Kochi Univ. Ser. A Math.* **3** (1982), 65–74.

[Og2] T. Ogoma, Existence of dualizing complexes, *J. Math. Kyoto Univ.* **24** (1984), 27–48.

[PS] C. Peskine, L. Szpiro, Dimension projective finie et cohomologie locale, *Publ. IHES* **42** (1973), 47–119.

[Pie] R. S. Pierce, *Associative Algebras*, Grad. Texts in Math. **88**, Springer, 1982.

[Qu] D. Quillen, Projective modules over polynomial rings, *Invent. Math.* **36** (1976), 167–171.

[R] D. Rees, Two classical theorems of ideal theory, *Proc. Cambridge Philos. Soc.* **52** (1956), 155–157.

[R2] D. Rees, $\mathfrak{a}$-transforms of local rings and a theorem on multiplicities of ideals, *Proc. Cambridge Philos. Soc.* **57** (1961), 8–17.

[Rei] G. A. Reisner, Cohen-Macaulay quotients of polynomial rings, *Adv. in Math.* **21** (1976), 30–49.

[Rob] P. C. Roberts, The homological conjectures, in *Progress in Commutative Algebra 1*, Walter de Gruyter & Co., Berlin (2012), 199–230.

[Sal] J. D. Sally, On the associated graded ring of a local Cohen-Macaulay ring, *J. Math. Kyoto Univ.* **17** (1977), 19–21.

[Salm] P. Salmon, Sulla fattorialita delle algebre graduate e degli annelli locali, *Rendiconti Sem. Nat. Padova*, **41** (1968), 119–138.

[Sam] Samuel, P., *Lectures on Unique Factorization Domains*, Notes by M. P. Murthy, Tata Inst. Fundam. Res. Lect. Math. **30**, Bombay, 1964.

[STC] P. Schenzel, N. V. Trung, and T. T. Cuong, Verallgemeinerte Cohen–Macaulay-Moduln (German), *Math. Nachr.* **85** (1978), 57–73.

[Se] J.-P. Serre, *Local Algebra*, Translated from the French by CheeWhye Chin and revised by the author. Springer Monographs in Mathematics, Springer-Verlag, Berlin, 2000.

[ST] G. C. Shepherd and J. A. Todd, Finite unitary reflection groups, *Canad. J. Math.* **6** (1954), 274–304.

[Sm] K. E. Smith, F-rational rings have rational singularities, *Amer. J. Math.* ¥textbf119 (1997), 159–180.

[St1] R. P. Stanley, Upper bound conjecture and Cohen-Macaulay rings, *Stud. Applied Math.* **54** (1975), 135–142.

[St1] R. P. Stanley, Hilbert functions of graded algebras, *Adv. Math.* **28** (1978), 57–83.

[St2] R. P. Stanley, Invariants of finite groups and its applications to combinatorics, *Bull. Amer. Math. Soc.* **1** (1979), 475–511.

[SV1] J. Stückrad and W. Vogel, Toward a theory of Buchsbaum singularities, *Amer. J. Math.* **100** (1978), 727–746.

[SV2] J. Stückrad and W. Vogel, *Buchsbaum Rings and Applications, An Interaction between Algebra, Geometry and Topology*, Springer-Verlag, Berlin, 1986.

[Su] A. A. Suslin, Projective modules over polynomial rings are free, *Dokl. Akad. Nauk SSSR* **229** (1976), 1063–1066.

[SwHu] I. Swanson and C. Hunete, Integral closure of ideals, rings, and modules. London Math. Soc. Lect. Note Ser **336**. Cambridge University Press.

[TW] S. Takagi and K.-i. Watanabe, F-singularities: applications of characteristic p methods to singularity theory, *Sugaku Expositions* **31** (2018), 1–42.

[Vel] J. D. Vélez, Openness of the F-rational locus and smooth base change, *J. Algebra* **172**, (1995), 425–453.

[WJ] J. Watanabe, A note on Gorenstein rings of embedding codimension three, *Nagoya Math. J.* **50** (1973), 227–232.

[Wa1] K.-i. Watanabe, Some examples of one dimensional Gorenstein domains, *Nagoya Math. J.* **49** (1973), 101–109.

[Wa2] K.-i. Watanabe, Certain invariant subrings are Gorenstein. I, *Osaka J. Math.* **11**, (1974), 1–8.

[Wa3] K.-i. Watanabe, Certain invariant subrings are Gorenstein. II, *Osaka J. Math.* **11**, (1974), 379–388.

[Wa4] K.-i. Watanabe, Invariant subrings which are complete intersections. I. Invariant subrings of finite abelian groups, *Nagoya Math. J.* **77** (1980), 89–98.

[Wa5] K.-i. Watanabe, Some remarks concerning Demazure's construction of normal graded rings, *Nagoya Math. J.* **83** (1981), 203–211.

[Wa6] K.-i. Watanabe, F-rationality of certain Rees algebras and counterexamples to "Boutot's theorem" for F-rational rings, *J. Pure Appl. Algebra* **122** (1997), 323–328.

[Wa7] K.-i. Watanabe, A short proof of Bresinski's theorem on Gorenstein semigroup rings generated by four elements, in *Numerical Semigroups: IMNS 2018*, Springer (2020), 369–373.

[Wei] C. A. Weibel, *An Introduction to Homological Algebra*, Cambridge University Press, 1994.

[Yo] Y. Yoshino, *Cohen–Macaulay Modules over Cohen–Macaulay Rings*, London Mathematical Society Lecture Note Series **146**, Cambridge University Press, 1990.

[ZS1] O. Zariski and P. Samuel, *Commutative Algebra. Vol. I.*, Grad. Texts in Math. **28**, Springer-Verlag, 1975.

[ZS2] O. Zariski and P. Samuel, *Commutative Algebra. Vol. II.*, Grad. Texts in Math. **29**, Springer-Verlag, 1975.

Notation

$\langle a_1, a_2, \ldots, a_m \rangle$ semigroup generated by $a_1, a_2, \ldots, a_m$. 252
$\overline{A}^B$ the integral closure of A in B. 86
A° stands for $A \setminus \bigcup_{P \in \mathrm{Min}\, A} P$. 329
$a \mid b$ a divides b. 18
A^G invariant subring of the action of G on A. 274
$A^{(I)}$ the direct sum of copies of A indexed by the set I. 359
$\alpha \not\sim \beta$ α and β are not comparable. 264
$(A, \mathfrak{m})$ We say that $(A, \mathfrak{m})$ is a local ring, if A is a local ring and $\mathfrak{m}$ is its unique maximal ideal. 10
$(A, \mathfrak{m}, k)$ We say that $(A, \mathfrak{m}, k)$ is a local ring, if $(A, \mathfrak{m})$ is a local ring, and $k = A/\mathfrak{m}$. 10
$\mathrm{Ann}_A(M)$ the annihilator of M ($= 0 :_A M$). 30
$A_\mathfrak{p}$ the localization of A at the prime ideal $\mathfrak{p}$. 58
$a(R)$ a-invariant of a graded ring R. 238
AS the A-(sub)module generated by S. 25
$A[S]$ the A-subalgebra generated by S. 48
$\mathrm{Ass}_A M$ the set of associated primes of the A-module M. 70
$\mathrm{Assh}_A M$ the set of minimal primes P of M such that $\dim A/P = \dim_A M$. 140
A_v the DVR defined by the (normalized) discrete valuation v. 97
$A[X]$ the polynomial ring over the ring A with the variable X. 13
Ax the cyclic module generated by x. 25
$A[X_1, X_2, \ldots, X_n]$ the polynomial ring over the ring A with several variables. 16

$\beta_i(M)$ the ith Betti number of M. 370
$B_n(P_\bullet)$ the module of n-boundaries of the chain complex $P_\bullet$. 361
$B * G$ twisted group algebra. 285

S. Goto, K. Watanabe, *Commutative Algebra*, Graduate Texts in Mathematics 309,
https://doi.org/10.1007/978-981-95-4227-7

$\mathrm{G}_{\mathfrak{a}}(A)$ the associated graded ring of A with respect to the ideal $\mathfrak{a}$. 121
$\mathrm{Gal}_k\, K$ the Galois group of the Galois extension K/k. 377
$\mathrm{G}_{\mathfrak{a}}(M)$ the associated graded module of M with respect to the $\mathfrak{a}$-adic topology. 121
$\Gamma_{\mathfrak{a}}(M)$ the A-submodule $\{x \in M \mid \exists n\ \mathfrak{a}^n x = 0\}$ of M. 178
$\mathrm{G}(M)$ the associated graded module of M. 121
$G(n, r)$ the homogeneous coordinate ring of the Grassmann variety $\mathrm{Grass}(n, r)$. 270

$\mathrm{H}^i_{\mathfrak{a}}(M)$ the ith local cohomology of M with respect to $\mathfrak{a}$. 179
$\mathrm{hd}_A\, M$ the projective dimension of the A-module M. 134
H-Max R the set of H-maximal ideals of the graded ring R. 379
$\mathrm{Hom}_A(M, N)$ the A-module of A-homomorphisms from M to N. 38
$\underline{\mathrm{Hom}}_R(M, N)$ the homogeneous $\mathrm{Hom}_R(M, N)$. 232
$\mathrm{H}_r(\underline{a})$ the rth homology group of $K_\bullet(\underline{a})$. 172
$\mathrm{Ht}_1(A)$ the set of height one prime ideals of A. 106
$\mathrm{ht}_A\, \mathfrak{p}$ the height of the (prime) ideal $\mathfrak{p}$ of A. 79

$I^{[q]}$ stands for $(a^q \mid a \in I)$. 330
$I + J$ the sum of the ideals I and J. 7
$\mathrm{I}(\varphi, M)$ stands for $\mathrm{I}_t(\varphi, M)$, where $t = \mathrm{rank}(\varphi, M)$. 184
IB the ideal of B generated by $\varphi(I)$ for a ring homomorphism $\varphi : A \to B$ and an ideal I of A. 8
$I : J$ the ideal quotient of I by J. 7
$\mathrm{id}_A\, M$ the injective dimension of the A-module M. 146
I_Δ the Stanley–Reisner ideal of Δ. 259
I^{-1} the A-submodule $\{x \in K \mid xI \subseteq A\}$ of $K = \mathrm{Q}(A)$. 104
IJ the product of the ideals I and J. 7
$\mathbb{I}(M)$ the Buchsbaum invariant of M. 318
$\mathrm{I}(\varphi)$ stands for $\mathrm{I}(\varphi, A)$. 184
i_S the map of localizatin $A \to A_S$ given by $i_S(a) = a/1$. 52
I^* the tight closure of the ideal I. 330
I^* the ideal generated by the homogeneous elements of I. 379
$\mathrm{I}_t(\varphi)$ the ideal generated by all the t-minors of the matrix of φ. 184

$\mathrm{J}(A)$ the Jacobson radical of the ring A. 10
$\mathrm{J}(P)$ the set of poset ideals of P. 270

$K(A)$ the homotopy category of the ring A. 362
K_A the canonical module of A. 199
$\kappa(\mathfrak{p})$ the residue field $A_{\mathfrak{p}}/\mathfrak{p}A_{\mathfrak{p}}$ of the local ring $A_{\mathfrak{p}}$. 59

R_+ the positive degree part of a graded ring R. 224
$R^{(r)}$ rth Veronese subring of R. 240
$R\sharp S$ the Segre product of R and S. 242
$R[[X]]$ the formal power series ring over R. 118
$R[[X_1, X_2, \ldots, X_n]]$ the formal power series ring over R in several variables. 118

(S) the ideal generated by S. 4
$\sum_{i\in I} N_i$ the sum of the submodules N_i. 25
$\mathrm{sgn}(\boldsymbol{a})$ Stands for $(\mathrm{sgn}(a_1), \ldots, \mathrm{sgn}(a_n))$. 261
$S^{-1}A$ the localization of A by the multiplicatively closed subset S. 52
$S^{-1}I$ the ideal of $S^{-1}A$ generated by $i_s(I)$. 53
$S^{-1}M$ the localization of M by the multiplicatively closed subset S. 54
(S_k) Serre's condition "$\forall P \in \operatorname{Spec} A$ [depth $P \geq \min(k, \operatorname{ht} P)$]". 99
$\operatorname{Soc}_A M$ the socle of the A-module M. 154
$\operatorname{Spec}(A)$ the prime spectrum of A. 56
$\operatorname{Supp}_A M$ the support of the A-module M. 59
$\operatorname{Sym} E$ the symmetric algebra of E. 371
$\operatorname{Sym}_n E$ the nth symmetric power of E. 372

$\tau \prec \sigma$ τ is a face of σ. 257
TE the tensor algebra of E. 371
$\widetilde{I}$ the divisorial hull of I. 105
$\mathrm{t}(M)$ the torsion part of M. 55
$\operatorname{Tor}_i^A(M, N)$ the ith torsion module of M and N. 366
$\operatorname{tr.deg}_k K$ the transcendence degree of K over k. 378

$U_M(N)$ the intersection of the primary components of N corresponding to the elements of $\operatorname{Assh}_A(M/N)$. 208
$\mathrm{U}(A)$ the set of units of the ring A. 3

v_A the normalized discrete valuation of the DVR A. 97
$v(A)$ the embedding dimension $\mu_A(\mathfrak{m})$ of A. 309
$v_{\mathfrak{p}}$ the normalized discrete valuation of $A_{\mathfrak{p}}$. 106
$\mathrm{V}(T)$ the set of prime that contains T. 54

$\bigwedge E$ the exterior algebra of E. 371
$\bigwedge^n E$ the nth exterior power of E. 371

$\mathbb{Z}$ the ring of integers. 1
$Z_n(P_\bullet)$ the module of n-cycles of the chain complex $P_\bullet$. 361

Index

S. Goto, K. Watanabe, *Commutative Algebra*, Graduate Texts in Mathematics 309,
https://doi.org/10.1007/978-981-95-4227-7

The manufacturer's authorised representative in the EU is Springer Nature Customer Service Centre GmbH, Europaplatz 3, 69115 Heidelberg, Germany. If you have any concerns regarding our products, please contact ProductSafety@springernature.com

Printed and bound by CPI Group (UK) Ltd, Croydon, CR0 4YY

05/10/2026

02228020-0005